AF856364

Dynamics of Flight

Stability and Control

THIRD EDITION

Dynamics of Flight
Stability and Control

BERNARD ETKIN
University Professor Emeritus
Institute for Aerospace Studies
University of Toronto

LLOYD DUFF REID
Professor
Institute for Aerospace Studies
University of Toronto

WILEY

Dynamics of Flight
Stability and Control, Third Edition

Authorized reprint by Wiley India Pvt. Ltd., 4435-36/7, Ansari Road, Daryaganj,
New Delhi – 110002.

Copyright © 1996 by John Wiley & Sons, Inc. All rights reserved.

No part of this book, including interior design, cover design, and icons, may be reproduced or transmitted in any form except with the permission of John Wiley & Sons, Inc., 111 River Street, Hoboken, NJ 07030, (201) 748-6011.

Limit of Liability/ Disclaimer of Warranty: The publisher and the author make no representations or warranties with respect to the accuracy or completeness of the contents of this work and specifically disclaim all warranties, including without limitation warranties of fitness for a particular purpose. No warranty may be created or extended by sales or promotional materials. The advice and strategies contained herein may not be suitable for every situation. This work is sold with the understanding that the publisher is not engaged in rendering legal, accounting, or other professional services. If professional assistance is required, the services of a competent professional person should be sought. Neither the publisher nor the author shall be liable for damages arising herefrom. The fact that an organization or Website is referred to in this work as a citation and/or a potential source of further information does not mean that the author or the publisher endorses the information the organization or Website may provide or recommendations it may make. Further, readers should be aware that Internet Websites listed in this work may have changed or disappeared between when this work was written and when it is read.

TRADEMARKS: Wiley and the Wiley logo are trademarks or registered trademarks of John Wiley & Sons Inc., and/or its affiliates, in the United States and other countries, and may not be used without written permission. All other trademarks are the property of their respective owners. John Wiley & Sons Inc., is not associated with any product or vendor mentioned in this book.

Wiley also publishes its books in a variety of electronic formats. Some content that appears in print may not be available in electronic books.

This edition is authorized for sale in the Indian Sub-continent only.

Reprint: 2015

Printed at: *Beekam Print & Pack (P) Ltd., Delhi*

ISBN : 978-81-265-2891-2

To the men and women of science and engineering whose contributions to aviation have made it a dominant force in shaping the destiny of mankind, and who, with sensitivity and concern, develop and apply their technological arts toward bettering the future.

PREFACE

The first edition of this book appeared in 1959—indeed before most students reading this were born. It was well received both by students and practicing aeronautical engineers of that era. The pace of development in aerospace engineering during the decade that followed was extremely rapid, and this was reflected in the subject of flight mechanics. The first author therefore saw the need at the time for a more advanced treatment of the subject that included the reality of the round rotating Earth and the real unsteady atmosphere, and hypersonic flight, and that reflected the explosive growth in computing power that was then taking place (and has not yet ended!). The result was the 1972 volume entitled *Dynamics of Atmospheric Flight.* That treatment made no concessions to the needs of undergraduate students, but attempted rather to portray the state of the art of flight mechanics as it was then. To meet the needs of students, a second edition of the 1959 book was later published in 1982. It is that volume that we have revised in the present edition, although in a number of details we have preferred the 1972 treatment, and used it instead.

We have retained the same philosophy as in the two preceding editions. That is, we have emphasized *basic principles rooted in the physics of flight, essential analytical techniques,* and *typical stability and control realities.* We continue to believe, as stated in the preface to the 1959 edition, that this is the preparation that students need to become aeronautical engineers who can face new and challenging situations with confidence.

This edition improves on its predecessors in several ways. It uses a real jet transport (the Boeing 747) for many numerical examples and includes exercises for students to work in most chapters. We learned from a survey of teachers of this subject that the latter was a sine qua non. Working out these exercises is an important part of acquiring skill in the subject. Moreover, some details in the theoretical development have been moved to the exercises, and it is good practice in analysis for the students to do these.

Students taking a course in this subject are assumed to have a good background in mathematics, mechanics, and aerodynamics, typical of a modern university course in aeronautical or aerospace engineering. Consequently, most of this basic material has been moved to appendices so as not to interrupt the flow of the text.

The content of Chapters 1 through 3 is very similar to that of the previous edition. Chapter 4, however, dealing with the equations of motion, contains two very significant changes. We have not presented the nondimensional equations of motion, but have left them in dimensional form to conform with current practice, and we have expressed the equations in the state vector form now commonly used. Chapter 5, on stability derivatives, is almost unchanged from the second edition, and Chapters 6 and 7 dealing with stability and open loop response, respectively, differ from their predecessors mainly in the use of the B747 as example and in the use of the dimensional equations. Chapter 8, on the other hand, on closed loop control, is very much expanded and almost entirely new. This is consistent with the much enhanced importance of automatic flight control systems in modern airplanes. We believe that the student who works through this chapter and does the exercises will have a good grasp of the basics of this subject.

The appendices of aerodynamic data have been retained as useful material for teachers and students. The same caveats apply as formerly. The data are not intended for design, but only to illustrate orders of magnitude and trends. They are provided to give students and teachers ready access to some data to use in problems and projects.

We acknowledge with thanks the assistance of our colleague, Dr. J. H. de Leeuw, who reviewed the manuscript of Chapter 8 and made a number of helpful suggestions.

On a personal note—as the first author is now in the 11th year of his retirement, this work would not have been undertaken had Lloyd Reid not agreed to collaborate in the task, and if Maya Etkin had not encouraged her husband to take it on and supported him in carrying it out.

In turn, the second author, having used the 1959 edition as a student (with the first author as supervisor), the 1972 text as a researcher, and the 1982 text as a teacher, was both pleased and honored to work with Bernard Etkin in producing this most recent version of the book.

Toronto
December, 1994

Bernard Etkin
Lloyd Duff Reid

CONTENTS

CHAPTER 1

Introduction

1.1 The Subject Matter of Dynamics of Flight

This book is about the *motion* of *vehicles* that *fly* in the atmosphere. As such it belongs to the branch of engineering science called applied mechanics. The three italicized words above warrant further discussion. To begin with *fly*—the dictionary definition is not very restrictive, although it implies motion through the air, the earliest application being of course to birds. However, we also say "a stone flies" or "an arrow flies," so the notion of sustention (lift) is not necessarily implied. Even the atmospheric medium is lost in "the flight of angels." We propose as a logical scientific definition that flying be defined as motion through a fluid medium or empty space. Thus a satellite "flies" through space and a submarine "flies" through the water. Note that a dirigible in the air and a submarine in the water are the same from a mechanical standpoint—the weight in each instance is balanced by buoyancy. They are simply separated by three orders of magnitude in density. By *vehicle* is meant any flying object that is made up of an arbitrary system of deformable bodies that are somehow joined together. To illustrate with some examples: (1) A rifle bullet is the simplest kind, which can be thought of as a single ideally rigid body. (2) A jet transport is a more complicated vehicle, comprising a main elastic body (the airframe and all the parts attached to it), rotating subsystems (the jet engines), articulated subsystems (the aerodynamic controls) and fluid subsystems (fuel in tanks). (3) An astronaut attached to his orbiting spacecraft by a long flexible cable is a further complex example of this general kind of system. Note that by the above definition a vehicle does not necessarily have to carry goods or passengers, although it usually does. The logic of the definitions is simply that the underlying engineering science is common to all these examples, and the methods of formulating and solving problems concerning the motion are fundamentally the same.

As is usual with definitions, we can find examples that don't fit very well. There are special cases of motion at an interface which we may or may not include in flying—for example, ships, hydrofoil craft and air-cushion vehicles (ACV's). In this connection it is worth noting that developments of hydrofoils and ACV's are frequently associated with the Aerospace industry. The main difference between these cases, and those of "true" flight, is that the latter is essentially three-dimensional, whereas the interface vehicles mentioned (as well as cars, trains, etc.) move approximately in a two-dimensional field. The underlying principles and methods are still the same however, with certain modifications in detail being needed to treat these "surface" vehicles.

Now having defined *vehicles* and *flying,* we go on to look more carefully at what we mean by *motion.* It is convenient to subdivide it into several parts:

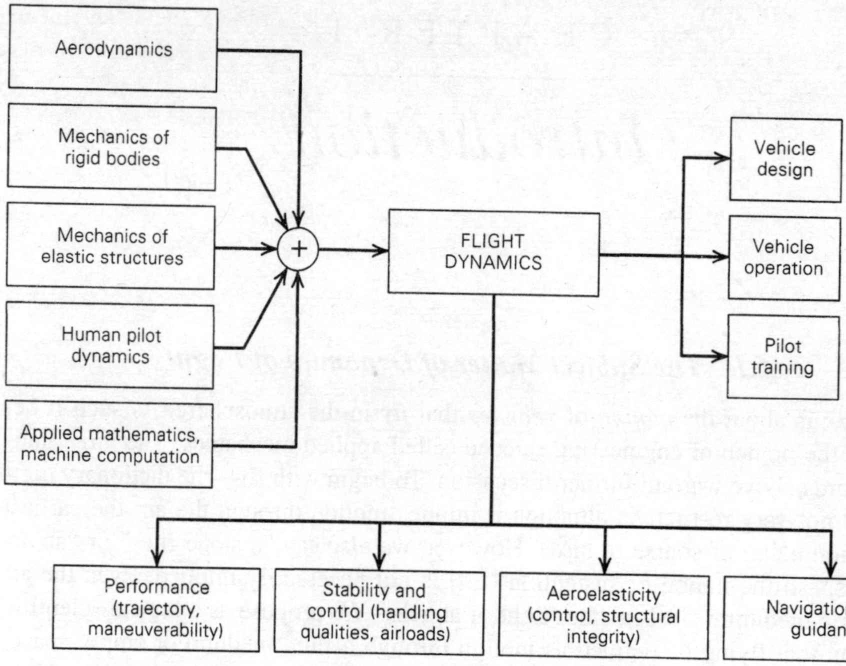

Figure 1.1 Block diagram of disciplines.

Gross Motion:

1. Trajectory of the vehicle mass center.[1]
2. "Attitude" motion, or rotations of the vehicle "as a whole."

Fine Motion:

3. Relative motion of rotating or articulated subsystems, such as engines, gyroscopes, or aerodynamic control surfaces.
4. Distortional motion of deformable structures, such as wing bending and twisting.
5. Liquid sloshing.

This subdivision is helpful both from the standpoint of the technical problems associated with the different motions, and of the formulation of their analysis. It is surely self-evident that studies of these motions must be central to the design and operation of aircraft, spacecraft, rockets, missiles, etc. To be able to formulate and solve the relevant problems, we must draw on several basic disciplines from engineering science. The relationships are shown on Fig. 1.1. It is quite evident from this figure that the practicing flight dynamicist requires intensive training in several branches of engineering science, and a broad outlook insofar as the practical ramifications of his work are concerned.

In the classes of vehicles, in the types of motions, and in the medium of flight, this book treats a very restricted set of all possible cases. It deals only with the flight

[1]It is assumed that gravity is uniform, and hence that the mass center and center of gravity (CG) are the same point.

of airplanes in the atmosphere. The general equations derived, and the methods of solution presented, are however readily modified and extended to treat many of the other situations that are embraced by the general problem.

All the *fundamental* science and mathematics needed to develop this subject existed in the literature by the time the Wright brothers flew. Newton, and other giants of the 18th and 19th centuries, such as Bernoulli, Euler, Lagrange, and Laplace, provided the building blocks in solid mechanics, fluid mechanics, and mathematics. The needed applications to aeronautics were made mostly after 1900 by workers in many countries, of whom special reference should be made to the Wright brothers, G. H. Bryan, F. W. Lanchester, J. C. Hunsaker, H. B. Glauert, B. M. Jones, and S. B. Gates. These pioneers introduced and extended the basis for analysis and experiment that underlies all modern practice.[2] This body of knowledge is well documented in several texts of that period, for example, Bairstow (1939). Concurrently, principally in the United States of America and Britain, a large body of aerodynamic data was accumulated, serving as a basis for practical design.

Newton's laws of motion provide the connection between environmental forces and resulting motion for all but relativistic and quantum-dynamical processes, including all of "ordinary" and much of celestial mechanics. What then distinguishes flight dynamics from other branches of applied mechanics? Primarily it is the special nature of the force fields with which we have to be concerned, the absence of the kinematical constraints central to machines and mechanisms, and the nature of the control systems used in flight. The external force fields may be identified as follows:

"Strong" Fields:

1. Gravity
2. Aerodynamic
3. Buoyancy

"Weak" Fields:

4. Magnetic
5. Solar radiation

We should observe that two of these fields, aerodynamic and solar radiation, produce important heat transfer to the vehicle in addition to momentum transfer (force). Sometimes we cannot separate the thermal and mechanical problems (Etkin and Hughes, 1967). Of these fields only the strong ones are of interest for atmospheric and oceanic flight, the weak fields being important only in space. It should be remarked that even in atmospheric flight the gravity force can not always be approximated as a constant vector in an inertial frame. Rotations associated with Earth curvature, and the inverse square law, become important in certain cases of high-speed and high-altitude flight (Etkin, 1972).

The prediction, measurement and representation of aerodynamic forces are the principal distinguishing features of flight dynamics. The size of this task is illustrated

[2]An excellent account of the early history is given in the 1970 von Kármán Lecture by Perkins (1970).

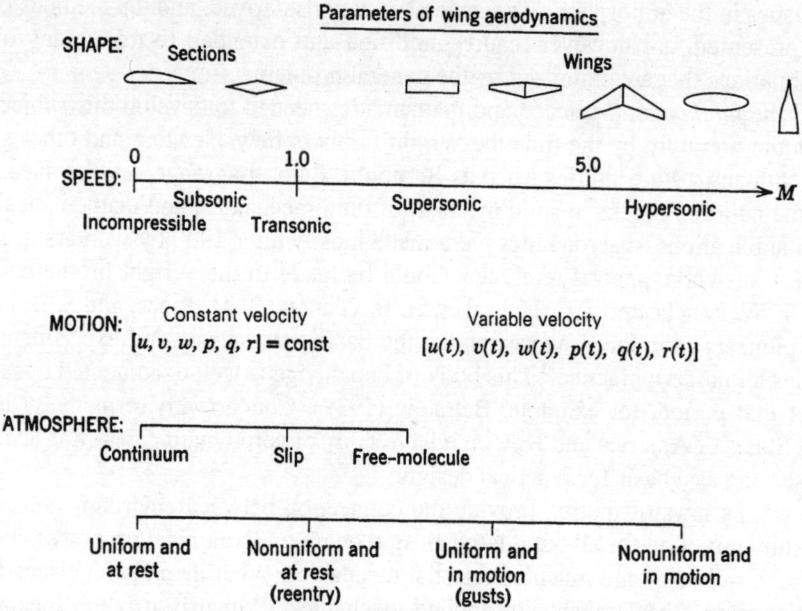

Figure 1.2 Spectrum of aerodynamic problems for wings.

by Fig. 1.2, which shows the enormous range of variables that need to be considered in connection with wings alone. To be added, of course, are the complications of propulsion systems (propellers, jets, rockets), compound geometries (wing + body + tail), and variable geometry (wing sweep, camber).

As remarked above, Newton's laws state the connection between force and motion. The commonest problem consists of finding the motion when the laws for the forces are given (all the numerical examples given in this book are of this kind). However, we must be aware of certain important variations:

1. Inverse problems of first kind—the system and the motion are given and the forces have to be calculated.
2. Inverse problems of the second kind—the forces and the motion are given and the system parameters have to be found.
3. Mixed problems—the unknowns are a mixture of variables from the force, system, and motion.

Examples of these inverse and mixed problems often turn up in research, when one is trying to deduce aerodynamic forces from the observed motion of a vehicle in flight or of a model in a wind tunnel. Another example is the deduction of harmonics of the Earth's gravity field from observed perturbations of satellite orbits. These problems are closely related to the "plant identification" or "parameter identification" problem of system theory. [Inverse problems were treated in Chap. 11 of Etkin (1959)].

TYPES OF PROBLEMS

The main types of flight dynamics problem that occur in engineering practice are:

1. Calculation of "performance" quantities, such as speed, height, range, and fuel consumption.
2. Calculation of trajectories, such as launch, reentry, orbital and landing.
3. Stability of motion.
4. Response of vehicle to control actuation and to propulsive changes.
5. Response to atmospheric turbulence, and how to control it.
6. Aeroelastic oscillations (flutter).
7. Assessment of human-pilot/machine combination (handling qualities).

It takes little imagination to appreciate that, in view of the many vehicle types that have to be dealt with, a number of subspecialties exist within the ranks of flight dynamicists, related to some extent to the above problem categories. In the context of the modern aerospace industry these problems are seldom simple or routine. On the contrary they present great challenges in analysis, computation, and experiment.

1.2 The Tools of Flight Dynamicists

The tools used by flight dynamicists to solve the design and operational problems of vehicles are of three kinds:

1. Analytical
2. Computational
3. Experimental

The analytical tools are essentially the same as those used in other branches of mechanics, that is the methods of applied mathematics. One important branch of applied mathematics is what is now known as system theory, including stability, automatic control, stochastic processes and optimization. Stability of the uncontrolled vehicle is neither a necessary nor a sufficient condition for successful controlled flight. Good airplanes have had slightly unstable modes in some part of their flight regime, and on the other hand, a completely stable vehicle may have quite unacceptable handling qualities. It is *dynamic performance* criteria that really matter, so to expend a great deal of analytical and computational effort on finding stability boundaries of nonlinear and time-varying systems may not be really worthwhile. On the other hand, the computation of stability of small disturbances from a steady state, that is, the linear eigenvalue problem that is normally part of the system study, is very useful indeed, and may well provide enough information about stability from a practical standpoint.

On the computation side, the most important fact is that the availability of machine computation has revolutionized practice in this subject over the past few decades. Problems of system performance, system design, and optimization that could not have been tackled at all in the past are now handled on a more or less routine basis.

The experimental tools of the flight dynamicist are generally unique to this field. First, there are those that are used to find the aerodynamic inputs. Wind tunnels and shock tubes that cover most of the spectrum of atmospheric flight are now available in the major aerodynamic laboratories of the world. In addition to fixed laboratory equipment, there are aeroballistic ranges for dynamic investigations, as well as rocket-boosted and gun-launched free-flight model techniques. Hand in hand with the development of these general facilities has gone that of a myriad of sensors and instruments, mainly electronic, for measuring forces, pressures, temperatures, acceleration, angular velocity, and so forth. The evolution of *computational fluid dynamics* (CFD) has sharply reduced the dependence of aerodynamicists on experiment. Many results that were formerly obtained in wind tunnel tests are now routinely provided by CFD analyses. The CFD codes themselves, of course, must be verified by comparison with experiment.

Second, we must mention the flight simulator as an experimental tool used directly by the flight dynamicist. In it he studies mainly the matching of the pilot to the machine. This is an essential step for radically new flight situations. The ability of the pilot to control the vehicle must be assured long before the prototype stage. This cannot yet be done without test, although limited progress in this direction is being made through studies of mathematical models of human pilots. Special simulators, built for most new major aircraft types, provide both efficient means for pilot training, and a research tool for studying handling qualities of vehicles and dynamics of human pilots. The development of high-fidelity simulators has made it possible to greatly reduce the time and cost of training pilots to fly new types of airplanes.

1.3 Stability, Control, and Equilibrium

It is appropriate here to define what is meant by the terms *stability* and *control.* To do so requires that we begin with the concept of equilibrium.

A body is in equilibrium when it is at rest or in uniform motion (i.e., has constant linear and angular momenta). The most familiar examples of equilibrium are the static ones; that is, bodies at rest. The equilibrium of an airplane in flight, however, is of the second kind; that is, uniform motion. Because the aerodynamic forces are dependent on the angular orientation of the airplane relative to its flight path, and because the resultant of them must exactly balance its weight, the equilibrium state is without rotation; that is, it is a motion of rectilinear translation.

Stability, or the lack of it, is a property of an equilibrium state.[3] The equilibrium is stable if, when the body is slightly disturbed in any of its degrees of freedom, it returns ultimately to its initial state. This is illustrated in Fig. 1.3*a*. The remaining sketches of Fig. 1.3 show neutral and unstable equilibrium. That in Fig. 1.3*d* is a more complex kind than that in Fig. 1.3*b* in that the ball is stable with respect to displacement in the y direction, but unstable with respect to x displacements. This has its counterpart in the airplane, which may be stable with respect to one degree of freedom and unstable with respect to another. Two kinds of instability are of interest in

[3]It is also possible to speak of the stability of a transient with prescribed initial condition.

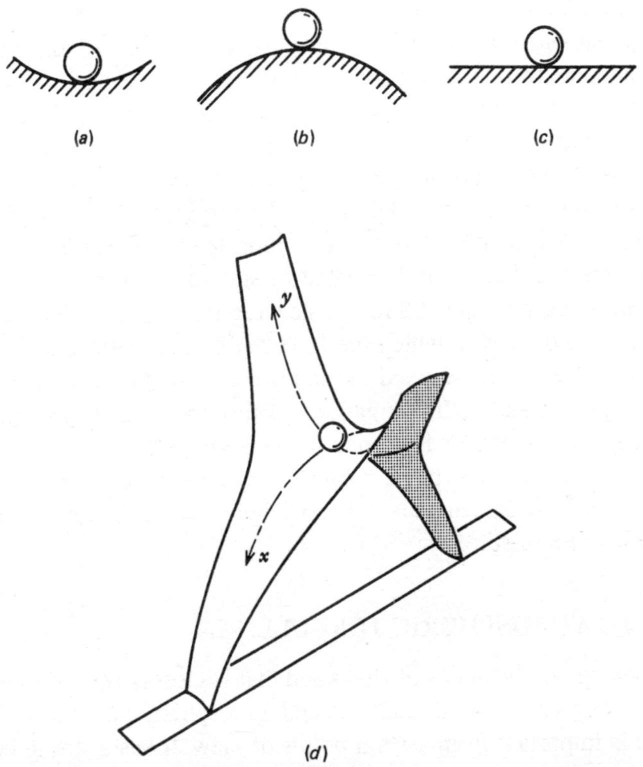

Figure 1.3 (*a*) Ball in a bowl—stable equilibrium. (*b*) Ball on a hill—unstable equilibrium. (*c*) Ball on a plane—neutral equilibrium. (*d*) Ball on a saddle surface—unstable equilibrium.

airplane dynamics. In the first, called *static instability,* the body departs continuously from its equilibrium condition. That is how the ball in Fig. 1.3*b* would behave if disturbed. The second, called *dynamic instability,* is a more complicated phenomenon in which the body oscillates about its equilibrium condition with ever-increasing amplitude.

When applying the concept of stability to airplanes, there are two classes that must be considered—*inherent stability* and *synthetic stability.* The discussion of the previous paragraph implicitly dealt with inherent stability, which is a property of the basic airframe with either fixed or free controls, that is, *control-fixed stability* or *control-free stability.* On the other hand, synthetic stability is that provided by an *automatic flight control system* (AFCS) and vanishes if the control system fails. Such automatic control systems are capable of stabilizing an inherently unstable airplane, or simply improving its stability with what is known as *stability augmentation systems* (SAS). The question of how much to rely on such systems to make an airplane flyable entails a trade-off among weight, cost, reliability, and safety. If the SAS works most of the time, and if the airplane can be controlled and landed after it has failed, albeit with diminished handling qualities, then poor inherent stability may be acceptable. Current aviation technology shows an increasing acceptance of SAS in all classes of airplanes.

If the airplane is controlled by a human pilot, some mild inherent instability can be tolerated, if it is something the pilot can control, such as a slow divergence. (Unstable bicycles have long been ridden by humans!). On the other hand, there is no margin for error when the airplane is under the control of an autopilot, for then the closed loop system *must be stable* in its response to atmospheric disturbances and to commands that come from a navigation system.

In addition to the role controls play in stabilizing an airplane, there are two others that are important. The first is to fix or to change the equilibrium condition (speed or angle of climb). An adequate control must be powerful enough to produce the whole range of equilibrium states of which the airplane is capable from a performance standpoint. The dynamics of the transition from one equilibrium state to another are of interest and are closely related to stability. The second function of the control is to produce nonequilibrium, or accelerated motions; that is, maneuvers. These may be steady states in which the forces and accelerations are constant when viewed from a reference frame fixed to the airplane (for example, a steady turn), or they may be transient states. Investigations of the transition from equilibrium to a nonequilibrium steady state, or from one maneuvering steady state to another, form part of the subject matter of airplane control. Very large aerodynamic forces may act on the airplane when it maneuvers—a knowledge of these forces is required for the proper design of the structure.

RESPONSE TO ATMOSPHERIC TURBULENCE

A topic that belongs in dynamics of flight and that is closely related to stability is the response of the airplane to wind gradients and atmospheric turbulence (Etkin, 1981). This response is important from several points of view. It has a strong bearing on the adequacy of the structure, on the safety of landing and take-off, on the acceptability of the airplane as a passenger transport, and on its accuracy as a gun or bombing platform.

1.4 The Human Pilot

Although the analysis and understanding of the dynamics of the airplane as an isolated unit is extremely important, one must be careful not to forget that for many flight situations it is the response of the total system, made up of the human pilot and the aircraft, that must be considered. It is for this reason that the designers of aircraft should apply the findings of studies into the human factors involved in order to ensure that the completed system is well suited to the pilots who must fly it.

Some of the areas of consideration include:

1. Cockpit environment; the occupants of the vehicle must be provided with oxygen, warmth, light, and so forth, to sustain them comfortably.
2. Instrument displays; instruments must be designed and positioned to provide a useful and unambiguous flow of information to the pilot.
3. Controls and switches; the control forces and control system dynamics must be acceptable to the pilot, and switches must be so positioned and designed as to prevent accidental operation. Tables 1.1 to 1.3 present some pilot data concerning control forces.
4. Pilot workload; the workload of the pilot can often be reduced through proper planning and the introduction of automatic equipment.

Table 1.1
Estimates of the Maximum Rudder Forces that Can Be Exerted for Various Positions of the Rudder Pedal (BuAer, 1954)

Rudder Pedal Position	*Distance from Back of Seat*		*Pedal Force*	
	(in)	*(cm)*	*(lb)*	*(N)*
Back	31.00	78.74	246	1,094
Neutral	34.75	88.27	424	1,886
Forward	38.50	97.79	334	1,486

Table 1.2
Hand-Operated Control Forces (From Flight Safety Foundation Human Engineering Bulletin 56-5H) (see figure in Table 1.3)

Direction of Movement		*180°*	*150°*	*120°*	*90°*	*60°*	
Pull	Rt. hand	52 (231)	56 (249)	42 (187)	37 (165)	24 (107)	
	Lft. hand	50 (222)	42 (187)	34 (151)	32 (142)	26 (116)	
Push	Rt. hand	50 (222)	42 (187)	36 (160)	36 (160)	34 (151)	
	Lft. hand	42 (187)	30 (133)	26 (116)	22 (98)	22 (98)	
Up	Rt. hand	14 (62)	18 (80)	24 (107)	20 (89)	20 (89)	Values given represent maximum exertable force in pounds (Newtons) by the 5 percentile man.
	Lft. hand	9 (40)	15 (67)	17 (76)	17 (76)	15 (67)	
Down	Rt. hand	17 (76)	20 (89)	26 (116)	26 (116)	20 (89)	
	Lft. hand	13 (58)	18 (80)	21 (93)	21 (93)	18 (80)	
Outboard	Rt. hand	14 (62)	15 (67)	15 (67)	16 (71)	17 (76)	
	Lft. hand	8 (36)	8 (36)	10 (44)	10 (44)	12 (53)	
Inboard	Rt. hand	20 (89)	20 (89)	22 (98)	18 (80)	20 (89)	
	Lft. hand	13 (58)	15 (67)	20 (89)	16 (71)	17 (76)	

Note: The above results are those obtained from unrestricted movement of the subject. Any force required to overcome garment restriction would reduce the effective forces by the same amount.

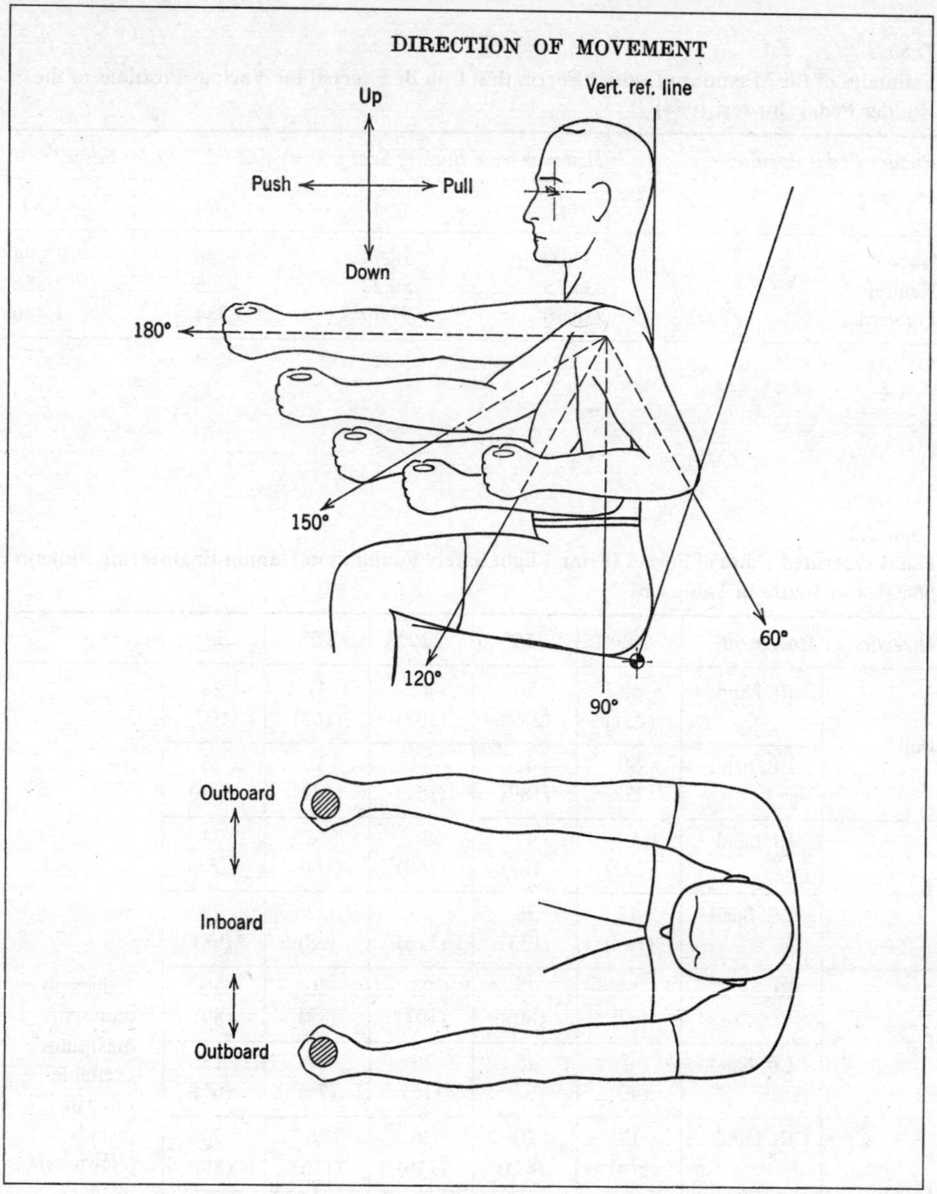

Table 1.3

Rates of Stick Movement in Flight Test Pull-ups Under Various Loads (BuAer, 1954)

Pull-up	*Maximum Stick Load* (lb)	(N)	*Average Rate of Stick Motion* (in/s)	(cm/s)	*Time for Full Deflection* (s)
1	35	156	51.85	131.70	0.162
2	74	329	15.58	39.57	0.475
3	77	343	11.00	27.94	0.600
4	97	431	10.27	26.09	0.750

The care exercised in considering the human element in the closed-loop system made up of pilot and aircraft can determine the success or failure of a given aircraft design to complete its mission in a safe and efficient manner.

Many critical tasks performed by pilots involve them in activities that resemble those of a servo control system. For example, the execution of a landing approach through turbulent air requires the pilot to monitor the aircraft's altitude, position, attitude, and airspeed and to maintain these variables near their desired values through the actuation of the control system. It has been found in this type of control situation that the pilot can be modeled by a linear control system based either on classical control theory or optimal control theory (Etkin, 1972; Kleinman et al., 1970; McRuer and Krendel, 1973).

1.5 Handling Qualities Requirements

As a result of the inability to carry out completely rational design of the pilot–machine combination, it is customary for the government agencies responsible for the procurement of military airplanes, or for licensing civil airplanes, to specify compliance with certain "handling (or flying) qualities requirements" (e.g., ICAO, 1991; USAF, 1980; USAF, 1990). *Handling qualities* refers to those qualities or characteristics of an aircraft that govern the ease and precision with which a pilot is able to perform the tasks required in support of an aircraft role (Cooper and Harper, 1969).

These requirements have been developed from extensive and continuing flight research. In the final analysis they are based on the opinions of research test pilots, substantiated by careful instrumentation. They vary from country to country and from agency to agency, and, of course, are different for different types of aircraft. They are subject to continuous study and modification in order to keep them abreast of the latest research and design information. Because of these circumstances, it is not feasible to present a detailed description of such requirements here. The following is intended to show the nature, not the detail, of typical handling qualities requirements.[4] Most of the specific requirements can be classified under one of the following headings.

CONTROL POWER

The term *control power* is used to describe the efficacy of a control in producing a range of steady equilibrium or maneuvering states. For example, an elevator control, which by taking positions between full up and full down can hold the airplane in equilibrium at all speeds in its speed range, for all configurations[5] and CG positions, is a powerful control. On the other hand, a rudder that is not capable at full deflection of maintaining equilibrium of yawing moments in a condition of one engine out and negligible sideslip is not powerful enough. The handling qualities requirements normally specify the specific speed ranges that must be achievable with full elevator de-

[4]For a more complete discussion, see AGARD (1959); Stevens and Lewis (1992).

[5]This word describes the position of movable elements of the airplane—for example, landing configuration means that landing flaps and undercarriage are down, climb configuration means that landing gear is up, and flaps are at take-off position, and so forth.

flection in the various important configurations and the asymmetric power condition that the rudder must balance. They may also contain references to the elevator angles required to achieve positive load factors, as in steady turns and pull-up maneuvers (see "elevator angle per *g*," Sec. 3.1).

CONTROL FORCES

The requirements invariably specify limits on the control forces that must be exerted by the pilot in order to effect specific changes from a given trimmed condition, or to maintain the trim speed following a sudden change in configuration or throttle setting. They frequently also include requirements on the control forces in pull-up maneuvers (see "control force per *g*," Sec. 3.1). In the case of light aircraft, the control forces can result directly from mechanical linkages between the aerodynamic control surfaces and the pilot's flight controls. In this case the hinge moments of Sec. 2.5 play a direct role in generating these forces. In heavy aircraft, systems such as partial or total hydraulic boost are used to counteract the aerodynamic hinge moments and a related or independent subsystem is used to create the control forces on the pilot's flight controls.

STATIC STABILITY

The requirement for static longitudinal stability (see Chap. 2) is usually stated in terms of the *neutral point.* The neutral point, defined more precisely in Sec. 2.3, is a special location of the center of gravity (CG) of the airplane. In a limited sense it is the boundary between stable and unstable CG positions. It is usually required that the relevant neutral point (stick free or stick fixed) shall lie some distance (e.g., 5% of the mean aerodynamic chord) behind the most aft position of the CG. This ensures that the airplane will tend to fly at a constant speed and angle of attack as long as the controls are not moved.

The requirement on static lateral stability is usually mild. It is simply that the spiral mode (see Chap. 6) if divergent shall have a time to double greater than some stated minimum (e.g., 4s).

DYNAMIC STABILITY

The requirement on dynamic stability is typically expressed in terms of the damping and frequency of a natural mode. Thus the USAF (1980) requires the damping and frequency of the lateral oscillation for various flight phases and stability levels to conform to the values in Table 1.4.

STALLING AND SPINNING

Finally, most requirements specify that the airplane's behavior following a stall or in a spin shall not include any dangerous characteristics, and that the controls must retain enough effectiveness to ensure a safe recovery to normal flight.

Table 1.4[1]
Minimum Dutch Roll Frequency and Damping

Level	Flight Phase Category	Class	Min ζ_d*	Min $\zeta_d\omega_{n_d}$,* rad/s	Min ω_{n_d}, rad/s
1	A	I, IV	0.19	0.35	1.0
		II, III	0.19	0.35	0.4
	B	All	0.08	0.15	0.4
	C	I, II-C, IV	0.08	0.15	1.0
		II-L, III	0.08	0.10	0.4
2	All	All	0.02	0.05	0.4
3	All	All	0	—	0.4

[1] *Level, Phase* and *Class* are defined in USAF, 1980.
**Note:* The damping coefficient ζ, and the undamped natural frequency ω_n, are defined in Chap. 6.

RATING OF HANDLING QUALITIES

To be able to assess aircraft handling qualities one must have a measuring technique with which any given vehicle's characteristics can be rated. In the early days of aviation, this was done by soliciting the comments of pilots after they had flown the aircraft. However, it was soon found that a communications problem existed with pilots using different adjectives to describe the same flight characteristics. These ambiguities have been alleviated considerably by the introduction of a uniform set of descriptive phrases by workers in the field. The most widely accepted set is referred to as the "Cooper-Harper Scale," where a numerical rating scale is utilized in conjunction with a set of descriptive phrases. This scale is presented in Fig. 1.4. To apply this rating technique it is necessary to describe accurately the conditions under which the results were obtained. In addition it should be realized that the numerical pilot rating (1–10) is merely a shorthand notation for the descriptive phrases and as such no mathematical operations can be carried out on them in a rigorous sense. For example, a vehicle configuration rated as 6 should not be thought to be "twice as bad" as one rated at 3. The comments from evaluation pilots are extremely useful and this information will provide the detailed reasons for the choice of a rating.

Other techniques have been applied to the rating of handling qualities. For example, attempts have been made to use the overall system performance as a rating parameter. However, due to the pilot's adaptive capability, quite often he can cause the overall system response of a bad vehicle to approach that of a good vehicle, leading to the same performance but vastly differing pilot ratings. Consequently system performance has not proved to be a good rating parameter. A more promising approach involves the measurement of the pilot's physiological and psychological state. Such methods lead to objective assessments of how the system is influencing the human controller. The measurement of human pilot describing functions is part of this technique (Kleinman et al., 1970; McRuer and Krendel, 1973; Reid, 1969).

Research into aircraft handling qualities is aimed in part at ascertaining which vehicle parameters influence pilot acceptance. It is obvious that the number of possi-

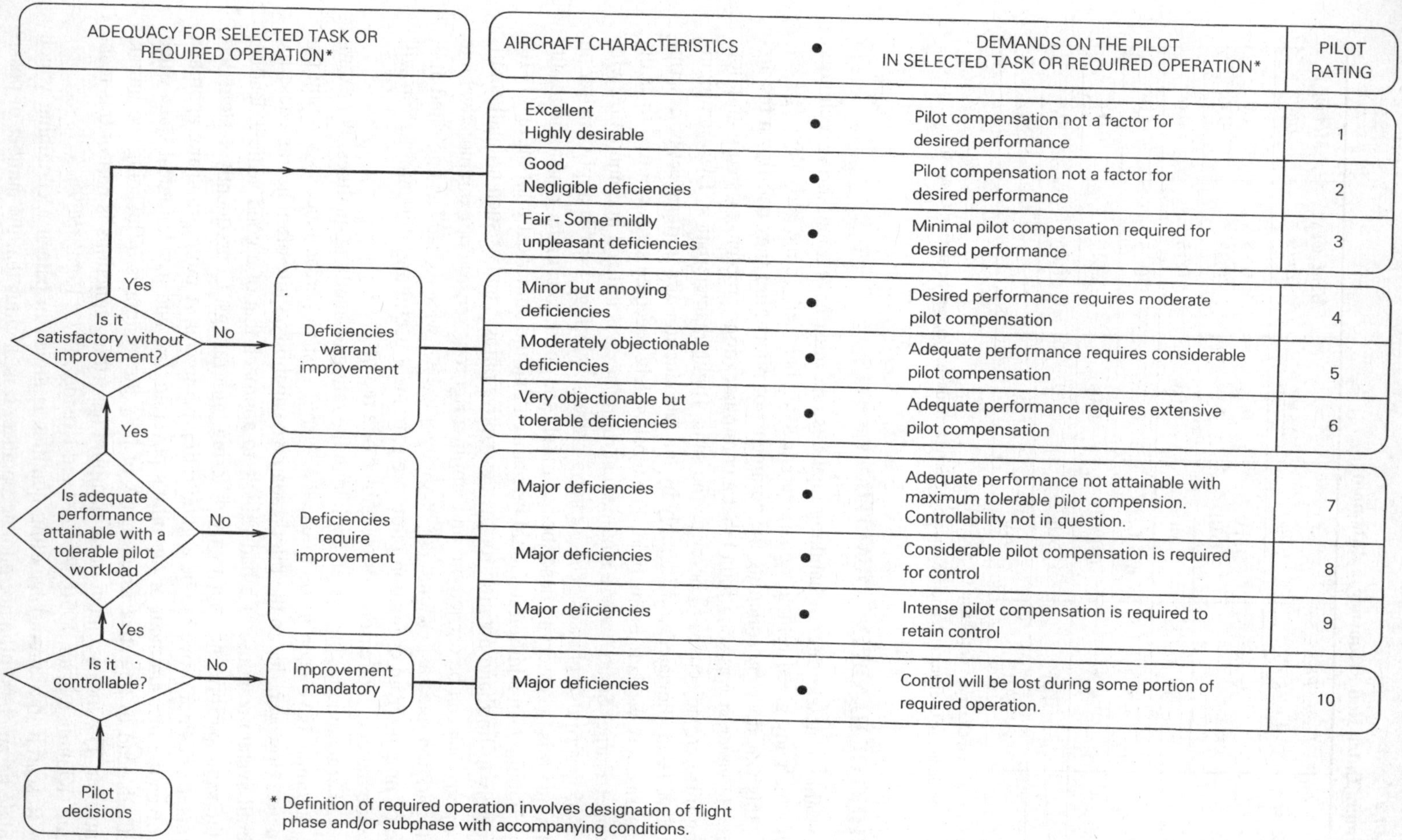

* Definition of required operation involves designation of flight phase and/or subphase with accompanying conditions.

Figure 1.4 **Handling qualities rating scale; Cooper/Harper scale (Cooper and Harper, 1969).**

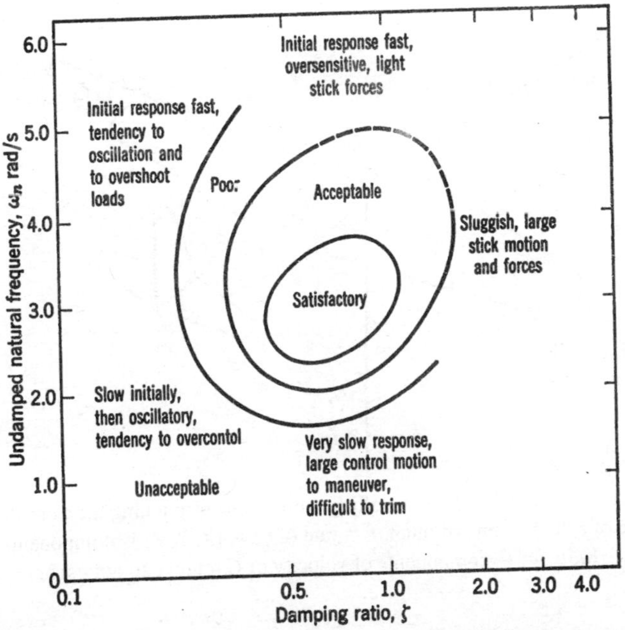

Figure 1.5 Longitudinal short-period oscillation—pilot opinion contours (O'Hara, 1967).

ble combinations of parameters is staggering, and consequently attempts are made to study one particular aspect of the vehicle while maintaining all others in a "satisfactory" configuration. Thus the task is formulated in a fashion that is amenable to study. The risk involved in this technique is that important interaction effects can be overlooked. For example, it is found that the degree of difficulty a pilot finds in controlling an aircraft's lateral-directional mode influences his rating of the longitudinal dynamics. Such facts must be taken into account when interpreting test results. Another possible bias exists in handling qualities results obtained in the past because most of the work has been done in conjunction with fighter aircraft. The findings from such research can often be presented as "isorating" curves such as those shown in Fig. 1.5.

1.6 *Axes and Notation*

In this book the Earth is regarded as flat and stationary in inertial space. Any coordinate system, or *frame of reference,* attached to the Earth is therefore an inertial system, one in which Newton's laws are valid. Clearly we shall need such a reference frame when we come to formulate the equations of motion of a flight vehicle. We denote that frame by $F_E(O_E, x_E, y_E, z_E)$. Its origin is arbitrarily located to suit the circumstances of the problem, the axis $O_E z_E$ points vertically downward, and the axis $O_E x_E$, which is horizontal, is chosen to point in any convenient direction, for example, North, or along a runway, or in some reference flight direction. It is additionally assumed that gravity is uniform, and hence that the mass center and center of gravity (CG) are the same point. The location of the CG is given by its Cartesian coordinates relative to F_E. Its velocity relative to F_E is denoted $\mathbf{V}^E$ and is frequently termed the *groundspeed.*

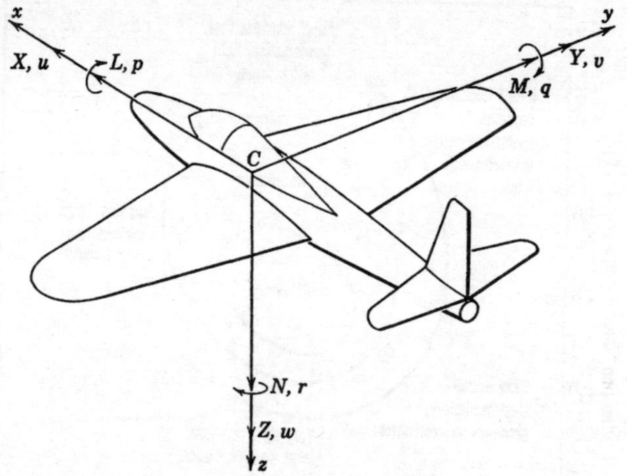

Figure 1.6 Notation for body axes. L = rolling moment, M = pitching moment, N = yawing moment, p = rate of roll, q = rate of pitch, r = rate of yaw. $[X, Y, Z]$ = components of resultant aerodynamic force. $[u, v, w]$ = components of velocity of C relative to atmosphere.

Aerodynamic forces, on the other hand, depend not on the velocity relative to F_E, but rather on the velocity relative to the surrounding air mass (the *airspeed*), which will differ from the groundspeed whenever there is a wind. If we denote the wind velocity vector relative to F_E by **W**, and that of the CG relative to the air by **V** then clearly

$$\mathbf{V}^E = \mathbf{V} + \mathbf{W} \tag{1.6,1}$$

The components of **W** in frame F_E, that is, relative to Earth, are given by

$$\mathbf{W}_E = [W_x \; W_y \; W_z]^T \tag{1.6,2}$$

V represents the magnitude of the airspeed (thus retaining the usual aerodynamics meaning of this symbol). For the most part we will have $\mathbf{W} = 0$, making the airspeed the same as the inertial velocity.

A second frame of reference will be needed in the development of the equations of motion. This frame is fixed to the airplane and moves with it, having its origin C at the CG, (see Fig. 1.6). It is denoted F_B and is commonly called body axes. Cxz is the plane of symmetry of the vehicle. The components of the aerodynamic forces and moments that act on the airplane, and of its linear and angular velocities relative to

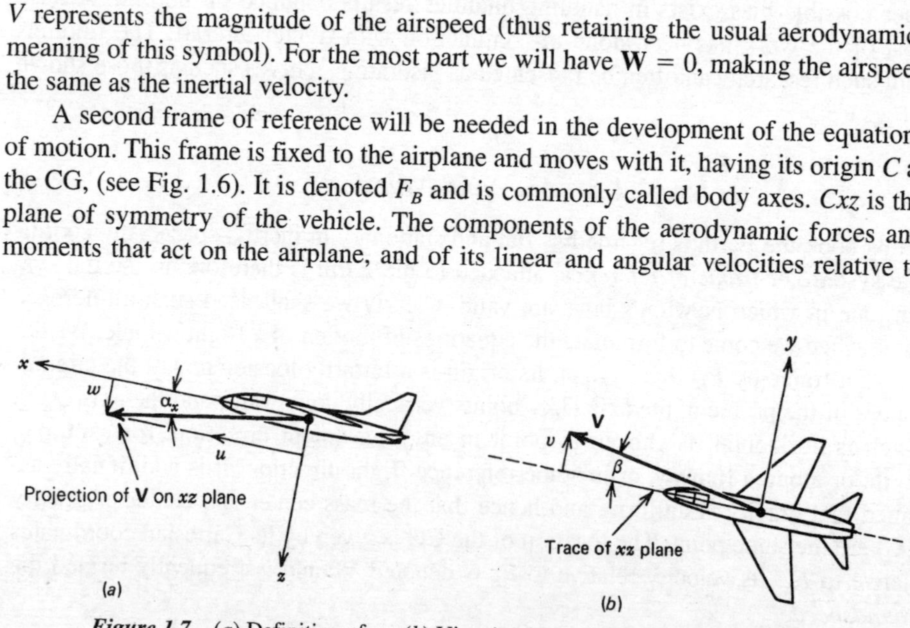

Figure 1.7 (*a*) Definition of α_x. (*b*) View in plane of y and **V**, definition of β.

the air are denoted by the symbols given in the figure. In the notation of Appendix A.1, this means, for example, that

$$\mathbf{V}_B = [u\ v\ w]^T \tag{1.6,3}$$

The vector **V** does not in general lie in any of the coordinate planes. Its orientation is defined by the two angles shown in Fig. 1.7:

$$\begin{aligned} &\text{Angle of attack,} \quad &\alpha_x = \tan^{-1}\frac{w}{u} \\ &\text{Angle of sideslip,} \quad &\beta = \sin^{-1}\frac{v}{V} \end{aligned} \tag{1.6,4}$$

With these definitions, the sideslip angle β is not dependent on the direction of Cx in the plane of symmetry.

The symbols used throughout the text correspond generally to current usage and are mainly used in a consistent manner.

CHAPTER 2

Static Stability and Control— Part 1

2.1 General Remarks

A general treatment of the stability and control of airplanes requires a study of the *dynamics* of flight, and this approach is taken in later chapters. Much useful information can be obtained, however, from a more limited view, in which we consider not the *motion* of the airplane, but only its *equilibrium states.* This is the approach in what is commonly known as *static* stability and control analysis.

The unsteady motions of an airplane can frequently be separated for convenience into two parts. One of these consists of the *longitudinal* or *symmetric* motions; that is, those in which the wings remain level, and in which the center of gravity moves in a vertical plane. The other consists of the *lateral* or *asymmetric* motions; that is, rolling, yawing, and sideslipping, while the angle of attack, the speed, and the angle of elevation of the x axis remains constant.

This separation can be made for both dynamic and static analyses. However, the results of greatest importance for static stability are those associated with the longitudinal analysis. Thus the principal subject matter of this and the following chapter is static longitudinal stability and control. A brief discussion of the static aspects of directional and rolling motions is contained in Secs. 3.9 and 3.11.

We shall be concerned with two aspects of the equilibrium state. Under the heading *stability* we shall consider the pitching moment that acts on the airplane when its angle of attack is changed from the equilibrium value, as by a vertical gust. We focus our attention on whether or not this moment acts in such a sense as to restore the airplane to its original angle of attack. Under the heading *control* we discuss the use of a longitudinal control (elevator) to change the equilibrium value of the angle of attack.

The restriction to angle of attack disturbances when dealing with stability must be noted, since the applicability of the results is thereby limited. When the aerodynamic characteristics of an airplane change with speed, owing to compressibility effects, structural distortion, or the influence of the propulsive system, then the airplane may be unstable with respect to disturbances in speed. Such instability is not predicted by a consideration of angle of attack disturbances only. (See Fig. 1.3*d*, and identify speed with x, angle of attack with y.) A more general point of view than that adopted in this chapter is required to assess that aspect of airplane stability. Such a viewpoint is taken in Chap. 6. To distinguish between true general static stability and the more limited version represented by C_m vs. α, we use the term *pitch stiffness* for the latter.

Although the major portion of this and the following chapter treats a rigid airplane, an introduction to the effects of airframe distortion is contained in Sec. 3.5.

THE BASIC LONGITUDINAL FORCES

The basic flight condition for most vehicles is symmetric steady flight. In this condition the velocity and force vectors are as illustrated in Fig. 2.1. All the nonzero forces and motion variables are members of the set defined as "longitudinal." The two main aerodynamic parameters of this condition are V and α.

Nothing can be said in general about the way the thrust vector varies with V and α, since it is so dependent on the type of propulsion unit—rockets, jet, propeller, or turboprop. Two particular idealizations are of interest, however,

1. T independent of V, that is, constant thrust; an approximation for rockets and pure jets.
2. TV independent of V, that is, constant power; an approximation for reciprocating engines with constant-speed propellers.

The variation of steady-state lift and drag with α for subsonic and supersonic Mach numbers ($\mathbf{M} <$ about 5) are characteristically as shown in Fig. 2.2 for the range of attached flow over the surfaces of the vehicle (McCormick, 1994; Miele, 1962; Schlichting and Truckenbrodt, 1979). Over a useful range of α (below the stall) the coefficients are given accurately enough by

$$C_L = C_{L_\alpha}\alpha \tag{2.1,1}$$

$$C_D = C_{D_{\min}} + KC_L^2 \tag{2.1,2}$$

The three constants C_{L_α}, $C_{D_{\min}}$, K are principally functions of the configuration shape, thrust coefficient, and Mach number.

Significant departure from the above idealizations may, of course, be anticipated in some cases. The minimum of C_D may occur at a value of $\alpha > 0$, and the curvature of the C_L vs. α relation may be an important consideration for flight at high C_L. When the vehicle is a "slender body," for example, a slender delta, or a slim wingless

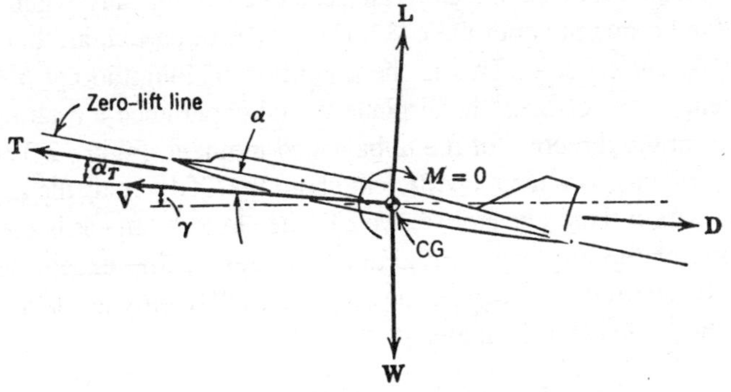

Figure 2.1 Steady symmetric flight.

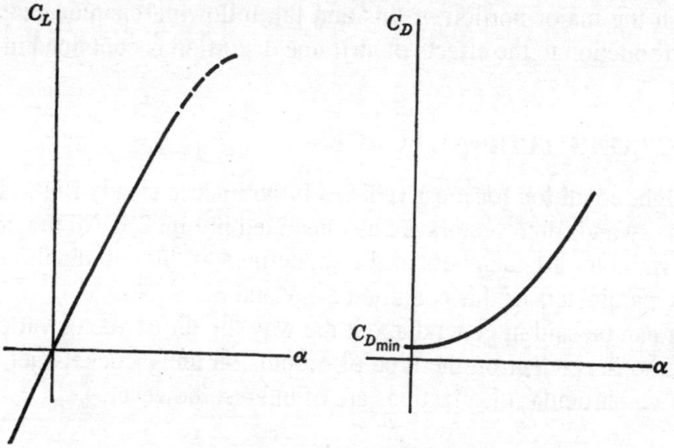

Figure 2.2 Lift and drag for subsonic and supersonic speeds.

body, the C_L curve may have a characteristic upward curvature even at small α (Flax and Lawrence, 1951). The upward curvature of C_L at small α is inherently present at hypersonic Mach numbers (Truitt, 1959). For the nonlinear cases, a suitable formulation for C_L is (USAF, 1978)

$$C_L = (\tfrac{1}{2}C_{N_\alpha} \sin 2\alpha + C_{N_{\alpha\alpha}} \sin \alpha \,|\sin \alpha|) \cos \alpha \qquad (2.1,3)$$

where C_{N_α} and $C_{N_{\alpha\alpha}}$ are coefficients (independent of α) that depend on the Mach number and configuration. [Actually C_N here is the coefficient of the aerodynamic force component normal to the wing chord, and C_{N_α} is the value of C_{L_α} at $\alpha = 0$, as can easily be seen by linearizing (2.1,3) with respect to α.] Equation (2.1,2) for the drag coefficient can serve quite well for flight dynamics applications up to hypersonic speeds ($\mathbf{M} > 5$) at which theory indicates that the exponent of C_L decreases from 2 to $\frac{3}{2}$. Miele (1962) presents in Chap. 6 a very useful and instructive set of typical lift and drag data for a wide range of vehicle types, from subsonic to hypersonic.

Balance, or Equilibrium

An airplane can continue in steady unaccelerated flight only when the resultant external force and moment about the CG both vanish. In particular, this requires that the pitching moment be zero. This is the condition of longitudinal *balance.* If the pitching moment were not zero, the airplane would experience a rotational acceleration component in the direction of the unbalanced moment. Figure 2.3 shows a typical graph of the pitching-moment coefficient about the CG[1] versus the angle of attack for an airplane with a fixed elevator (curve *a*). The angle of attack is measured from the zero-lift line of the airplane. The graph is a straight line except near the stall. Since zero C_m is required for balance, the airplane can fly only at the angle of attack marked *A*, for the given elevator angle.

[1]Unless otherwise specified, C_m always refers to moment about the CG.

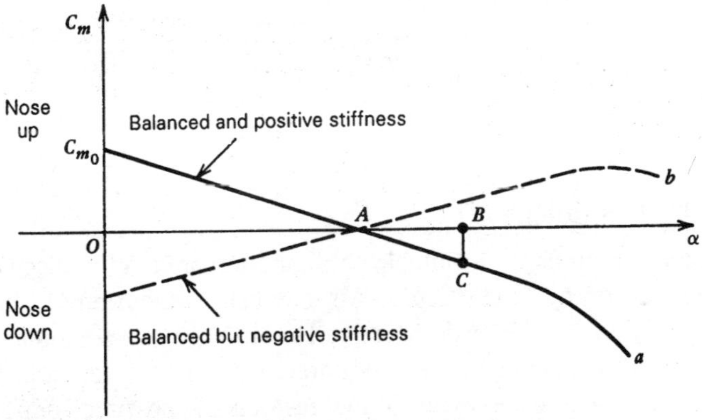

Figure 2.3 Pitching moment of an airplane about the CG.

Pitch Stiffness

Suppose that the airplane of curve *a* on Fig. 2.3 is disturbed from its equilibrium attitude, the angle of attack being increased to that at *B* while its speed remains unaltered. It is now subject to a negative, or nose-down, moment, whose magnitude corresponds to *BC*. This moment tends to reduce the angle of attack to its equilibrium value, and hence is a *restoring* moment. In this case, the airplane has positive pitch stiffness, obviously a desirable characteristic.

On the other hand, if C_m were given by the curve *b*, the moment acting when disturbed would be positive, or nose-up, and would tend to rotate the airplane still farther from its equilibrium attitude. We see that the pitch stiffness is determined by the sign and magnitude of the slope $\partial C_m/\partial \alpha$. *If the pitch stiffness is to be positive at the equilibrium* α, C_m *must be zero, and* $\partial C_m/\partial \alpha$ *must be negative.* It will be appreciated from Fig. 2.3 that an alternative statement is "C_{m_0} *must be positive, and* $\partial C_m/\partial \alpha$ *negative* if the airplane is to meet this (limited) condition for stable equilibrium." The various possibilities corresponding to the possible signs of C_{m_0} and $\partial C_m/\partial \alpha$ are shown in Figs. 2.3 and 2.4.

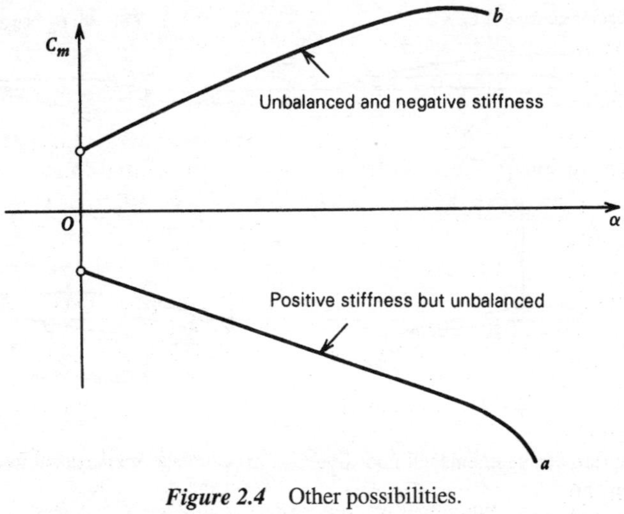

Figure 2.4 Other possibilities.

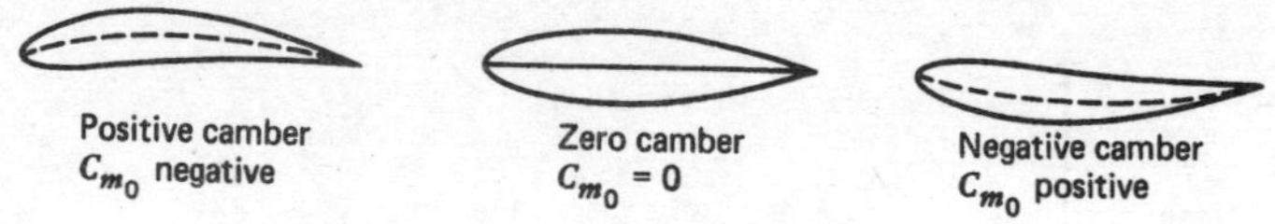

Figure 2.5 C_{m_0} of airfoil sections.

Possible Configurations

The possible solutions for a suitable configuration are readily discussed in terms of the requirements on C_{m_0} and $\partial C_m/\partial\alpha$. We state here without proof (this is given in Sec. 2.3) that $\partial C_m/\partial\alpha$ can be made negative for virtually any combination of lifting surfaces and bodies by placing the center of gravity far enough forward. Thus it is not the stiffness requirement, taken by itself, that restricts the possible configurations, but rather the requirement that the airplane must be *simultaneously* balanced and have positive pitch stiffness. Since a proper choice of the CG location can ensure a negative $\partial C_m/\partial\alpha$, *then any configuration with a positive* C_{m_0} *can satisfy the (limited) conditions for balanced and stable flight.*

Figure 2.5 shows the C_{m_0} of conventional airfoil sections. If an airplane were to consist of a straight wing alone (flying wing), then the wing camber would determine the airplane characteristics as follows:

Negative camber—flight possible at $\alpha > 0$; i.e., $C_L > 0$ (Fig. 2.3*a*).

Zero camber—flight possible only at $\alpha = 0$, or $C_L = 0$.

Positive camber—flight not possible at any positive α or C_L.

For straight-winged tailless airplanes, only the negative camber satisfies the conditions for stable, balanced flight. Effectively the same result is attained if a flap, deflected upward, is incorporated at the trailing edge of a symmetrical airfoil. A conventional low-speed airplane, with essentially straight wings and positive camber, could fly upside down without a tail, provided the CG were far enough forward (ahead of the wing mean aerodynamic center). Flying wing airplanes based on a straight wing with negative camber are not in general use for three main reasons:

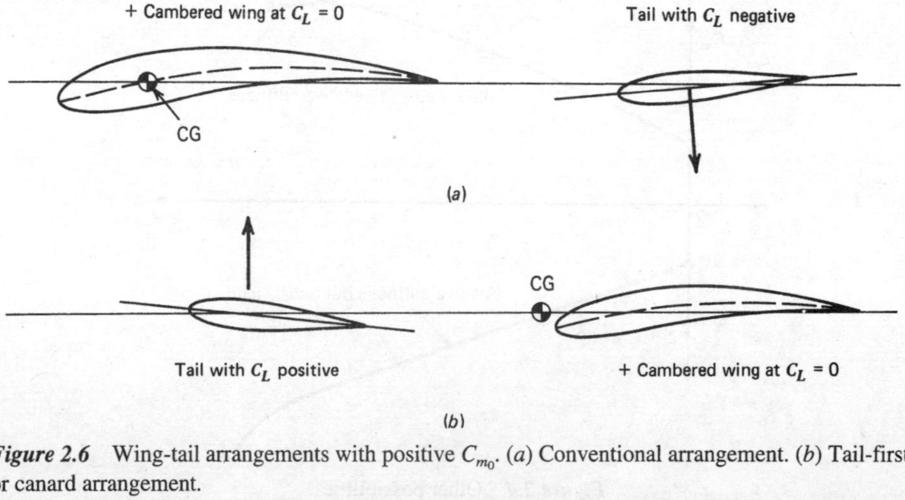

Figure 2.6 Wing-tail arrangements with positive C_{m_0}. (*a*) Conventional arrangement. (*b*) Tail-first or canard arrangement.

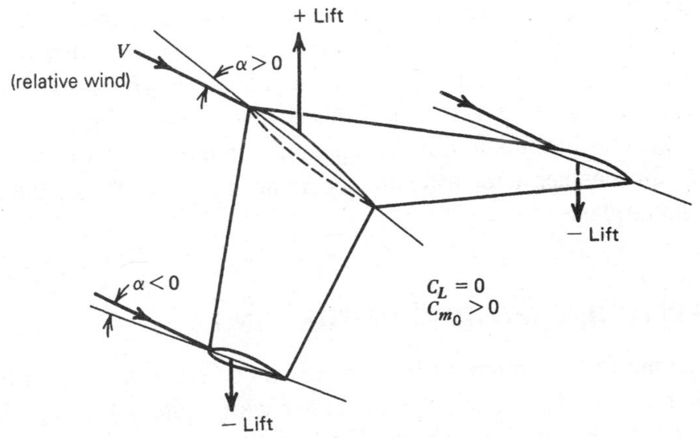

Figure 2.7 Swept-back wing with twisted tips.

1. The dynamic characteristics tend to be unsatisfactory.
2. The permissible CG range is too small.
3. The drag and $C_{L_{max}}$ characteristics are not good.

The positively cambered straight wing can be used only in conjunction with an auxiliary device that provides the positive C_{m_0}. The solution adopted by experimenters as far back as Samuel Henson (1842) and John Stringfellow (1848) was to add a tail behind the wing. The Wright brothers (1903) used a tail ahead of the wing (canard configuration). Either of these alternatives can supply a positive C_{m_0}, as illustrated in Fig. 2.6. When the wing is at zero lift, the auxiliary surface must provide a nose-up moment. The conventional tail must therefore be at a negative angle of attack, and the canard tail at a positive angle.

An alternative to the wing-tail combination is the swept-back wing with twisted tips (Fig. 2.7). When the net lift is zero, the forward part of the wing has positive lift, and the rear part negative. The result is a positive couple, as desired.

A variant of the swept-back wing is the delta wing. The positive C_{m_0} can be achieved with such planforms by twisting the tips, by employing negative camber, or by incorporating an upturned tailing edge flap.

2.2 *Synthesis of Lift and Pitching Moment*

The total lift and pitching moment of an airplane are, in general, functions of angle of attack, control-surface angle(s), Mach number, Reynolds number, thrust coefficient, and dynamic pressure.[2] (The last-named quantity enters because of aeroelastic effects. Changes in the dynamic pressure ($\frac{1}{2}\rho V^2$), when all the other parameters are constant, may induce enough distortion of the structure to alter C_m significantly.) An accurate determination of the lift and pitching moment is one of the major tasks in a static stability analysis. Extensive use is made of wind-tunnel tests, supplemented by aerodynamic and aeroelastic analyses.

[2]When partial derivatives are taken in the following equations with respect to one of these variables, for example, $\partial C_m/\partial\alpha$, it is to be understood that all the others are held constant.

For purposes of estimation, the total lift and pitching moment may be synthesized from the contributions of the various parts of the airplane, that is, wing, body, nacelles, propulsive system, and tail, and their mutual interferences. Some data for estimating the various aerodynamic parameters involved are contained in Appendix B, while the general formulation of the equations, in terms of these parameters, follows here. In this chapter aeroelastic effects are not included. Hence the analysis applies to a *rigid* airplane.

LIFT AND PITCHING MOMENT OF THE WING

The aerodynamic forces on any lifting surface can be represented as a lift and drag acting at the mean aerodynamic center, together with a pitching couple independent of the angle of attack (Fig. 2.8). The pitching moment of this force system about the CG is given by (Fig. 2.9)[3]

$$\begin{aligned} M_w = M_{ac_w} &+ (L_w \cos \alpha_w + D_w \sin \alpha_w)(h - h_{n_w})\bar{c} \\ &+ (L_w \sin \alpha_w - D_w \cos \alpha_w)z \end{aligned} \tag{2.2,1}$$

We assume that the angle of attack is sufficiently small to justify the approximations

$$\cos \alpha_w = 1, \qquad \sin \alpha_w = \alpha_w$$

and the equation is made nondimensional by dividing through by $\frac{1}{2}\rho V^2 S\bar{c}$. It then becomes

$$C_{m_w} = C_{m_{ac_w}} + (C_{L_w} + C_{D_w}\alpha_w)(h - h_{n_w}) + (C_{L_w}\alpha_w - C_{D_w})z/\bar{c} \tag{2.2,2}$$

Although it may occasionally be necessary to retain all the terms in (2.2,2), experience has shown that the last term is frequently negligible, and that $C_{D_w}\alpha_w$ may be neglected in comparison with C_{L_w}. With these simplifications, we obtain

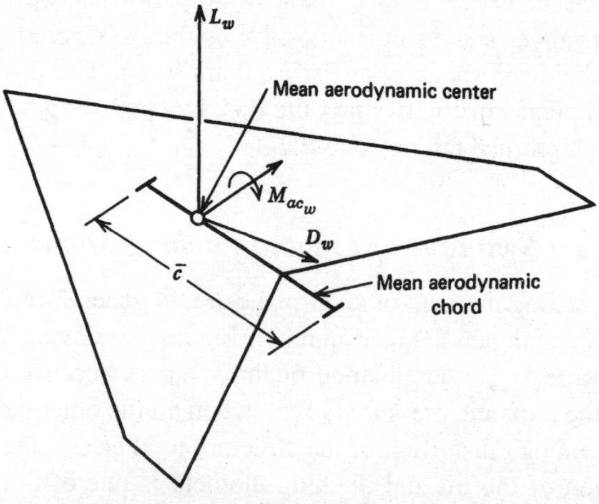

Figure 2.8 Aerodynamic forces on the wing.

[3]The notation h_{n_w} indicates that the mean aerodynamic center of the wing is also the *neutral point* of the wing. *Neutral point* is defined in Sec. 2.3.

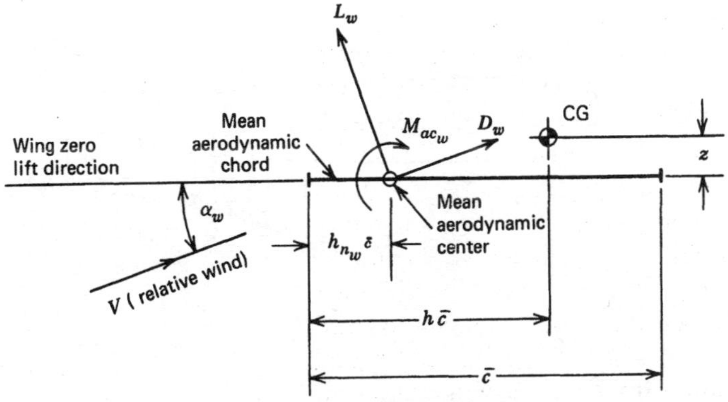

Figure 2.9 Moment about the CG in the plane of symmetry.

$$\begin{aligned} C_{m_w} &= C_{m_{ac_w}} + C_{L_w}(h - h_{n_w}) \\ &= C_{m_{ac_w}} + \alpha_w a_w(h - h_{n_w}) \end{aligned} \tag{2.2,3}$$

where $a_w = C_{L_{\alpha_w}}$ is the lift-curve-slope of the wing.

Equation 2.2,3 will be used to represent the wing pitching moment in the discussions that follow.

LIFT AND PITCHING MOMENT OF THE BODY AND NACELLES

The influences of the body and nacelles are complex. A body alone in an airstream is subjected to aerodynamic forces. These, like those on the wing, may be represented over moderate ranges of angle of attack by lift and drag forces at an aerodynamic center, and a pitching couple independent of α. Also as for a wing alone, the lift-α relation is approximately linear. When the wing and body are put together, however, a simple superposition of the aerodynamic forces that act upon them separately does not give a correct result. Strong interference effects are usually present, the flow field of the wing affecting the forces on the body, and vice versa.

These interference flow fields are illustrated for subsonic flow in Fig. 2.10. Part (*a*) shows the pattern of induced velocity along the body that is caused by the wing vortex system. This induced flow produces a positive moment that increases with wing lift or α. Hence a positive (destabilizing) contribution to C_{m_α} results. Part (*b*) shows an effect of the body on the wing. When the body axis is at angle α to the stream, there is a cross-flow component $V \sin \alpha$. The body distorts this flow locally, leading to cross-flow components that can be of order $2V \sin \alpha$ at the wing-body intersection. There is a resulting change in the wing lift distribution.

The result of adding a body and nacelles to a wing may usually be interpreted as a shift (forward) of the mean aerodynamic center, an increase in the lift-curve slope, and a negative increment in $C_{m_{ac}}$. The equation that corresponds to (2.2,3) for a wing-body-nacelle combination is then of the same form as (2.2,3), but with different values of the parameters. The subscript *wb* is used to denote these values.

$$\begin{aligned} C_{m_{wb}} &= C_{m_{ac_{wb}}} + C_{L_{wb}}(h - h_{n_{wb}}) \\ &= C_{m_{ac_{wb}}} + a_{wb}\alpha_{wb}(h - h_{n_{wb}}) \end{aligned} \tag{2.2,4}$$

where a_{wb} is the lift-curve-slope of the wing-body-nacelle combination.

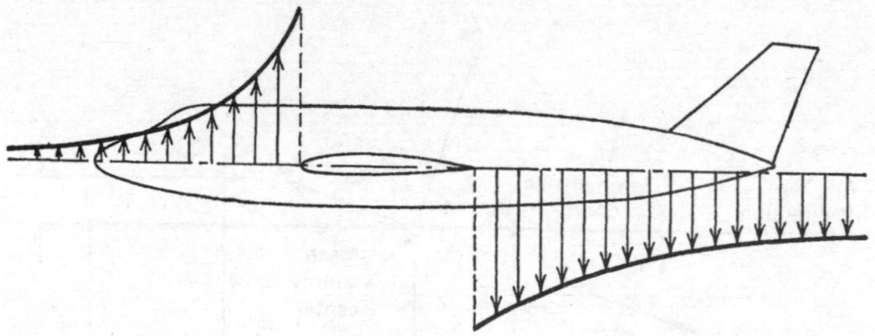

(a)

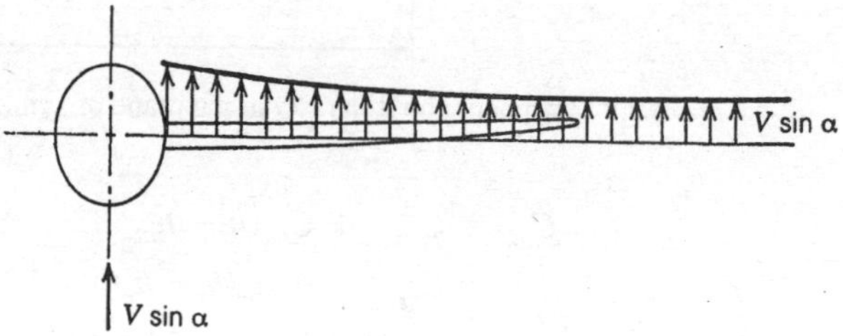

(b)

Figure 2.10 Example of mutual interference flow fields of wing and body—subsonic flow. (*a*) Qualitative pattern of upwash and downwash induced along the body axis by the wing vorticity. (*b*) Qualitative pattern of upwash induced along wing by the cross-flow past the body.

LIFT AND PITCHING MOMENT OF THE TAIL

The forces on an isolated tail are represented just like those on an isolated wing. When the tail is mounted on an airplane, however, important interferences occur. The most significant of these, and one that is usually predictable by aerodynamic theory, is a downward deflection of the flow at the tail caused by the wing. This is characterized by the mean downwash angle ϵ. Blanking of part of the tail by the body is a second effect, and a reduction of the relative wind when the tail lies in the wing wake is the third.

Figure 2.11 depicts the forces acting on the tail showing the relative wind vector of the airplane. $\mathbf{V}'$ is the average or effective relative wind at the tail. The tail lift and drag forces are, respectively, perpendicular and parallel to $\mathbf{V}'$. The reader should note

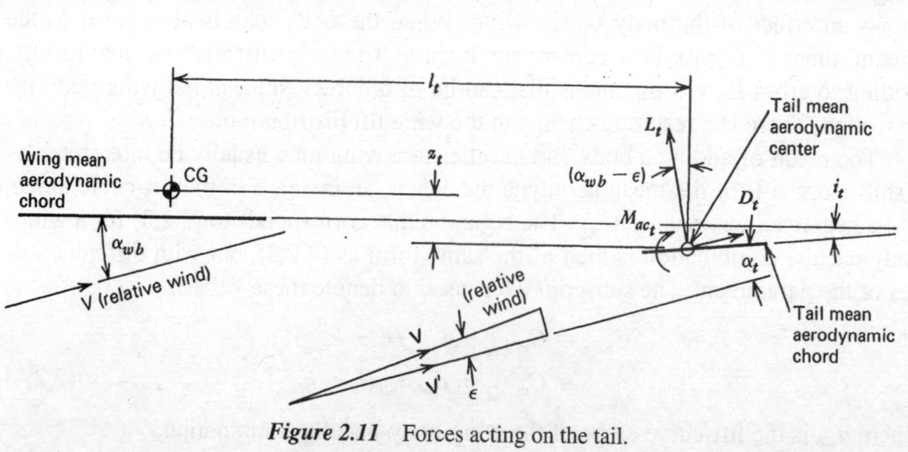

Figure 2.11 Forces acting on the tail.

the tail angle i_t, which must be positive as shown for equilibrium. This is sometimes referred to as *longitudinal dihedral.*

The contribution of the tail to the airplane lift, which by definition is perpendicular to **V**, is

$$L_t \cos \epsilon - D_t \sin \epsilon$$

ϵ is always a small angle, and we assume that $D_t\epsilon$ may be neglected compared with L_t. The contribution of the tail to the airplane lift then becomes simply L_t. We introduce the symbol C_{L_t} to represent the lift coefficient of the tail, based on the airplane dynamic pressure $\frac{1}{2}\rho V^2$ and the tail area S_t.

$$C_{L_t} = \frac{L_t}{\frac{1}{2}\rho V^2 S_t} \tag{2.2,5}$$

The total lift of the airplane is

$$L = L_{wb} + L_t,$$

or in coefficient form

$$C_L = C_{L_{wb}} + \frac{S_t}{S} C_{L_t} \tag{2.2,6}$$

The reader should note that the lift coefficient of the tail is often based on the local dynamic pressure at the tail, which differs from $\frac{1}{2}\rho V^2$ when the tail lies in the wing wake. This practice entails carrying the ratio V'/V in many subsequent equations. The definition employed here amounts to incorporating V'/V into the tail lift-curve slope a_t. This quantity is in any event different from that for the isolated tail, owing to the interference effects previously noted. This circumstance is handled in various ways in the literature. Sometimes a tail efficiency factor η_t is introduced, the isolated tail lift slope being multiplied by η_t. In other treatments, η_t is used to represent $(V'/V)^2$. In the convention adopted here, a_t is the lift-curve slope of the tail, as measured in situ on the airplane, and based on the dynamic pressure $\frac{1}{2}\rho V^2$. This is the quantity that is directly obtained in a wind-tunnel test.

From Fig. 2.11 we find the pitching moment of the tail about the CG to be

$$\begin{aligned} M_t = &-l_t[L_t \cos(\alpha_{wb} - \epsilon) + D_t \sin(\alpha_{wb} - \epsilon)] \\ &- z_t[D_t \cos(\alpha_{wb} - \epsilon) - L_t \sin(\alpha_{wb} - \epsilon)] + M_{ac_t} \end{aligned} \tag{2.2,7}$$

Experience has shown that in the majority of instances the dominant term in this equation is the first one, and that all others are negligible by comparison. Only this case will be dealt with here. The reader is left to extend the analysis to cases in which this approximation is not valid. With the above approximation, and that of small angles,

$$M_t = -l_t L_t = -l_t C_{L_t} \tfrac{1}{2}\rho V^2 S_t$$

Upon conversion to coefficient form, we obtain

$$C_{m_t} = \frac{M_t}{\frac{1}{2}\rho V^2 S\bar{c}} = -\frac{l_t}{\bar{c}}\frac{S_t}{S} C_{L_t} \tag{2.2,8}$$

The combination $l_t S_t / S\bar{c}$ is the ratio of two volumes characteristic of the airplane's

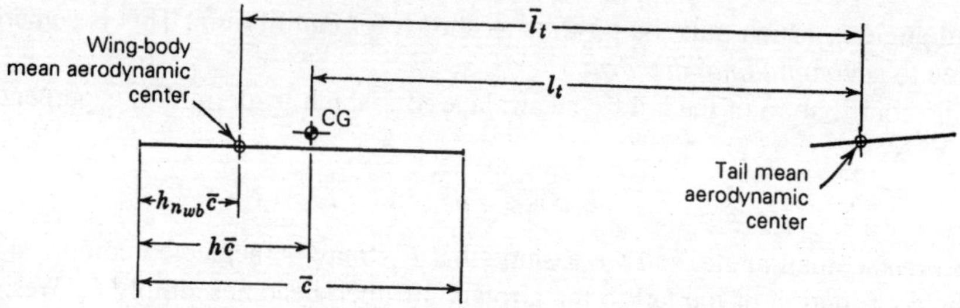

Figure 2.12 Wing-body and tail mean aerodynamic centers.

geometry. It is commonly called the "horizontal-tail volume ratio," or more simply, the "tail volume." It is denoted here by V_H. Thus

$$C_{m_t} = -V_H C_{L_t} \tag{2.2,9}$$

Since the center of gravity is not a fixed point, but varies with the loading condition and fuel consumption of the vehicle, V_H in (2.2,9) is not a constant (although it does not vary much). It is a little more convenient to calculate the moment of the tail about a fixed point, the mean aerodynamic center of the wing-body combination, and to use this moment in the subsequent algebraic manipulations. Figure 2.12 shows the relevant relationships, and we define

$$\bar{V}_H = \frac{\bar{l}_t S_t}{\bar{c} S} \tag{2.2,10}$$

which leads to

$$V_H = \bar{V}_H - \frac{S_t}{S}(h - h_{n_{wb}}) \tag{2.2,11}$$

The moment of the tail about the wing-body mean aerodynamic center is then [cf. (2.2,9)]

$$\bar{C}_{m_t} = -\bar{V}_H C_{L_t} \tag{2.2,12}$$

and its moment about the CG is, from substitution of (2.2,11) into (2.2,9)

$$C_{m_t} = -\bar{V}_H C_{L_t} + C_{L_t}\frac{S_t}{S}(h - h_{n_{wb}}) \tag{2.2,13}$$

PITCHING MOMENT OF A PROPULSIVE SYSTEM

The moment provided by a propulsive system is in two parts: (1) that coming from the forces acting on the unit itself, for example, the thrust and in-plane force acting on a propeller, and (2) that coming from the interaction of the propulsive slipstream with the other parts of the airplane. These are discussed in more detail in Sec. 3.4. We assume that the interference part is included in the moments already given for the wing, body, and tail, and denote by C_{m_p} the remaining moment from the propulsion units.

2.3 Total Pitching Moment and Neutral Point

On summing the first of (2.2,4) and (2.2,13) making use of (2.2,6) and adding the contribution C_{m_p} for the propulsive system, we obtain the total pitching moment about the CG

$$C_m = C_{m_{ac_{wb}}} + C_L(h - h_{n_{wb}}) - \bar{V}_H C_{L_t} + C_{m_p} \tag{2.3,1}$$

It is worthwhile repeating that no assumptions about thrust, compressibility, or aeroelastic effects have been made in respect of (2.3,1). The pitch stiffness $(-C_{m_\alpha})$ is now obtained from (2.3,1). Recall that the mean aerodynamic centers of the wing-body combination and of the tail are fixed points, so that

$$C_{m_\alpha} = \frac{\partial C_{m_{ac_{wb}}}}{\partial \alpha} + C_{L_\alpha}(h - h_{n_{wb}}) - \bar{V}_H \frac{\partial C_{L_t}}{\partial \alpha} + \frac{\partial C_{m_p}}{\partial \alpha} \tag{2.3,2}$$

If a true mean aerodynamic center in the classical sense exists, then $\partial C_{m_{ac_{wb}}}/\partial \alpha$ is zero and

$$C_{m_\alpha} = C_{L_\alpha}(h - h_{n_{wb}}) - \bar{V}_H \frac{\partial C_{L_t}}{\partial \alpha} + \frac{\partial C_{m_p}}{\partial \alpha} \tag{2.3,3}$$

C_{m_α} as given by (2.3,2) or (2.3,3) depends linearly on the CG position, h. Since C_{L_α} is usually large, the magnitude and sign of C_{m_α} depend strongly on h. This is the basis of the statement in Sec. 2.2 that C_{m_α} can always be made negative by a suitable choice of h. The CG position h_n for which C_{m_α} is zero is of particular significance, since this represents a boundary between positive and negative pitch stiffness. In this book we define h_n as the *neutral point,* NP. It has the same significance for the vehicle as a whole as does the mean aerodynamic center for a wing alone, and indeed the term *vehicle aerodynamic center* is an acceptable alternative to "neutral point."

The location of the NP is readily calculated from (2.3,2) by setting the left-hand side to zero leading to

$$h_n = h_{n_{wb}} - \frac{1}{C_{L_\alpha}}\left(\frac{\partial C_{m_{ac_{wb}}}}{\partial \alpha} - \bar{V}_H \frac{\partial C_{L_t}}{\partial \alpha} + \frac{\partial C_{m_p}}{\partial \alpha}\right) \tag{2.3,4}$$

Substitution of (2.3,4) back into (2.3,2) simplifies the latter to

$$C_{m_\alpha} = C_{L_\alpha}(h - h_n) \tag{2.3,5}$$

which is valid whether $C_{m_{ac_{wb}}}$ and C_{m_p} vary with α or not. Equation (2.3,5) clearly provides an excellent way of finding h_n from test results, that is from measurements of C_{m_α} and C_{L_α}. The difference between the CG position and the NP is sometimes called the *static margin,*

$$K_n = (h_n - h) \tag{2.3,6}$$

Since the criterion to be satisfied is $C_{m_\alpha} < 0$, that is, positive pitch stiffness, then we see that we must have $h < h_n$, or $K_n > 0$. In other words the CG must be forward of the NP. The farther forward the CG the greater is K_n, and in the sense of "static stability" the more stable the vehicle.

The neutral point has sometimes been defined as the CG location at which the derivative $dC_m/dC_L = 0$. When this definition is applied to the gliding flight of a rigid airplane at low Mach number, the neutral point obtained is identical with that defined in this book. This is so because under these restricted conditions C_L is a unique function of α, and $dC_m/dC_L = (\partial C_m/\partial\alpha)/(\partial C_L/\partial\alpha)$. Then dC_m/dC_L and $\partial C_m/\partial\alpha$ are simultaneously zero. In general, however, C_m and C_L are both functions of several variables, as pointed out at the beginning of Sec. 2.2. For fixed values of δ_e and h, and neglecting Reynolds number effects (these are usually very small), we may write

$$C_L = f(\alpha, \mathbf{M}, C_T, \tfrac{1}{2}\rho V^2), \qquad C_m = g(\alpha, \mathbf{M}, C_T, \tfrac{1}{2}\rho V^2) \tag{2.3,7}$$

where C_T is the thrust coefficient, defined in Sec. 2.11.

Mathematically speaking, the derivative dC_m/dC_L does not exist unless $\mathbf{M}$, C_T, and $\frac{1}{2}\rho V^2$ are functions of C_L. When that is the case, then

$$\frac{dC_m}{dC_L} = \frac{\partial C_m}{\partial\alpha}\frac{\partial\alpha}{\partial C_L} + \frac{\partial C_m}{\partial\mathbf{M}}\frac{\partial\mathbf{M}}{\partial C_L} + \frac{\partial C_m}{\partial C_T}\frac{\partial C_T}{\partial C_L} + \frac{\partial C_m}{\partial(\frac{1}{2}\rho V^2)}\frac{\partial(\frac{1}{2}\rho V^2)}{\partial C_L} \tag{2.3,8}$$

Equation 2.3,8 has meaning only when a specific kind of flight is prescribed: e.g., horizontal unaccelerated flight, or rectilinear climbing flight at full throttle. When a condition of this kind is imposed, then $\mathbf{M}$, C_T, and the dynamic pressure are definite functions of C_L, dC_m/dC_L exists, and a neutral point may be calculated. The neutral point so found is not an index of stability with respect to angle of attack disturbances, and the question arises as to what it does relate to. It can be shown that it relates to the trim curves of the airplane. A plot of the elevator angle to trim versus speed will have a zero slope when dC_m/dC_L is zero, and a negative slope when the CG lies aft of the neutral point so defined. As shown in Sec. 2.4, this reversal of slope indicates a *tendency* toward instability with respect to speed, but only a dynamic analysis can show whether or not the airplane is stable in this condition. There are cases when the application of the "trim-slope" criterion can be definitely misleading as to stability. One such is level unaccelerated flight, during which the throttle must be adjusted every time the flight speed or C_L is altered.

It can be seen from the foregoing remarks that the "trim-slope" criterion for the neutral point does not lead to any definite and clear-cut conclusions, either about the stability with respect to angle of attack disturbances, or about the general static stability involving both speed and angle of attack disturbances. It is mainly for this reason that the neutral point has been defined herein on the basis of $\partial C_m/\partial\alpha$.

EFFECT OF LINEAR LIFT AND MOMENT ON NEUTRAL POINT

When the forces and moments on the wing, body, tail, and propulsive system are linear in α, as may be near enough the case in reality, some additional useful relations can be obtained. We then have

$$C_{L_{wb}} = a_{wb}\alpha_{wb} \tag{2.3,9}$$

$$C_{L_t} = a_t\alpha_t \tag{2.3,10}$$

and

$$C_{m_p} = C_{m0_p} + \frac{\partial C_{m_p}}{\partial\alpha}\cdot\alpha \tag{2.3,11}$$

If $C_{m_{wb}}$ is linear in $C_{L_{wb}}$, it can be shown (see exercise 2.3) that $C_{m_{ac_{wb}}}$ does not vary with $C_{L_{wb}}$, i.e. that a true mean aerodynamic center exists. Figure 2.11 shows that the tail angle of attack is

$$\alpha_t = \alpha_{wb} - i_t - \epsilon \tag{2.3,12}$$

and hence

$$C_{L_t} = a_t(\alpha_{wb} - i_t - \epsilon) \tag{2.3,13}$$

The downwash ϵ can usually be adequately approximated by

$$\epsilon = \epsilon_0 + \frac{\partial \epsilon}{\partial \alpha}\alpha_{wb} \tag{2.3,14}$$

The downwash ϵ_0 at $\alpha_{wb} = 0$ results from the induced velocity field of the body and from wing twist; the latter produces a vortex wake and downwash field even at zero total lift. The constant derivative $\partial\epsilon/\partial\alpha$ occurs because the main contribution to the downwash at the tail comes from the wing trailing vortex wake, the strength of which is, in the linear case, proportional to C_L.

The tail lift coefficient then is

$$C_{L_t} = a_t\left[\alpha_{wb}\left(1 - \frac{\partial \epsilon}{\partial \alpha}\right) - i_t - \epsilon_0\right] \tag{2.3,15}$$

and the total lift, from (2.2,6) and (2.3,9) is

$$C_L = a_{wb}\alpha_{wb}\left[1 + \frac{a_t S_t}{a_{wb} S}\left(1 - \frac{\partial \epsilon}{\partial \alpha}\right)\right] - a_t \frac{S_t}{S}(i_t + \epsilon_0) \quad (a)$$

or

$$C_L = (C_L)_0 + a\alpha_{wb} \quad (b) \quad (2.3,16)$$

or since α_{wb} and α differ by a constant

$$C_L = a\alpha \quad (c)$$

where

$$(C_L)_0 = -a_t \frac{S_t}{S}(i_t + \epsilon_0) \tag{2.3,17}$$

is the coefficient of lift when $\alpha_{wb} = 0$;

$$a = \frac{\partial C_L}{\partial \alpha} = a_{wb}\left[1 + \frac{a_t}{a_{wb}}\frac{S_t}{S}\left(1 - \frac{\partial \epsilon}{\partial \alpha}\right)\right] \tag{2.3,18}$$

is the lift-curve slope of the whole configuration; and α is the angle of attack of the zero-lift line of the whole configuration (see Fig. 2.13). Note that, since i_t is positive,

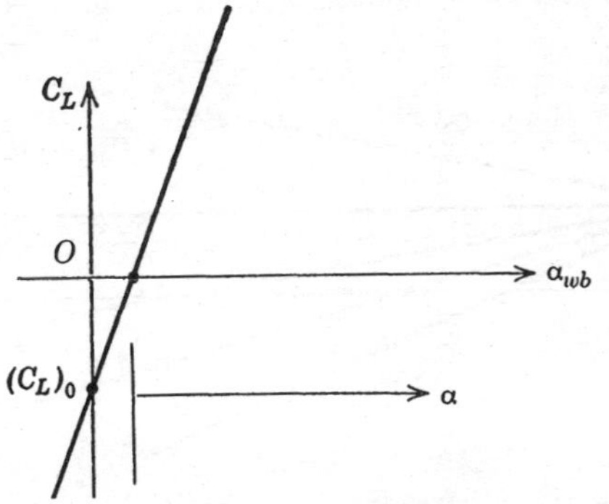

Figure 2.13 Graph of total lift.

then $(C_L)_0$ is negative. The difference between α and α_{wb} is found by equating (2.3,16*b* and *c*) to be

$$\alpha - \alpha_{wb} = -\frac{a_t}{a}\frac{S_t}{S}(i_t + \epsilon_0) \tag{2.3,19}$$

When the linear relations for C_L, C_{L_t} and C_{m_p} are substituted into (2.3,1) the following results can be obtained after some algebraic reduction:

$$\begin{aligned} C_m &= C_{m_0} + C_{m_\alpha}\alpha & (a)\\ C_m &= \bar{C}_{m_0} + C_{m_\alpha}\alpha_{wb} & (b) \end{aligned} \qquad (2.3,20)$$

where
$$C_{m_\alpha} = a(h - h_{n_{wb}}) - a_t\bar{V}_H\left(1 - \frac{\partial\epsilon}{\partial\alpha}\right) + \frac{\partial C_{m_p}}{\partial\alpha} \qquad (a)$$

or
$$C_{m_\alpha} = a_{wb}(h - h_{n_{wb}}) - a_tV_H\left(1 - \frac{\partial\epsilon}{\partial\alpha}\right) + \frac{\partial C_{m_p}}{\partial\alpha} \qquad (b) \qquad (2.3,21)$$

and
$$C_{m_0} = C_{m_{ac_{wb}}} + C_{m_{0_p}} + a_t\bar{V}_H(\epsilon_0 + i_t) \times\left[1 - \frac{a_t}{a}\frac{S_t}{S}\left(1 - \frac{\partial\epsilon}{\partial\alpha}\right)\right] \qquad (a)$$

$$\bar{C}_{m_0} = C_{m_{ac_{wb}}} + \bar{C}_{m_{0_p}} + a_tV_H(\epsilon_0 + i_t) \qquad (b) \qquad (2.3,22)$$

where
$$\bar{C}_{m_{0_p}} = C_{m_{0_p}} + (\alpha - \alpha_{wb})\frac{\partial C_{m_p}}{\partial\alpha} \qquad (c)$$

$$h_n = h_{n_{wb}} + \frac{a_t}{a}V_H\left(1 - \frac{\partial\epsilon}{\partial\alpha}\right) - \frac{1}{a}\frac{\partial C_{m_p}}{\partial\alpha} \tag{2.3,23}$$

Note that since $\bar{C}_{m_0}$ is the pitching moment at zero α_{wb}, not at zero total lift, its value depends on h (via V_H), whereas C_{m_0}, being the moment at zero total lift, represents a couple and is hence independent of CG position. All the above relations apply to tailless aircraft by putting $\bar{V}_H = 0$. Another useful relation comes from integrating (2.3,5), i.e.

$$\begin{aligned} C_m &= C_{m_0} + C_L(h - h_n) & (a)\\ C_m &= C_{m_0} + a\alpha(h - h_n) & (b)\\ \text{or}\quad C_{m_\alpha} &= a(h - h_n) & (c) \end{aligned} \qquad (2.3,25)$$

Figure 2.14 shows the linear C_m vs. α relation, and Fig. 2.15 shows the resultant system of lift and moment that corresponds to (2.3,25), that is a force C_L and a couple

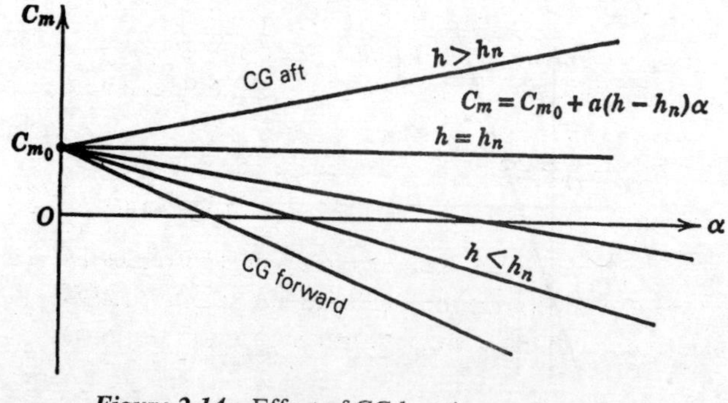

Figure 2.14 Effect of CG location on C_m curve.

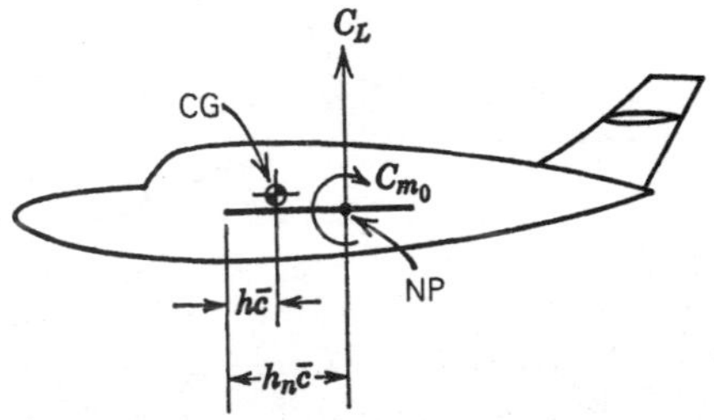

Figure 2.15 Total lift and moment acting on vehicle.

C_{m0} at the NP. Figure 2.15 is a very important result that the student should fix in his mind.

2.4 Longitudinal Control

In this section we discuss the longitudinal control of the vehicle from a *static* point of view. That is, we concern ourselves with how the equilibrium state of steady rectilinear flight is governed by the available controls. Basically there are two kinds of changes that can be made by the pilot or automatic control system—a change of propulsive thrust, or a change of configuration. Included in the latter are the operation of aerodynamic controls—elevators, wing flaps, spoilers, and horizontal tail rotation. Since the equilibrium state is dominated by the requirement $C_m = 0$, the most powerful controls are those that have the greatest effect on C_m.

Figure 2.14 shows that another theoretically possible way of changing the trim condition is to move the CG, which changes the value of α at which $C_m = 0$. Moving it forward reduces the trim α or C_L, and hence produces an increase in the trim speed. This method was actually used by Lilienthal, a pioneer of aviation, in gliding flights during 1891/1896, in which he shifted his body to move the CG. It has the inherent disadvantage, apart from practical difficulties, of changing C_{m_α} at the same time, reducing the pitch stiffness and hence stability, when the trim speed is reduced.

The longitudinal control now generally used is aerodynamic. A variable pitching moment is provided by moving the elevator, which may be all or part of the tail, or a trailing-edge flap in a tailless design. Deflection of the elevator through an angle δ_e produces increments in both the C_m and C_L of the airplane. The ΔC_L caused by the elevator of aircraft with tails is small enough to be neglected for many purposes. This is not so for tailless aircraft, where the ΔC_L due to elevators is usually significant. We shall assume that the lift and moment increments for both kinds of airplane are linear in δ_e, which is a fair representation of the characteristics of typical controls at high Reynolds number. Therefore,

$$\Delta C_L = C_{L_{\delta_e}} \delta_e \qquad (a)$$

$$C_L = C_L(\alpha) + C_{L_{\delta_e}} \delta_e \qquad (b)$$

$$\Delta C_m = C_{m_{\delta_e}} \delta_e \qquad (c) \qquad (2.4,1)$$

and

$$C_m = C_m(\alpha) + C_{m_{\delta_e}} \delta_e \qquad (d)$$

where $C_{L_{\delta_e}} = \partial C_L/\partial \delta_e$, $C_{m_{\delta_e}} = \partial C_m/\partial \delta_e$, and $C_L(\alpha)$, $C_m(\alpha)$ are the "basic" lift and moment when $\delta_e = 0$. The usual convention is to take down elevator as positive (Fig. 2.16*a*). This leads to positive $C_{L_{\delta_e}}$ and negative $C_{m_{\delta_e}}$. The deflection of the elevator

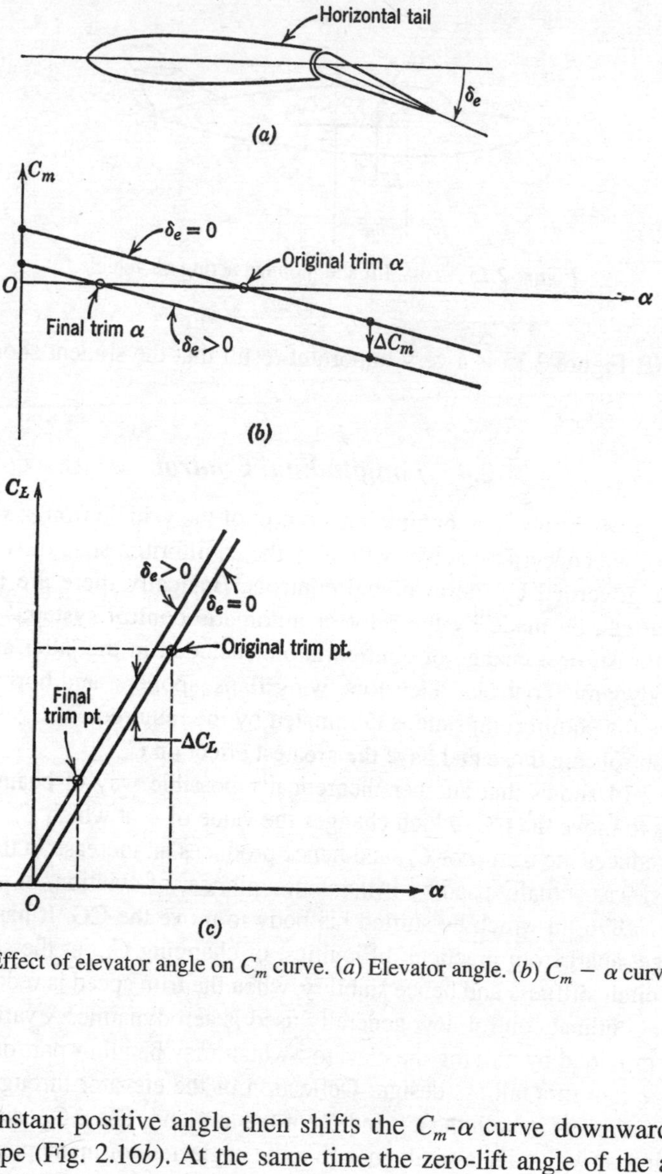

Figure 2.16 Effect of elevator angle on C_m curve. (*a*) Elevator angle. (*b*) $C_m - \alpha$ curve. (*c*) $C_L - \alpha$ curve.

through a constant positive angle then shifts the C_m-α curve downward, without change of slope (Fig. 2.16*b*). At the same time the zero-lift angle of the airplane is slightly changed (Fig. 2.16*c*).

In the case of linear lift and moment, we have

$$C_L = C_{L_\alpha}\alpha + C_{L_{\delta_e}}\delta_e \qquad (a)$$
$$C_m = C_{m_0} + C_{m_\alpha}\alpha + C_{m_{\delta_e}}\delta_e \qquad (b) \qquad (2.4,2)$$

THE DERIVATIVES $C_{L_{\delta_e}}$ AND $C_{m_{\delta_e}}$

Equation (2.2,6) gives the vehicle lift. Hence

$$C_{L_{\delta_e}} = \frac{\partial C_L}{\partial \delta_e} = \frac{\partial C_{L_{wb}}}{\partial \delta_e} + \frac{S_t}{S}\frac{\partial C_{L_t}}{\partial \delta_e} \qquad (2.4,3)$$

in which only the first term applies for tailless aircraft and the second for conventional tail elevators or all moving tails (when i_t is used instead of δ_e). We define the elevator lift effectiveness as

$$a_e = \frac{\partial C_{L_t}}{\partial \delta_e} \tag{2.4,4}$$

so that (2.4,3) becomes

$$C_{L_{\delta_e}} = \frac{\partial C_{L_{wb}}}{\partial \delta_e} + a_e \frac{S_t}{S} \qquad (a) \quad (2.4,5)$$

and the lift coefficient of the tail is

$$C_{L_t} = a_t\alpha_t + a_e\delta_e \qquad (b)$$

The total vehicle C_m is given for both tailed and tailless types by (2.3,1). For the latter, of course, $\bar{V}_H = 0$. Taking the derivative with respect to δ_e gives

$$C_{m_{\delta_e}} = \frac{\partial C_{m_{ac_{wb}}}}{\partial \delta_e} + C_{L_{\delta_e}}(h - h_{n_{wb}}) - \bar{V}_H \frac{\partial C_{L_t}}{\partial \delta_e} + \frac{\partial C_{m_p}}{\partial \delta_e} \tag{2.4,6}$$

We may usually neglect the last term, since there is unlikely to be any propulsive-elevator interaction that cannot be included in a_e. Then (2.4,6) becomes

$$C_{m_{\delta_e}} = \frac{\partial C_{m_{ac_{wb}}}}{\partial \delta_e} + C_{L_{\delta_e}}(h - h_{n_{wb}}) - a_e\bar{V}_H \tag{2.4,7}$$

Summarizing for both types of vehicle, we have (retaining only the dominant terms)

Tailed aircraft:

$$C_{L_{\delta_e}} = a_e \frac{S_t}{S} \qquad (a)$$
$$C_{m_{\delta_e}} = -a_e\bar{V}_H + C_{L_{\delta_e}}(h - h_{n_{wb}}) \qquad (b) \quad (2.4,8)$$

Tailless aircraft:

$$C_{L_{\delta_e}} = \frac{\partial C_L}{\partial \delta_e} \qquad (a)$$
$$C_{m_{\delta_e}} = \frac{\partial C_{m_{ac}}}{\partial \delta_e} + C_{L_{\delta_e}}(h - h_n) \qquad (b) \quad (2.4,9)$$

In the last case, the subscript *wb* is, of course, redundant and has been dropped. The primary parameters to be predicted or measured are a_e for tailed aircraft, and $\partial C_L/\partial \delta_e$, $\partial C_{m_{ac}}/\partial \delta_e$ for tailless.

ELEVATOR ANGLE TO TRIM

The trim condition is $C_m = 0$, whence from (2.4,1*d*)

$$\delta_{e_{\text{trim}}} = -\frac{C_m(\alpha)}{C_{m_{\delta_e}}} \tag{2.4,10}$$

and the corresponding lift coefficient is

$$C_{L_{\text{trim}}} = C_L(\alpha) + C_{L_{\delta_e}} \delta_{e_{\text{trim}}}$$

$$= C_L(\alpha) - \frac{C_{L_{\delta_e}}}{C_{m_{\delta_e}}} C_m(\alpha) \qquad (2.4,11)$$

When the linear lift and moment relations (2.4,2) apply the equations for trim are

$$\begin{bmatrix} C_{L_\alpha} & C_{L_{\delta_e}} \\ C_{m_\alpha} & C_{m_{\delta_e}} \end{bmatrix} \begin{bmatrix} \alpha_{\text{trim}} \\ \delta_{e_{\text{trim}}} \end{bmatrix} = \begin{bmatrix} C_{L_{\text{trim}}} \\ -C_{m_0} \end{bmatrix} \qquad (2.4,12)$$

These equations are solved for α and δ_e to give

$$\alpha_{\text{trim}} = \frac{C_{m_0} C_{L_{\delta_e}} + C_{m_{\delta_e}} C_{L_{\text{trim}}}}{det} \qquad (a)$$

$$\delta_{e_{\text{trim}}} = -\frac{C_{m_0} C_{L_\alpha} + C_{m_\alpha} C_{L_{\text{trim}}}}{det} \qquad (b) \quad (2.4,13)$$

$$\frac{d\delta_{e_{\text{trim}}}}{dC_{L_{\text{trim}}}} = -\frac{C_{m_\alpha}}{det} = -\frac{C_{L_\alpha}}{det}(h - h_n) \qquad (c)$$

where

$$det = C_{L_\alpha} C_{m_{\delta_e}} - C_{L_{\delta_e}} C_{m_\alpha} \qquad (d)$$

is the determinant of the square matrix in (2.4,12) and is normally negative. The values of *det* for the two types of airplane are readily calculated from (2.4,8 and 9) together with (2.3,5) to give

Tailed aircraft:

$$det = C_{L_\alpha}[C_{L_{\delta_e}}(h_n - h_{n_{wb}}) - a_e \bar{V}_H] \qquad (a)$$

Tailless aircraft:

$$det = C_{L_\alpha} \frac{\partial C_{m_{ac}}}{\partial \delta_e} \qquad (b) \quad (2.4,14)$$

and both are independent of h. From (2.4,13*a*) we get the *trimmed lift curve:*

$$C_{L_{\text{trim}}} = -\frac{C_{m_0} C_{L_{\delta_e}}}{C_{m_{\delta_e}}} + \frac{det}{C_{m_{\delta_e}}} \alpha_{\text{trim}} \qquad (2.4,15)$$

and the slope is given by

$$\left(\frac{dC_L}{d\alpha}\right)_{\text{trim}} = C_{L_\alpha} - \frac{C_{L_{\delta_e}}}{C_{m_{\delta_e}}} C_{m_\alpha} \qquad (2.4,16)$$

The trimmed lift-curve slope is seen to be less than C_{L_α} by an amount that depends on C_{m_α}, i.e., on the static margin, and that vanishes when $h = h_n$. The difference is only a few percent for tailed airplanes at normal CG position, but may be appreciable for tailless vehicles because of their larger C_{L_δ}. The relation between the basic and trimmed lift curves is shown in Fig. 2.17.

Equation (2.4,13*b*) is plotted on Fig. 2.18, showing how $\delta_{e_{\text{trim}}}$ varies with $C_{L_{\text{trim}}}$ and CG position when the aerodynamic coefficients are constant.

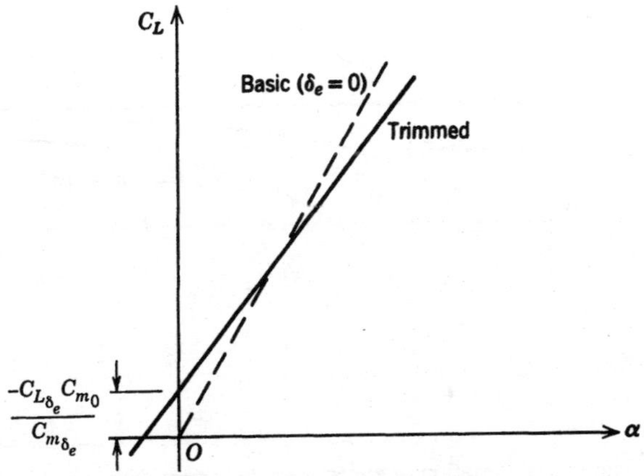

Figure 2.17 Trimmed lift curve.

VARIATION OF $\delta_{e_{trim}}$ WITH SPEED

When, in the absence of compressibility, aeroelastic effects, and propulsive system effects, the aerodynamic coefficients of (2.4,13) are constant, the variation of $\delta_{e_{trim}}$ with speed is simple. Then $\delta_{e_{trim}}$ is a unique function of $C_{L_{trim}}$ for each CG position. Since $C_{L_{trim}}$ is in turn fixed by the equivalent airspeed,[4] for horizontal flight

$$C_{L_{trim}} = \frac{W}{\frac{1}{2}\rho_0 V_E{}^2 S} \tag{2.4,17}$$

then $\delta_{e_{trim}}$ becomes a unique function of V_E. The form of the curves is shown in Fig. 2.19 for representative values of the coefficients.

The variation of $\delta_{e_{trim}}$ with $C_{L_{trim}}$ or speed shown on Figs. 2.18 and 2.19 is the normal and desirable one. For any CG position, an increase in trim speed from any initial value to a larger one requires a downward deflection of the elevator (a forward

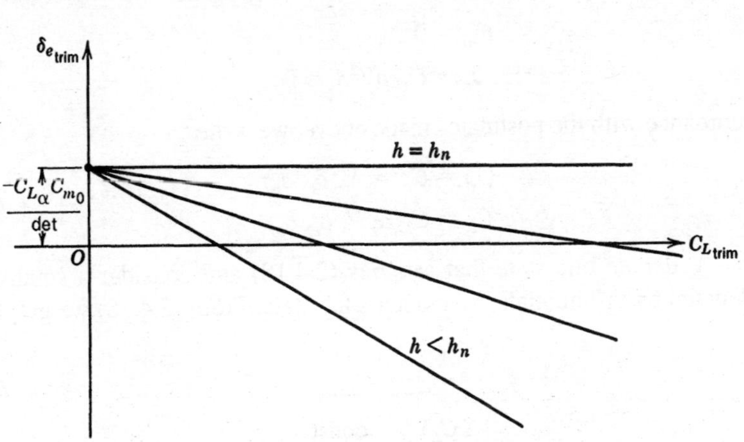

Figure 2.18 Elevator angle to trim at various CG positions.

[4]Equivalent airspeed (EAS) is $V_E = V\sqrt{\rho/\rho_0}$ where ρ_0 is standard sea-level density.

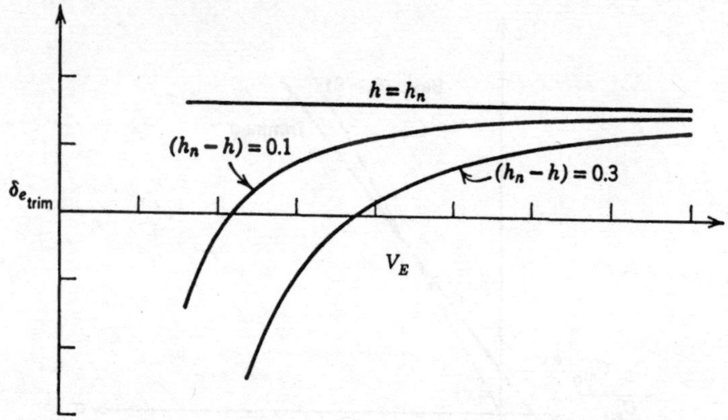

Figure 2.19 Example of variation of elevator angle to trim with speed and CG position.

movement of the pilot's control). The "gradient" of the movement $\partial\delta_{e_{\rm trim}}/\partial V_E$ is seen to decrease with rearward movement of the CG until it vanishes altogether at the NP. In this condition the pilot in effect has no control over trim speed, and control of the vehicle becomes very difficult. For even more rearward positions of the CG the gradient reverses, and the controllability deteriorates still further.

When the aerodynamic coefficients vary with speed, the above simple analysis must be extended. In order to be still more general, we shall in the following explicitly include propulsive effects as well, by means of the parameter δ_p, which stands for the state of the pilot's propulsion control (e.g., throttle position). δ_p = constant therefore denotes fixed-throttle and, of course, for horizontal flight at varying speed, δ_p must be a function of V that is compatible with $T = D$. For angles of climb or descent in the normal range of conventional airplanes $L \doteq W$ is a reasonable approximation, and we adopt it in the following. When nonhorizontal flight is thus included, δ_p becomes an independent variable, with the angle of climb γ then becoming a function of δ_p, V, and altitude.

The two basic conditions then, for trimmed steady flight on a straight line are

$$\begin{aligned} C_m &= 0 \\ L &= C_L\tfrac{1}{2}\rho V^2 S = W \end{aligned} \qquad (2.4,18)$$

and in accordance with the postulates made above, we write

$$\begin{aligned} C_m &= C_m(\alpha, V, \delta_e, \delta_p) \\ C_L &= C_L(\alpha, V, \delta_e, \delta_p) \end{aligned} \qquad (2.4,19)$$

Now let $(\ )_e$ denote one state that satisfies (2,4,18) and consider a small change from it, denoted by differentials, to another such state. From (2.4,18) we get, for ρ = constant,

$$dC_m = 0 \qquad (2.4,20)$$

and

$$C_L V^2 = \text{const}$$

or

$$2V_e C_{L_e} dV + V_e^2 dC_L = 0$$

so that

$$dC_L = -2C_{L_e}\frac{dV}{V_e} = -2C_{L_e}d\hat{V} \qquad (2.4,21)$$

where $\hat{V}$ is V/V_e. Taking the differentials of (2.4,19) and equating to (2.4,20 and 21) we get

$$\begin{aligned} C_{L_\alpha} d\alpha + C_{L_{\delta_e}} d\delta_e &= -C_{L_{\delta_p}} d\delta_p - (C_{L_V} + 2C_{L_e}) d\hat{V} \\ C_{m_\alpha} d\alpha + C_{m_{\delta_e}} d\delta_e &= -C_{m_{\delta_p}} d\delta_p - C_{m_V} d\hat{V} \end{aligned} \tag{2.4,22}$$

where $C_{L_V} = \partial C_L/\partial \hat{V}$ and $C_{m_V} = \partial C_m/\partial \hat{V}$. From (2.4,22) we get the solution for $d\delta_e$ as

$$d\delta_e = \frac{1}{det}\left\{[(C_{L_V} + 2C_{L_e})C_{m_\alpha} - C_{L_\alpha}C_{m_V}]d\hat{V} + (C_{L_{\delta_p}}C_{m_\alpha} - C_{L_\alpha}C_{m_{\delta_p}})d\delta_p\right\} \tag{2.4,23}$$

There are two possibilities, δ_p constant and δ_p variable. In the first case (fixed throttle), $d\delta_p = 0$ and

$$\left(\frac{d\delta_{e_{\text{trim}}}}{d\hat{V}}\right)_{\delta_p} = \frac{(C_{L_V} + 2C_{L_e})C_{m_\alpha} - C_{L_\alpha}C_{m_V}}{det} \tag{2.4,24}$$

It will be shown in Chap. 6 that the vanishing of this quantity is a true criterion of stability, that is it must be >0 for a stable airplane. In the second case, for example exactly horizontal flight, $\delta_p = \delta_p(V)$ and the δ_p term on the right-hand side of (2.4,23) remains. For such cases the gradient $(d\delta_{e_{\text{trim}}}/d\hat{V})$ is not necessarily related to stability. For purposes of calculating the propulsion contributions, the terms $C_{L_{\delta_p}}\,d\delta_p$ and $C_{m_{\delta_p}}\,d\delta_p$ in (2.4,23) would be evaluated as dC_{L_p} and dC_{m_p} [see the notation of (2.3,1)]. These contributions to the lift and moment are discussed in Sec. 3.4.

The derivatives C_{L_V} and C_{m_V} may be quite large owing to slipstream effects on STOL airplanes, aeroelastic effects, or Mach number effects near transonic speeds. These variations with **M** can result in reversal of the slope of $\delta_{e_{\text{trim}}}$ as illustrated on Fig. 2.20. The negative slope at *A*, according to the stability criterion referred to above, indicates that the airplane is unstable at *A*. This can be seen as follows. Let the airplane be in equilibrium flight at the point *A*, and be subsequently perturbed so that its speed increases to that of *B* with no change in α or δ_e. Now at *B* the elevator angle is too positive for trim: that is there is an unbalanced nose-down moment on the airplane. This puts the airplane into a dive and increases its speed still further. The speed will continue to increase until point *C* is reached, when the δ_e is again the correct value for trim, but here the slope is positive and there is no tendency for the speed to change any further.

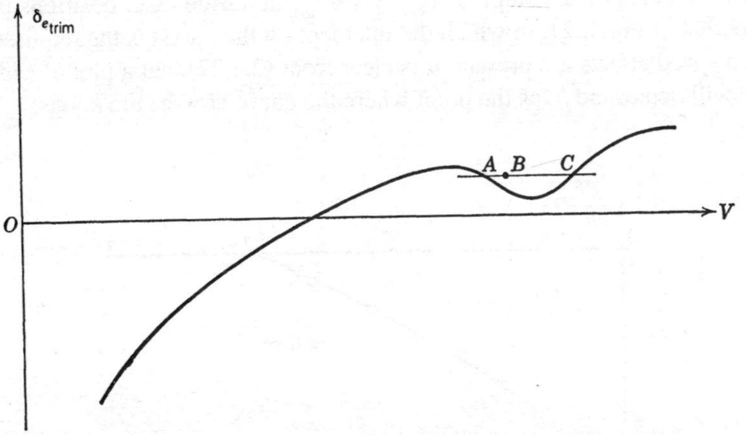

Figure 2.20 Reversal of $\delta_{e_{\text{trim}}}$ slope at transonic speeds, δ_p = const.

STATIC STABILITY LIMIT, h_s

The critical CG position for zero elevator trim slope (i.e. for stability) can be found by setting (2.4,24) equal to zero. Recalling that $C_{m_\alpha} = C_{L_\alpha}(h - h_n)$, this yields

$$h - h_n - \frac{C_{mV}}{C_{L_V} + 2C_{L_e}} = 0 \tag{2.4,25}$$

or

$$h = h_s$$

where

$$h_s = h_n + \frac{C_{mV}}{C_{L_V} + 2C_{L_e}} \tag{2.4,26}$$

Depending on the sign of C_{mV}, h_s may be greater or less than h_n. In terms of h_s, (2.4,24) can be rewritten as

$$\left(\frac{d\delta_{e_{\text{trim}}}}{d\hat{V}}\right)_{\delta_p} = \frac{C_{L_\alpha}}{det}(C_{L_V} + 2C_{L_e})(h - h_s) \tag{2.4,27}$$

$(h - h_s)$ is the "stability margin," which may be greater or less than the static margin.

FLIGHT DETERMINATION OF h_n AND h_s

For the general case, (2.3,5) suggests that the measurement of h_n requires the measurement of C_{m_α} and C_{L_α}. Flight measurements of aerodynamic derivatives such as these can be made by dynamic techniques. However, in the simpler case when the complications presented by propulsive, compressibility, or aeroelastic effects are absent, then the relations implicit in Figs. 2.18 and 2.19 lead to a means of finding h_n from the elevator trim curves. In that case all the coefficients of (2.4,13) are constants, and

$$\frac{d\delta_{e_{\text{trim}}}}{dC_{L_{\text{trim}}}} = -\frac{C_{m_\alpha}}{det} \tag{2.4,28}$$

or

$$\frac{d\delta_{e_{\text{trim}}}}{dC_{L_{\text{trim}}}} = -\frac{C_{L_\alpha}}{det}(h - h_n) \tag{2.4,29}$$

Thus measurements of the slope of $\delta_{e_{\text{trim}}}$ vs. $C_{L_{\text{trim}}}$ at various CG positions produce a curve like that of Fig. 2.21, in which the intercept on the h axis is the required NP.

When speed effects are present, it is clear from (2.4,27) that a plot of $(d\delta_{e_{\text{trim}}}/d\hat{V})_{\delta_p}$ against h will determine h_s as the point where the curve crosses the h axis.

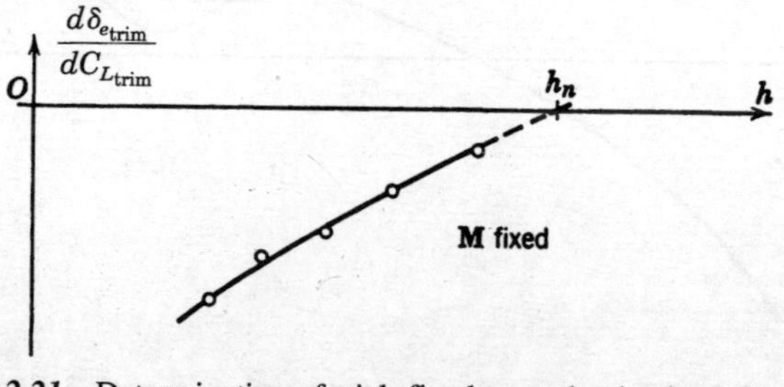

Figure 2.21 Determination of stick-fixed neutral point from flight test.

2.5 The Control Hinge Moment

To rotate any of the aerodynamic control surfaces, elevator, aileron, or rudder, about its hinge, it is necessary to apply a force to it to overcome the aerodynamic pressures that resist the motion. This force may be supplied entirely by a human pilot through a mechanical system of cables, pulleys, rods, and levers; it may be provided partly by a powered actuator; or the pilot may be altogether mechanically disconnected from the control surface ("fly-by-wire" or "fly-by-light"). In any case, the force that has to be applied to the control surface must be known with precision if the control system that connects the primary controls in the cockpit to the aerodynamic surface is to be designed correctly. The range of control system options is so great that it is not feasible in this text to present a comprehensive coverage of them. We have therefore limited ourselves in this and the following chapter to some material related to elevator control forces when the human pilot supplies all of the actuation, or when a power assist relieves the pilot of a fixed fraction of the force required. This treatment necessarily begins with a discussion of the aerodynamics; that is, of the *aerodynamic hinge moment.*

The aerodynamic forces on any control surface produce a moment about the hinge. Figure 2.22 shows a typical tail surface incorporating an elevator with a tab. The tab usually exerts a negligible effect on the lift of the aerodynamic surface to which it is attached, although its influence on the hinge moment is large.

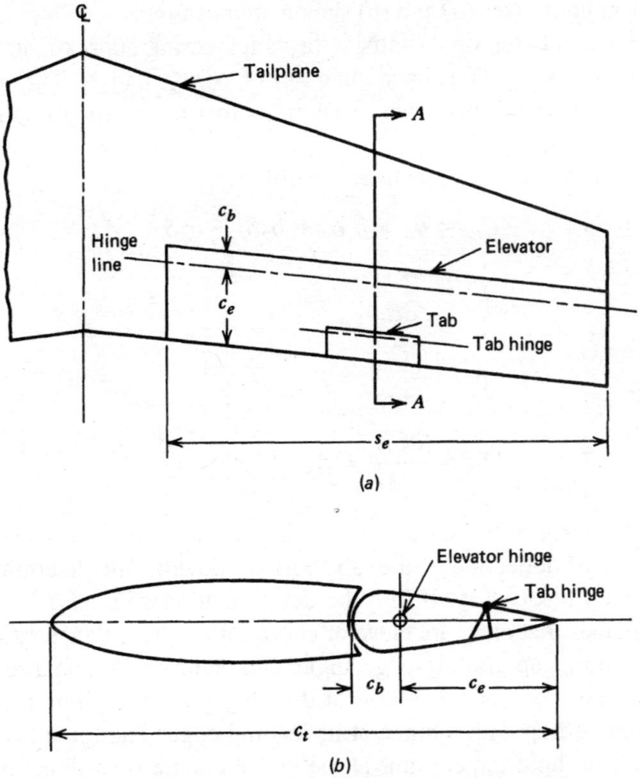

Figure 2.22 Elevator and tab geometry. (*a*) Plan view. (*b*) Section *A*—*A*.

The coefficient of elevator hinge moment is defined by

$$C_{he} = \frac{H_e}{\frac{1}{2}\rho V^2 S_e \bar{c}_e}$$

Here H_e is the moment, about the elevator hinge line, of the aerodynamic forces on the elevator and tab, S_e is the area of that portion of the elevator and tab that lies *aft of the elevator hinge line,* and $\bar{c}_e$ is a mean chord of the same portion of the elevator and tab. Sometimes $\bar{c}_e$ is taken to be the geometric mean value, that is, $\bar{c}_e = S_e/2s_e$, and other times it is the root-mean square of c_e. The taper of elevators is usually slight, and the difference between the two values is generally small. The reader is cautioned to note which definition is employed when using reports on experimental measurements of C_{he}.

Of all the aerodynamic parameters required in stability and control analysis, the hinge-moment coefficients are most difficult to determine with precision. A large number of geometrical parameters influence these coefficients, and the range of design configurations is wide. Scale effects tend to be larger than for many other parameters, owing to the sensitivity of the hinge moment to the state of the boundary layer at the trailing edge. Two-dimensional airfoil theory shows that the hinge moment of simple flap controls is linear with angle of attack and control angle in both subsonic and supersonic flow.

The normal-force distributions typical of subsonic flow associated with changes in α and δ_e are shown qualitatively in Fig. 2.23. The force acting on the movable flap has a moment about the hinge that is quite sensitive to its location. Ordinarily the hinge moments in both cases (*a*) and (*b*) shown are negative.

In many practical cases it is a satisfactory engineering approximation to assume that for finite surfaces C_{he} is a linear function of α_s, δ_e, and δ_t. The reader should note however that there are important exceptions in which strong nonlinearities are present.

We assume therefore that C_{he} is linear, as follows,

$$C_{he} = b_0 + b_1\alpha_s + b_2\delta_e + b_3\delta_t \tag{2.5,1}$$

where

$$b_1 = \frac{\partial C_{he}}{\partial \alpha_s} = C_{he_{\alpha_s}}$$

$$b_2 = \frac{\partial C_{he}}{\partial \delta_e} = C_{he_{\delta_e}}$$

$$b_3 = \frac{\partial C_{he}}{\partial \delta_t} = C_{he_{\delta_t}}$$

α_s is the angle of attack of the surface to which the control is attached (wing or tail), and δ_t is the angle of deflection of the tab (positive down). The determination of the hinge moment then resolves itself into the determination of b_0, b_1, b_2, and b_3. The geometrical variables that enter are elevator chord ratio c_e/c_t, balance ratio c_b/c_e, nose shape, hinge location, gap, trailing-edge angle, and planform. When a set-back hinge is used, some of the pressure acts ahead of the hinge, and the hinge moment is less than that of a simple flap with a hinge at its leading edge. The force that the control system must exert to hold the elevator at the desired angle is in direct proportion to the hinge moment.

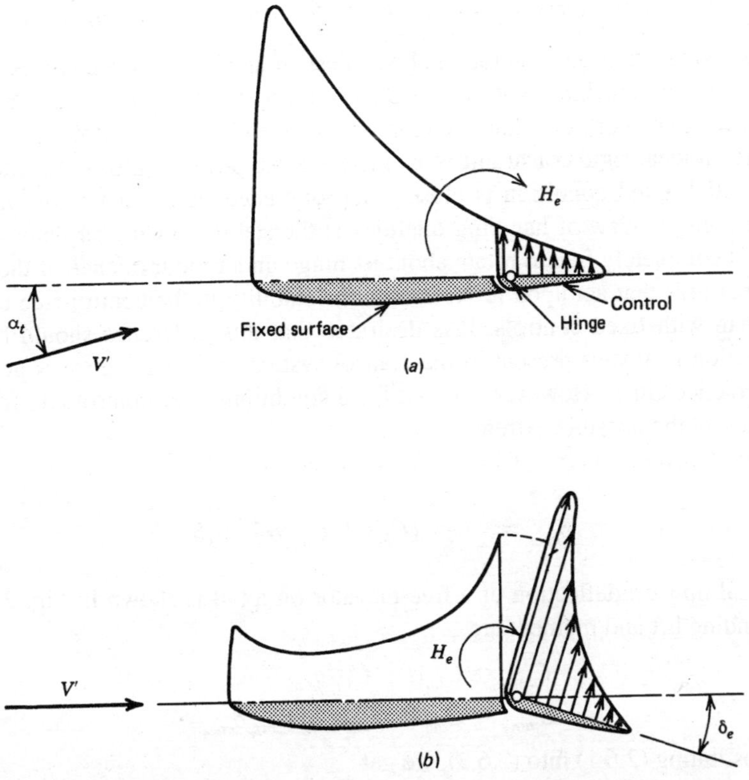

Figure 2.23 Normal-force distribution over control surface at subsonic speed. (*a*) Force distribution over control associated with α_t at $\delta_e = 0$. (*b*) Force distribution over control associated with δ_e at zero α_t.

We shall find it convenient subsequently to have an equation like (2.5,1) with α instead of α_s. For tailless aircraft, α_s is equal to α, but for aircraft with tails, $\alpha_s = \alpha_t$. Let us write for both types

$$C_{he} = C_{he_0} + C_{he_\alpha}\alpha + b_2\delta_e + b_3\delta_t \tag{2.5,2}$$

where for tailless aircraft $C_{he_0} = b_0$, $C_{he_\alpha} = b_1$. For aircraft with tails, the relation between α and α_t is derived from (2.3,12) and (2.3,19), that is,

$$\alpha_t = \alpha\left(1 - \frac{\partial\epsilon}{\partial\alpha}\right) - (\epsilon_0 + i_t)\left[1 - \frac{a_t}{a}\frac{S_t}{S}\left(1 - \frac{\partial\epsilon}{\partial\alpha}\right)\right] \tag{2.5,3}$$

whence it follows that for tailed aircraft, with symmetrical airfoil sections in the tail, for which $b_0 = 0$,

$$C_{he_0} = -b_1(i_t + \epsilon_0)\left[1 - \frac{a_t}{a}\frac{S_t}{S}\left(1 - \frac{\partial\epsilon}{\partial\alpha}\right)\right] \qquad (a)$$

$$C_{he_\alpha} = b_1\left(1 - \frac{\partial\epsilon}{\partial\alpha}\right) \qquad (b) \tag{2.5,4}$$

2.6 Influence of a Free Elevator on Lift and Moment

In Sec. 2.3 we have dealt with the pitch stiffness of an airplane the controls of which are fixed in position. Even with a completely rigid structure, which never exists, a manually operated control cannot be regarded as fixed. A human pilot is incapable of supplying an ideal rigid constraint. When irreversible power controls are fitted, however, the stick-fixed condition is closely approximated. A characteristic of interest from the point of view of handling qualities is the stability of the airplane when the elevator is completely free to rotate about its hinge under the influence of the aerodynamic pressures that act upon it. Normally, the stability in the control-free condition is less than with fixed controls. It is desirable that this difference should be small. Since friction is always present in the control system, the free control is never realized in practice either. However, the two ideal conditions, free control and fixed control, represent the possible extremes.

When the control is free, then $C_{he} = 0$, so that from (2.5,2)

$$\delta_{e_{\text{free}}} = -\frac{1}{b_2}(C_{he_0} + C_{he_\alpha}\alpha + b_3\delta_t) \tag{2.6,1}$$

The typical upward deflection of a free-elevator on a tail is shown in Fig. 2.24. The corresponding lift and moment are

$$\begin{aligned} C_{L_{\text{free}}} &= C_{L_\alpha}\alpha + C_{L_{\delta_e}}\delta_{e_{\text{free}}} \\ C_{m_{\text{free}}} &= C_{m_0} + C_{m_\alpha}\alpha + C_{m_{\delta_e}}\delta_{e_{\text{free}}} \end{aligned} \tag{2.6,2}$$

After substituting (2.6,1) into (2.6,2), we get

$$\begin{aligned} C_{L_{\text{free}}} &= C'_{L_0} + C_{L_\alpha}{}'\alpha & (a) \\ C_{m_{\text{free}}} &= C'_{m_0} + C'_{m_\alpha}\alpha & (b) \end{aligned} \tag{2.6,3}$$

where

$$\begin{aligned} C'_{L_0} &= -\frac{C_{L_{\delta_e}}}{b_2}(C_{he_0} + b_3\delta_t) & (a) \\ a' = C'_{L_\alpha} &= C_{L_\alpha} - \frac{C_{L_{\delta_e}}C_{he_\alpha}}{b_2} & (b) \end{aligned} \tag{2.6,4}$$

$$\begin{aligned} C'_{m_0} &= C_{m_0} - \frac{C_{m_{\delta_e}}}{b_2}(C_{he_0} + b_3\delta_t) & (a) \\ C'_{m_\alpha} &= C_{m_\alpha} - \frac{C_{m_{\delta_e}}C_{he_\alpha}}{b_2} & (b) \end{aligned} \tag{2.6,5}$$

When due consideration is given to the usual signs of the coefficients in these equations, we see that the two important gradients C_{L_α} and C_{m_α} are reduced in absolute

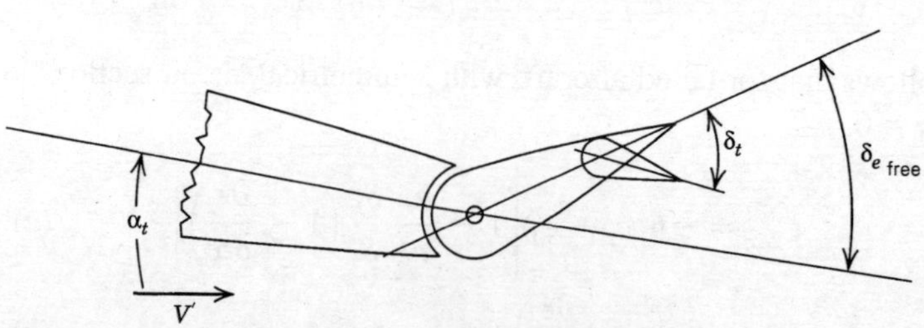

Figure 2.24 Elevator floating angle.

magnitude when the control is released. This leads, broadly speaking, to a reduction of stability.

FREE-ELEVATOR FACTOR

For a tailless aircraft with a free elevator, the lift-curve slope is (cf 2.6,4*b*)

$$a' = a\left(1 - \frac{C_{L_{\delta_e}} b_1}{a b_2}\right) \tag{2.6,6}$$

The factor in parentheses is the *free elevator factor,* and normally has a value less than unity. When the elevator is part of the tail, the floating angle can be related to α_t, viz for $b_0 = 0$

$$C_{he} = b_1\alpha_t + b_2\delta_{e_{\text{free}}} + b_3\delta_t = 0$$

or

$$\delta_{e_{\text{free}}} = -\frac{1}{b_2}(b_1\alpha_t + b_3\delta_t) \tag{2.6,7}$$

and the tail lift coefficient is

$$\begin{aligned} C'_{L_t} &= a_t\alpha_t + a_e\delta_{e_{\text{free}}} \\ &= a_t\left(1 - \frac{a_e}{a_t}\frac{b_1}{b_2}\right)\alpha_t - \frac{a_e b_3}{b_2}\delta_t \end{aligned} \tag{2.6,8}$$

The effective lift-curve slope is

$$\frac{\partial C'_{L_t}}{\partial \alpha_t} = F a_t \tag{2.6,9}$$

where $F = \left(1 - \frac{a_e}{a_t}\frac{b_1}{b_2}\right)$ is the *free elevator factor* for a tail. If Fa_t is used in place of a_t and a' in place of a, then all the equations given in Sec. 2.3 hold for aircraft with a free elevator.

ELEVATOR-FREE NEUTRAL POINT

It is evident from the preceding comment that the NP of a tailed aircraft when the elevator is free is given by (2.3,23) as

$$h'_n = h_{n_{wb}} + \frac{Fa_t}{a'}\bar{V}_H\left(1 - \frac{\partial\epsilon}{\partial\alpha}\right) - \frac{1}{a'}\frac{\partial C_{m_p}}{\partial\alpha} \tag{2.6,10}$$

Alternatively, we can derive the NP location from (2.6,5*b*), for we know from (2.3,5) that

$$C'_{m_\alpha} = C'_{L_\alpha}(h - h'_n) \tag{a}$$

or

$$\begin{aligned} (h - h'_n) &= \frac{C'_{m_\alpha}}{C'_{L_\alpha}} = \frac{1}{a'}\left(C_{m_\alpha} - \frac{C_{m_{\delta_e}} C_{he_\alpha}}{b_2}\right) \\ &= \frac{1}{a'}\left[a(h - h_n) - \frac{C_{m_{\delta_e}} C_{he_\alpha}}{b_2}\right] \end{aligned} \tag{2.6,11}$$

(*b*)

Since $C_{m_{\delta_e}}$ is of different form for the two main types of aircraft, we proceed separately below.

TAILED AIRCRAFT

$C_{m_{\delta_e}}$ is given by (2.4,8), so (2.6,11) becomes for this case

$$(h - h_n') = \frac{a}{a'}(h - h_n) - \frac{C_{he_\alpha}}{a'b_2} C_{L_{\delta_e}}(h - h_{n_{wb}}) + \frac{C_{he_\alpha}}{a'b_2} a_e \bar{V}_H$$

Using (2.6,4*b*) this becomes

$$h - h_n' = h - \frac{1}{a'}\left(ah_n - \frac{C_{he_\alpha} C_{L_{\delta_e}}}{b_2} h_{n_{wb}}\right) + \frac{C_{he_\alpha}}{a'b_2} a_e \bar{V}_H$$

We replace $h_{n_{wb}}$ by $(h_{n_{wb}} - h_n) + h_n$ to get

$$h_n' = h_n + \frac{C_{he_\alpha} C_{L_{\delta_e}}}{b_2 a'}(h_n - h_{n_{wb}}) - \frac{a_e C_{he_\alpha}}{a'b_2} \bar{V}_H \tag{2.6,12}$$

Finally, using (2.4,8) for $C_{L_{\delta_e}}$, and (2.5,4) for C_{he_α}, we get

$$h_n' = h_n - \frac{a_e}{a'}\frac{b_1}{b_2}\left(1 - \frac{\partial \epsilon}{\partial \alpha}\right)\left(-(h_n - h_{n_{wb}})\frac{S_t}{S} + \bar{V}_H\right) \tag{2.6,13}$$

TAILLESS AIRCRAFT

$C_{m_{\delta_e}}$ is given by (2.4,9) and $C_{he_\alpha} = b_1$. When these are substituted into (2.6,11) the result is

$$\begin{aligned} h - h_n' &= \frac{a}{a'}(h - h_n) - \frac{b_1}{a'b_2} C_{L_{\delta_e}}(h - h_n) - \frac{b_1}{a'b_2}\frac{\partial C_{m_{ac}}}{\partial \delta_e} \\ &= \frac{(h - h_n)}{a'}\left(a - \frac{b_1}{b_2} C_{L_{\delta_e}}\right) - \frac{b_1}{a'b_2}\frac{\partial C_{m_{ac}}}{\partial \delta_e} \end{aligned}$$

By virtue of (2.6,6) this becomes

$$h - h_n' = h - h_n - \frac{b_1}{a'b_2}\frac{\partial C_{m_{ac}}}{\partial \delta_e}$$

or

$$h_n' = h_n + \frac{b_1}{a'b_2}\frac{\partial C_{m_{ac}}}{\partial \delta_e} \tag{2.6,14}$$

The difference $(h_n' - h)$ is called the *control-free static margin, K_n'*. When representative numerical values are used in (2.6,13) one finds that $h_n - h_n'$ may be typically about 0.08. This represents a substantial forward movement of the NP, with consequent reduction of static margin, pitch stiffness, and stability.

2.7 The Use of Tabs

TRIM TABS

In order to fly at a given speed, or C_L, it has been shown in Sec. 2.4 that a certain elevator angle $\delta_{e_{\text{trim}}}$ is required. When this differs from the free-floating angle $\delta_{e_{\text{free}}}$, a force is required to hold the elevator. When flying for long periods at a constant speed, it is very fatiguing for the pilot to maintain such a force. The trim tabs are used to relieve the pilot of this load by causing $\delta_{e_{\text{trim}}}$ and $\delta_{e_{\text{free}}}$ to coincide. The trim-tab angle required is calculated below.

When C_{he} and C_m are both zero, the tab angle is obtained from (2.5,2) as

$$\delta_{t_{\text{trim}}} = -\frac{1}{b_3}(C_{he_0} + C_{he_\alpha}\alpha_{\text{trim}} + b_2\delta_{e_{\text{trim}}}) \tag{2.7,1}$$

On substituting from (2.4,13) (which implies neglecting $\partial C_m/\partial\delta_t$), we get

$$\delta_{t_{\text{trim}}} = -\frac{1}{b_3}\left[C_{he_0} + \frac{C_{m_0}}{det}(C_{he_\alpha}C_{L_{\delta_e}} - b_2C_{L_\alpha}) + \frac{C_{L_{\text{trim}}}}{det}(C_{he_\alpha}C_{m_{\delta_e}} - b_2C_{m_\alpha})\right]$$

which is linear in $C_{L_{\text{trim}}}$ for constant h, as shown in Fig. 2.25. The dependence on h is simple, since from (2.6,11) we find that

$$(C_{he_\alpha}C_{m_{\delta_e}} - b_2C_{m_\alpha}) = -a'b_2(h - h_n')$$

and hence

$$\delta_{t_{\text{trim}}} = -\frac{1}{b_3}\left[C_{he_0} + \frac{C_{m_0}}{det}(C_{he_\alpha}C_{L_{\delta_e}} - b_2C_{L_\alpha}) - \frac{a'b_2}{det}(h - h_n')C_{L_{\text{trim}}}\right] \tag{2.7,2}$$

This result applies to both tailed and tailless aircraft, provided only that the appropriate values of the coefficients are used. It should be realized, of course, in reference to Fig. 2.25, that each different $C_{L_{\text{trim}}}$ in a real flight situation corresponds to a different set of values of $\mathbf{M}$, $\frac{1}{2}\rho V^2$, and C_T, so that in general the coefficients of (2.7,2) vary with C_L, and the graphs will depart from straight lines.

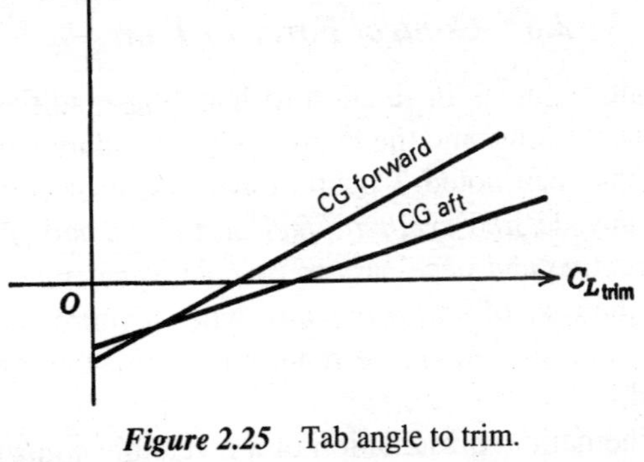

Figure 2.25 Tab angle to trim.

Equation (2.7,2) shows that the slope of the $\delta_{t_{\text{trim}}}$ vs $C_{L_{\text{trim}}}$ curve is proportional to the control-free static margin. When the coefficients are constants, we have

$$\frac{d\delta_{t_{\text{trim}}}}{dC_{L_{\text{trim}}}} = \frac{b_2}{b_3}\frac{a'}{det}(h - h_n') \tag{2.7,3}$$

The similarity between (2.7,3) and (2.4,13*c*) is noteworthy, that is the trim-tab slope bears the same relation to the control-free NP as the elevator angle slope does to the control-fixed NP. It follows that flight determination of h_n' from measurements of $d\delta_{t_{\text{trim}}}/dC_{L_{\text{trim}}}$ is possible subject to the same restrictions as discussed in relation to the measurement of h_n in Sec. 2.4.

OTHER USES

Tabs are used for purposes other than trimming, especially for manually actuated controls. Three of the main types are as follows:

Geared Tabs. Tabs connected to the main surface by a mechanical linkage that causes the tab to deflect automatically when the main surface is deflected, but in the opposite direction. The hinge moment produced by the tab then assists the rotation of the main surface. These have the effect of reducing the b_2 of the surface.

Spring Tabs. Tabs connected to the main surface by an elastic element. The design is such that the deflection of the tab depends on the dynamic pressure in a way that mitigates the effect of the speed-squared law on the control force.

Servo Tabs. An arrangement in which the pilot controls the tab directly through a mechanical linkage. It is then the tab, not the pilot, that provides the hinge moment needed to rotate the main surface.

Both spring tabs and servo tabs are effective devices for reducing control forces on large high-speed airplanes. However, both add an additional degree of freedom to the control system dynamics, and this is a potential source of trouble due to vibration or flutter.

For further details of how these tabs function, see Etkin (1972).

2.8 *Control Force to Trim*

The importance of control forces in relation to handling qualities has already been emphasized in previous sections, and the many options available to designers of powered control systems has been noted. Cockpit devices can of course be designed to produce more or less any desired *synthetic feel* on the primary flight controls. It is nevertheless both instructive and necessary to be able to calculate the control forces that will be present in the case of *natural feel,* or when a simple power assist is present. A case in point is the elevator force required to trim the airplane, and how it varies with flight speed.

Figure 2.26 is a schematic representation of a reversible control system. The box denoted "control system linkage" represents any assemblage of levers, rods, pulleys,

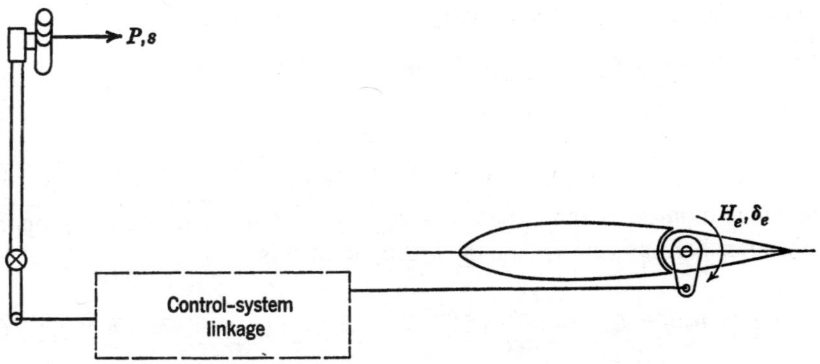

Figure 2.26 Schematic diagram of an elevator control system.

cables, and power-boost elements that comprise a general control system. We assume that the elements of the linkage and the structure to which it is attached are ideally rigid, so that no strain energy is stored in them, and we neglect friction. The system then has one degree of freedom. P is the force applied by the pilot, (positive to the rear) s is the displacement of the hand grip, and the work done by the power boost system is W_b. Considering a small quasistatic displacement from equilibrium (i.e., no kinetic energy appears in the control system), conservation of energy gives

$$Pds + dW_b + H_e d\delta_e = 0 \tag{2.8,1}$$

or

$$P = -\frac{dW_b}{ds} - \frac{d\delta_e}{ds}H_e$$

Now the nature of ratio or power boost controls is such that dW_b/ds is proportional to P or H_e. Hence we can write

$$P = (G_1 - G_2)H_e \tag{2.8,2}$$

where $\quad G_1 = -\dfrac{d\delta_e}{ds} > 0,\quad$ the *elevator gearing* (rad/ft or rad/m)

and $\quad G_2 = \dfrac{dW_b/ds}{H_e},\quad$ the *boost gearing* (ft^{-1} or m^{-1})

(2.8,2) is now rewritten as

$$P = GH_e \tag{2.8,3}$$

where $G = G_1 - G_2$. For fixed G_1, i.e., *for a given movement of the control surface to result from a given displacement of the pilot's control,* then the introduction of power boost is seen to reduce G and hence P. G may be designed to be constant over the whole range of δ_e, or it may, by the use of special linkages and power systems, be made variable in almost any desired manner.

Introduction of the hinge-moment coefficient gives the expression for P as

$$P = GC_{he}S_e\bar{c}_e\tfrac{1}{2}\rho V^2 \tag{2.8,4}$$

and the variation of P with flight speed depends on both V^2 and on how C_{he} varies with speed.

The value of C_{he} at trim for arbitrary tab angle is given by

$$C_{he} = C_{he_0} + C_{he_\alpha}\alpha_{\text{trim}} + b_2\delta_{e_{\text{trim}}} + b_3\delta_t \tag{2.8,5}$$

(2.8,5) in combination with (2.7,1) yields

$$C_{he} = b_3(\delta_t - \delta_{t_{\text{trim}}}) \tag{2.8,6}$$

i.e., the hinge moment is zero when $\delta_t = \delta_{t_{\text{trim}}}$ as expected, and linearly proportional to the difference. From (2.7,2) then the hinge moment is

$$C_{he} = b_3\delta_t + C_{he_0} + \frac{C_{m_0}}{det}(C_{he_\alpha}C_{L_{\delta_e}} - b_2C_{L_\alpha}) - \frac{a'b_2}{det}(h - h'_n)C_{L_{\text{trim}}} \tag{2.8,7}$$

Lift equals the weight in horizontal flight, so that

$$C_{L_{\text{trim}}} = \frac{w}{\frac{1}{2}\rho V^2} \tag{2.8,8}$$

where $w = W/S$ is the wing loading. When (2.8,7 and 8) are substituted into (2.8,4) the result obtained is

$$P = A + B\tfrac{1}{2}\rho V^2 \tag{2.8,9}$$

where

$$A = -GS_e\bar{c}_e w\,\frac{a'b_2}{det}(h - h'_n)$$

$$B = GS_e\bar{c}_e\left[b_3\delta_t + C_{he_0} + \frac{C_{m_0}}{det}(C_{he_\alpha}C_{L_{\delta_e}} - b_2C_{L_\alpha})\right]$$

The typical parabolic variation of P with V when the aerodynamic coefficients are all constant, is shown in Fig. 2.27. The following conclusions may be drawn.

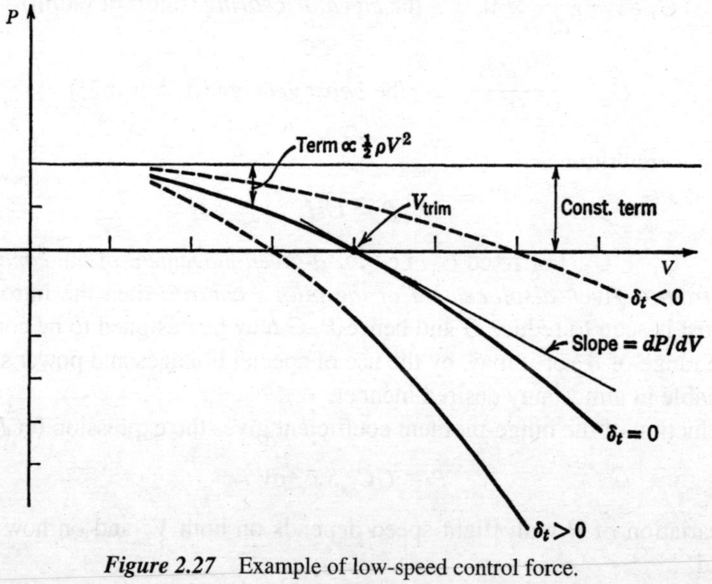

Figure 2.27 Example of low-speed control force.

1. Other things remaining equal, $P \propto S_e\bar{c}_e$, i.e., to the cube of the airplane size. This indicates a very rapid increase in control force with size.
2. P is directly proportional to the gearing G.
3. The CG position only affects the constant term (apart from a second-order influence on $C_{m_{\delta_e}}$). A forward movement of the CG produces an upward translation of the curve.
4. The weight of the airplane enters only through the wing loading, a quantity that tends to be constant for airplanes serving a given function, regardless of weight. An increase in wing loading has the same effect as a forward shift of the CG.
5. The part of P that varies with $\frac{1}{2}\rho V^2$ decreases with height, and increases as the speed squared.
6. Of the terms contained in B, none can be said in general to be negligible. All of them are "built-in" constants except for δ_t.
7. The effect of the trim tab is to change the coefficient of $\frac{1}{2}\rho V^2$, and hence the curvature of the parabola in Fig. 2.27. Thus it controls the intercept of the curve with the V axis. This intercept is denoted V_{trim}; it is the speed for zero control force.

2.9 Control Force Gradient

It was pointed out in Sec. 2.7 how the trim tabs can be used to reduce the control force to zero. A significant handling characteristic is the gradient of P with V at $P = 0$. The manner in which this changes as the CG is moved aft is illustrated in Fig. 2.28. The trim tab is assumed to be set so as to keep V_{trim} the same. The gradient dP/dV is seen to decrease in magnitude as the CG moves backward. When it is at the control-free neutral point, $A = 0$ for aircraft with or without tails, and, under the stated conditions, the P/V graph becomes a straight line lying on the V axis. This is an important characteristic of the control-free NP; that is, when the CG is at that point, no force is required to change the trim speed.

A quantitative analysis of the control-force gradient follows.

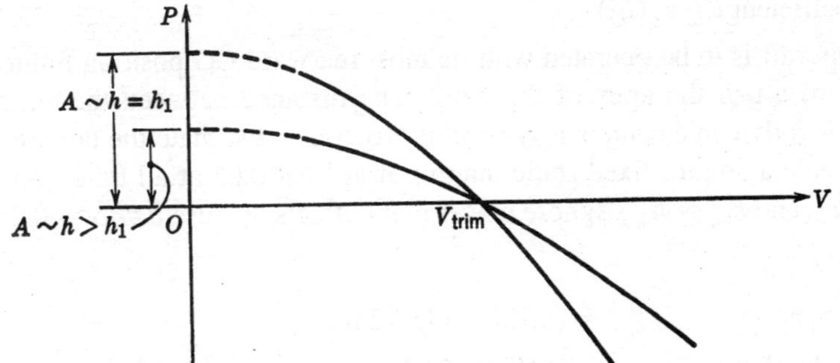

Figure 2.28 Effect of CG location on control-force gradient at fixed trim speed.

The force is given by (2.8,9). From it we obtain the derivative

$$\frac{\partial P}{\partial V} = B\rho V$$

At the speed V_{trim}, $P = 0$, and $B = -A/\frac{1}{2}\rho V^2_{\text{trim}}$, whence

$$\frac{\partial P}{\partial V} = -\frac{2A}{V_{\text{trim}}} \tag{2.9,1}$$

A is given following (2.8,9). Substituting the value into (2.9,1) we get

$$\frac{\partial P}{\partial V} = 2GS_e\bar{c}_e\frac{a'b_2}{det}\frac{w}{V_{\text{trim}}}(h - h'_n) \tag{2.9,2}$$

From (2.9,2) we deduce the following:

1. The control-force gradient is proportional to $S_e\bar{c}_e$; that is, to the cube of airplane size.
2. It is inversely proportional to the trim speed; i.e. it increases with decreasing speed. This effect is also evident in Fig. 2.27.
3. It is directly proportional to wing loading.
4. It is *independent* of height for a given true airspeed, but *decreases* with height for a fixed V_E.
5. It is directly proportional to the control-free static margin.

Thus, in the absence of Mach number effects, the elevator control will be "heaviest" at sea-level, low-speed, forward CG, and maximum weight.

2.10 Exercises

2.1 A subsonic transport aircraft has a tapered, untwisted sweptback wing with straight leading and trailing edges. The wing tips are straight and parallel to the root chord. In the following, use the data of Appendix C and assume that the airfoil section local aerodynamic center is at the $\frac{1}{4}$-chord point.

(a) Make an accurate three-view drawing of the wing chord plane.

(b) Calculate wing area S, aspect ratio A, taper ratio $\lambda = c_t/c_r$ and the mean aerodynamic chord $\bar{c}$.

(c) Calculate the location of the wing's mean aerodynamic center, and locate it and $\bar{c}$ on the side view of the wing (with dimensions). (Assume a uniform additional lift coefficient $C_{l_a} = C_L$.)

(d) The aircraft is to be operated with its most rearward CG position limited to 25 ft (7.62 m) aft of the apex of the wing. The distance between the wing and tail mean aerodynamic centers is $\bar{l}_t = 55$ ft (16.76 m). Estimate the tail area required to provide a control-fixed static margin of at least 0.05 at all times. Assume that $a_t = a_{wb}$ and $h_{n_w} = h_{n_{wb}}$. Ignore power plant effects and use $\partial\epsilon/\partial\alpha = 0.25$.

Geometric Data

Wing Span, b	150 ft (45.72 m)
Root Chord, c_r	25 ft (7.62 m)
Tip Chord, c_t	12 ft (3.66 m)

Leading edge sweep, Λ_0	26°
Dihedral angle, γ	4°

2.2 Evaluate the validity of the approximation made in going from (2.2,2) to (2.2,3) by using the data for the airplane of Exercise (2.1) and calculating C_{m_w} from both equations. Assume $C_{D_w} = C_{D_{\min_w}} + KC_{L_w}^2$ and $L = L_w$. The following additional data are provided.

Geometric Data

Weight, W	207,750 lb (924,488 N)
$z/\bar{c}$	0.15

Aerodynamic Data

a_w	0.080/deg
$C_{m_{ac_w}}$	−0.05
$C_{D_{\min_w}}$	0.013
K	0.054
V	350 kts (180 m/s)
ρ	2.377×10^{-3} slugs/ft³ (1.225 kg/m³)

2.3 Show that if $C_{m_{wb}}$ is a linear function of $C_{L_{wb}}$ then $C_{m_{ac_{wb}}}$ is a constant.

2.4 Beginning with (2.3,1) perform the reductions to derive (2.3,20) to (2.3,23).

2.5 The following data apply to a $\frac{1}{25}$ scale wind tunnel model of a transport airplane. The full-scale mass of the aircraft is 1,552.80 slugs (22,680 kg). Assume that the aerodynamic data can be applied at full-scale. For level unaccelerated flight at $V = 239$ knots (123 m/s) of the full-scale aircraft, under the assumption that propulsion effects can be ignored,

(a) Find the limits on tail angle i_t and CG position h imposed by the conditions $C_{m_0} > 0$ and $C_{m_\alpha} < 0$.

(b) For trimmed flight with $\delta_e = 0$, plot i_t vs. h for the aircraft and indicate where this line meets the boundaries of part (a).

Geometric Data

Wing area, S	1.50 ft² (0.139 m²)
Wing mean aerodynamic chord, $\bar{c}$	6.145 in (15.61 cm)
$\bar{l}_t$	15.29 in (38.84 cm)
Tail area, S_t	0.368 ft² (0.0342 m²)

Aerodynamic Data

a_{wb}	0.077/deg
a_t	0.064/deg
ϵ_0	0.72°
$\dfrac{\partial \epsilon}{\partial \alpha}$	0.30
$C_{m_{ac_{wb}}}$	−0.018

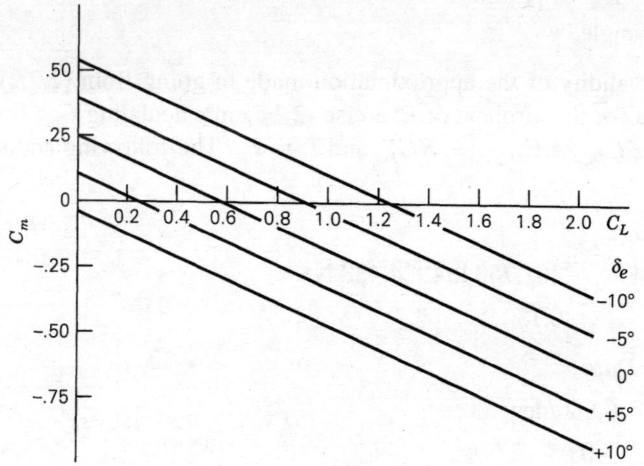

Figure 2.29 Data for Exercise 2.6.

$h_{n_{wb}}$	0.25
ρ	2.377×10^{-3} slugs/ft^3 (1.225 kg/m^3)

2.6* The McDonnell Douglas C-17 is a four-engined jet STOL transport airplane.

(a) Find A and $\bar{c}$ for the wing using the geometrical data and Appendix C.

(b) Use Appendix B to estimate a_w, the wing lift curve slope, assuming that $\beta = 1$ and $\kappa = 1$.

(c) If $a_t = 0.068$/deg and $a_{wb} = a_w$, find the lift curve slope, a, of the aircraft. Assume $\dfrac{\partial \epsilon}{\partial \alpha} = \dfrac{2a_w}{\pi A}$ (with a_w expressed in rad^{-1}).

(d) Find C_{m_α} for the case where $l_t = \bar{l}_t = 92$ ft (28.04 m). Ignore propulsion effects.

*Problem courtesy of Professor E. K. Parks, University of Arizona.

Figure 2.30 Trim data for Exercise 2.6.

(e) From the experimental curves of Figs. 2.29 and 2.30 and the given geometry, find $C_{m_{\delta_e}}$ and h_n. Find C_{m_α} for $h = 0.30$.

Geometric Data

Wing area, S	3,800 ft^2 (353.0 m^2)
Wing span, b	165 ft (50.29 m)
Root chord, c_r	37.3 ft (11.37 m)
Tip chord, c_t	8.8 ft (2.68 m)
$\frac{1}{4}$ chord line sweep, Λ	25°
$\frac{1}{2}$ chord line sweep, $\Lambda_{c/2}$	22°
Tail area, S_t	870 ft^2 (80.83 m^2)

2.7 Consider an aircraft with its tail identical to its wing (i.e., the same span, area, chord, etc.). Neglect body and wing-body interaction effects [i.e., in general $(\ )_{wb} \equiv (\ )_w$], neglect propulsion effects and assume zero elevator and tab deflections. Assume (2.2,7) in this instance is approximated by $M_t = -l_t L_t + M_{ac_t}$.

(a) What changes should be made in the expressions for a (2.3,18), C_{m_α} (2.3,21*a*), and C_{m_0} (2.3,22*a*)?

(b) What would $\dfrac{\partial \epsilon}{\partial \alpha}$ have to be numerically in order that the neutral point h_n lies midway between the mean aerodynamic centers of the wing and tail?

(c) For trimmed level flight, derive an expression for the ratio of the lift generated by the wing to the lift generated by the tail as a function of the tail angle i_t. Assume $\epsilon_0 = 0$, $\dfrac{\partial \epsilon}{\partial \alpha} = 0.2$, $a = 5$ (rad^{-1}), $C_{m_0} = 0.2$ and $(h - h_n) = -0.3$.

2.8* The following data were taken from a flight test of a PA-32R-300 Cherokee-6 airplane.

Altitude (ft)	(km)	V_E (mph)	(m/s)	*Mass* (slugs)	(kg)	i_t (deg)	x_{CG} (in)	(cm)
4540	1.384	91.0	40.7	113.4	1656	1.5	93.89	238.5
4560	1.390	109	48.7	113.0	1650	0	93.89	238.5
4700	1.433	126	56.3	112.9	1649	−1.0	93.89	238.5
4580	1.396	155	69.3	112.7	1646	−2.0	93.89	238.5
5320	1.622	89.0	39.8	100.4	1466	4.5	86.82	220.5
4620	1.408	105	46.9	100.2	1463	2.0	86.82	220.5
4740	1.445	123	55.0	100.0	1461	0.3	86.82	220.5
4900	1.494	151	67.5	99.84	1458	−1.0	86.82	220.5
4880	1.487	87.0	38.9	88.51	1293	7.2	80.43	204.3
4820	1.469	103	46.0	88.35	1290	3.5	80.43	204.3
4880	1.487	122	54.5	88.20	1288	1.5	80.43	204.3
4740	1.445	152	68.0	88.04	1286	0	80.43	204.3

*Problem courtesy of Professor E. K. Parks, University of Arizona.

The data were taken in trimmed level flight. x_{CG} is the distance of the CG aft of the nose of the aircraft. The aircraft has an all-moving tail and thus i_t is used instead of δ_e to trim the aircraft. The wing area is $S = 174.5$ ft^2 (16.21 m^2).

(a) Plot tail-setting angle, i_t, versus the lift coefficient of the aircraft for each of the three CG locations.

(b) Curve fit the data points in (a) with three straight lines having a common intercept (refer to Fig. 2.18).

(c) Use a graphical technique to find the location of the neutral point (controls fixed) relative to the nose of the aircraft (refer to Fig. 2.21).

2.9 Starting with (2.6,11*b*), derive (2.6,13).

2.10 The elevator control force to trim a particular airplane at a speed of 300 kts (154 m/s) is zero. Using the following data estimate the force required to change the trim speed to 310 kts (159 m/s). Assume that $C_{L_{\delta_e}}$ is sufficiently small that $C_{L_{\delta_e}} = 0$ can be used in the expression for control force.

Geometric Data

Elevator gearing, G	3°/in (1.18°/cm)
Elevator area aft of hinge line, S_e	40 ft^2 (3.72 m^2)
Mean elevator chord, $\bar{c}_e$	2.0 ft (0.61 m)
$\bar{V}_H$	0.56
CG location, h	0.38
Wing loading, w	50 psf (2,395 Pa)

Aerodynamic Data

Elevator hinge moment coefficient, $\dfrac{\partial C_{h_e}}{\partial \delta_e}$	−0.005/deg
a_e	0.025/deg
Neutral point, elevator free, h'_n	0.45

2.11 A fatal airplane accident has led to a civil court case. It is alleged that the airplane in question was unstable—"that the neutral point was ahead of the center of gravity."

You are called as an expert witness to explain to the court (i.e., to the judge and the attorneys) the meaning and importance of *stability, center of gravity,* and *neutral point.* You know that your audience is composed of intelligent persons educated in the humanities and law, but you must assume that they have only a rudimentary knowledge of science and mathematics.

Write an essay describing how you would meet this challenge. You are free to use diagrams and simple models, but *you must avoid any use of mathematics.* You should keep your language simple, avoiding any technical jargon. Even the word *moment* should be avoided. Your goal is to clarify, not to impress the court with your superior technical knowledge.

2.11 Additional Symbols Introduced in Chapter 2

$a = C_{L\alpha}$	airplane lift-curve slope, $\partial C_L/\partial\alpha$, elevator fixed
$a' = C'_{L\alpha}$	airplane lift-curve slope, $\partial C'_L/\partial\alpha$, elevator free
$a_e = C_{L_t\delta_e}$	$\partial C_{L_t}/\partial\delta_e$
a_w	wing lift-curve slope, $\partial C_{L_w}/\partial\alpha$
a_{wb}	wing-body lift-curve slope, $\partial C_{L_{wb}}/\partial\alpha$
a_t	tail lift-curve slope, $\partial C_{L_t}/\partial\alpha_t$
$b_0 = C_{he_0}$	see Eq. 2.5,1
$b_1 = C_{he_{\alpha_t}}$	$\partial C_{he}/\partial\alpha_t$
$b_2 = C_{he_{\delta_e}}$	$\partial C_{he}/\partial\delta_e$
$b_3 = C_{he_{\delta_t}}$	$\partial C_{he}/\partial\delta_t$
$\bar{c}$	length of mean aerodynamic chord
$\bar{c}_e$	mean elevator chord (see Sec. 2.5)
C_D	$D/\frac{1}{2}\rho V^2 S$
C_{he}	elevator hinge-moment coefficient, $H_e/\frac{1}{2}\rho V^2 S_e\bar{c}_e$
C_L	$L/\frac{1}{2}\rho V^2 S$
$C_{L_{wb}}$	$L_{wb}/\frac{1}{2}\rho V^2 S$
C_{L_t}	tail lift coefficient, $L_t/\frac{1}{2}\rho V^2 S_t$
$C_{L_{\delta_e}}$	$\partial C_L/\partial\delta_e$
C_{L_V}	$\partial C_L/\partial\hat{V}$
C_m	$M/\frac{1}{2}\rho V^2 S\bar{c}$
$C_{m_{ac_w}}$	$M_{ac_w}/\frac{1}{2}\rho V^2 S\bar{c}$
$C_{m_{ac_{wb}}}$	$M_{ac_{wb}}/\frac{1}{2}\rho V^2 S\bar{c}$
C_{m_o}	airplane pitching-moment coefficient at zero α
$\bar{C}_{m_o}$	airplane pitching-moment coefficient at zero α_{wb}
C_{m_p}	pitching-moment coefficient of the propulsion units
C_{m_t}	$M_t/\frac{1}{2}\rho V^2 S\bar{c}$
C_{m_V}	$\partial C_m/\partial\hat{V}$
C_{m_α}	$\partial C_m/\partial\alpha$
$C_{m_{\delta_e}}$	$\partial C_m/\partial\delta_e$
C_T	$T/\frac{1}{2}\rho V^2 S$
det	see (2.4,14)
D_w	wing drag
D_t	drag of the tail
F	free-elevator factor $(1 - a_e b_1/a_t b_2)$ for aircraft with tails
G	elevator gearing
H_e	elevator hinge moment

h	CG position, fraction of mean chord (see Fig. 2.9)
h_n	neutral point of airplane, fraction of mean chord, elevator fixed
h'_n	neutral point of airplane, fraction of mean chord, elevator free
h_{n_w}	neutral point of wing, fraction of mean chord
$h_{n_{wb}}$	neutral point of the wing-body combination
h_s	static stability limit
i_t	tail-setting angle (see Fig. 2.11)
K_n	static margin, see (2.3,6)
L	airplane lift
L_w	wing lift
L_{wb}	lift of wing-body combination
L_t	lift of the tail
l_t	distance between CG and tail mean aerodynamic center
$\bar{l}_t$	see Fig. 2.12
$\mathbf{M}$	Mach number
M	pitching moment about the CG
M_w	pitching moment of the wing about the CG
M_{ac_w}	pitching moment of the wing about its mean aerodynamic center
$M_{ac_{wb}}$	pitching moment of the wing-body combination about its mean aerodynamic center
M_{wb}	pitching moment, about the CG, of the wing-body combination
M_t	pitching moment of tail about CG
P	control force, positive to the rear
S	wing area
s_e	span of elevator
S_e	area of elevator aft of hinge line
S_t	area of tail
T	thrust
V	true airspeed
$\hat{V}$	V/V_e
V_e	reference equilibrium airspeed
V_E	equivalent airspeed (EAS), $V\sqrt{\rho/\rho_0}$
V_H	horizontal tail volume, $S_t l_t/S\bar{c}$
$\bar{V}_H$	$S_t\bar{l}_t/S\bar{c}$
W	aircraft weight
w	wing loading (W/S)
α	angle of attack of the zero lift line of the airplane (elevator angle zero)

α_w	angle of attack of the zero lift line of the wing
α_{wb}	angle of attack of the zero lift line of the wing-body combination
α_t	angle of attack of the tail
α_T	angle of attack of the thrust line
δ_e	elevator angle
δ_t	tab angle
ϵ	downwash angle
ϵ_0	downwash when $\alpha_{wb} = 0$
ρ	air density
ρ_0	standard sea-level value of ρ (see Appendix D)

CHAPTER 3

Static Stability and Control—Part 2

3.1 Maneuverability—Elevator Angle per g

In this and the following sections, we investigate the elevator angle and control force required to hold the airplane in a steady pull-up with load factor[1] n (Fig. 3.1). The concepts discussed here were introduced by S. B. Gates (1942). The flight-path tangent is horizontal at the point under analysis, and hence the net normal force is $L - W = (n - 1)W$ vertically upward. The normal acceleration is therefore $(n - 1)g$.

When the airplane is in straight horizontal flight at the same speed and altitude, the elevator angle and control force to trim are δ_e and P, respectively. When in the pull-up, these are changed to $\delta_e + \Delta\delta_e$ and $P + \Delta P$. The ratios $\Delta\delta_e/(n - 1)$ and $\Delta P/(n - 1)$ are known, respectively, as the *elevator angle per g*, and the *control force per g*. These two quantities provide a measure of the maneuverability of the airplane; the smaller they are, the more maneuverable it is.

The angular velocity of the airplane is fixed by the speed and normal acceleration (Fig. 3.1).

$$q = \frac{(n - 1)g}{V} \tag{3.1,1}$$

As a consequence of this angular velocity, the field of the relative air flow past the airplane is curved. It is as though the aircraft were attached to the end of a whirling arm pivoted at O (Fig. 3.1). This curvature of the flow field alters the pressure distribution and the aerodynamic forces from their values in translational flight. The change is large enough that it must be taken into account in the equations describing the motion.

We assume that q and the increments $\Delta\alpha$, $\Delta\delta_e$ etc. between the rectilinear and curved flight conditions are small, so that the increments in lift and moment may be written

$$\Delta C_L = C_{L_\alpha}\Delta\alpha + C_{L_q}\hat{q} + C_{L_{\delta_e}}\Delta\delta_e \tag{3.1,2}$$

$$\Delta C_m = C_{m_\alpha}\Delta\alpha + C_{m_q}\hat{q} + C_{m_{\delta_e}}\Delta\delta_e \tag{3.1,3}$$

where, in order to maintain a nondimensional form of equations, we have introduced the dimensionless pitch rate $\hat{q} = q\bar{c}/2V$, and $C_{L_q} = \partial C_L/\partial\hat{q}$, $C_{m_q} = \partial C_m/\partial\hat{q}$. The q de-

[1]The load factor is the ratio of lift to weight, $n = L/W$. It is unity in straight horizontal flight.

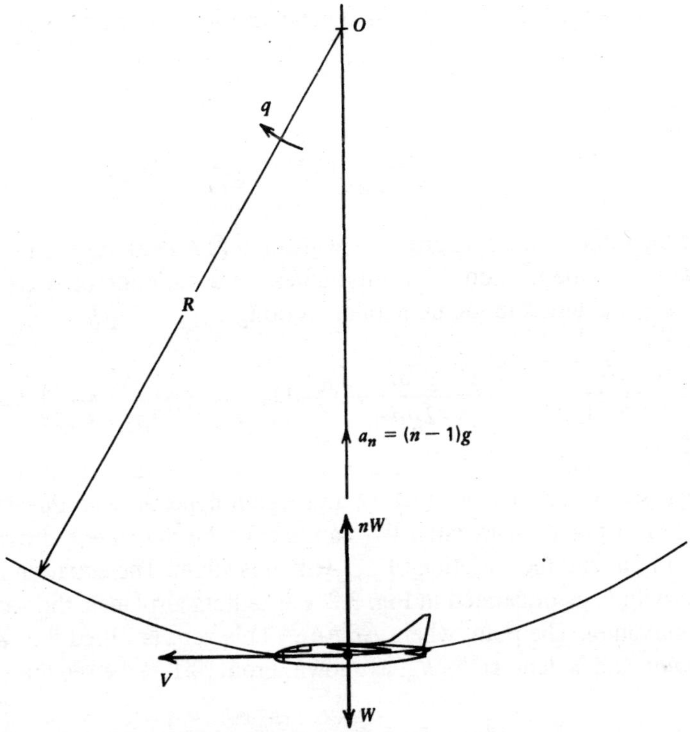

Figure 3.1 Airplane in a pull-up.

rivatives are discussed in Sec. 5.4. In this form, these equations apply to any configuration. From (3.1,1) we get

$$\hat{q} = (n-1)\frac{g\bar{c}}{2V^2}$$

which is more conveniently expressed in terms of *the weight coefficient* C_W and *the mass ratio* μ (see Sec. 3.15), that is,

$$\hat{q} = (n-1)\frac{C_W}{2\mu} \tag{3.1,4}$$

Since the curved flight condition is also assumed to be steady, that is, without angular acceleration, then $\Delta C_m = 0$. Finally, we can relate ΔC_L to n thus:

$$\Delta C_L = \frac{nW - W}{\frac{1}{2}\rho V^2 S} = (n-1)C_W \tag{3.1,5}$$

Equations (3.1,2 and 3) therefore become

$$(n-1)C_W = C_{L_\alpha}\Delta\alpha + (n-1)C_{L_q}\frac{C_W}{2\mu} + C_{L_{\delta_e}}\Delta\delta_e$$

$$0 = C_{m_\alpha}\Delta\alpha + (n-1)C_{m_q}\frac{C_W}{2\mu} + C_{m_{\delta_e}}\Delta\delta_e$$

which are readily solved for $\Delta\alpha$ and $\Delta\delta_e$ to yield the elevator angle per g

$$\frac{\Delta\delta_e}{n-1} = -\frac{C_W}{det}\left[C_{m_\alpha} - \frac{1}{2\mu}(C_{L_q}C_{m_\alpha} - C_{L_\alpha}C_{m_q})\right] \qquad (a)$$

and

$$\frac{\Delta\alpha}{n-1} = \frac{1}{C_{L_\alpha}}\left(C_W - C_{L_q}\frac{C_W}{2\mu} - C_{L_{\delta_e}}\frac{\Delta\delta_e}{n-1}\right) \qquad (b) \qquad (3.1,6)$$

where *det* is the same expression previously given in (2.4,13*d*). As has been shown in Sec. 2.4 *det* does not depend on CG position, hence the variation of $\Delta\delta_e/(n-1)$ with h is provided by the terms in the numerator. Writing $C_{m_\alpha} = C_{L_\alpha}(h - h_n)$ (3.1,6*a*) becomes

$$\frac{\Delta\delta_e}{n-1} = -\frac{C_W C_{L_\alpha}(2\mu - C_{L_q})}{2\mu det}\left(h - h_n + \frac{C_{m_q}}{2\mu - C_{L_q}}\right) \qquad (3.1,7)$$

The derivatives C_{L_q} and C_{m_q} both in general vary with h, the former linearly, the latter quadratically, (see Sec. 5.4). Thus (3.1,7), although it appears to be linear in h, is not exactly so. For airplanes with tails, C_{L_q} can usually be neglected altogether when compared with 2μ, and the variation of C_{m_q} with h is slight. The equation is then very nearly linear with h, as illustrated in Fig. 3.2. For tailless airplanes, the variation may show more curvature. The point where $\Delta\delta_e/(n-1)$ is zero is called the *control-fixed maneuver point,* and is denoted by h_m, as shown. From (3.1,7) we see that

$$h_m = h_n - \frac{C_{m_q}(h_m)}{2\mu - C_{L_q}(h_m)} \qquad (3.1,8)$$

where $C_{m_q}(h_m)$ and $C_{L_q}(h_m)$ are the values of these two derivatives evaluated for $h = h_m$. When C_{m_q} and C_{L_q} can be assumed to be independent of h, (3.1,7) reduces to

$$\frac{\Delta\delta_e}{n-1} = -\frac{C_W C_{L_\alpha}(2\mu - C_{L_q})}{2\mu det}(h - h_m) \qquad (3.1,9)$$

The difference $(h_m - h)$ is known as the *control-fixed maneuver margin.*

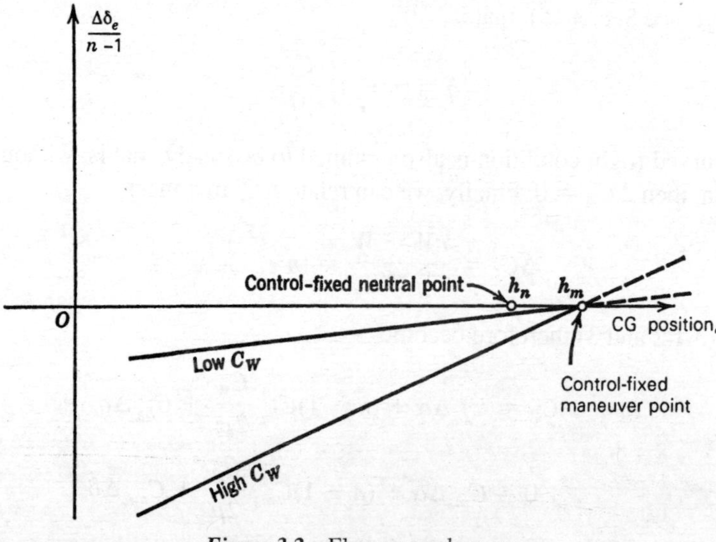

Figure 3.2 Elevator angle per g.

3.2 Control Force per g

From (2.8,4) we get the incremental control force

$$\Delta P = GS_e\bar{c}_e\tfrac{1}{2}\rho V^2 \,\Delta C_{he} \tag{3.2,1}$$

C_{he} is given for rectilinear flight by (2.5,2). Since it too will in general be influenced by q, we write for the incremental value ($\Delta\delta_t = 0$)

$$\Delta C_{he} = C_{he_\alpha}\Delta\alpha + C_{he_q}\hat{q} + b_2\Delta\delta_e \tag{3.2,2}$$

The derivative C_{he_q} is discussed in Sec. 5.4. Using (3.1,4) and (3.1,6*b*), (3.2,2) is readily expanded to give

$$\frac{\Delta C_{he}}{n-1} = \frac{C_W}{2\mu C_{L_\alpha}}[(2\mu - C_{L_q})C_{he_\alpha} + C_{h_q}C_{L_\alpha}] + \frac{\Delta\delta_e}{n-1}\left(b_2 - \frac{C_{L_{\delta_e}}C_{he_\alpha}}{C_{L_\alpha}}\right) \tag{3.2,3}$$

From (2.6,4*b*) we note that the last parenthetical factor is $b_2C'_{L_\alpha}/C_{L_\alpha}$ or b_2a'/a. For $\Delta\delta_e$ we use the approximation (3.1,9) in the interest of simplicity and the result for ΔC_{he} after some algebraic reduction is

$$\frac{\Delta C_{he}}{n-1} = -\frac{C_W}{2\mu}\frac{a'b_2}{det}(2\mu - C_{L_q})(h - h'_m) \tag{3.2,4}$$

where

$$h'_m = h_m + \frac{det}{a'b_2}\left(\frac{C_{he_\alpha}}{C_{L_\alpha}} + \frac{C_{he_q}}{2\mu - C_{L_q}}\right) \tag{3.2,5}$$

In keeping with earlier nomenclature, h'_m is the *control-free maneuver point* and $(h'_m - h)$ is the corresponding *margin*. On noting that $C_W\frac{1}{2}\rho V^2$ is the wing loading w, we find the control force per g is given by

$$Q = \frac{\Delta P}{n-1} = -GS_e\bar{c}_e w\frac{a'b_2}{2\mu det}(2\mu - C_{L_q})(h - h'_m) \tag{3.2,6}$$

Note that this result applies to both tailed and tailless aircraft provided that the appropriate derivatives are used. The following conclusions may be drawn from (3.2,6).

1. The control force per g increases linearly from zero as the CG is moved forward from the control-free maneuver point, and reverses sign for $h > h'_m$.
2. It is directly proportional to the wing loading. High wing loading produces "heavier" controls.
3. For similar aircraft of different size but equal wing loading, $Q \propto S_e\bar{c}_e$; i.e. to the cube of the linear size.
4. Neither C_L nor V enters the expression for Q explicitly. Thus, apart from **M** and Reynolds number effects, Q is independent of speed.
5. The factor μ which appears in (3.2,5) causes the separation of the control-free neutral and maneuver points to vary with altitude, size, and wing loading, in the same manner as the interval $(h_m - h_n)$.

Figure 3.3 shows a typical variation of Q with CG position. The statement made above that the control force per g is "reversed" when $h > h'_m$ must be interpreted cor-

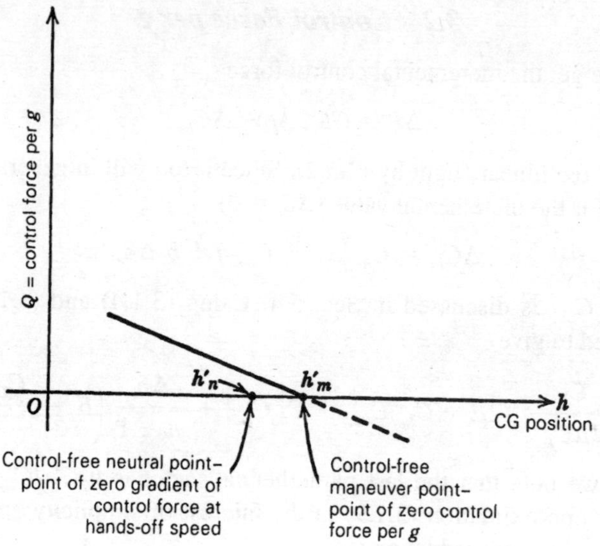

Figure 3.3 Control force per g.

rectly. In the first place this does not necessarily mean a reversal of control movement per g, for this is governed by the elevator angle per g. If $h'_m < h < h_m$, then there would be reversal of Q without reversal of control movement. In the second place, the analysis given applies only to the *steady state* at load factor n, and throws no light whatsoever on the transition between unaccelerated flight and the pull-up condition. No matter what the value of h, the *initial* control force and movement required to start the maneuver will be in the normal direction (backward for a pull-up), although one or both of them may have to be reversed before the final steady state is reached.

CONTROL-FORCE GADGETS

The control forces on a manually controlled airplane can be made to deviate from the "natural" pattern that flows from the size of the airplane, the aerodynamic design, and the speed and altitude of flight without necessarily using either powered controls or aerodynamic tabs (see Sec. 2.7). Some "gadgets" that can be used to this end are springs, weights, and variable-ratio sprockets and linkages. These can have the effect of modifying the control-force to trim and the control-force per g, giving the pilot the same feel as if the control-free neutral and maneuver points were moved. Some details of these effects are given in Sec. 7.1 of Etkin (1972).

3.3 Influence of High-Lift Devices on Trim and Pitch Stiffness

Conventional airplanes utilize a wide range of aerodynamic devices for increasing $C_{L_{max}}$. These include various forms of trailing edge elements (plain flaps, split flaps, slotted flaps, etc.), leading edge elements (drooped nose, slats, slots, etc.) and purely fluid mechanical solutions such as boundary layer control by blowing. Each of these has its own characteristic effects on the lift and pitching moment curves, and it is not feasible to go into them in depth here. The specific changes that result from the "configuration-type" devices, i.e. flaps, slots, etc., can always be incorporated by making the appropriate changes to $h_{n_{wb}}$, $C_{m_{ac_{wb}}}$ and $C_{L_{wb}}$ in (2.2,4) and following through the consequences. Consider for example the common case of part-span trailing edge flaps on a conventional tailed airplane. The main aerodynamic effects of such flaps are illustrated in Fig. 3.4.[2]

1. Their deflection distorts the shape of the spanwise distribution of lift on the wing, increasing the vorticity behind the flap tips, as in (*a*).
2. They have the same effect locally as an increase in the wing-section camber, that is, a negative increment in $C_{m_{ac}}$ and a positive increment in $C_{L_{wb}}$.
3. The downwash at the tail is increased; both ϵ_0 and $\partial\epsilon/\partial\alpha$ will in general change.

The change in wing-body C_m is obtained from (2.2,4) as

$$\Delta C_{m_{wb}} = \Delta C_{m_{ac_{wb}}} + \Delta C_{L_{wb}}(h - h_{n_{wb}}) \tag{3.3,1}$$

The change in airplane C_L is

$$\Delta C_L = \Delta C_{L_{wb}} - a_t \frac{S_t}{S} \Delta\epsilon \tag{3.3,2}$$

and the change in tail pitching moment is

$$\Delta C_{m_t} = a_t V_H \Delta\epsilon \tag{3.3,3}$$

When the increments $\Delta C_{m_{ac_{wb}}}$ and $\Delta C_{L_{wb}}$ are constant with α and $\Delta h_{n_{wb}}$ is negligible, then the only effect on C_{L_α} and C_{m_α} is that of $\partial\epsilon/\partial\alpha$, and from (2.3,18) and (2.3,21*a*) these are

$$\Delta a = \Delta C_{L_\alpha} = -a_t \frac{S_t}{S} \Delta \frac{\partial\epsilon}{\partial\alpha} \tag{3.3,4}$$

$$\Delta C_{m_\alpha} = (h - h_{n_{wb}})\Delta a + a_t \bar{V}_H \Delta \frac{\partial\epsilon}{\partial\alpha} \tag{3.3,5}$$

The net result on the C_L and C_m curves is obviously very much configuration dependent. If the $C_m - \alpha$ relation were as in Fig. 3.4*e*, then the trim change would be very large, from α_1 at $\delta_f = 0$ to α_2 after flap deflection. The C_L at α_2 is much larger than at α_1 and hence if the flap operation is to take place without change of trim speed, a down-elevator deflection would be needed to reduce α_{trim} to α_3 (Fig. 3.4*c*). This would result in a nose-down rotation of the aircraft.

[2]Note that α is still the angle of attack of the zero-lift line of the basic configuration, and that the lift with flap deflected is not zero at zero α.

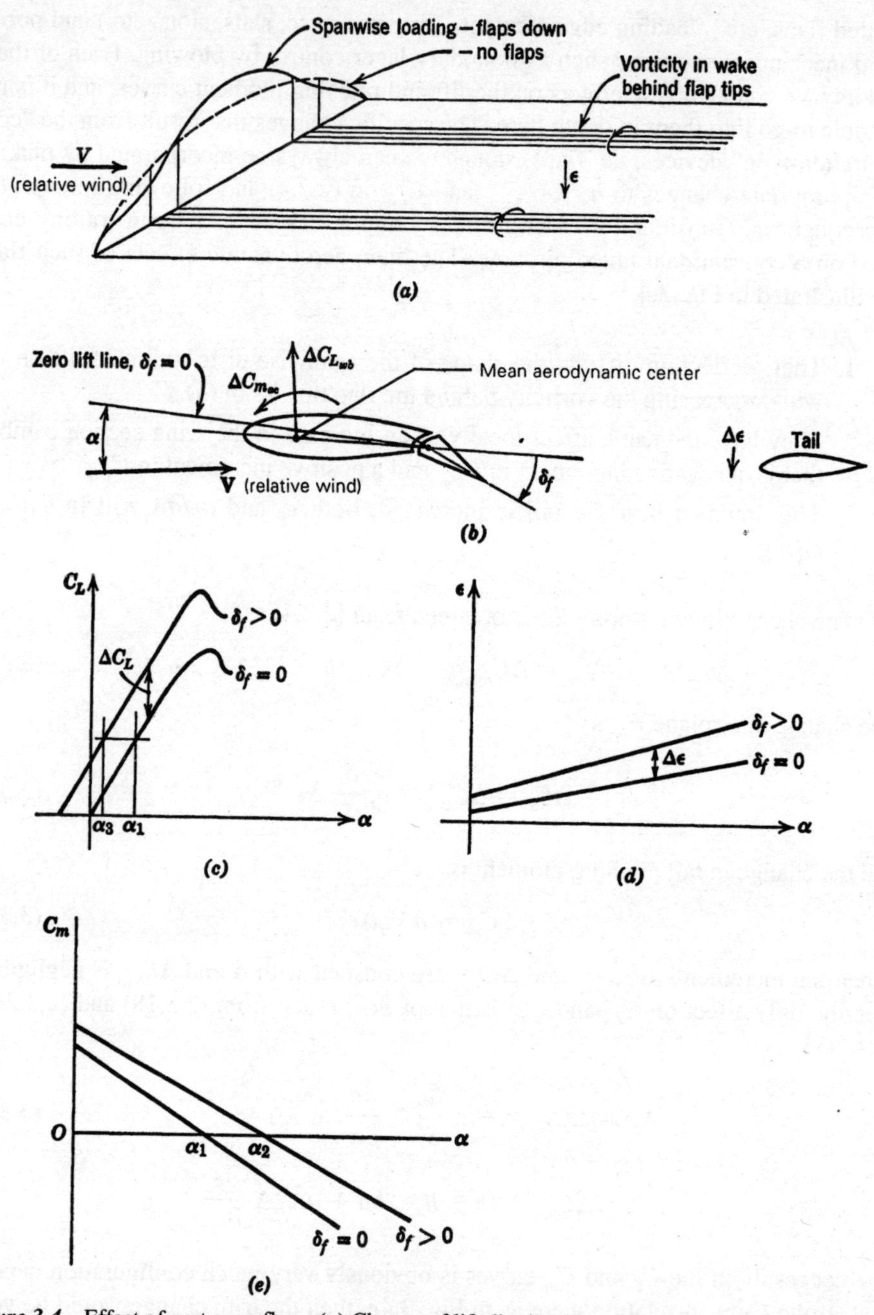

Figure 3.4 Effect of part-span flaps. (*a*) Change of lift distribution and vorticity. (*b*) Changes in forces and moments. (*c*) Change in C_L. (*d*) Change in downwash. (*e*) Change in C_m.

3.4 Influence of the Propulsive System on Trim and Pitch Stiffness

The influences of the propulsive system upon trim and stability may be both important and complex. The range of conditions to be considered in this connection is extremely wide. There are several types of propulsive units in common use—reciprocating-engine-driven propellers, turbojets, turboprops, and rockets, and the variations in engine-plus-vehicle geometry are very great. The analyst may have to deal with such widely divergent cases as a high-aspect-ratio straight-winged airplane with six wing-mounted counterrotating propellers or a low-aspect-ratio delta with buried jet engines. Owing to its complexity, a comprehensive treatment of propulsive system influences on stability is not feasible. There does not exist sufficient theoretical or empirical information to enable reliable predictions to be made under all the above-mentioned conditions. However, certain of the major effects of propellers and propulsive jets are sufficiently well understood to make it worth while to discuss them, and this is done in the following.

In a purely formal sense, of course, it is only necessary to add the appropriate direct effects, C_{m0_p} and $\partial C_{m_p}/\partial\alpha$ in (2.3,21 and 22), together with the indirect effects on the various wing-body and tail coefficients in order to calculate all the results with power on.

When calculating the trim curves (i.e., elevator angle, tab angle, and control force to trim) the thrust must be that required to maintain equilibrium at the condition of speed and angle of climb being investigated (see Sec. 2.4). For example (see Fig. 2.1), assuming that $\alpha_T \ll 1$

$$\begin{aligned} C_T &= C_D + C_W \sin\gamma && (a) \\ C_W \cos\gamma &= C_L + C_T\alpha_T && (b) \end{aligned} \qquad (3.4,1)$$

Solving for C_T, we get

$$C_T = \frac{C_D + C_L \tan\gamma}{1 - \alpha_T \tan\gamma} \qquad (3.4,2)$$

Except for very steep climb angles, $\alpha_T \tan\gamma \ll 1$, and we may write approximately,

$$C_T = C_D + C_L \tan\gamma \qquad (3.4,3)$$

Let the thrust line be offset by a distance z_p from the CG (as in Fig. 3.5) and neglecting for the moment all other thrust contributions to the pitching moment except Tz_p, we have

$$\begin{aligned} C_{m_p} &= C_T \frac{z_p}{\bar{c}} \\ &= (C_D + C_L \tan\gamma)\frac{z_p}{\bar{c}} \end{aligned} \qquad (3.4,4)$$

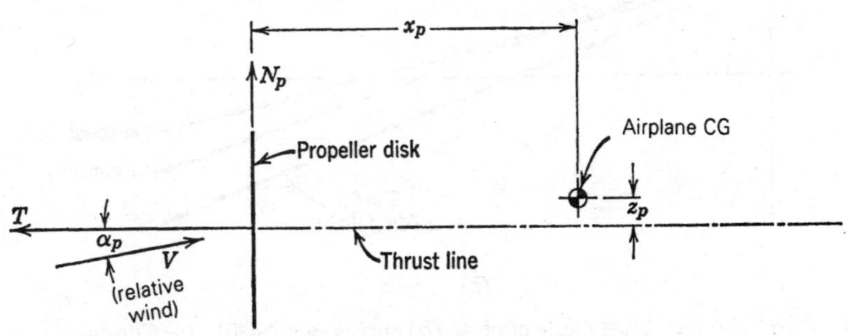

Figure 3.5 Forces on a propeller.

Now let C_D be given by the parabolic polar (2.1,2), so that

$$C_{m_p} = (C_{D_{\min}} + KC_L^2 + C_L \tan \gamma) \frac{z_p}{\bar{c}} \tag{3.4,5}$$

Strictly speaking, the values of C_D and C_L in (3.4,4 and 5) are those for trimmed flight, i.e. with $\delta_e = \delta_{e_{\text{trim}}}$. For the purposes of this discussion of propulsion effects we shall neglect the effects of δ_e on C_D and C_L, and assume that the values in (3.4,5) are those corresponding to $\delta_e = 0$. The addition of this propulsive effect to the C_m curve for rectilinear gliding flight in the absence of aeroelastic and compressibility effects might then appear as in Fig. 3.6*a*. We note that the gradient $-dC_m/dC_L$ for any value of $\gamma > 0$ is less than for unpowered flight. If dC_m/dC_L is used uncritically as a criterion for stability an *entirely erroneous conclusion* may be drawn from such curves.

1. Within the assumptions made above, the thrust moment Tz_p is *independent of* α, hence $\partial C_{m_p}/\partial\alpha = 0$ and there is *no change* in the NP from that for unpowered flight.
2. A true analysis of stability when *both* speed and α are changing requires that the propulsive system controls (e.g., the throttle) be kept *fixed,* whereas each point on the curves of Fig. 3.6*a* corresponds to a *different* throttle setting. This parallels exactly the argument of Sec. 2.4 concerning the elevator trim slope.

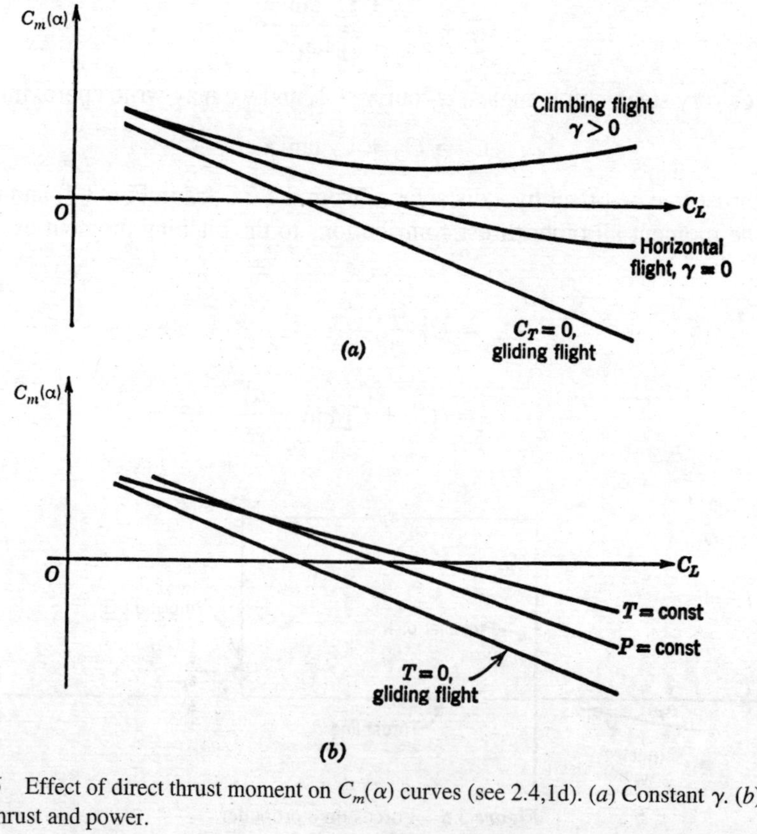

Figure 3.6 Effect of direct thrust moment on $C_m(\alpha)$ curves (see 2.4,1d). (*a*) Constant γ. (*b*) Constant thrust and power.

For in fact, under the stated conditions, the $C_m - C_L$ curve is transformed into a curve of $\delta_{e_{\text{trim}}}$ vs. V by using the relations $\delta_{e_{\text{trim}}} = -C_m(\alpha)/C_{m_{\delta_e}}$ and $C_L = W/\frac{1}{2}\rho V^2 S$. The slopes of C_m vs. C_L and $\delta_{e_{\text{trim}}}$ vs. V will vanish together.

If a graph of C_m vs. C_L be prepared for fixed throttle, then γ will be a variable along it, and its gradient dC_m/dC_L *is* an index of stability, as shown in Sec. 6.4. The two idealized cases of constant thrust and constant power are of interest. If the thrust at fixed throttle does not change with speed, then we easily find

$$C_{m_p} = \frac{T}{W} C_L \frac{z_p}{\bar{c}} \qquad (a)$$

and (3.4,6)

$$\frac{dC_{m_p}}{dC_L} = \frac{Tz_p}{W\bar{c}} \qquad (b)$$

If the power P is invariant, instead of the thrust, then $T = P/V$ and we find

$$C_{m_p} = \frac{P}{W}\sqrt{\frac{\rho}{2w}}\frac{z_p}{\bar{c}} C_L^{3/2} \qquad (c)$$

(3.4,6)

whence

$$\frac{dC_{m_p}}{dC_L} = \frac{3}{2}\frac{P}{W}\sqrt{\frac{\rho}{2w}}\frac{z_p}{\bar{c}} C_L^{1/2} \qquad (d)$$

Thus in the constant thrust case, the power-off $C_m - C_L$ graph simply has its slope changed by the addition of thrust, and in the constant power case the shape is changed as well. The form of these changes is illustrated in Fig. 3.6*b* and it is evident by comparison with 3.6*a* that the behavior of dC_m/dC_L is quite different in these two situations.

THE INFLUENCE OF RUNNING PROPELLERS

The forces on a single propeller are illustrated in Fig. 3.5, where α_p is the angle of attack of the local flow at the propeller. It is most convenient to resolve the resultant into the two components T along the axis, and N_p in the plane of the propeller. The moment associated with T has already been treated above, and does not affect C_{m_α}. That due to N_p is

$$\Delta C_m = C_{N_p}\frac{x_p}{\bar{c}}\frac{S_p}{S} \qquad (3.4,7)$$

where $C_{N_p} = N_p/\frac{1}{2}\rho V^2 S_p$ and S_p is the propeller disk area. To get the total ΔC_m for several propellers, increments such as (3.4,7) must be calculated for each and summed. Theory shows (Ribner, 1945) that for small angles C_{N_p} is proportional to α_p. Hence N_p contributes to both C_{m0_p} and $\partial C_{m_p}/\partial\alpha$. The latter is

$$\frac{\partial C_{m_p}}{\partial\alpha} = \frac{S_p}{S}\frac{x_p}{\bar{c}}\frac{\partial C_{N_p}}{\partial\alpha_p}\frac{\partial\alpha_p}{\partial\alpha} \qquad (3.4,8)$$

If the propeller were situated far from the flow field of the wing, then $\partial\alpha_p/\partial\alpha$ would be unity. However, for the common case of wing-mounted tractor propellers with the propeller plane close to the wing, there is a strong upwash ϵ_p at the propeller. Thus

$$\alpha_p = \alpha + \epsilon_p + \text{const} \quad (a)$$

and

$$\frac{\partial \alpha_p}{\partial \alpha} = 1 + \frac{\partial \epsilon_p}{\partial \alpha} \quad (b) \tag{3.4,9}$$

where the constant in (3.4,9*a*) is the angle of attack of the propeller axis relative to the airplane zero-lift line. Finally,

$$\frac{\partial C_{m_p}}{\partial \alpha} = \frac{S_p}{S}\frac{x_p}{\bar{c}}\left(1 + \frac{\partial \epsilon_p}{\partial \alpha}\right)\frac{\partial C_{N_p}}{\partial \alpha_p} \tag{3.4,10}$$

Increase of Wing Lift

When a propeller is located ahead of a wing, the high-velocity slipstream causes a distortion of the lift distribution, and an increase in the total lift. This is a principal mechanism in obtaining high lift on so-called deflected slipstream STOL airplanes. For accurate results that allow for the details of wing and flap geometry powered-model testing is needed. However, for some cases there are available theoretical results (Ellis, 1971; Kuhn, 1959; Priestly, 1953) suitable for estimates. Both theory and experiment show that the lift increment tends to be linear in α for constant C_T, and hence has the effect of increasing a_{wb}, the lift-curve slope for the wing-body combination. From (2.3,23) this is seen to reduce the effect of the tail on the NP location, and can result in a decrease of pitch stiffness.

Effects on the Tail

The propeller slipstream can affect the tail principally in two ways. (1) Depending on how much if any of the tail lies in it, the effective values of a_t and a_e will experience some increase. (2) The downwash values ϵ_0 and $\partial\epsilon/\partial\alpha$ may be appreciably altered in any case. Methods of estimating these effects are at best uncertain, and powered-model testing is needed to get results with engineering precision for most new configurations. However, some empirical methods (Smelt and Davies, 1937; Millikan, 1940; Weil and Sleeman, 1949) are available that are suitable for some cases.

THE INFLUENCE OF JET ENGINES

The direct thrust moment of jet engines is treated as shown at the beginning of this section, the constant-thrust idealization given in (3.4,6) often being adequate. In addition, however, there is a normal force on jet engines as well as on propellers.

Jet Normal Force

The air that passes through a propulsive duct experiences, in general, changes in both the direction and magnitude of its velocity. The change in magnitude is the principal source of the thrust, and the direction change entails a force normal to the thrust line. The magnitude and line of action of this force can be found from momentum considerations. Let the mass flow through the duct be m' and the velocity vectors at the inlet and outlet be $\mathbf{V}_i$ and $\mathbf{V}_j$. Application of the momentum principle then shows that the reaction on the airplane of the air flowing through the duct is

$$\mathbf{F} = -m'(\mathbf{V}_j - \mathbf{V}_i) + \mathbf{F}'$$

where $\mathbf{F}'$ is the resultant of the pressure forces acting across the inlet and outlet areas. For the present purpose, $\mathbf{F}'$ may be neglected, since it is approximately in the direction of the thrust T. The component of $\mathbf{F}$ normal to the thrust line is then found as in Fig. 3.7. It acts through the intersection of $\mathbf{V}_i$ and $\mathbf{V}_j$. The magnitude is given by

$$N_j = m'V_i \sin \theta$$

or, for small angles,

$$N_j = m'V_i\theta \tag{3.4,11}$$

In order to use this relation, both V_i and θ are required. It is assumed that V_i has that direction which the flow would take in the absence of the engine; that is, θ equals the angle of attack of the thrust line α_j plus the upwash angle due to wing induction ϵ_j.

$$\theta = \alpha_j + \epsilon_j \tag{3.4,12}$$

It is further assumed that the magnitude V_i is determined by the mass flow and inlet area; thus

$$V_i = \frac{m'}{A_i\rho_i} \tag{3.4,13}$$

where A_i is the inlet area, and ρ_i the air density in the inlet. We then get for N_j the expression

$$N_j = \frac{m'^2}{A_i\rho_i}(\alpha_j + \epsilon_j)$$

The corresponding pitching-moment coefficient is

$$\Delta C_m = \frac{m'^2}{A_i\rho_i}\frac{x_j}{\frac{1}{2}\rho V^2 S\bar{c}}(\alpha_j + \epsilon_j) \tag{3.4,14}$$

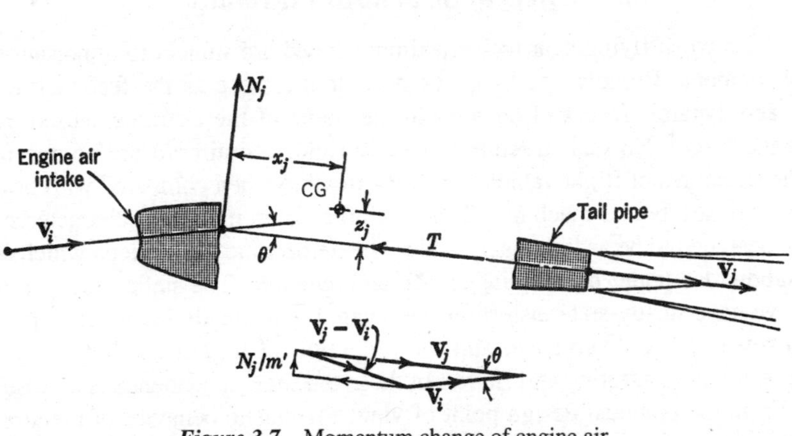

Figure 3.7 Momentum change of engine air.

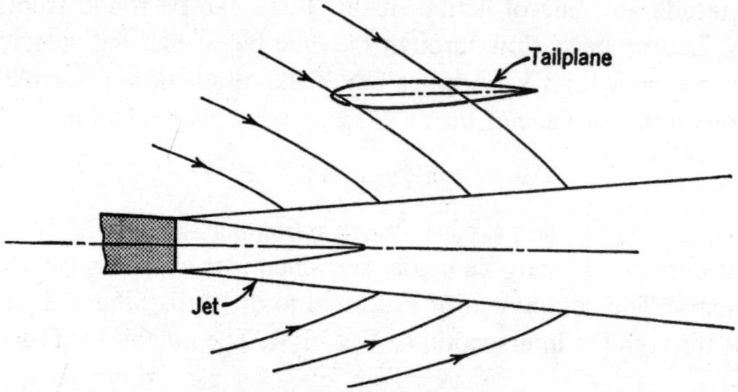

Figure 3.8 Jet-induced inflow.

Since the pitching moment given by (3.4,14) varies with α at constant thrust, then there is a change in C_{m_α} given by

$$\Delta C_{m_\alpha} = \frac{m'^2}{A_i \rho_i} \frac{1}{\frac{1}{2}\rho V^2 S\bar{c}} \left[x_j \left(1 + \frac{\partial \epsilon_j}{\partial \alpha}\right) + \theta \frac{\partial x_j}{\partial \alpha} \right] \tag{3.4,15}$$

The quantities m' and ρ_i can be determined from the engine performance data, and for subsonic flow, $\partial \epsilon_j / \partial \alpha$ is the same as the value $\partial \epsilon_p / \partial \alpha$ used for propellers. $\partial x_j / \partial \alpha$ can be calculated from the geometry.

Jet Induced Inflow

A spreading jet entrains the air that surrounds it, as illustrated in Fig. 3.8, thereby inducing a flow toward the jet axis. If a tailplane is placed in the induced flow field, the angle of attack will be modified by this inflow. A theory of this phenomenon which allows for the curvature of the jet due to angle of attack has been formulated by Ribner (1946). This inflow at the tail may vary with α sufficiently to reduce the stability by a significant amount.

3.5 *Effect of Structural Flexibility*

Many vehicles when flying near their maximum speed are subject to important aeroelastic phenomena. Broadly speaking, we may define these as the feedback effects upon the aerodynamic forces of changes in the shape of the airframe caused by the aerodynamic forces. No real structure is ideally rigid, and aircraft are no exception. Indeed the structures of flight vehicles are very flexible when compared with bridges, buildings, and earthbound machines. This flexibility is an inevitable characteristic of structures designed to be as light as possible. The aeroelastic phenomena which result may be subdivided under the headings static and dynamic. The static cases are those in which we have steady-state distortions associated with steady loads. Examples are aileron reversal, wing divergence, and the reduction of longitudinal stability. Dynamic cases include buffeting and flutter. In these the time dependence is an essential element. From the practical design point of view, the elastic behavior of the airplane affects all three of its basic characteristics: namely performance, stability, and structural integrity. This subject occupies a well-established position as a separate branch of aeronautical engineering. For further information the reader is referred to one of the books devoted to it (Bisplinghoff, 1962; Dowell, 1994).

In this section we take up by way of example a relatively simple aeroelastic effect; namely, the influence of fuselage flexibility on longitudinal stiffness and control. Assume that the tail load L_t bends the fuselage so that the tail rotates through the angle $\Delta\alpha_t = -kL_t$ (Fig. 3.9) while the wing angle of attack remains unaltered. The net angle of attack of the tail will then be

$$\alpha_t = \alpha_{wb} - \epsilon - i_t - kL_t$$

and the tail lift coefficient at $\delta_e = 0$ will be

$$C_{L_t} = a_t\alpha_t = a_t(\alpha_{wb} - \epsilon - i_t - kL_t)$$

But $L_t = C_{L_t}\frac{1}{2}\rho V^2 S_t$, from which

$$C_{L_t} = a_t(\alpha_{wb} - \epsilon - i_t - kC_{L_t}\tfrac{1}{2}\rho V^2 S_t) \tag{3.5,1}$$

Solving for C_{L_t}, we get

$$C_{L_t} = \frac{a_t}{1 + ka_tS_t\dfrac{\rho}{2}V^2}(\alpha_{wb} - \epsilon - i_t) \tag{3.5,2}$$

Comparison of (3.5,2) with (2.3,13) shows that the tail effectiveness has been reduced by the factor $1/[1 + ka_t(\rho/2)V^2S_t]$. The main variable in this expression is V, and it is seen that the reduction is greatest at high speeds. From (2.3,23) we find that the reduction in tail effectiveness causes the neutral point to move forward. The shift is given by

$$\Delta h_n = \frac{\Delta a_t}{a}\bar{V}_H\left(1 - \frac{\partial \epsilon}{\partial \alpha}\right) \tag{3.5,3}$$

where

$$\Delta a_t = a_t\left(\frac{1}{1 + ka_t\frac{1}{2}\rho V^2 S_t} - 1\right) \tag{3.5,4}$$

The elevator effectiveness is also reduced by the bending of the fuselage. For, if we consider the case when δ_e is different from zero, then (3.5,1) becomes

$$C_{L_t} = a_t(\alpha_{wb} - \epsilon - i_t - kC_{L_t}\tfrac{1}{2}\rho V^2 S_t) + a_e\delta_e$$

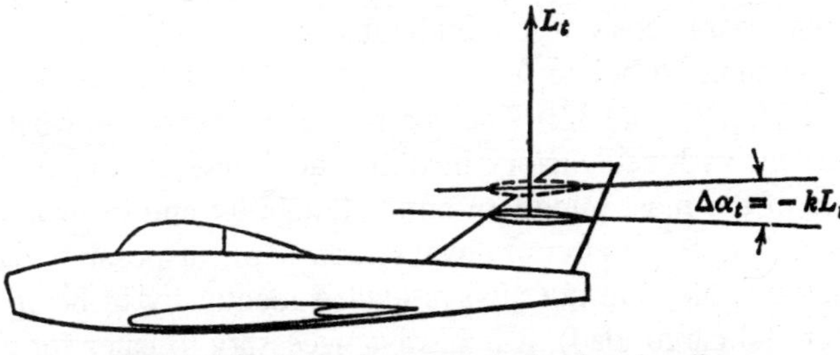

Figure 3.9 Tail rotation due to fuselage bending.

and (3.5,2) becomes

$$C_{L_t} = \frac{a_t(\alpha_{wb} - \epsilon - i_t) + a_e\delta_e}{1 + ka_t\frac{1}{2}\rho V^2 S_t}$$

Thus the same factor $1/(1 + ka_t\rho/2V^2S_t)$ that operates on the tail lift slope a_t also multiplies the elevator effectiveness a_e.

3.6 Ground Effect

At landing and takeoff airplanes fly for very brief (but none the less extremely important) time intervals close to the ground. The presence of the ground modifies the flow past the airplane significantly, so that large changes can take place in the trim and stability. For conventional airplanes, the takeoff and landing cases provide some of the governing design criteria.

The presence of the ground imposes a boundary condition that inhibits the downward flow of air normally associated with the lifting action of the wing and tail. The reduced downwash has three main effects, being in the usual order of importance:

1. A reduction in ϵ, the downwash angle at the tail.
2. An increase in the wing-body lift slope a_{wb}.
3. An increase in the tail lift slope a_t.

The problem of calculating the stability and control near the ground then resolves itself into estimating these three effects. When appropriate values of $\partial\epsilon/\partial\alpha$, a_{wb}, and a_t have been found, their use in the equations of the foregoing sections will readily yield the required information. The most important items to be determined are the elevator angle and control force required to maintain $C_{L_{max}}$ in level flight close to the ground. It will usually be found that the ratio a_t/a is decreased by the presence of the ground. (2.3,23) shows that this would tend to move the neutral point forward. However, the reduction in $\partial\epsilon/\partial\alpha$ is usually so great that the net effect is a large rearward shift of the neutral point. Since the value of C_{m_0} is only slightly affected, it turns out that the elevator angle required to trim at $C_{L_{max}}$ is much larger than in flight remote from the ground. It commonly happens that this is the critical design condition on the elevator, and it will govern the ratio S_e/S_t, or the forward CG limit (see Sec. 3.7).

3.7 CG Limits

One of the dominant parameters of longitudinal stability and control has been shown in the foregoing sections to be the fore-and-aft location of the CG (see Figs. 2.14, 2.18, 2.19, 2.25, 2.27, 2.28, and 3.2). The question now arises as to what range of CG position is consistent with satisfactory handling qualities. This is a critical design problem, and one of the most important aims of stability and control analysis is to provide the answer to it. Since aircraft always carry some disposable load (e.g., fuel, armaments), and since they are not always loaded identically to begin with (variations in passenger and cargo load), it is always necessary to cater for a variation in the CG position. The range to be provided for is kept to a minimum by proper location of the items of variable load, but still it often becomes a difficult matter to keep the handling qualities acceptable over the whole CG range. Sometimes the problem is not solved, and the airplane must be subjected to restrictions on the fore-and-aft distribution of its variable load when operating at part load.

THE AFT LIMIT

The most rearward allowable location of the CG is determined by considerations of longitudinal stability and control sensitivity. The behavior of the five principal control gradients are summarized in Fig. 3.10 for the case when the aerodynamic coefficients are independent of speed. From the handling qualities point of view, none of the gradients should be "reversed," that is, they should have the signs associated with low values of h. When the controls are reversible, this requires that $h < h'_n$. If the controls are irreversible, and if the artificial feel system is suitably designed, then the control force gradient $\partial P/\partial V$ can be kept negative to values of $h > h'_n$, and the rear limit can be somewhat farther back than with reversible controls. The magnitudes of the gradients are also important. If they are allowed to fall to very small values the vehicle will be too sensitive to the controls. When the coefficients do not depend on speed, as assumed for Fig. 3.10, the NP also gives the stability boundary (this is proved in Chap. 6), the vehicle becoming unstable for $h > h'_n$ with free controls or $h > h_n$ with fixed controls. If the coefficients dependent on speed, for example, $C_m = C_m(\mathbf{M})$, then the CG boundary for stability will be different and may be forward of the NP.

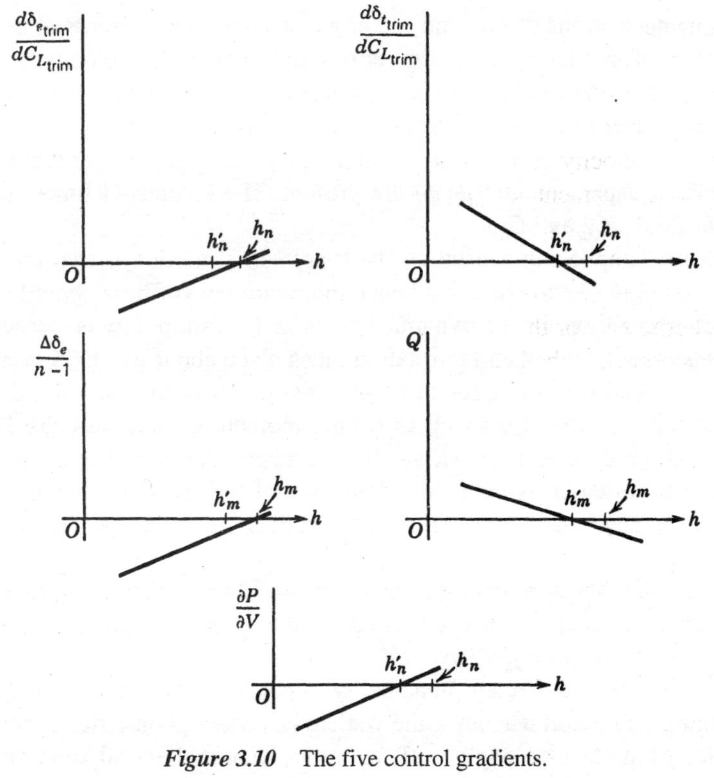

Figure 3.10 The five control gradients.

As noted in Chap. 1, it is possible to increase the inherent stability of a flight vehicle. *Stability augmentation systems* (SAS) are in widespread use on a variety of airplanes and rotorcraft. If such a system is added to the longitudinal controls of an airplane, it permits the use of more rearward CG positions than otherwise, but the risk of failure must be reckoned with, for then the airplane is reduced to its "inherent" stability, and would still need to be manageable by a human pilot.

THE FORWARD LIMIT

As the CG moves forward, the stability of the airplane increases, and larger control movements and forces are required to maneuver or change the trim. The forward CG limit is therefore based on control considerations and may be determined by any one of the following requirements:

1. The control force per g shall not exceed a specified value.
2. The control-force gradient at trim, $\partial P/\partial V$, shall not exceed a specified value.
3. The control force required to land, from trim at the approach speed, shall not exceed a specified value.
4. The elevator angle required to land shall not exceed maximum up elevator.
5. The elevator angle required to raise the nose-wheel off the ground at takeoff speed shall not exceed the maximum up elevator.

3.8 *Lateral Aerodynamics*

In the preceding sections of this and the previous chapter we discussed aerodynamic characteristics of symmetrical configurations flying with the velocity vector in the plane of symmetry. As a result the only nonzero motion variables were V, α, and q, and the only nonzero forces and moments were T, D, L, and M. We now turn to the cases in which the velocity vector is not in the plane of symmetry, and in which yawing and rolling displacements (β, ϕ) are present. The associated force and moment coefficients are C_y, C_l, and C_n.

One of the simplifying aspects of the longitudinal motion is that the rotation is about one axis only (the y axis), and hence the rotational stiffness about that axis is a very important criterion for the dynamic behavior. This simplicity is lost when we go to the lateral motions, for then the rotation takes place about two axes (x and z). The moments associated with these rotations are cross-coupled, that is, roll rotation p produces a yawing moment C_n as well as rolling moment C_l, and yaw displacement β and rate r both produce rolling *and* yawing moments. Furthermore, the roll and yaw controls are also often cross-coupled—deflection of the ailerons can produce significant yawing moments, and deflection of the rudder can produce significant rolling moments.

Another important difference between the two cases is that in "normal" flight—that is, steady rectilinear symmetric motion, all the lateral motion and force variables are zero. Hence there is no *fundamental* trimming problem—the ailerons and rudder would be nominally undeflected. In actuality of course, these controls do have a secondary trimming function whenever the vehicle has either geometric or inertial asymmetries—for example, one engine off, or multiple propellers all rotating the same way. Because the gravity vector in normal flight also lies in the plane of symmetry, the CG position is not a dominant parameter for the lateral characteristics as it is for the longitudinal. Thus the CG limits, (see Sec. 3.7) are governed by considerations deriving from the longitudinal characteristics.

3.9 Weathercock Stability (Yaw Stiffness)

Application of the static stability principle to rotation about the z axis suggests that a stable airplane should have "weathercock" stability. That is, when the airplane is at an angle of sideslip β relative to its flight path (see Fig. 3.11), the yawing moment produced should be such as to tend to restore it to symmetric flight. The yawing moment N is positive as shown. Hence the requirement for yaw stiffness is that $\partial N/\partial\beta$ must be positive. The nondimensional coefficient of N is

$$C_n = \frac{N}{\frac{1}{2}\rho V^2 Sb}$$

and hence for positive yaw stiffness $\partial C_n/\partial\beta$ must be positive. The usual notation for this derivative is

$$\frac{\partial C_n}{\partial\beta} = C_{n\beta} \tag{3.9,1}$$

This quantity is analogous in some respects to the longitudinal stability parameter C_{m_α}. It is estimated in a similar way by synthesis of the contributions of the various components of the airplane. The principal contributions are those of the body and the vertical-tail surface. By contrast with C_{m_α}, the wing has little influence in most cases, and the CG location is a weak parameter. Whether or not a positive value of C_{n_β} will produce lateral stability can only be determined by a full dynamic analysis such as is done in Chap. 6.

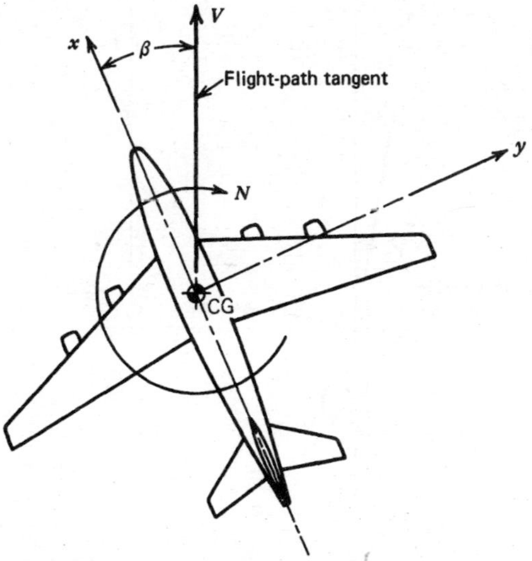

Figure 3.11 Sideslip angle and yawing moment.

In Fig. 3.12 are shown the relevant geometry and the lift force L_F acting on the vertical tail surface. If the surface were alone in an airstream, the velocity vector $\mathbf{V}_F$ would be that of the free stream, so that (cf. Fig. 3.11) α_F would be equal to $-\beta$. When installed on an airplane, however, changes in both magnitude and direction of the local flow at the tail take place. These changes may be caused by the propeller slipstream, and by the wing and fuselage when the airplane is yawed. The angular deflection is allowed for by introducing the sidewash angle σ, analogous to the downwash angle ϵ. σ is positive when it corresponds to a flow in the y direction; that is, when it tends to increase α_F. Thus the angle of attack is

$$\alpha_F = -\beta + \sigma \tag{3.9,2}$$

and the lift coefficient of the vertical-tail surface is

$$C_{L_F} = a_F(-\beta + \sigma) + a_r\delta_r \tag{3.9,3}$$

The lift is then

$$L_F = C_{L_F} \frac{\rho}{2} V_F^2 S_F \tag{3.9,4}$$

and the yawing moment is

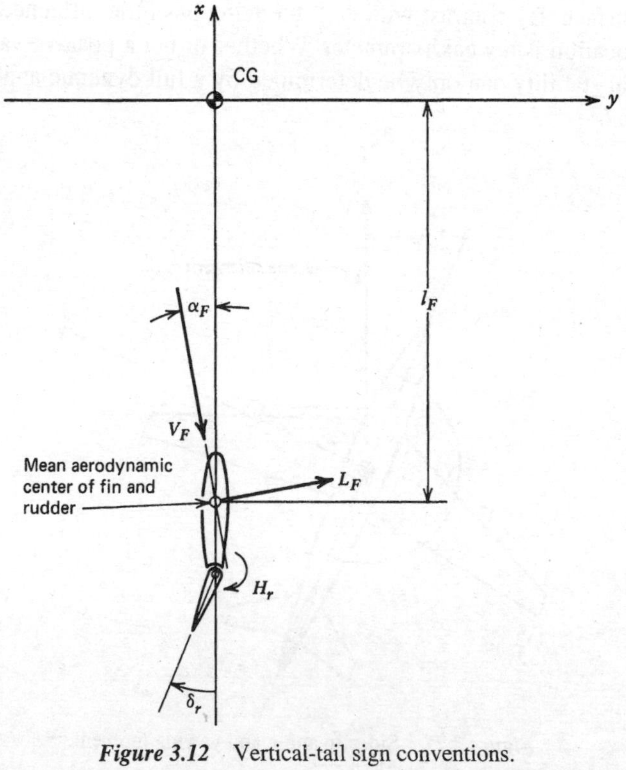

Figure 3.12 Vertical-tail sign conventions.

$$N_F = -C_{L_F}\frac{\rho}{2}V_F^2 S_F l_F$$

thus

$$C_{n_F} = -C_{L_F}\frac{S_F l_F}{Sb}\left(\frac{V_F}{V}\right)^2 \tag{3.9,5}$$

The ratio $S_F l_F/Sb$ is analogous to the horizontal-tail volume ratio, and is therefore called the *vertical-tail volume ratio,* denoted here by V_V. Equation 3.9,5 then reads

$$C_{n_F} = -V_V C_{L_F}\left(\frac{V_F}{V}\right)^2 \tag{3.9,6}$$

and the corresponding contribution to the weathercock stability is

$$\frac{\partial C_{n_F}}{\partial \beta} = -V_V\left(\frac{V_F}{V}\right)^2\frac{\partial C_{L_F}}{\partial \beta} = V_V a_F\left(\frac{V_F}{V}\right)^2\left(1-\frac{\partial \sigma}{\partial \beta}\right) \tag{3.9,7}$$

THE SIDEWASH FACTOR $\partial\sigma/\partial\beta$

Generally speaking, the sidewash is difficult to estimate with engineering precision. Suitable wind-tunnel tests are required for this purpose. The contribution from the fuselage arises through its behavior as a lifting body when yawed. Associated with the side force that develops is a vortex wake which induces a lateral-flow field at the tail. The sidewash from the propeller is associated with the side force which acts upon it when yawed, and may be estimated by the method of (Ribner, 1944). The contribution from the wing is associated with the asymmetric structure of the flow which develops when the airplane is yawed. This phenomenon is especially pronounced with low-aspect-ratio swept wings. It is illustrated in Fig. 3.13.

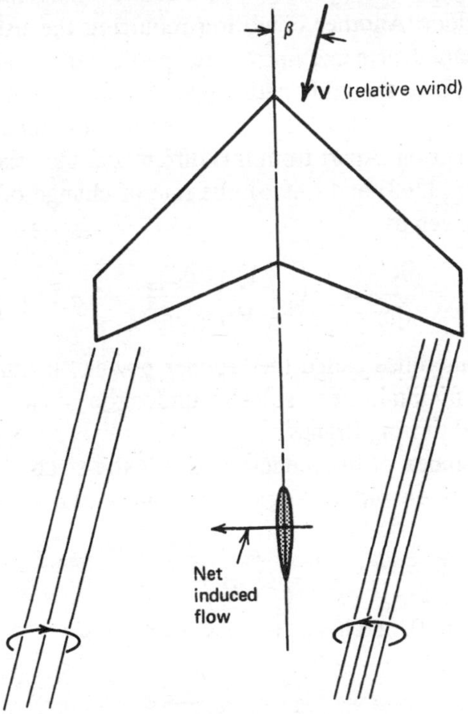

Figure 3.13 Vortex wake of yawed wing.

THE VELOCITY RATIO V_F/V

When the vertical tail is not in a propellor slipstream, V_F/V is unity. When it is in a slipstream, the effective velocity increment may be dealt with as for a horizontal tail.

CONTRIBUTION OF PROPELLER NORMAL FORCE

The yawing moment produced by the normal force that acts on the yawed propeller is calculated in the same way as the pitching-moment increment dealt with in Sec. 3.4. The result is similar to (3.4,8):

$$\Delta\frac{\partial C_n}{\partial\beta} = -\frac{x_p}{b}\frac{S_p}{S}\frac{\partial C_{N_p}}{\partial\alpha_p}. \tag{3.9,8}$$

This is known as the propeller fin effect and is negative (i.e., destabilizing) when the propeller is forward of the CG, but is usually positive for pusher propellers. There is a similar yawing moment effect for jet engines (see Exercise 3.7).

3.10 Yaw Control

In most flight conditions it is desired to maintain the sideslip angle at zero. If the airplane has positive yaw stiffness, and is truly symmetrical, then it will tend to fly in this condition. However, yawing moments may act upon the airplane as a result of unsymmetrical thrust (e.g., one engine inoperative), slipstream rotation, or the unsymmetrical flow field associated with turning flight. Under these circumstances, β can be kept zero only by the application of a control moment. The control that provides this is the rudder. Another condition requiring the use of the rudder is the steady sideslip, a maneuver sometimes used, particularly with light aircraft, to increase the drag and hence the glide path angle. A major point of difference between the rudder and the elevator is that for the former trimming the airplane is a secondary and not a primary function. Apart from this difference, the treatment of the two controls is similar. From (3.9,3) and (3.9,6), the rate of change of yawing moment with rudder deflection is given by

$$C_{n_{\delta_r}} = \frac{\partial C_n}{\partial\delta_r} = -V_V\left(\frac{V_F}{V}\right)^2\frac{\partial C_{L_F}}{\partial\delta_r} = -a_r V_V\left(\frac{V_F}{V}\right)^2 \tag{3.10,1}$$

This derivative is sometimes called the "rudder power." It must be large enough to make it possible to maintain zero sideslip under the most extreme conditions of asymmetric thrust and turning flight.

A second useful index of the rudder control is the steady sideslip angle that can be maintained by a given rudder angle. The total yawing moment during steady sideslip may be written

$$C_n = C_{n_\beta}\beta + C_{n_{\delta_r}}\delta_r \tag{3.10,2}$$

For steady motion, $C_n = 0$, and hence the desired ratio is

$$\frac{\beta}{\delta_r} = -\frac{C_{n_{\delta_r}}}{C_{n_\beta}} \tag{3.10,3}$$

The rudder hinge moment and control force are also treated in a manner similar to that employed for the elevator. Let the rudder hinge-moment coefficient be given by

$$C_{hr} = b_1\alpha_F + b_2\delta_r \tag{3.10,4}$$

The rudder pedal force will then be given by

$$P = G\frac{\rho}{2} V_F^2 S_r\bar{c}_r(b_1\alpha_F + b_2\delta_r)$$

$$= G\frac{\rho}{2} V_F^2 S_r\bar{c}_r[b_1(-\beta + \sigma) + b_2\delta_r] \tag{3.10,5}$$

where G is the rudder system gearing.

The effect of a free rudder on the directional stability is found by setting $C_{hr} = 0$ in (3.10,4). Then the rudder floating angle is

$$\delta_{r_{\text{free}}} = -\frac{b_1}{b_2}\alpha_F \tag{3.10,6}$$

The vertical-tail lift coefficient with rudder free is found from (3.9,3) to be

$$C'_{L_F} = a_F\alpha_F - a_r\frac{b_1}{b_2}\alpha_F$$

$$= a_F\alpha_F\left(1 - \frac{a_r}{a_F}\frac{b_1}{b_2}\right) \tag{3.10,7}$$

The free control factor for the rudder is thus seen to be of the same form as that for the elevator (see Sec. 2.6) and to have a similar effect.

3.11 Roll Stiffness

Consider a vehicle constrained, as on bearings in a wind tunnel, to one degree of freedom-rolling about the x axis. The forces and moments resulting from a fixed displacement ϕ are fundamentally different in character from those associated with the rotations α and β. In the first place if the x axis coincides with the velocity vector $\mathbf{V}$, *no aerodynamic change whatsoever* follows from the fixed rotation ϕ (see Fig. 3.14). The aerodynamic field remains symmetrical with respect to the plane of symmetry, the resultant aerodynamic force remains in that plane, and no changes occur in any of the aerodynamic coefficients. Thus the roll stiffness $\partial C_l/\partial\phi = C_{l_\phi}$ is zero in that case.

If the x axis does not coincide with $\mathbf{V}$, then a second-order roll stiffness results through the medium of the derivative $\partial C_l/\partial\beta = C_{l_\beta}$. Let the angle of attack of the x axis be α_x (see Fig. 1.7), then the velocity vector when $\phi = 0$ is

$$\mathbf{V}_1 = \begin{bmatrix} V\cos\alpha_x \\ 0 \\ V\sin\alpha_x \end{bmatrix} \tag{3.11,1}$$

After rolling through angle ϕ about Ox, the x component of the velocity vector remains unchanged, but the component $V\sin\alpha_x$ has projections on both of the new y

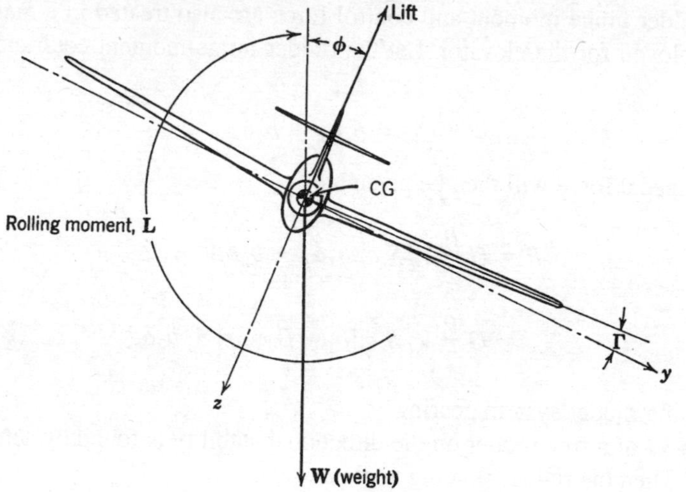

Figure 3.14 Rolled airplane.

and z axes. Thus there is now a *sideslip,* and hence, an angle β and a resulting rolling moment. Using the notation of Appendix A.4, we get for the velocity vector in the new reference frame after the rotation ϕ

$$\mathbf{V}_2 = \mathbf{L}_1(\phi)\mathbf{V}_1 = \begin{bmatrix} V\cos\alpha_x \\ V\sin\alpha_x\sin\phi \\ V\sin\alpha_x\cos\phi \end{bmatrix} \tag{3.11,2}$$

Thus the sideslip component is $v = V\sin\alpha_x\sin\phi$, and the sideslip angle is

$$\beta = \sin^{-1}\frac{v}{V} = \sin^{-1}(\sin\alpha_x\sin\phi) \tag{3.11,3}$$

As a result of this positive β, and the usually negative C_{l_β} there is a restoring rolling moment $C_{l_\beta}\beta$, that is,

$$\Delta C_l = C_{l_\beta}\sin^{-1}(\sin\alpha_x\sin\phi) \qquad (a)$$

For small α_x, we get the approximate result

$$\Delta C_l \doteq C_{l_\beta}\sin^{-1}(\alpha_x\sin\phi) \doteq C_{l_\beta}\alpha_x\sin\phi \qquad (b) \quad (3.11,4)$$

and if ϕ also is small,

$$\Delta C_l \doteq C_{l_\beta}\alpha_x\phi \qquad (c)$$

The stiffness derivative for rolling about Ox is then from (3.11,4a)

$$\frac{\partial C_l}{\partial\phi} = C_{l_\beta}\frac{\sin\alpha_x\cos\phi}{(1-\sin^2\alpha_x\sin^2\phi)^{1/2}} \qquad (a)$$

or for $\alpha_x \ll 1$,

$$\frac{\partial C_l}{\partial\phi} \doteq C_{l_\beta}\alpha_x\cos\phi \qquad (b) \quad (3.11,5)$$

or for $\alpha_x, \phi \ll 1$

$$\frac{\partial C_l}{\partial \phi} \doteq C_{l_\beta}\alpha_x \qquad (c)$$

Thus there is a roll stiffness that resists rolling if α_x is >0, and would tend to keep the wings level. If rolling occurs about the wind vector, the stiffness is zero and the vehicle has no preferred roll angle. If $\alpha_x < 0$, then the stiffness is negative and the vehicle would roll to the position $\phi = 180°$, at which point $C_l = 0$ and $C_{l_\phi} < 0$.

The above discussion applies to a vehicle constrained, as stated, to one degree of freedom. It does not, by any means, give the full answer for an unconstrained airplane to the question: "What happens when the airplane rolls away from a wings-level attitude—does it tend to come back or not?" That answer can only be provided by a full dynamic analysis like the kind given in Chaps. 6 and 7. The roll stiffness argument given above, however, does help in understanding the behavior of slender airplanes, ones with very low aspect ratio and hence small roll inertia. These tend, in response to aileron deflection when at angle of attack, to rotate about the x axis, not the velocity vector, and hence experience the roll stiffness effect at the beginning of the response.

Even though airplanes have no first-order aerodynamic roll stiffness, stable airplanes do have an inherent tendency to fly with wings level. They do so because of what is known as the *dihedral effect.* This is a complex pattern involving gravity and the derivative C_{l_β}, which owes its existence largely to the wing dihedral (see Sec. 3.12). When rolled to an angle ϕ, there is a weight component $mg \sin \phi$ in the y direction (Fig. 3.14). This induces a sideslip velocity to the right, with consequent $\beta > 0$, and a rolling moment $C_{l_\beta}\beta$ that tends to bring the wings level. The rolling and yawing motions that result from such an initial condition are, however, strongly coupled, so no significant conclusions can be drawn about the behavior except by a dynamic analysis (see Chap. 6).

3.12 The Derivative C_{l_β}

The derivative C_{l_β} is of paramount importance. We have already noted its relation to roll stiffness and to the tendency of airplanes to fly with wings level. The primary contribution to C_{l_β} is from the wing—its dihedral angle, aspect ratio, and sweep all being important parameters.

The effect of wing dihedral is illustrated in Fig. 3.15. With the coordinate system shown, the normal velocity component V_n on the right wing panel (R) is, for small dihedral angle Γ,

$$\begin{aligned} V_n &= w \cos \Gamma + v \sin \Gamma \\ &\doteq w + v\Gamma \end{aligned}$$

and that on the other panel is $w - v\Gamma$. The terms $\pm v\Gamma/V = \pm\beta\Gamma$ represent opposite changes in the angle of attack of the two panels resulting from sideslip. The "upwind" panel has its angle of attack and therefore its lift increased, and vice versa. The result is a rolling moment approximately linear in both β and Γ, and hence a fixed

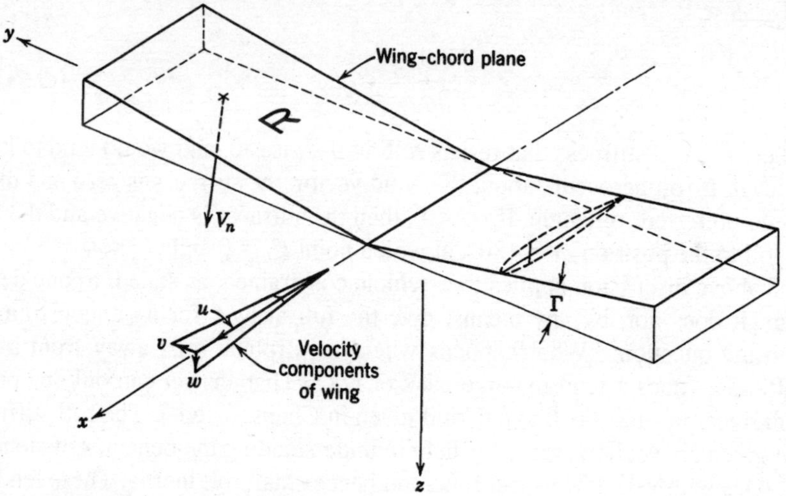

Figure 3.15 Dihedral effect. V_n = normal velocity of panel $R = w \cos \Gamma + v \sin \Gamma \doteqdot w + v\Gamma$ ∴ $\Delta\alpha$ of R due to dihedral $\doteqdot \dfrac{v\Gamma}{V} = \dfrac{V\beta\Gamma}{V} = \beta\Gamma$.

value of C_{l_β} for a given Γ. This part of C_{l_β} is essentially independent of wing angle of attack so long as the flow remains attached.

Even in the absence of dihedral, a flat lifting wing panel has a C_{l_β} proportional to C_L. Consider the case of Fig. 3.16. The vertical induced velocity (downwash) of the vortex wake is greater at L than at R simply by virtue of the geometry of the wake. Hence the local wing angle of attack and lift are less at L than at R, and a negative C_l results. Since this effect depends, essentially linearly, on the strength of the vortex wake, which is itself proportional to the wing C_L, then the result is $\Delta C_{l_\beta} \propto C_L$.

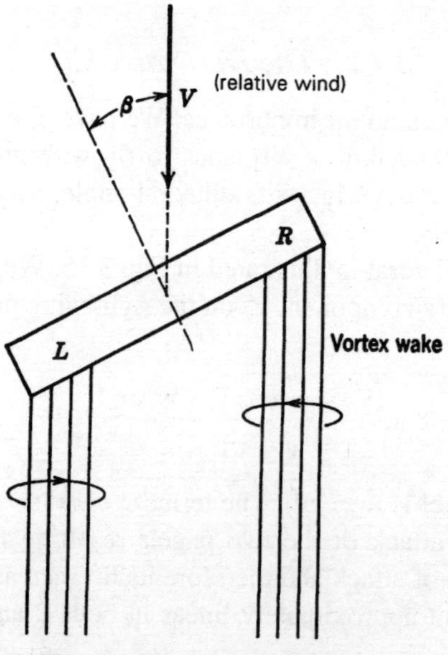

Figure 3.16 Yawed lifting wing.

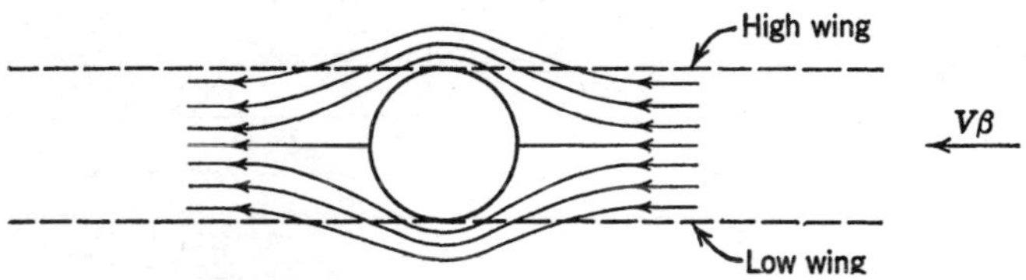

Figure 3.17 Influence of body on C_{l_β}.

INFLUENCE OF FUSELAGE ON C_{l_β}

The flow field of the body interacts with the wing in such a way as to modify its dihedral effect. To illustrate this, consider a long cylindrical body, of circular cross section, yawed with respect to the main stream. Consider only the cross-flow component of the stream, of magnitude $V\beta$, and the flow pattern which it produces about the body. This is illustrated in Fig. 3.17. It is clearly seen that the body induces vertical velocities which, when combined with the mainstream velocity, alter the local angle of attack of the wing. When the wing is at the top of the body (high-wing), then the angle of attack distribution is such as to produce a negative rolling moment; that is, the dihedral effect is enhanced. Conversely, when the airplane has a low wing, the dihedral effect is diminished by the fuselage interference. The magnitude of the effect is dependent upon the fuselage length ahead of the wing, its cross-section shape, and the planform and location of the wing. Generally, this explains why high-wing airplanes usually have less wing dihedral than low-wing airplanes.

INFLUENCE OF SWEEP ON C_{l_β}

Wing sweep is a major parameter affecting C_{l_β}. Consider the swept yawed wing illustrated in Fig. 3.18. According to simple sweep theory it is the velocity V_n normal to a wing reference line ($\frac{1}{4}$ chord line for subsonic, leading edge for supersonic) that determines the lift. It follows obviously that the lift is greater on the right half of the

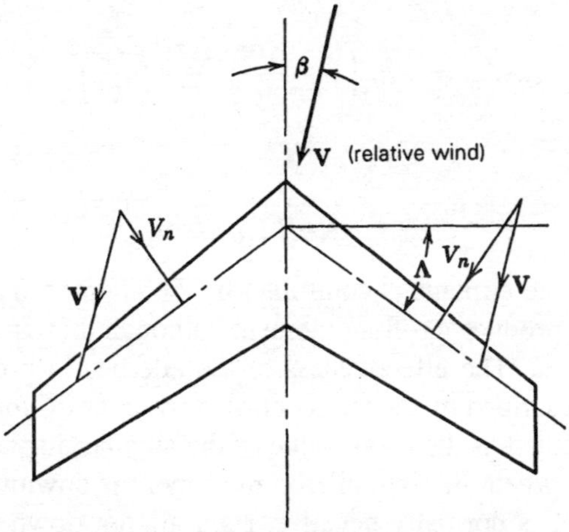

Figure 3.18 Dihedral effect of a swept wing.

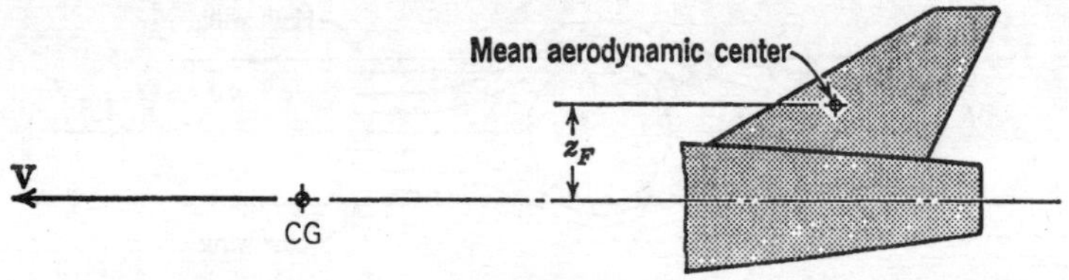

Figure 3.19 Dihedral effect of the vertical tail.

wing shown than on the left half, and hence that there is a negative rolling moment. The rolling moment would be expected for small β to be proportional to

$$C_L[(V_n^2)_{\text{right}} - (V_n^2)_{\text{left}}] = C_L V^2[\cos^2(\Lambda - \beta) - \cos^2(\Lambda + \beta)]$$
$$\doteq 2C_L\beta V^2 \sin 2\Lambda$$

The proportionality with C_L and β is correct, but the sin 2Λ factor is not a good approximation to the variation with Λ. The result is a $C_{l_\beta} \propto C_L$, that can be calculated by the methods of linear wing theory.

INFLUENCE OF FIN ON C_{l_β}

The sideslipping airplane gives rise to a side force on the vertical tail (see Sec. 3.9). When the mean aerodynamic center of the vertical surface is appreciably offset from the rolling axis (see Fig. 3.19) then this force may produce a significant rolling moment. We can calculate this moment from (3.9,3). When the rudder angle is zero, that is, $\delta_r = 0$, the moment is found to be

$$a_F(-\beta + \sigma)\frac{\rho}{2}V_F^2 S_F z_F$$

thus

$$\Delta C_l = a_F(-\beta + \sigma)\frac{S_F z_F}{Sb}\left(\frac{V_F}{V}\right)^2$$

and

$$\Delta C_{l_\beta} = -a_F\left(1 - \frac{\partial\sigma}{\partial\beta}\right)\frac{S_F z_F}{Sb}\left(\frac{V_F}{V}\right)^2 \tag{3.12,1}$$

3.13 Roll Control

The angle of bank of the airplane is controlled by the ailerons. The primary function of these controls is to produce a rolling moment, although they frequently introduce a yawing moment as well. The effectiveness of the ailerons in producing rolling and yawing moments is described by the two control derivatives $\partial C_l/\partial\delta_a$ and $\partial C_n/\partial\delta_a$. The aileron angle δ_a is defined as the mean value of the angular displacements of the two ailerons. It is positive when the right aileron movement is downward (see Fig. 3.20). The derivative $\partial C_l/\partial\delta_a$ is normally negative, right aileron down producing a roll to the left.

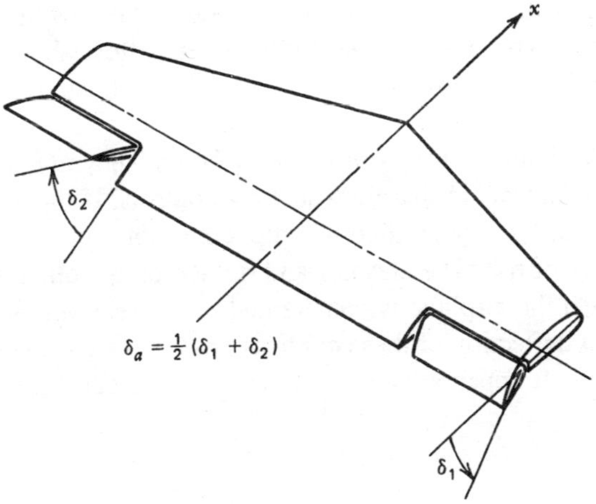

Figure 3.20 Aileron angle.

For simple flap-type ailerons, the increase in lift on the right side and the decrease on the left side produce a drag differential that gives a positive (nose-right) yawing moment. Since the normal reason for moving the right aileron down is to initiate a turn to the left, then the yawing moment is seen to be in just the wrong direction. It is therefore called *aileron adverse yaw.* On high-aspect-ratio airplanes this tendency may introduce decided difficulties in lateral control. Means for avoiding this particular difficulty include the use of spoilers and Frise ailerons. Spoilers are illustrated in Fig. 3.21. They achieve the desired result by reducing the lift and increasing the drag on the side where the spoiler is raised. Thus the rolling and yawing moments developed are mutually complementary with respect to turning. Frise ailerons diminish the adverse yaw or eliminate it entirely by increasing the drag on the side of the upgoing aileron. This is achieved by the shaping of the aileron nose and the choice of hinge location. When deflected upward, the gap between the control surface and the wing is increased, and the relatively sharp nose protrudes into the stream. Both these geometrical factors produce a drag increase.

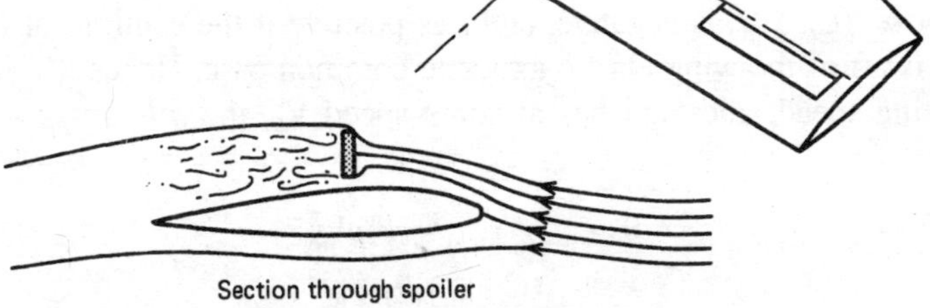

Figure 3.21 Spoilers.

The deflection of the ailerons leads to still additional yawing moments once the airplane starts to roll. These are caused by the altered flow about the wing and tail. These effects are discussed in Sec. 5.7 (C_{n_p}), and are illustrated in Figs. 5.12 and 5.15.

A final remark about aileron controls is in order. They are functionally distinct from the other two controls in that they are *rate* controls. If the airplane is restricted only to rotation about the x axis, then the application of a constant aileron angle results in a steady *rate* of roll. The elevator and rudder, on the other hand, are *displacement* controls. When the airplane is constrained to the relevant single axis degree of freedom, a constant deflection of these controls produces a constant angular *displacement* of the airplane. It appears that both rate and displacement controls are acceptable to pilots.

AILERON REVERSAL

There is an important aeroelastic effect on roll control by ailerons that is significant on most conventional airplanes at both subsonic and supersonic speeds. This results from the elastic distortion of the wing structure associated with the aerodynamic load increment produced by the control. As illustrated in Fig. 2.23, the incremental load caused by deflecting a control flap at subsonic speeds has a centroid somewhere near the middle of the wing chord. At supersonic speeds the control load acts mainly on the deflected surface itself, and hence has its centroid even farther to the rear. If this load centroid is behind the elastic axis of the wing structure, then a nose-down twist of the main wing surface results. The reduction of angle of attack corresponding to $\delta_a > 0$ causes a reduction in lift on the surface as compared with the rigid case, and a consequent reduction in the control effectiveness. The form of the variation of $C_{l_{\delta_a}}$ with $\frac{1}{2}\rho V^2$ for example can be seen by considering an idealized model of the phenomenon. Let the aerodynamic torsional moment resulting from equal deflection of the two ailerons be $T(y) \propto \frac{1}{2}\rho V^2 \delta_a$ and let $T(y)$ be antisymmetric, $T(y) = -T(-y)$. The twist distribution corresponding to $T(y)$ is $\theta(y)$, also antisymmetric, such that $\theta(y)$ is proportional to T at any reference station, and hence proportional to $\frac{1}{2}\rho V^2 \delta_a$. Finally, the antisymmetric twist causes an antisymmetric increment in the lift distribution, giving a proportional rolling moment increment $\Delta C_l = k\frac{1}{2}\rho V^2 \delta_a$. The total rolling moment due to aileron deflection is then

$$\Delta C_l = (C_{l_{\delta_a}})_{\text{rigid}} \delta_a + k\tfrac{1}{2}\rho V^2 \delta_a \tag{3.13,1}$$

and the control effectiveness is

$$C_{l_{\delta_a}} = (C_{l_{\delta_a}})_{\text{rigid}} + k\tfrac{1}{2}\rho V^2 \tag{3.13,2}$$

As noted above, $(C_{l_{\delta_a}})_{\text{rigid}}$ is negative, and k is positive if the centroid of the aileron-induced lift is aft of the wing elastic axis, the common case. Hence $|C_{l_{\delta_a}}|$ diminishes with increasing speed, and vanishes at some speed V_R, the *aileron reversal speed.* Hence

$$0 = (C_{l_{\delta_a}})_{\text{rigid}} + k\tfrac{1}{2}\rho V_R^2$$

or

$$k = -(C_{l_{\delta_a}})_{\text{rigid}} / \tfrac{1}{2}\rho V_R^2 \tag{3.13,3}$$

Substitution of (3.13,3) into (3.13,2) yields

$$C_{l_{\delta_a}} = (C_{l_{\delta_a}})_{\text{rigid}}\left(1 - \frac{V^2}{V_R^2}\right) \tag{3.13,4}$$

This result, of course, applies strictly only if the basic aerodynamics are not Mach-number dependent, i.e. so long as V_R is at a value of $\mathbf{M}$ appreciably below 1.0. Otherwise k and $(C_{l_{\delta_a}})_{\text{rigid}}$ are both functions of $\mathbf{M}$, and the equation corresponding to (3.13,4) is

$$C_{l_{\delta_a}}(\mathbf{M}) = (C_{l_{\delta_a}})_{\text{rigid}}(\mathbf{M}) - \frac{k(\mathbf{M})}{k(\mathbf{M}_R)}\,(C_{l_{\delta_a}})_{\text{rigid}}(\mathbf{M}_R)\,\frac{V^2}{V_R^2} \tag{3.13,5}$$

where $\mathbf{M}_R$ is the reversal Mach number.

It is evident from (3.13,4) that the torsional stiffness of the wing has to be great enough to keep V_R appreciably higher than the maximum flight speed if unacceptable loss of aileron control is to be avoided.

3.14 Exercises

3.1 Derive (3.1,4).

3.2 Derive an expression for the elevator angle per g in dimensional form. Denote the derivatives of L and M with respect to α and q by $\partial L/\partial q = L_q$, and so on. There are two choices: (1) do the derivation in dimensional form from the beginning, or (2) convert the nondimensional result (3.1,6) to dimensional form. Do it both ways and check that they agree.

3.3 Calculate the variation of the control force per g with altitude from the following data. Ignore propulsion effects.

Geometric Data

Weight, W	50,000 lb (222,500 N)
Wing area, S	937.5 ft^2 (87.10 m^2)
Wing mean aerodynamic chord, $\bar{c}$	12.80 ft (3.90 m)
$\bar{l}t$	31.85 ft (9.71 m)
Tail area, S_t	230 ft^2 (21.4 m^2)
S_e	71.30 ft^2 (6.62 m^2)
Mean elevator chord, $\bar{c}_e$	2.21 ft (0.674 m)
G	30°/ft (98.4°/m)

Aerodynamic Data

a	0.088/deg
a_e	0.044/deg
a_t	0.064/deg
b_o	0

b_1	−0.17/rad
b_2	−0.48/rad
$C_{h_{eq}}$	−0.846
C_{L_q}	0
C_{m_q}	−22.9
$(h - h_n)$	−0.10
$\partial\epsilon\backslash\partial\alpha$	0.30
δ_t	0

3.4 A small manually controlled airplane has an undesirable handling characteristic—the control force per g is too large. List some design changes that could reduce it, and describe the other consequences that each such change would entail.

3.5 The range of elevator motion on an airplane is from 20° down to 30° up. Use Table 1.3 as a guide, a fellow student as a model, and a tape measure to arrive at a reasonable value for the elevator gearing ratio G.

3.6 Two airplanes are similar, but one is jet-propelled and the other has a piston engine and propeller. The thrust line in each case is well below the CG with $z_p/\bar{c} = 0.4$. The power-off pitching moment at $\delta_e = 0$ is $C_m = 0.1 - 0.2\ C_L$. The throttle is set to give level flight with $C_L = 0.4$ and $L/D = 12$. Consider several steady rectilinear flight conditions having the same throttle setting but different elevator settings, C_L values and flight-path angles. Find $dC_m(\alpha)/dC_L$ (for $\delta_e = 0$) for the two airplanes when passing through the altitude corresponding to the level flight conditions. As indicated in Sec. 3.4 and (6.4,10) dC_m/dC_L is an index of static longitudinal stability under certain conditions. Assuming that these conditions are met in this problem, how will the static longitudinal stability of the two aircraft change as the aircraft slow down?

3.7 Derive an expression for the increment $\Delta \dfrac{\partial C_n}{\partial \beta}$ attributable to a jet engine. (Hint, refer to (3.4,15).)

3.8 Suppose that as a result of an accident in flight the rear fuselage of an airplane is damaged, so that the flexibility parameter k in (3.5,1) et seq. is suddenly increased. The effect is large enough that the pilot notices a loss in longitudinal stability and control. Bearing in mind that the integrity of the fuselage structure depends on the tail load L_t and the stability and control on the factor in parentheses in (3.5,4), analyze how the situation changes as the pilot slows down and descends to an emergency landing. Consider two cases; (1) C_{L_t} initially positive, (2) C_{L_t} initially negative.

3.9 Derive (3.9,8). Explain clearly each step in the development and justify any assumptions you make.

3.10 Use Appendix B to determine the elevator hinge moment parameters b_1 and b_2 for a NACA 0009 airfoil (a symmetric airfoil with a thickness-to-chord ratio of $t/c = 0.09$). The elevator has an elliptic nose, a sealed gap and a balance ratio of 0.2. In using the curves assume that transition is at the leading edge; $R = 10^7$; $\tan\left(\dfrac{\tau}{2}\right) = \tan\left(\dfrac{\phi'_{\mathrm{TE}}}{2}\right) = 0.12$; $F_3 = 1$; $c_f/c = 0.325$; $A = 4.84$; $\mathbf{M} = 0$.

3.15 Additional Symbols Introduced in Chapter 3

a_F	$\partial C_{L_F}/\partial\alpha_F$
a_r	$\partial C_{L_F}/\partial\alpha_r$
b	span of airplane
$\bar{c}_r$	mean rudder chord
C_{hr}	rudder hinge-moment coefficient
C_l	rolling-moment coefficient, $L/\frac{1}{2}\rho V^2 Sb$
C_{l_β}	$\partial C_l/\partial\beta$
C_{L_F}	vertical-tail lift coefficient
C_{l_q}	$\partial C_L/\partial\hat{q}$
C_{m_q}	damping in pitch, $\partial C_m/\partial\hat{q}$
C_n	yawing-moment coefficient, $N/\frac{1}{2}\rho V^2 Sb$
C_{n_β}	$\partial C_n/\partial\beta$
$C_{n_{\delta_r}}$	$\partial C_n/\partial\delta_r$
C_T	$T/\frac{1}{2}\rho V^2 S$
C_W	$W/\frac{1}{2}\rho V^2 S$
d	diameter of propeller or jet
g	acceleration due to gravity
h_m	maneuver point, stick fixed (see 3.1,8)
h'_m	maneuver point, stick free (see 3.2,5)
L	rolling moment
L_F	vertical-tail lift
l_F	see Fig. 3.12
m	airplane mass
M_q	$\partial M/\partial q$
n	load factor, L/W
N	yawing moment
q	angular velocity in pitch, rad/s
$\hat{q}$	$q\bar{c}/2V$
S_F	vertical-tail area
S_p	propeller disk area
S_r	area of rudder
T	thrust of one propulsion unit
$\mathbf{V}_F$	effective velocity vector at the fin
V_V	vertical-tail volume ratio, $S_F l_F/Sb$
w	wing loading, W/S
z_F	see Fig. 3.19

α_F	effective angle of attack of the fin, (see Fig. 3.12)
β	sideslip angle
Γ	dihedral angle
ϵ_p	upwash at propeller
δ_a	aileron angle (see Fig. 3.20)
δ_r	rudder angle (see Fig. 3.12)
Λ	sweepback angle
σ	sidewash angle
μ	$2m/\rho S\bar{c}$
ϕ	bank angle

CHAPTER 4

General Equations of Unsteady Motion

4.1 General Remarks

The basis for analysis, computation, or simulation of the unsteady motions of a flight vehicle is the mathematical model of the vehicle and its subsystems. An airplane in flight is a very complicated dynamic system. It consists of an aggregate of elastic bodies so connected that both rigid and elastic relative motions can occur. For example, the propeller or jet-engine rotor rotates, the control surfaces move about their hinges, and bending and twisting of the various aerodynamic surfaces occur. The external forces that act on the airplane are also complicated functions of its shape and its motion. It seems clear that realistic analyses of engineering precision are not likely to be accomplished with a very simple mathematical model. The model that is developed in the following has been found by aeronautical engineers and researchers to be very useful in practise. We begin by first treating the vehicle as a single rigid body with six degrees of freedom. This body is free to move in the atmosphere under the actions of gravity and aerodynamic forces—it is primarily the nature and complexity of aerodynamic forces that distinguish flight vehicles from other dynamic systems. Next we add the gyroscopic effects of spinning rotors and then continue with a discussion of structural distortion (aeroelastic effects).

As was noted in Chap. 1, the Earth is treated as flat and stationary in inertial space. These assumptions simplify the model enormously, and are quite acceptable for dealing with most problems of airplane flight. The effects of a round rotating Earth are treated at some length in Etkin (1972).

Extensive use is made in the developments that follow of linear algebra, with which the reader is assumed to be familiar. Appendix A.1 contains a brief review of some pertinent material.

4.2 The Rigid-Body Equations

In the interest of completeness, the rigid-body equations are derived from first principles, that is to say, we apply Newton's laws to an element *dm* of the airplane, and then integrate over all elements. The velocities and accelerations must of course be relative to an inertial, or Newtonian, frame of reference. As we noted in Sec. 1.6 the frame F_E, fixed to the Earth, is assumed to be such a frame. We also noted there that velocities relative to F_E are identified by a superscript E. In order to avoid the carrying of the cumbersome superscript throughout the following development, we shall temporarily assume that $\mathbf{W} = 0$ in (1.6,1), so that $\mathbf{V}^E = \mathbf{V}$, and make an appropriate adjustment at the end. In the frame F_B

$$\mathbf{V}_B = [u \quad v \quad w]^T \tag{4.2,1}$$

The position vector of dm relative to the origin of F_E is $\mathbf{r}_c + \mathbf{r}$ (see Fig. 4.1). In the frame F_E,

$$\mathbf{r}_{c_E} = [x_E \quad y_E \quad z_E]^T \tag{4.2,2}$$

and in the frame F_B

$$\mathbf{r}_B = [x \quad y \quad z]^T \tag{4.2,3}$$

The inertial velocity of dm is

$$\mathbf{v}_E = (\dot{\mathbf{r}}_{c_E} + \dot{\mathbf{r}}_E) = \mathbf{V}_E + \dot{\mathbf{r}}_E \tag{4.2,4}$$

The momentum of dm is $\mathbf{v}dm$, and of the whole airplane is

$$\int \mathbf{v}_E dm = \int (\mathbf{V}_E + \dot{\mathbf{r}}_E) dm = \mathbf{V}_E \int dm + \int \dot{\mathbf{r}}_E dm \tag{4.2,5}$$

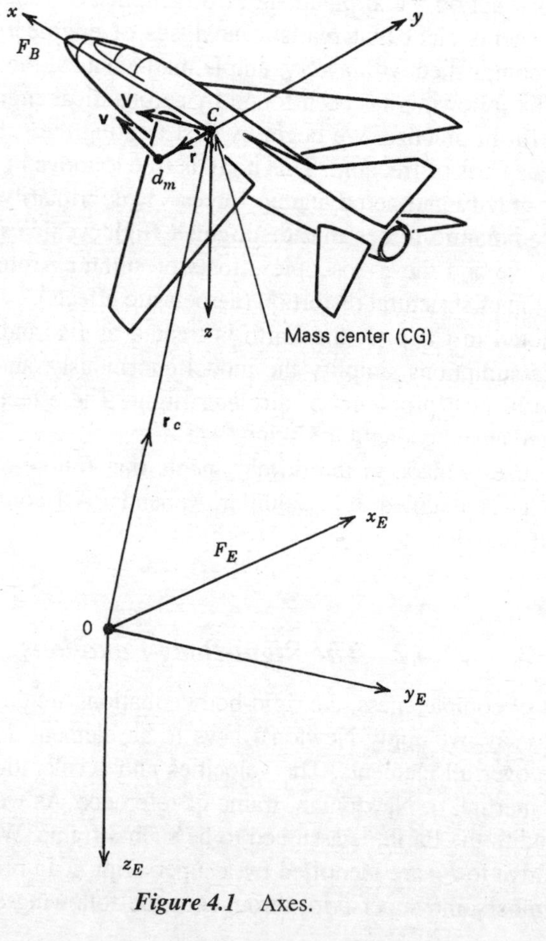

Figure 4.1 Axes.

Since C is the mass center, the last integral in (4.2,5) is zero and

$$\int \mathbf{v}_E dm = m\mathbf{V}_E \tag{4.2,6}$$

where m is the total mass of the airplane. Newton's second law applied to dm is

$$d\mathbf{f}_E = \dot{\mathbf{v}}_E dm \tag{4.2,7}$$

where $d\mathbf{f}_E$ is the resultant of all forces acting on dm. The integral of (4.2,7) is

$$\mathbf{f}_E = \int d\mathbf{f}_E = \int \dot{\mathbf{v}}_E dm$$

or, from (4.2,6)

$$\mathbf{f}_E = m\dot{\mathbf{V}}_E \tag{4.2,8}$$

The quantity $\int d\mathbf{f}_E$ is a summation of all the forces that act upon all the elements. The internal forces, that is, those exerted by one element upon another, all occur in equal and opposite pairs, by Newton's third law, and hence contribute nothing to the summation. Thus $\mathbf{f}_E$ is the resultant *external* force acting upon the airplane.

This equation relates the external force on the airplane to the motion of the CG. We need also the relation between the external moment and the rotation of the airplane. It is obtained from a consideration of the moment of momentum. The moment of momentum of dm with respect to C is by definition $d\mathbf{h} = \mathbf{r} \times \mathbf{v} dm$. It is convenient in the following to use the matrix form of the cross product (see Appendix A.1) so that

$$d\mathbf{h}_E = \tilde{\mathbf{r}}_E \mathbf{v}_E dm$$

Consider

$$\frac{d}{dt}(d\mathbf{h}_E) = \dot{\tilde{\mathbf{r}}}_E \mathbf{v}_E dm + \tilde{\mathbf{r}}_E \dot{\mathbf{v}}_E dm \tag{4.2,9}$$

Now from (4.2,4),

$$\dot{\tilde{\mathbf{r}}}_E = \tilde{\mathbf{v}}_E - \tilde{\mathbf{V}}_E$$

and the *moment* of $d\mathbf{f}$ about C is

$$d\mathbf{G} = \mathbf{r} \times d\mathbf{f}$$

so that from (4.2,7)

$$d\mathbf{G}_E = \tilde{\mathbf{r}}_E d\mathbf{f}_E = \tilde{\mathbf{r}}_E \dot{\mathbf{v}}_E dm \tag{4.2,10}$$

Equation (4.2,9) then becomes

$$d\mathbf{G}_E = \frac{d}{dt}(d\mathbf{h}_E) - (\tilde{\mathbf{v}}_E - \tilde{\mathbf{V}}_E)\mathbf{v}_E dm \tag{4.2,11}$$

Since $\mathbf{v} \times \mathbf{v} = 0$, (4.2,11) becomes

$$d\mathbf{G}_E = \frac{d}{dt}(d\mathbf{h}_E) + \tilde{\mathbf{V}}_E \mathbf{v}_E dm \tag{4.2,12}$$

Equation (4.2,12) is now integrated as was (4.2,7), and becomes

$$\int d\mathbf{G}_E = \frac{d}{dt}\int d\mathbf{h}_E + \tilde{\mathbf{V}}_E \int \mathbf{v}_E dm \tag{4.2,13}$$

By an argument similar to that for $\int d\mathbf{f}$, $\int d\mathbf{G}$ is shown to be the resultant external moment about C, denoted $\mathbf{G}$. $\int d\mathbf{h}$ is called the *moment of momentum,* or *angular momentum* of the airplane and is denoted $\mathbf{h}$. Formulas for $\mathbf{h}$ are derived in Sec. 4.3. Using (4.2,6) and noting that $\mathbf{V} \times \mathbf{V} = 0$, (4.2,13) reduces to

$$\mathbf{G}_E = \dot{\mathbf{h}}_E \tag{4.2,14}$$

where

$$\mathbf{h}_E = \int \tilde{\mathbf{r}}_E \mathbf{v}_E dm$$

The reader should note that, in (4.2,14), both $\mathbf{G}$ and $\mathbf{h}$ are referred to a moving point, the mass center. For a moving reference point other than the mass center, the equation does *not* in general apply. The reader should also note that (4.2,8) and (4.2,14) are both valid when there is relative motion of parts of the airplane.

The two vector equations of motion of the airplane, equivalent to six scalar equations, are (4.2,8) and (4.2,14)

$$\mathbf{f}_E = m\dot{\mathbf{V}}_E$$
$$\mathbf{G}_E = \dot{\mathbf{h}}_E$$

When the wind vector $\mathbf{W}$ is not zero, the velocity $\mathbf{V}_E$ in (4.2,5) is that of the CG relative to F_E. The angular momentum $\mathbf{h}$ is the same whether $\mathbf{W}$ is zero or not (see Exercise 4.1). Hence in the general case when wind is present the equations of motion are:

$$\begin{aligned} \mathbf{f}_E &= m\dot{\mathbf{V}}_E^E \\ \mathbf{G}_E &= \dot{\mathbf{h}}_E \end{aligned} \tag{4.2,15}$$

4.3 Evaluation of the Angular Momentum h

We shall want the angular momentum components in F_B. Now

$$\mathbf{h} = \int d\mathbf{h} = \int \mathbf{r} \times \mathbf{v}\, dm$$

So in F_B we have

$$\mathbf{h}_B = \int \tilde{\mathbf{r}}_B \mathbf{v}_B\, dm \tag{4.3,1}$$

where

$$\tilde{\mathbf{r}}_B = \begin{bmatrix} 0 & -z & y \\ z & 0 & -x \\ -y & x & 0 \end{bmatrix}$$

Let the angular velocity of the airplane relative to inertial space[1] be

$$\boldsymbol{\omega}_B = [p \quad q \quad r]^T$$

where p, q, r are the rates of roll, pitch and yaw respectively (see Fig. 1.6).

Now the velocity of a point in a rigid rotating body is given by[2]

$$\mathbf{v}_B = \mathbf{V}_B + \tilde{\boldsymbol{\omega}}_B \mathbf{r}_B \qquad (a)$$

where

$$\tilde{\boldsymbol{\omega}}_B = \begin{bmatrix} 0 & -r & q \\ r & 0 & -p \\ -q & p & 0 \end{bmatrix} \qquad (b) \tag{4.3,2}$$

so that

$$\begin{aligned} \mathbf{h}_B &= \int \tilde{\mathbf{r}}_B(\mathbf{V}_B + \tilde{\boldsymbol{\omega}}_B \mathbf{r}_B)dm \\ &= \int \tilde{\mathbf{r}}_B \mathbf{V}_B \, dm + \int \tilde{\mathbf{r}}_B \tilde{\boldsymbol{\omega}}_B \mathbf{r}_B \, dm \\ &= \left(\int \tilde{\mathbf{r}}_B \, dm\right) \mathbf{V}_B + \int \tilde{\mathbf{r}}_B \tilde{\boldsymbol{\omega}}_B \mathbf{r}_B \, dm \end{aligned} \tag{4.3,3}$$

The first integral in (4.3,3) vanishes since the origin of $\mathbf{r}$ is the CG. When the triple matrix product of the second integral is expanded (see Exercise 4.2) we get the result for $\mathbf{h}_B$:

$$\mathbf{h}_B = \mathbf{I}_B \boldsymbol{\omega}_B \tag{4.3,4}$$

where

$$\mathbf{I}_B = \begin{bmatrix} I_x & -I_{xy} & -I_{xz} \\ -I_{yx} & I_y & -I_{yz} \\ -I_{zx} & -I_{zy} & I_z \end{bmatrix} \tag{4.3,5}$$

and

$$\begin{aligned} I_x &= \int (y^2 + z^2)dm; \quad I_y = \int (x^2 + z^2)dm; \quad I_z = \int (x^2 + y^2)dm \\ I_{xy} &= I_{yx} = \int xy \, dm; \quad I_{xz} = I_{zx} = \int xz \, dm; \quad I_{yz} = I_{zy} = \int yz \, dm \end{aligned} \tag{4.3,6}$$

$\mathbf{I}_B$ is the *inertia matrix,* its elements being the moments and products of inertia of the airplane. When the xz plane is a plane of symmetry, which is the usual assumption, then

$$I_{xy} = I_{yz} = 0$$

and the only off-diagonal term remaining is I_{zx}. If the direction of the x axis is so chosen that this product of inertia also vanishes, which is always possible in principle, then the axes are *principal axes.*

[1]Since there is no need to use the angular velocity relative to any other frame of reference the distinguishing superscript E is not needed on $\boldsymbol{\omega}$.

[2]See Appendix A.6.

4.4 *Orientation and Position of the Airplane*

The position and orientation of the airplane are given relative to the Earth-fixed frame F_E. The CG has position vector $\mathbf{r}_c$ (see Fig. 4.1), with coordinates (x_E, y_E, z_E).

The orientation of the airplane is given by a series of three consecutive rotations, *the Euler angles,* whose order is important. The airplane is imagined first to be oriented so that its axes are parallel to those of F_E, (see Fig. 4.2). It is then in the position $Cx_1y_1z_1$. The following rotations are then applied.

1. A rotation $\mathbf{\Psi}$ about oz_1, carrying the axes to $Cx_2y_2z_2$ (bringing Cx to its final azimuth).
2. A rotation $\mathbf{\Theta}$ about oy_2, carrying the axes to $Cx_3y_3z_3$ (bringing Cx to its final elevation).
3. A rotation $\mathbf{\Phi}$ about ox_3, carrying the axes to their final position $Cxyz$ (giving the final angle of bank to the wings).

In order to avoid ambiguities which can otherwise result in the set of angles (ψ, θ, ϕ) the ranges are limited to

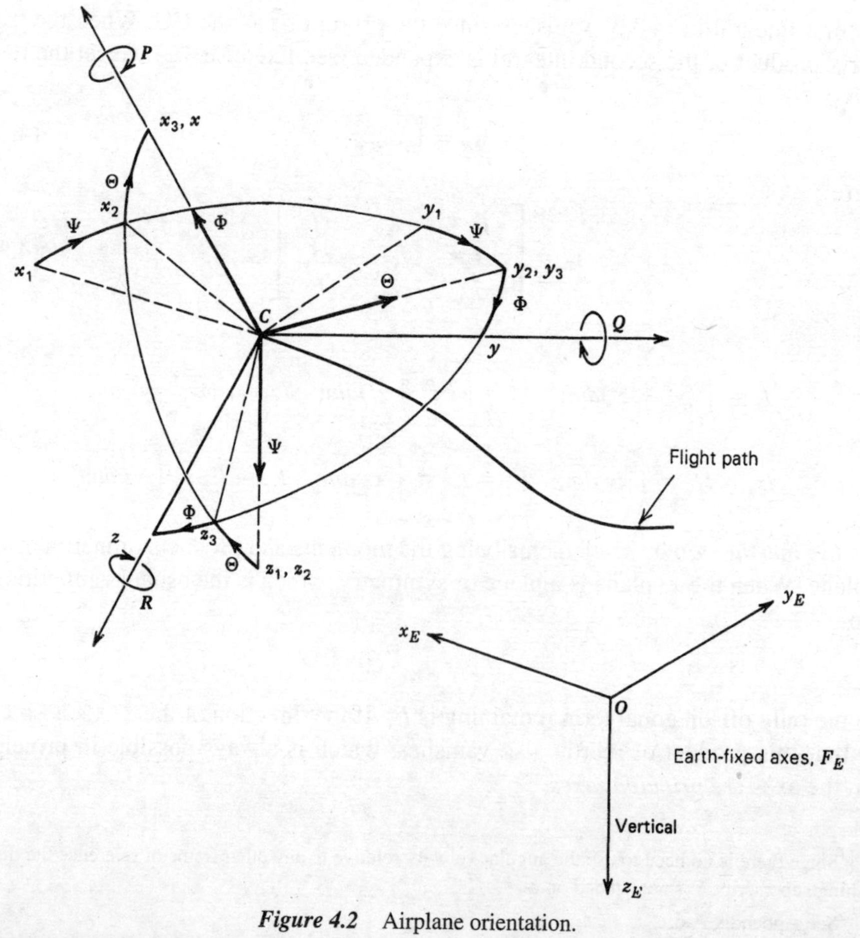

Figure 4.2 Airplane orientation.

$$-\pi \leq \psi < \pi \quad \text{or} \quad 0 \leq \psi \leq 2\pi$$

$$-\frac{\pi}{2} \leq \theta \leq \frac{\pi}{2}$$

$$-\pi \leq \phi < \pi \quad \text{or} \quad 0 \leq \phi \leq 2\pi$$

The Euler angles are then unique for most orientations of the vehicle. It should be noted that in a continuous steady rotation, such as rolling, the time variation of ϕ for example is a discontinuous sawtooth function, and that another exception occurs in a vertical climb or dive, when $\theta = \pm\pi/2$. For then $(\psi, \theta, \phi) = (\psi + a, \pm\pi/2, -a)$ gives the same final orientation regardless of the value of a. The above difficulties can be avoided by using direction cosines[3] or quaternions[4] to define the orientation of the airplane instead of Euler angles. We use the Euler angles because they give a more physical picture of the airplane attitude than the other alternatives. For a more complete discussion of methods of describing vehicle orientation the reader is referred to Hughes (1986).

THE FLIGHT PATH

To track the flight path relative to F_E, we need the velocity components in the directions of the axes of F_E. These we get by transforming the velocity vector $\mathbf{V}_B^E$ into $\mathbf{V}_E^E$ as shown in Appendix A.4.[5]

$$\mathbf{V}_E^E = \mathbf{L}_{EB}\mathbf{V}_B^E \tag{4.4,1}$$

Here $\mathbf{L}_{EB}$ is the matrix of direction cosines that corresponds to the *reverse* of the sequence of rotations given above, which are for a transformation from F_E to F_B. Thus

$$\mathbf{L}_{EB} = \mathbf{L}_z(-\Psi)\mathbf{L}_y(-\Theta)\mathbf{L}_x(-\Phi) \tag{4.4,2}$$

where L_x, L_y, L_z are respectively L_1, L_2, L_3 of Appendix A.4. Using the rotation matrices given in Appendix A.4, and carrying out the multiplication, we get the final result (4.4,3).

$$L_{EB} = \begin{bmatrix} \cos\theta\cos\psi & \sin\phi\sin\theta\cos\psi - \cos\phi\sin\psi & \cos\phi\sin\theta\cos\psi + \sin\phi\sin\psi \\ \cos\theta\sin\psi & \sin\phi\sin\theta\sin\psi + \cos\phi\cos\psi & \cos\phi\sin\theta\sin\psi - \sin\phi\cos\psi \\ -\sin\theta & \sin\phi\cos\theta & \cos\phi\cos\theta \end{bmatrix} \tag{4.4,3}$$

The differential equations for the coordinates of the flight path are then

$$\begin{bmatrix} \dot{x}_E \\ \dot{y}_E \\ \dot{z}_E \end{bmatrix} = \mathbf{L}_{EB}\mathbf{V}_B^E$$

The position of the vehicle CG is obtained by integrating the preceding equation.

[3]Sec. 5.2 of Etkin (1972).

[4]Appendix E of ANSI/AIAA (1992).

[5]Recall that the superscript signifies velocity relative to F_E, and that the subscript identifies the reference frame in which the components are given.

ORIENTATION OF THE AIRPLANE

We now need a set of differential equations from which the Euler angles can be calculated. These are obtained as follows: Let $(\mathbf{i}, \mathbf{j}, \mathbf{k})$ be unit vectors, with subscripts 1, 2, 3 denoting directions (x_1, y_1, z_1), and so on of Fig. 4.2.

Let the airplane experience, in time Δt, an infinitesimal rotation from the position defined by Ψ, Θ, Φ to that corresponding to $(\Psi + \Delta\Psi)$, $(\Theta + \Delta\Theta)$, $(\Phi + \Delta\Phi)$. The vector representing this rotation is approximately

$$\Delta\mathbf{n} \doteqdot \mathbf{i}_3\,\Delta\Phi + \mathbf{j}_2\,\Delta\Theta + \mathbf{k}_1\,\Delta\Psi$$

and the angular velocity is exactly

$$\boldsymbol{\omega} = \lim_{\Delta t \to 0} \frac{\Delta\mathbf{n}}{\Delta t} = \mathbf{i}_3\dot{\Phi} + \mathbf{j}_2\dot{\Theta} + \mathbf{k}_1\dot{\Psi} \tag{4.4,4}$$

By using (A.1,4), (A.4,3*a*) and (A.4,10) the unit vectors are found to be in frame F_B

$$\mathbf{i}_{3_B} = \begin{bmatrix} 1 \\ 0 \\ 0 \end{bmatrix} \quad \mathbf{j}_{2_B} = \begin{bmatrix} 0 \\ \cos\phi \\ -\sin\phi \end{bmatrix} \quad \mathbf{k}_{1_B} = \begin{bmatrix} -\sin\theta \\ \cos\theta\sin\phi \\ \cos\theta\cos\phi \end{bmatrix} \tag{4.4,5}$$

Recalling that $\boldsymbol{\omega}_B = [p \ \ q \ \ r]^T$, we get

$$\begin{bmatrix} p \\ q \\ r \end{bmatrix} = \mathbf{R} \begin{bmatrix} \dot{\phi} \\ \dot{\theta} \\ \dot{\psi} \end{bmatrix} \tag{a}$$

where (4.4,6)

$$\mathbf{R} = \begin{bmatrix} 1 & 0 & -\sin\theta \\ 0 & \cos\phi & \sin\phi\cos\theta \\ 0 & -\sin\phi & \cos\phi\cos\theta \end{bmatrix} \tag{b}$$

Inverting (4.4,6) we get the Euler angle rates as

$$\begin{bmatrix} \dot{\phi} \\ \dot{\theta} \\ \dot{\psi} \end{bmatrix} = \mathbf{T} \begin{bmatrix} p \\ q \\ r \end{bmatrix} \tag{a}$$

where (4.4,7)

$$\mathbf{T} = \begin{bmatrix} 1 & \sin\phi\tan\theta & \cos\phi\tan\theta \\ 0 & \cos\phi & -\sin\phi \\ 0 & \sin\phi\sec\theta & \cos\phi\sec\theta \end{bmatrix} \tag{b}$$

4.5 Euler's Equations of Motion

We come now to the reason why body axes are so important. Equation (4.2,15) shows that we need the time derivative of the angular momentum, $\mathbf{h}$, which contains the moments and products of inertia with respect to whatever axes are chosen. If those axes are fixed relative to inertial space, then the inertias will be variables in the equations. This is most undesirable and can be avoided by writing the equations in the frame F_B, in which all the inertias are constant.

We begin with the force equation of (4.2,15):

$$\mathbf{f}_E = m\dot{\mathbf{V}}_E^E \tag{4.5,1}$$

Both vectors in (4.5,1) are now expressed in F_B components; thus:

$$\mathbf{L}_{EB}\mathbf{f}_B = m\frac{d}{dt}(\mathbf{L}_{EB}\mathbf{V}_B^E) = m(\dot{\mathbf{L}}_{EB}\mathbf{V}_B^E + \mathbf{L}_{EB}\dot{\mathbf{V}}_B^E) \tag{4.5,2}$$

The derivative of the transformation matrix is obtained from (A.4) as

$$\dot{\mathbf{L}}_{EB} = \mathbf{L}_{EB}\tilde{\boldsymbol{\omega}}_B \tag{4.5,3}$$

With (4.5,3), (4.5,2) becomes

$$\mathbf{L}_{EB}\mathbf{f}_B = m(\mathbf{L}_{EB}\tilde{\boldsymbol{\omega}}_B\mathbf{V}_B^E + \mathbf{L}_{EB}\dot{\mathbf{V}}_B^E)$$

Now premultiply by $\mathbf{L}_{BE}$ to get

$$\mathbf{f}_B = m(\dot{\mathbf{V}}_B^E + \tilde{\boldsymbol{\omega}}_B\mathbf{V}_B^E) \tag{4.5,4}$$

A similar procedure applied to the moment equation of (4.2,15) leads to

$$\mathbf{G}_B = \dot{\mathbf{h}}_B + \tilde{\boldsymbol{\omega}}_B\mathbf{h}_B \tag{4.5,5}$$

The force vector **f** is the sum of the aerodynamic force **A** and the gravitational force $m\mathbf{g}$, that is,

$$\mathbf{f} = m\mathbf{g} + \mathbf{A} \tag{4.5,6}$$

where

$$\mathbf{A}_B = [X \quad Y \quad Z]^T$$

and

$$m\mathbf{g}_B = m\mathbf{L}_{BE}\mathbf{g}_E = m\mathbf{L}_{BE}[0 \quad 0 \quad g]^T \tag{4.5,7}$$

We denote $\mathbf{V}_B^E = [u^E \ v^E \ w^E]^T$ and use (4.3,2*b*) for $\tilde{\boldsymbol{\omega}}_B$. Equations (4.5,4 and 5) are then expanded using (4.3,4) to yield the desired equations. In doing so we note that $\mathbf{G}_B = [L \ M \ N]^T$ and that, because the airplane is assumed to be rigid, $\dot{\mathbf{I}}_B = 0$.

$$\begin{aligned} X - mg\sin\theta &= m(\dot{u}^E + qw^E - rv^E) && (a)\\ Y + mg\cos\theta\sin\phi &= m(\dot{v}^E + ru^E - pw^E) && (b) \quad (4.5,8)\\ Z + mg\cos\theta\cos\phi &= m(\dot{w}^E + pv^E - qu^E) && (c)\end{aligned}$$

$$\begin{aligned} L &= I_x\dot{p} - I_{yz}(q^2 - r^2) - I_{zx}(\dot{r} + pq) - I_{xy}(\dot{q} - rp) - (I_y - I_z)qr && (a)\\ M &= I_y\dot{q} - I_{zx}(r^2 - p^2) - I_{xy}(\dot{p} + qr) - I_{yz}(\dot{r} - pq) - (I_z - I_x)rp && (b) \quad (4.5,9)\\ N &= I_z\dot{r} - I_{xy}(p^2 - q^2) - I_{yz}(\dot{q} + rp) - I_{zx}(\dot{p} - qr) - (I_x - I_y)pq && (c)\end{aligned}$$

CHOICE OF BODY AXES

The equations derived in the preceding sections are valid for any orthogonal axes fixed in the airplane, with origin at the CG, and known as *body axes.* Since most aircraft are very nearly symmetrical, it is usual to assume exact symmetry, and to let Cxz be the plane of symmetry. Then Cx points "forward," Cz "downward," and Cy to

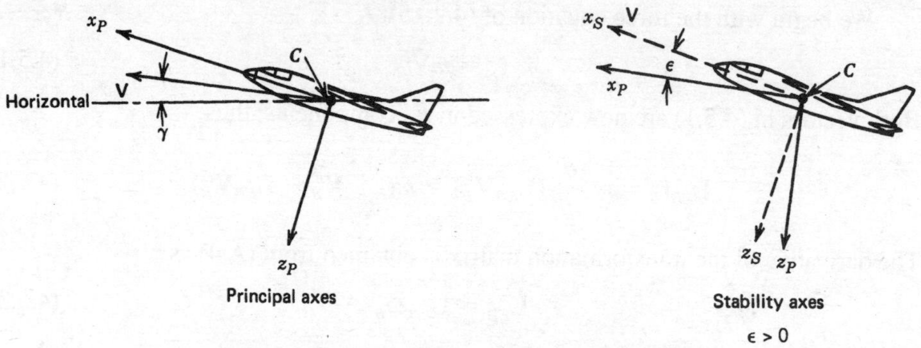

Figure 4.3 Illustrating two choices of body axes.

the right. In this case, the two products of inertia, I_{xy} and I_{yz}, are zero, and (4.5,9) are consequently simplified.

The directions of Cx and Cz in the plane of symmetry are conventionally fixed in one of three ways (see Fig. 4.3).

Principal Axes

These are chosen to coincide with the principal axes of the vehicle, so that the remaining product of inertia I_{zx} vanishes; (4.3,4 and 5) then yield

$$\begin{aligned} h_x &= I_x p \\ h_y &= I_y q \\ h_z &= I_z r \end{aligned} \tag{4.5,10}$$

Stability Axes

These are chosen so that Cx is aligned with **V** in a reference condition of steady symmetric flight. In this case, the reference values of v and w are zero, and the axes are termed *stability axes.* These axes are commonly used, owing to the simplifications that result in the equations of motion, and in the expressions for the aerodynamic forces.

With this choice, it should be noted that for different initial flight conditions the axes are differently oriented in the airplane, and hence the values of I_x, I_z, and I_{zx} will vary from problem to problem. The "stability axes," just as the principal axes, are body axes that remain fixed to the airplane during the motion considered in any one problem.

The following formulas are convenient for computing I_x, I_z, I_{zx} when the values I_{x_p} and I_{z_p} are known for principal axes (see Exercise 4.4).

$$\begin{aligned} I_x &= I_{x_p} \cos^2 \epsilon + I_{z_p} \sin^2 \epsilon \\ I_z &= I_{x_p} \sin^2 \epsilon + I_{z_p} \cos^2 \epsilon \\ I_{zx} &= \frac{1}{2} (I_{z_p} - I_{x_p}) \sin 2\epsilon \end{aligned} \tag{4.5,11}$$

where ϵ = angle between x_p (principal axis) and x_s (stability axis), positive as shown (see Fig. 4.3).

Body Axes

When the axes are neither principal axes nor stability axes, they are usually called simply *body axes.* In this case the x axis is usually fixed to a longitudinal reference line in the airplane. These axes may be the most convenient ones to use if the aerodynamic data have been measured by a wind-tunnel balance that resolves the forces and moments into body-fixed axes rather than tunnel-fixed axes.

4.6 Effect of Spinning Rotors on the Euler Equations

In evaluating the angular momentum **h** (see Sec. 4.3) it was tacitly assumed that the airplane is a single rigid body. This is implied in the equation for the velocity of an element (4.3,2). Let us now imagine that some portions of the airplane mass are spinning relative to the body axes; for example, rotors of jet engines, or propellers. Each such rotor has an angular momentum relative to the body axes. This can be computed from (4.3,4) by interpreting the moments and products of inertia therein as those of the rotor with respect to axes parallel to $Cxyz$ and origin at the rotor mass center. The angular velocities in (4.3,4) are interpreted as those of the rotor relative to the airplane body axes. Let the resultant relative angular momentum of all rotors be $\mathbf{h}'$, with components (h'_x, h'_y, h'_z) in F_B, which are assumed to be constant. It can be shown that the total angular momentum of an airplane with spinning rotors is obtained simply by adding $\mathbf{h}'$ to the **h** previously obtained in Sec. 4.3 (see Exercise 4.5). The equation that corresponds to (4.3,4) is then[6]

$$\mathbf{h}_B = \mathbf{I}_B\boldsymbol{\omega}_B + \mathbf{h}'_B \tag{4.6,1}$$

Because of the additional terms in the angular momentum, certain extra terms appear in the right-hand side of the moment equations, (4.5,9). Those additional terms, known as the gyroscopic couples, are

$$\begin{aligned} &\text{In the } L \text{ equation:} \quad qh'_z - rh'_y \\ &\text{In the } M \text{ equation:} \quad rh'_x - ph'_z \\ &\text{In the } N \text{ equation:} \quad ph'_y - qh'_x \end{aligned} \tag{4.6,2}$$

As an example, suppose the rotor axes are parallel to Cx, with angular momentum $\mathbf{h}' = \mathbf{i}I\Omega$. Then the gyroscopic terms in the three equations are, respectively, 0, $I\Omega r$, and $-I\Omega q$.

4.7 The Equations Collected

The kinematical and dynamical equations derived in the foregoing are collected below for convenience. The assumption that Cxz is a plane of symmetry is used, so that $I_{xy} = I_{yz} = 0$, and (4.6,2) are added to (4.5,9) to give (4.7,2).

[6]Note that the inertias of the rotors are also included in $\mathbf{I}_B$.

$$\begin{aligned} X - mg\sin\theta &= m(\dot{u}^E + qw^E - rv^E) && (a)\\ Y + mg\cos\theta\sin\phi &= m(\dot{v}^E + ru^E - pw^E) && (b) \quad (4.7,1)\\ Z + mg\cos\theta\cos\phi &= m(\dot{w}^E + pv^E - qu^E) && (c) \end{aligned}$$

$$\begin{aligned} L &= I_x\dot{p} - I_{zx}\dot{r} + qr(I_z - I_y) - I_{zx}pq + qh'_z - rh'_y && (a)\\ M &= I_y\dot{q} + rp(I_x - I_z) + I_{zx}(p^2 - r^2) + rh'_x - ph'_z && (b) \quad (4.7,2)\\ N &= I_z\dot{r} - I_{zx}\dot{p} + pq(I_y - I_x) + I_{zx}qr + ph'_y - qh'_x && (c) \end{aligned}$$

$$\begin{aligned} p &= \dot{\phi} - \dot{\psi}\sin\theta && (a)\\ q &= \dot{\theta}\cos\phi + \dot{\psi}\cos\theta\sin\phi && (b)\\ r &= \dot{\psi}\cos\theta\cos\phi - \dot{\theta}\sin\phi && (c)\\ \dot{\phi} &= p + (q\sin\phi + r\cos\phi)\tan\theta && (d) \quad (4.7,3)\\ \dot{\theta} &= q\cos\phi - r\sin\phi && (e)\\ \dot{\psi} &= (q\sin\phi + r\cos\phi)\sec\theta && (f) \end{aligned}$$

$$\begin{aligned} \dot{x}_E &= u^E\cos\theta\cos\psi + v^E(\sin\phi\sin\theta\cos\psi - \cos\phi\sin\psi)\\ &\quad + w^E(\cos\phi\sin\theta\cos\psi + \sin\phi\sin\psi) && (a)\\ \dot{y}_E &= u^E\cos\theta\sin\psi + v^E(\sin\phi\sin\theta\sin\psi + \cos\phi\cos\psi) && (4.7,4)\\ &\quad + w^E(\cos\phi\sin\theta\sin\psi - \sin\phi\cos\psi) && (b)\\ \dot{z}_E &= -u^E\sin\theta + v^E\sin\phi\cos\theta + w^E\cos\phi\cos\theta && (c) \end{aligned}$$

$$\begin{aligned} u^E &= u + W_x && (a)\\ v^E &= v + W_y && (b) \quad (4.7,5)\\ w^E &= w + W_z && (c) \end{aligned}$$

4.8 Discussion of the Equations

The above equations are quite general and contain few assumptions. These are:

1. The airplane is a rigid body, which may have attached to it any number of rigid spinning rotors.
2. Cxz is a plane of mirror symmetry.
3. The axes of any spinning rotors are fixed in direction relative to the body axes, and the rotors have constant angular speed relative to the body axes.

The equations of Sec. 4.7 are many and complex. They consist of 15 coupled nonlinear ordinary differential equations in the independent variable t, and three algebraic equations. Before we can identify the true dependent variables, however, we must first consider the aerodynamic forces (X,Y,Z) and moments (L,M,N). It is clear that these must depend in some manner on the relative motion of the airplane with respect to the air, given by the linear and angular velocities **V** and $\boldsymbol{\omega}$, on the control variables that fix the angles of any moveable surfaces and on the settings of any propulsion controls that determine the thrust vector. Thus it is universally assumed in flight dynamics that the six forces and moments are functions of the six linear and angular velocities (u,v,w,p,q,r) and of a *control vector.* The latter clearly depends to some extent on the particular airplane, but we can, with adequate generality, write it as

$$\mathbf{c} = [\delta_a \quad \delta_e \quad \delta_r \quad \delta_p]^T$$

of which the first three are the familiar aileron, elevator, and rudder angles, and the last is the throttle control. Other components can be added to the control vector as needed to meet special requirements—for example, direct lift control. The control variables, from the standpoint of the mathematical system, are arbitrary functions of time. How they are determined is the subject of later sections. The wind vector whose components appear in (4.7,5) would ordinarily be a known function of position $\mathbf{r}_c$ with its components given in frame F_E. Its components in F_B are $\mathbf{W}_B = \mathbf{L}_{BE}\mathbf{W}_E$.

The true implicit dependent variables of the system are thus seen to be 12 in number:

CG position: x_E, y_E, z_E
Attitude: ψ, θ, ϕ
Velocity: u^E, v^E, w^E
Angular velocity: p, q, r

Of the 15 differential equations we note that 3 of (4.7,3) are not independent, so the number of independent equations is actually 12, the same as the number of dependent variables. The mathematical system is therefore complete.

A useful view of the equations is given in the block diagram of Fig. 4.4, which is specialized to the case of zero wind. Each block represents one set of equations, with *inputs* and *outputs* (the dependent variables). All the inputs needed for the left-hand side are generated as outputs on the right, except for the control inputs, which remain to be specified. The nature of the mathematical problem that ensues is very much

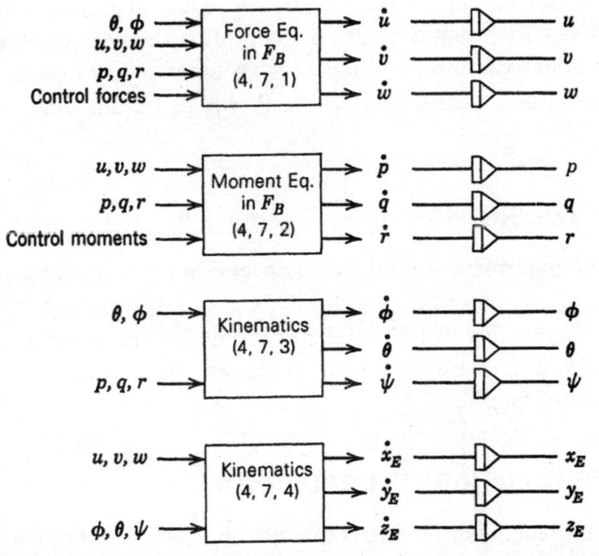

Figure 4.4 Block diagram of equations for vehicle with plane of symmetry. Body axes. Flat-Earth approximation. No wind.

governed by the specifics of the control inputs. In the following paragraphs, we discuss the various cases that commonly occur in engineering practise and in research.

STABILITY PROBLEMS—CONTROLS FIXED

In these problems the airplane is considered to be disturbed from an initially steady flight condition, with the controls locked in position. Thus, **c** is zero or a known constant. The equations are nonlinear and, consequently, extremely difficult to deal with analytically. In many problems of practical importance, it is satisfactory to linearize the equations by dealing only with small perturbations from the reference condition. In that case we obtain a set of homogeneous linear differential equations with constant coefficients, a type that is readily solved. Problems of this class are treated in Chap. 6.

STABILITY PROBLEMS—CONTROLS FREE

The free-control case is of interest primarily only for manually controlled airplanes. In that case, one or more of the primary control systems is presumed to be freed as in "hands off" by the pilot. The variation of the control angles with time, which is of course needed for the aerodynamic force and moment inputs, is then the result of an interaction between the dynamics and aerodynamics of the vehicle and those of the control system itself, which is usually simplified as a system with one degree of freedom relative to F_B. For a derivation of these equations see Etkin (1972, Sec. 11.3). Each such free control adds one dependent variable and one equation to the mathematical system.

STABILITY PROBLEMS—AUTOMATIC CONTROLS

When the airplane is under the control of an AFCS the controls are neither fixed nor free. The control vector **c**, which fixes the control inputs of Fig. 4.4, are then determined by the *feedback loop* that activates the control systems in response to the values of the 12 dependent variables and inputs from other sources such as navigation, guidance, or fire control systems, and so on. Problems of this class are studied in Chap. 8.

RESPONSE TO CONTROLS

The effectiveness of an airplane's controls is conventionally studied by specifying the variation of $(\delta_e, \delta_r, \delta_a, \delta_p)$ with time arbitrarily, e.g., a step-function input of aileron angle. The airplane equations of motion then become inhomogeneous equations for (u, v, w), (p, q, r), (θ, ϕ, ψ) with the control angles as forcing functions. Problems of this type are treated in Chap. 7.

RESPONSE TO ATMOSPHERIC TURBULENCE

The motion of an airplane, and the forces that act on it, as a consequence of the turbulent motion of the atmosphere, are very important for both design and operation. The associated mathematical problems are treated with the same general equations as given above. (u^E, v^E, w^E) then have to be sums of (u, v, w) and the velocity of the atmosphere at the CG, and additional complications arise from the fact that the relative wind varies, in general, from point to point on the airplane. This case is treated in depth in Chap. 13 of Etkin (1972) and in Etkin (1981).

INVERSE PROBLEMS

A class of problems that has not received much attention in the past, but that is nevertheless both interesting and useful, is that in which some of the 12 variables usually regarded as dependent are prescribed in advance as functions of time. An equal number of equations must then be dropped in order to maintain a complete system. This is the kind of problem that occurs when we ask questions of the type "Given the airplane motion, what pilot action is required to produce it?" Such questions may be relevant to problems of control design and maneuvering loads.

The mathematical problem that results is generally simpler than those of stability and control. The equations to be solved are sometimes algebraic, sometimes differential. A decided advantage is the ability of this approach to cope with the nonlinear equations of large disturbances.

Another category is the mathematical problem that arises in flight testing when time records are available of some control variables and some of the 12 dependent variables. The question then is "What must the airplane parameters be to produce the measured response from the measured input?" (See Etkin, 1959, Chap. 11; AGARD 1991; Maine and Iliff, 1986.) This is an example of the important "plant identification" problem of system theory.

4.9 The Small-Disturbance Theory

As remarked in Sec. 4.8, the equations of motion are frequently linearized for use in stability and control analysis. It is assumed that the motion of the airplane consists of small deviations from a reference condition of steady flight. The use of the small-disturbance theory has been found in practice to give good results. It predicts with satisfactory precision the stability of unaccelerated flight, and it can be used, with sufficient accuracy for engineering purposes, for response calculations where the disturbances are not infinitesimal. There are, of course, limitations to the theory. It is not suitable for solutions of problems in which large disturbance angles occur, for example $\Phi \doteq \pi/2$.

The reasons for the success of the method are twofold: (1) In many cases, the major aerodynamic effects are nearly linear functions of the disturbances, and (2) disturbed flight of considerable violence can occur with quite small values of the linear- and angular-velocity disturbances.

NOTATION FOR SMALL DISTURBANCES

The reference values of all the variables are denoted by a subscript zero, and the small perturbations by prefix Δ. When the reference value is zero, the Δ may be omitted. All the disturbance quantities and their derivatives are assumed to be small, so that their squares and products are negligible compared to first-order quantities.

The reference flight condition is assumed for convenience to be symmetric and with no angular velocity. Thus $v_0 = p_0 = q_0 = r_0 = \phi_0 = \psi_0 = 0$. Furthermore, stability axes are selected as standard in this book, and thus $w_0 = 0$ for all the problems considered. u_0 is then equal to the reference flight speed, and θ_0 to the reference angle of climb (not assumed to be small). In dealing with the trigonometric functions in the equations the following relations are used:

$$\begin{aligned}\sin(\theta_0 + \Delta\theta) &= \sin\theta_0 \cos\Delta\theta + \cos\theta_0 \sin\Delta\theta \\ &\doteqdot \sin\theta_0 + \Delta\theta\cos\theta_0 \\ \cos(\theta_0 + \Delta\theta) &= \cos\theta_0\cos\Delta\theta - \sin\theta_0\sin\Delta\theta \\ &\doteqdot \cos\theta_0 - \Delta\theta\sin\theta_0\end{aligned} \tag{4.9,1}$$

FURTHER ASSUMPTIONS

The small-disturbance equations will be slightly restricted by the adoption of two more assumptions, which correspond to current practice. These are

1. The effects of spinning rotors are negligible. This is the case when the airplane is in gliding flight with power off, when the symmetrical engines have opposite rotation, or when the rotor angular momentum is small.
2. The wind velocity is zero, so that $\mathbf{V}^E = \mathbf{V}$

LINEARIZATION

When the small-disturbance notation is introduced into the equations of Sec. 4.7, the additional assumptions noted above are incorporated, and only the first-order terms in disturbance quantities are kept, then the following linear equations are obtained.

$$\begin{aligned}X_0 + \Delta X - mg(\sin\theta_0 + \Delta\theta\cos\theta_0) &= m\Delta\dot{u} && (a)\\ Y_0 + \Delta Y + mg\phi\cos\theta_0 &= m(\dot{v} + u_0 r) && (b)\\ Z_0 + \Delta Z + mg(\cos\theta_0 - \Delta\theta\sin\theta_0) &= m(\dot{w} - u_0 q) && (c)\end{aligned} \tag{4.9,2}$$

$$\begin{aligned}L_0 + \Delta L &= I_x\dot{p} - I_{zx}\dot{r} && (a)\\ M_0 + \Delta M &= I_y\dot{q} && (b)\\ N_0 + \Delta N &= -I_{zx}\dot{p} + I_z\dot{r} && (c)\end{aligned} \tag{4.9,3}$$

$$\begin{aligned}&\Delta\dot{\theta} = q && (a)\\ &\dot{\phi} = p + r\tan\theta_0, \qquad p = \dot{\phi} - \dot{\psi}\sin\theta_0 && (b)\\ &\dot{\psi} = r\sec\theta_0 && (c)\end{aligned} \tag{4.9,4}$$

$$\begin{aligned}\dot{x}_E &= (u_0 + \Delta u)\cos\theta_0 - u_0\Delta\theta\sin\theta_0 + w\sin\theta_0 && (a)\\ \dot{y}_E &= u_0\psi\cos\theta_0 + v && (b)\\ \dot{z}_E &= -(u_0 + \Delta u)\sin\theta_0 - u_0\Delta\theta\cos\theta_0 + w\cos\theta_0 && (c)\end{aligned} \tag{4.9,5}$$

REFERENCE STEADY STATE

If all the disturbance quantities are set equal to zero in the foregoing equations, then they apply to the reference flight condition. When this is done, we get the following relations, which may be used to eliminate all the reference forces and moments from the equations:

$$
\begin{aligned}
X_0 - mg \sin \theta_0 &= 0 \\
Y_0 &= 0 \\
Z_0 + mg \cos \theta_0 &= 0 \\
L_0 = M_0 = N_0 &= 0
\end{aligned} \tag{4.9,6}
$$

$$\dot{x}_{E_0} = u_0 \cos \theta_0, \qquad \dot{y}_{E0} = 0, \qquad \dot{z}_{E0} = -u_0 \sin \theta_0$$

We further postulate that in the reference steady state the aileron and rudder angles are zero. When (4.9,6) are substituted into (4.9,2 to 4.9,5), (4.9,3) are solved for $\dot{p}$ and $\dot{r}$, and the equations are rearranged, the result is (4.9,7 to 4.9,10).

$$
\begin{aligned}
\Delta\dot{u} &= \frac{\Delta X}{m} - g\Delta\theta \cos \theta_0 && (a) \\
\dot{v} &= \frac{\Delta Y}{m} + g\phi \cos \theta_0 - u_0 r && (b) \\
\dot{w} &= \frac{\Delta Z}{m} - g\Delta\theta \sin \theta_0 + u_0 q && (c)
\end{aligned} \tag{4.9,7}
$$

$$
\begin{aligned}
\dot{p} &= (I_x I_z - I_{xz}^2)^{-1} (I_z \Delta L + I_{xz} \Delta N) && (a) \\
\dot{q} &= \frac{\Delta M}{I_y} && (b) \\
\dot{r} &= (I_x I_z - I_{zx}^2)^{-1} (I_{zx} \Delta L + I_x \Delta N) && (c)
\end{aligned} \tag{4.9,8}
$$

$$
\begin{aligned}
\Delta\dot{\theta} &= q && (a) \\
\dot{\phi} &= p + r \tan \theta_0; \qquad p = \dot{\phi} - \dot{\psi} \sin \theta_0 && (b) \\
\dot{\psi} &= r \sec \theta_0 && (c)
\end{aligned} \tag{4.9,9}
$$

$$
\begin{aligned}
\Delta\dot{x}_E &= \Delta u \cos \theta_0 - u_0 \Delta\theta \sin \theta_0 + w \sin \theta_0 && (a) \\
\Delta\dot{y}_E &= u_0 \psi \cos \theta_0 + v && (b) \\
\Delta\dot{z}_E &= -\Delta u \sin \theta_0 - u_0 \Delta\theta \cos \theta_0 + w \cos \theta_0 && (c)
\end{aligned} \tag{4.9,10}
$$

THE LINEAR AIR REACTIONS

At the heart of the subject of atmospheric flight mechanics lies the problem of determining and describing the aerodynamic forces and moments that act on a given body in arbitrary motion. It is primarily this aerodynamic ingredient that distinguishes it from other branches of mechanics. Aerodynamic forces and moments are strictly speaking *functionals* of the state variables. Consider for example the time-dependent lift $L(t)$ on a wing with variable angle of attack $\alpha(t)$. Because the wing leaves behind it a vortex wake that in general generates an induced velocity field at the wing, and because hysteresis is present in flow separation processes, the aerodynamic field that fixes the lift at any given moment is actually dependent not only on the instantaneous value of α but strictly speaking on its entire past history. This functional relation is expressed by

$$L(t) = L[\alpha(\tau)] \qquad -\infty \le \tau \le t \tag{4.9,11}$$

When $\alpha(\tau)$ can be expressed as a convergent Taylor series around t, i.e.

$$\alpha(\tau) = \alpha(t) + (\tau - t)\dot{\alpha}(t) + \tfrac{1}{2}(\tau - t)^2\ddot{\alpha}(t) + \cdots \tag{4.9,12}$$

then the infinite series $\alpha(t)$, $\dot{\alpha}(t)$, $\ddot{\alpha}(t)$ $\cdots$ can replace $\alpha(\tau)$ in (4.9,11). For time $t = t_0$ this becomes

$$L(t_0) = L(\alpha, \dot{\alpha}, \ddot{\alpha} \cdots) \tag{4.9,13}$$

where α, $\dot{\alpha}$ $\cdots$ are values at time t_0. Thus the lift at time t_0 is in this case determined by α and all its derivatives at time t_0. A further series expansion of the right-hand side of (4.9,13) around $\alpha(t_0)$, $\dot{\alpha}(t_0)$, etc. yields

$$\Delta L(t_0) = L_\alpha \Delta\alpha + \tfrac{1}{2}L_{\alpha\alpha}(\Delta\alpha)^2 + \cdots + L_{\dot{\alpha}}\Delta\dot{\alpha} + \tfrac{1}{2}L_{\dot{\alpha}\dot{\alpha}}(\Delta\dot{\alpha})^2 + \cdots \tag{4.9,14}$$

in which all the products and powers of $\Delta\alpha$, $\Delta\dot{\alpha}$ $\cdots$ appear, and where

$$L_\alpha = \left.\frac{\partial L}{\partial \alpha}\right|_{\alpha_0(t_0)} \quad \text{etc.} \tag{4.9,15}$$

The classical assumption of linear aerodynamic theory, due to Bryan (1911) is to accept the linear reduction of (4.9,14) as a representation of the aerodynamic force, even when $\Delta\alpha(t)$ is not an analytic function as implied by (4.9,12), i.e.

$$\Delta L(t_0) = L_\alpha \Delta\alpha + L_{\dot{\alpha}}\Delta\dot{\alpha} + L_{\ddot{\alpha}}\Delta\ddot{\alpha} + \cdots \tag{4.9,16}$$

Derivatives such as L_α in (4.9,16) are known as the *stability derivatives,* or more generally as *aerodynamic derivatives.* For most forces and state variables, only the first term of (4.9,16) is kept, but in some cases, terms up to the second derivative must be retained for sufficient accuracy. This assumption has been found to work extremely well over a wide range of practical applications. Occasionally the addition of nonlinear terms such as $\frac{1}{2}L_{\alpha\alpha}(\Delta\alpha)^2 \equiv L_{\alpha^2}(\Delta\alpha)^2$ can extend the useful range considerably. Another way of including nonlinear effects is to treat the derivatives as functions of the variables, for example, $L_\alpha = L_\alpha(\alpha)$.

A major fraction of the total effort in aerodynamic research in the past has been devoted to the determination, by theoretical and experimental means, of the aerodynamic derivatives needed for application to flight mechanics. A great mass of information about these parameters has now been accumulated and Chap. 5 is devoted to this topic.

For a truly symmetric configuration, it is evident that the side force Y, the rolling moment L, and the yawing moment N will all be exactly zero in any condition of symmetric flight, that is, when the plane of symmetry remains in a fixed vertical plane. In that case, β, p, r, ϕ, ψ, and y_E are all identically zero. Thus the derivatives of the *asymmetric* or *lateral* forces and moments, Y, L, N with respect to the *symmetric* or *longitudinal* motion variables u, w, q are zero. In writing out the complete linear expression for the aerodynamic forces and moments, we use this fact, and in addition make the further approximations:

1. We may neglect as well all the derivatives of the symmetric forces and moments with respect to the asymmetric motion variables.
2. We may neglect all derivatives with respect to rates of change of motion variables except for $Z_{\dot{w}}$ and $M_{\dot{w}}$.

3. The derivative X_q is also negligibly small.
4. The density of the atmosphere is assumed not to vary with altitude (see Sec. 6.5).

It should be emphasized that none of these assumptions is basically necessary for the solution of airplane dynamics problems. They are made as a matter of experience and convenience. When it appears necessary to do so, any of the terms dropped can be restored into the equations. With these assumptions, however, the linear forces and moments are:

$$\begin{aligned}
\Delta X &= X_u \Delta u + X_w w + \Delta X_c & (a)\\
\Delta Y &= Y_v v + Y_p p + Y_r r + \Delta Y_c & (b)\\
\Delta Z &= Z_u \Delta u + Z_w w + Z_{\dot{w}} \dot{w} + Z_q q + \Delta Z_c & (c)\\
\Delta L &= L_v v + L_p p + L_r r + \Delta L_c & (d)\\
\Delta M &= M_u \Delta u + M_w w + M_{\dot{w}} \dot{w} + M_q q + \Delta M_c & (e)\\
\Delta N &= N_v v + N_p p + N_r r + \Delta N_c & (f)
\end{aligned} \qquad (4.9,17)$$

In the above equations, the terms on the right with subscript c are control forces and moments that result from the control vector $\mathbf{c}$. Explicit forms for the controls will be introduced as they are needed in the following.

Aerodynamic Transfer Functions

The preceding equations are subject to the theoretical objection (not of great practical importance) that the Bryan formulation for the aerodynamics is not quite sound even within the restriction of linearity. This is readily illustrated by considering the lift on a wing following a step change in angle of attack. Let $\Delta\alpha$ be given by $\Delta\alpha(t) = \alpha_0 \mathbf{1}(t)$ where α_0 is a constant. For $t > 0$, the Bryan formula (4.9,16) gives

$$\Delta L(t) = L_\alpha \alpha_0 = \text{const}$$

whereas in fact the lift undergoes a transient approach to the asymptote $L_\alpha \alpha_0$, the details of which depend on the wing shape and the Mach number. Equation (4.9,16) fails in this case because $\Delta\alpha$ is not an analytic function, having a discontinuity at $t = 0$. Now the transient process is often well approximated as a linear one, and as such is subject to exact representation by linear mathematics in the form of an *indicial function* (Tobak, 1954), or an *aerodynamic transfer function* (Etkin, 1956). The implementation of these alternative representations of aerodynamic force is described in Etkin (1972, Sec. 5.11).

THE LINEAR EQUATIONS OF MOTION

When (4.9,17) are substituted into (4.9,7 to 4.9,10) two of the equations [(4.9,7c) and (4.9,8b)], contain $\dot{w}$ terms on the right-hand side. In order to retain the desired form with first derivatives of the dependent variables on the left, we have to solve these two equations simultaneously for $\dot{w}$ and $\dot{q}$. The result is presented in matrix form in (4.9,18 and 4.9,19). Here the equations are divided into two groups, termed *longitudinal* and *lateral,* for reasons that are explained as follows.

Longitudinal Equations, Eq. (4.9,18):

$$
\begin{bmatrix} \Delta\dot{u} \\ \dot{w} \\ \dot{q} \\ \Delta\dot{\theta} \end{bmatrix} =
\begin{bmatrix}
\dfrac{X_u}{m} & \dfrac{X_w}{m} & 0 & -g\cos\theta_o \\
\dfrac{Z_u}{m - Z_{\dot{w}}} & \dfrac{Z_w}{m - Z_{\dot{w}}} & \dfrac{Z_q + mu_o}{m - Z_{\dot{w}}} & \dfrac{-mg\sin\theta_o}{m - Z_{\dot{w}}} \\
\dfrac{1}{I_y}\left[M_u + \dfrac{M_{\dot{w}} Z_u}{(m - Z_{\dot{w}})}\right] & \dfrac{1}{I_y}\left[M_w + \dfrac{M_{\dot{w}} Z_w}{(m - Z_{\dot{w}})}\right] & \dfrac{1}{I_y}\left[M_q + \dfrac{M_{\dot{w}}(Z_q + mu_o)}{(m - Z_{\dot{w}})}\right] & -\dfrac{M_{\dot{w}}\, mg\sin\theta_o}{I_y(m - Z_{\dot{w}})} \\
0 & 0 & 1 & 0
\end{bmatrix}
\begin{bmatrix} \Delta u \\ w \\ q \\ \Delta\theta \end{bmatrix} +
\begin{bmatrix}
\dfrac{\Delta X_c}{m} \\
\dfrac{\Delta Z_c}{m - Z_{\dot{w}}} \\
\dfrac{\Delta M_c}{I_y} + \dfrac{M_{\dot{w}}}{I_y}\dfrac{\Delta Z_c}{(m - Z_{\dot{w}})} \\
0
\end{bmatrix}
$$

$$\Delta\dot{x}_E = \Delta u\cos\theta_o + w\sin\theta_o - u_o\,\Delta\theta\sin\theta_o$$

$$\Delta\dot{z}_E = -\Delta u\sin\theta_o + w\cos\theta_o - u_o\,\Delta\theta\cos\theta_o$$

Lateral Equations, Eq. (4.9,19):

$$
\begin{bmatrix} \dot{v} \\ \dot{p} \\ \dot{r} \\ \dot{\phi} \end{bmatrix} =
\begin{bmatrix}
\dfrac{Y_v}{m} & \dfrac{Y_p}{m} & \left(\dfrac{Y_r}{m} - u_o\right) & g\cos\theta_0 \\
\left(\dfrac{L_v}{I'_x} + I'_{zx}N_v\right) & \left(\dfrac{L_p}{I'_x} + I'_{zx}N_p\right) & \left(\dfrac{L_r}{I'_x} + I'_{zx}N_r\right) & 0 \\
\left(I'_{zx}L_v + \dfrac{N_v}{I'_z}\right) & \left(I'_{zx}L_p + \dfrac{N_p}{I'_z}\right) & \left(I'_{zx}L_r + \dfrac{N_r}{I'_z}\right) & 0 \\
0 & 1 & \tan\theta_0 & 0
\end{bmatrix}
\begin{bmatrix} v \\ p \\ r \\ \phi \end{bmatrix} +
\begin{bmatrix}
\dfrac{\Delta Y_c}{m} \\
\dfrac{\Delta L_c}{I'_x} + I'_{zx}N_c \\
I'_{zx}\Delta L_c + \dfrac{\Delta N_c}{I'_z} \\
0
\end{bmatrix}
$$

$$\dot{\psi} = r \sec\theta_o$$

$$\Delta\dot{y}_E = u_o\psi\cos\theta_o + v$$

$$I'_x = (I_xI_z - I_{zx}^2)/I_z$$

$$I'_z = (I_xI_z - I_{zx}^2)/I_x$$

$$I'_{zx} = I_{zx}/(I_xI_z - I_{xz}^2)$$

As a consequence of the simplifying assumptions made in their derivation, the preceding equations have the exceedingly useful property of splitting into two independent groups. Suppose that ϕ, v, p, r, ΔY_c, ΔL_c and ΔN_c are identically zero. Then (4.9,19) are all identically satisfied. The remaining equations (4.9,18) form a complete set for the six homogeneous variables Δu, w, q, $\Delta\theta$, Δx_E, Δz_E. Thus we may conclude that modes of motion are possible in which *only* these variables differ from zero. Such motions are called *longitudinal* or *symmetric,* and the corresponding equations and variables are likewise named. Conversely, if the longitudinal variables are set equal to zero, the remaining six equations (4.9,19) form a complete set for the determination of the variables ϕ, ψ, v, p, r, y_E. These are known as the *lateral* variables, the corresponding equations and motions being likewise named.

It is worthwhile recording here the specific assumptions upon which this separation depends. A study of the various steps that have led to the final equations reveals these facts—the existence of the pure longitudinal motions depends on only two assumptions:

1. The existence of a plane of symmetry.
2. The absence of rotor gyroscopic effects.

The existence of the pure lateral motions, however, depends on more restrictive approximations; namely

1. The linearization of the equations.
2. The absence of rotor gyroscopic effects.
3. The neglect of all aerodynamic cross-coupling (approximation 1 p. 110).

If the equations were not linearized, then there would be inertial cross-coupling between the longitudinal and lateral modes, as evidenced by terms such as mpv^E in (4.7,1*c*) and $rp(I_x - I_z)$ in (4.7,2*b*). That is, motion in the lateral modes would induce longitudinal motion but not the reverse.

Equations (4.9,18 and 4.9,19) are both in the desired first-order form, commonly referred to as *state vector* form, conventionally written in vector/matrix notation as

$$\dot{\mathbf{x}} = \mathbf{A}\mathbf{x} + \mathbf{B}\mathbf{c} \tag{4.9,20}$$

Here $\mathbf{x}$ is the state vector, $\mathbf{c}$ is the control vector, and $\mathbf{A}$ and $\mathbf{B}$ are system matrices. The state vectors for the longitudinal and lateral systems are, respectively:

$$\begin{aligned} \mathbf{x} &= [\Delta u \quad w \quad q \quad \Delta\theta]^T \\ \mathbf{x} &= [v \quad p \quad r \quad \phi]^T \end{aligned} \tag{4.9,21}$$

and the matrices $\mathbf{A}$ for the two cases can be inferred from the full equations. The dependent variables x_E, y_E, z_E and ψ are not included in the state vectors because they do not appear on the right-hand side of the equations. The matrix $\mathbf{B}$ will be discussed later when we come to the analysis of controlled motions.

4.10 Nondimensional System

The reader will already be familiar with the great advantage of using nondimensional coefficients for aerodynamic forces and moments such as lift, drag, and pitching moment. In this way the major effects of speed, size, and air density are automatically accounted for. Similarly we need nondimensional coefficients for the many derivatives—X_u and so on—that occur in (4.9,18 and 4.9,19). Unfortunately, there is no universally accepted standard for these coefficients, although attempts have been made to devise one (e.g., ANSI/AIAA, 1992). The student, and indeed the practising engineer, should be sure to note carefully the exact notation and definitions employed in any reference material or data sources being used. The notation and definitions used in this book are essentially the NASA system, which is widely used.

Before presenting this system, we digress briefly to a dimensional analysis of the general flight dynamics problem. This helps to provide insight into what the true underlying variables are, and provides a basis for what follows. Imagine a class of geometrically similar airplanes of various sizes and masses in steady unaccelerated flight at various heights and speeds. Suppose that one of these airplanes is subjected to a disturbance. After the disturbance, some typical nondimensional variable π varies with time. For example, π may be the angle of yaw, the load factor, or the helix angle in roll. Thus, for this one airplane, under one particular set of conditions we shall have

$$\pi = f(t) \tag{4.10,1}$$

Let it be assumed that this equation can be generalized to cover the whole class of airplanes, under all flight conditions. That is, we shall assume that π is a function not of t alone, but also of

$$u_0, \rho, m, l, g, \mathbf{M}, RN$$

where m is the airplane mass and l is a characteristic length. Instead of (4.10,1), then, we write

$$\pi = f(u_0, \rho, m, l, g, \mathbf{M}, RN, t) \tag{4.10,2}$$

Buckingham's π theorem (Langhaar, 1951) tells us that, since there are nine quantities in (4.10,2) containing three fundamental dimensions, L, M, and T, then there are $9 - 3 = 6$ independent dimensionless combinations of the nine quantities. These six so-called π functions are to be regarded as the meaningful physical variables of the equation, instead of the original nine. Two systems of the same class are *dynamically similar* when all the π functions of one are numerically equal to those of the other. By inspection, we can easily form the following six independent nondimensional combinations:

$$\pi, \mathbf{M}, RN, \frac{m}{\rho l^3}, \frac{u_0 t}{l}, \frac{u_0^2}{lg}$$

Following the π theorem, we write as the symbolic solution to our problem

$$\pi = f\left(\mathbf{M}, RN, \frac{m}{\rho l^3}, \frac{u_0 t}{l}, \frac{u_0^2}{lg}\right) \tag{4.10,3}$$

The effects of the six variables m, ρ, l, g, u_0, and t are thus seen to be compressed into the three combinations: $m/\rho l^3$, u_0^2/lg, and u_0t/l. We replace l^3 by Sl, where S is a characteristic area, without changing its dimensions, and denote the resulting nondimensional quantity $m/\rho Sl$ by μ. The quantity l/u_0 has the dimensions of time and is denoted t^*. The quantity u_0^2/lg is the Froude number (FN). Equation 4.10,3 then becomes

$$\pi = f(\mathbf{M}, RN, FN, \mu, t/t^*) \tag{4.10,4}$$

The significance of (4.10,4) is that it shows π to be a function of only five variables, instead of the original eight. The result is of sufficient importance that it is customary to elaborate on it still further. Since μ is the ratio of the airplane mass to the mass of a volume Sl of air, it is called the *relative mass parameter* or *relative density parameter.* It is smallest at sea level and increases with altitude.

The main symbols for which nondimensional forms are wanted are listed in Table 4.1. The nondimensional item in column 3 is obtained by dividing the corresponding dimensional item of column 1 by the divisor in column 2. In the small-disturbance case, $\hat{v}$ and $\hat{w}$ are aerodynamic angles, for then

$$V = [(u_0 + \Delta u)^2 + v^2 + w^2)]^{1/2}$$

From (1.6,4)

$$\alpha_x = \tan^{-1}(w/u) = \tan^{-1}[w/(u_0 + \Delta u)]$$

To first order in v, w, and Δu these are

$$V \doteq u_0 + \Delta u$$

$$\alpha_x \doteq \frac{w}{u_0} = \hat{w}$$

Table 4.1
The Nondimensional System

(1)	*(2)*		*(3)*
Dimensional Quantity	*Divisor—General Case*	*Divisor—Small-Disturbance Case*	*Nondimensional Quantity*
X, Y, Z	$\frac{1}{2}\rho V^2 S$	$\frac{1}{2}\rho V^2 S$	C_x, C_y, C_z
mg	$\frac{1}{2}\rho V^2 S$	$\frac{1}{2}\rho V^2 S$	C_W
M	$\frac{1}{2}\rho V^2 S\bar{c}$	$\frac{1}{2}\rho V^2 S\bar{c}$	C_m
L, N	$\frac{1}{2}\rho V^2 Sb$	$\frac{1}{2}\rho V^2 Sb$	C_l, C_n
u, v, w	V	u_0	$\hat{u}, \hat{v}, \hat{w}$ (1)
$\dot{\alpha}, q$	$2V/\bar{c}$	$2u_0/\bar{c}$	$\hat{\dot{\alpha}}, \hat{q}$
$\dot{\beta}, p, r$	$2V/b$	$2u_0/b$	$\hat{\dot{\beta}}, \hat{p}, \hat{r}$
m	$\rho S\bar{c}/2$	$\rho S\bar{c}/2$	μ
I_y	$\rho S(\bar{c}/2)^3$	$\rho S(\bar{c}/2)^3$	$\hat{I}_y$
I_x, I_z, I_{zx}	$\rho S(b/2)^3$	$\rho S(b/2)^3$	$\hat{I}_x, \hat{I}_z, \hat{I}_{zx}$
t	—	$t^* = \bar{c}/(2u_0)$	$\hat{t}$

Note: (1) β and α_x are used interchangeably with $\hat{v}$ and $\hat{w}$, respectively, in the small disturbance case.

Table 4.2
Longitudinal Nondimensional Derivatives

	C_x	C_z	C_m
$\hat{u}$	C_{x_u}	C_{z_u}	C_{m_u}
α	C_{x_α}	C_{z_α}	C_{m_α}
$\hat{q}$	C_{x_q}	C_{z_q}	C_{m_q}
$\hat{\dot{\alpha}}$	$C_{x_{\dot{\alpha}}}$	$C_{z_{\dot{\alpha}}}$	$C_{m_{\dot{\alpha}}}$

and similarly

$$\beta = \sin^{-1}\frac{v}{V} \doteq \frac{v}{u_0} = \hat{v}$$

NONDIMENSIONAL STABILITY DERIVATIVES

The nondimensional stability derivatives are the partial derivatives of the force and moment coefficients in lines 1, 3, and 4 of Table 4.1 with respect to the nondimensional motion variables in lines 5, 6, and 7. The notation for these is displayed in Tables 4.2 and 4.3. Each entry in the tables represents the derivative of the column heading with respect to the row variable.

Since α_x differs from α only by a constant (the angle between the zero-lift line and the x axis), then $\Delta\alpha_x = \Delta\alpha$, $\partial/\partial\alpha_x \equiv \partial/\partial\alpha$, and no distinction need be made between these two derivatives.

NONDIMENSIONAL EQUATIONS

It is possible with the definitions given in Tables 4.1–4.3 to make the equations of motion entirely nondimensional, and such equations have been widely used in the past, especially for analytical work (see Etkin, 1972 and 1982). The prevailing current practise in design and research, however, is to use the dimensional equations and program them for calculation on a digital computer. We are therefore not including the nondimensional equations in this book. There is no real loss in so doing however, since any analytical results that are obtained with the dimensional equations can subsequently be expressed, for maximum generality, in nondimensional form. Examples of this are contained in Chaps. 6 and 7.

Table 4.3
Lateral Nondimensional Derivatives

	C_y	C_l	C_n
β	C_{y_β}	C_{l_β}	C_{n_β}
$\hat{p}$	C_{y_p}	C_{l_p}	C_{n_p}
$\hat{r}$	C_{y_r}	C_{l_r}	C_{n_r}
$\hat{\dot{\beta}}$	$C_{y_{\dot{\beta}}}$	$C_{l_{\dot{\beta}}}$	$C_{n_{\dot{\beta}}}$

Table 4.4
Longitudinal Dimensional Derivatives

	X	Z	M
u	$\rho u_0 SC_{w_0} \sin\theta_0 + \frac{1}{2}\rho u_0 SC_{x_u}$	$-\rho u_0 SC_{w_0}\cos\theta_0 + \frac{1}{2}\rho u_0 SC_{z_u}$	$\frac{1}{2}\rho u_0 \bar{c} SC_{m_u}$
w	$\frac{1}{2}\rho u_0 SC_{x_\alpha}$	$\frac{1}{2}\rho u_0 SC_{z_\alpha}$	$\frac{1}{2}\rho u_0 \bar{c} SC_{m_\alpha}$
q	$\frac{1}{4}\rho u_0 \bar{c} SC_{x_q}$	$\frac{1}{4}\rho u_0 \bar{c} SC_{z_q}$	$\frac{1}{4}\rho u_0 \bar{c}^2 SC_{m_q}$
$\dot{w}$	$\frac{1}{4}\rho \bar{c} SC_{x_{\dot\alpha}}$	$\frac{1}{4}\rho \bar{c} SC_{z_{\dot\alpha}}$	$\frac{1}{4}\rho \bar{c}^2 SC_{m_{\dot\alpha}}$

4.11 Dimensional Stability Derivatives

We now need expressions for the derivatives that appear in (4.9,18) and (4.9,19) in terms of the nondimensional derivatives. A few examples of these are derived as follows to illustrate the procedure, and the whole set needed for (4.9,17) is displayed in Tables 4.4 and 4.5. Derivatives with respect to $\dot{v}$ or $\dot{\beta}$ are usually negligible and are not included.

THE Z DERIVATIVES

From Table 4.1, $Z = C_z\frac{1}{2}\rho V^2 S$ where $V^2 = u^2 + v^2 + w^2$ and $u = u_0 + \Delta u$. Hence

$$Z_u = \left(\frac{\partial Z}{\partial u}\right)_0 = C_{z0}\rho u_0 S\left(\frac{\partial V}{\partial u}\right)_0 + \tfrac{1}{2}\rho u_0{}^2 S\left(\frac{\partial C_z}{\partial u}\right)_0$$

where the subscript zero indicates the reference flight condition. But $2V(\partial V/\partial u) = 2u$, and hence $(\partial V/\partial u)_0 = 1$. Also

$$\left(\frac{\partial C_z}{\partial u}\right)_0 = \frac{1}{u_0}\left(\frac{\partial C_z}{\partial \hat{u}}\right)_0 = \frac{1}{u_0}C_{z_u}$$

From (4.9,6)

$$Z_0 = -mg\cos\theta_0$$

hence

$$C_{z0} = -C_{w0}\cos\theta_0$$

so that

$$Z_u = -\rho u_0 SC_{w0}\cos\theta_0 + \frac{1}{2}\rho u_0 SC_{z_u}$$

Table 4.5
Lateral Dimensional Derivatives

	Y	L	N
v	$\frac{1}{2}\rho u_0 SC_{y_\beta}$	$\frac{1}{2}\rho u_0 bSC_{l_\beta}$	$\frac{1}{2}\rho u_0 bSC_{n_\beta}$
p	$\frac{1}{4}\rho u_0 bSC_{y_p}$	$\frac{1}{4}\rho u_0 b^2 SC_{l_p}$	$\frac{1}{4}\rho u_0 b^2 SC_{n_p}$
r	$\frac{1}{4}\rho u_0 bSC_{y_r}$	$\frac{1}{4}\rho u_0 b^2 SC_{l_r}$	$\frac{1}{4}\rho u_0 b^2 SC_{n_r}$

Also, since $(\partial V/\partial w)_0 = 0$, then

$$Z_w = \left(\frac{\partial Z}{\partial w}\right)_0 = \frac{1}{2}\rho u_0^2 S\left(\frac{\partial C_z}{\partial w}\right)_0$$

But $w = u_0\alpha_x$, so that

$$Z_w = \frac{1}{2}\rho u_0 S\left(\frac{\partial C_z}{\partial \alpha_x}\right)_0 = \frac{1}{2}\rho u_0 S C_{z_\alpha}$$

In a similar way,

$$Z_{\dot{w}} = \left(\frac{\partial Z}{\partial \dot{w}}\right)_0 \frac{1}{2}\rho u_0^2 S\left(\frac{\partial C_z}{\partial \dot{w}}\right)_0 = \frac{1}{2}\rho u_0 S\left(\frac{\partial C_z}{\partial \dot{\alpha}_x}\right)_0$$

But

$$\left(\frac{\partial C_z}{\partial \dot{\alpha}_x}\right)_0 = \frac{\bar{c}}{2u_0} C_{z\dot{\alpha}}$$

Hence

$$Z_{\dot{w}} = \frac{1}{4}\rho\bar{c}SC_{z_{\dot{\alpha}}}$$

Also,

$$Z_q = \left(\frac{\partial Z}{\partial q}\right)_0 = \frac{1}{2}\rho u_0^2 S\left(\frac{\partial C_z}{\partial q}\right)_0$$

But

$$q = \frac{2u_0}{\bar{c}}\hat{q}$$

Hence

$$Z_q = \frac{1}{4}\rho u_0 \bar{c} S C_{z_q}$$

THE X DERIVATIVES

These are found in a manner similar to the Z derivatives. In this instance from (4.9,6)

$$C_{x0} = C_{w_0}\sin\theta_0$$

THE M DERIVATIVES

These are also found in a manner similar to the Z derivatives. In this case we start with $M = C_m\frac{1}{2}\rho V^2 S\bar{c}$ and note from (4.9,6) that $C_{m_0} = 0$.

THE L DERIVATIVES

From Table 4.1, $L = C_l \rho V^2 S \dfrac{b}{2}$. Hence

$$L_v = \left(\frac{\partial L}{\partial v}\right)_0 = \rho u_0^2 S \frac{b}{2}\left(\frac{\partial C_l}{\partial v}\right)_0 = \rho u_0 S \frac{b}{2}\left(\frac{\partial C_l}{\partial \beta}\right)_0$$
$$= \rho u_0 S \frac{b}{2} C_{l_\beta}$$

Also

$$L_p = \left(\frac{\partial L}{\partial p}\right)_0 = \rho u_0^2 S \frac{b}{2}\left(\frac{\partial C_l}{\partial p}\right)_0 = \rho u_0 S \frac{b^2}{4}\left(\frac{\partial C_l}{\partial \hat{p}}\right)_0$$
$$= \frac{1}{4}\rho u_0 b^2 S C_{l_p}$$

Similarly

$$L_r = \left(\frac{\partial L}{\partial r}\right)_0 = \rho u_0^2 S \frac{b}{2}\left(\frac{\partial C_l}{\partial r}\right)_0$$
$$= \frac{1}{4}\rho u_0 b^2 S C_{l_r}$$

THE N DERIVATIVES

These are found in a manner similar to the L derivatives.

THE Y DERIVATIVES

These are also found in a manner similar to the L derivatives. In this case we start with $Y = C_y \frac{1}{2}\rho V^2 S$.

4.12 Elastic Degrees of Freedom

In the preceding sections we have presented the "main" equations of motion, that is, those associated with the six rigid-body degrees of freedom. Now it is known that the stability and control characteristics of flight vehicles may be profoundly influenced by the elastic distortions of the structure under aerodynamic load (AGARD, 1970; Milne, 1964; McLaughlin, 1956; Rodden, 1956). Additionally, there are phenomena not primarily related to stability and control, but rather to structural integrity, in which elastic deformation is a primary element—i.e., structural divergence and flutter. In order to understand and analyze all these effects, one needs the equations that govern the elastic deformations, and as well the changes that such deformation introduces into the six main dynamical equations.

A full treatment of this branch of flight mechanics—aeroelasticity and structural vibration—is beyond the scope of this text, and the reader is referred to (Bisplinghoff, 1962 and Dowell, 1994) for comprehensive treatises on it. Here we content ourselves with presenting the framework of the analysis, but omit most of the structural and aerodynamic details. Enough material is given, however, to show how the static and dynamic deformations are integrated into the preceding mathematical model of the "gross" vehicle motion.

The deformation analysis is almost invariably treated by a linear theory, even when the rigid-body motion is not. We shall therefore assume at the outset that the distortional motions are "small" and that all the associated aerodynamic forces are linear functions.

THE METHOD OF QUASISTATIC DEFLECTIONS

Many of the important effects of distortion can be accounted for simply by altering the aerodynamic derivatives. The assumption is made that the changes in aerodynamic loading take place so slowly that the structure is at all times in static equilibrium. (This is equivalent to assuming that the natural frequencies of vibration of the structure are much higher than the frequencies of the rigid-body motions.) Thus a change in load produces a proportional change in the shape of the vehicle, which in turn influences the load. Examples of this kind of analysis are given in Sec. 3.5 (effect of fuselage bending on the location of the neutral point), and Secs. 5.3 and 5.10.

THE METHOD OF NORMAL MODES

When the separation in frequency between the elastic degrees of freedom and the rigid-body motions is not large, then significant inertial coupling can occur between the two. In that case a dynamic analysis is required, which takes account of the time dependence of the elastic motions.

The method that is described here for accomplishing this purpose is based upon the representation of the deformation of the elastic vehicle in terms of its *normal modes of free vibration.* Imagine that the vehicle is at rest under the action of *no* external forces, aerodynamic, gravity, or other, and that a frame of reference with origin at the mass center, but otherwise arbitrary, is attached to it. The position of mass element δm is then (x_0, y_0, z_0). Now let the structure be deformed by a self-equilibrating set of external forces and couples, so that it takes a new form, stationary with respect to the coordinate system. Upon instantaneous release of this force system, a *free* vibration ensues, that is, one in which external forces play no part, and in which the position of δm at time t is (x, y, z). Since there is zero net force, and zero net moment, the linear and angular momenta of the elastic motion must vanish, whatever the initial distorted shape. In particular this is true for each and every undamped normal mode of free vibration. Any small arbitrary elastic motion of the vehicle can, therefore, relative to the chosen axes (transients as well as steady oscillations), be represented by a superposition of free undamped normal modes as follows:

$$
\begin{aligned}
x'(t) &= \sum_{1}^{\infty} f_n(x_0, y_0, z_0)\epsilon_n(t) \\
y'(t) &= \sum_{1}^{\infty} g_n(x_0, y_0, z_0)\epsilon_n(t) \qquad (4.12,1) \\
z'(t) &= \sum_{1}^{\infty} h_n(x_0, y_0, z_0)\epsilon_n(t)
\end{aligned}
$$

where (x', y', z') are the elastic displacements, $(x - x_0)$ etc., (f_n, g_n, h_n) are the mode shape functions,[7] and $\epsilon_n(t)$ are the generalized coordinates giving the magnitudes of the modal displacements.

We have specified idealized undamped modes, as opposed to the true modes of a real physical structure with internal and external damping, because the latter may not be "simple" modes with fixed nodes, describable by a single set of three functions. More generally they each consist of a superposition of two "submodes" 90° out of phase. Because of this, the equations of motion for the elastic degrees of freedom of the real structure are not perfectly uncoupled from one another, but contain intercoupling damping terms that would usually be negligible in practical applications.

The use of the free undamped normal modes is seen to ensure that the linear and angular momenta of the distortional motion vanishes. Consequently the elastic motions have no inertial coupling with the rigid-body motions except through the moments and products of inertia. However, it can be shown that this coupling is second-order and negligible in the small-perturbation theory. The determination of the shapes and frequencies ω_n of the normal modes is a major task, and the methods for finding them are beyond the scope of this text. For treatises on this subject the reader should refer to (Bisplinghoff et al., 1955; Fung, 1955). As indicated in (4.12,1), there are actually an infinite number of normal modes of vehicle structures. In practice, of course, only a finite number N of those at the low-frequency end of the set need be retained, and the summations in (4.12,1) are approximated by finite series of N terms. Some judgment and experience is needed to decide just how many modes are needed in any application, but a general rule that is helpful is to discard those whose frequencies are substantially higher than the significant ones present in the spectral representation of inputs arising from control action or atmospheric turbulence.

MODIFICATION OF THE RIGID BODY EQUATIONS

Although the inertia terms of the previous equations, for example, (4.9,18) and (4.9,19), remain unchanged to first order by the presence of elastic motions, the elastic and rigid-body motions are not nevertheless entirely uncoupled.

The deformations of the structure in general cause perturbations in the aerodynamic forces and moments. These may be introduced into the linearized equations of motion by the addition of appropriate derivatives to the expressions for the aerodynamic forces given by (4.9,17). For example, the added terms in the pitching moment associated with the nth elastic degree of freedom would be

$$M_{\epsilon_n}\epsilon_n + M_{\dot{\epsilon}_n}\dot{\epsilon}_n + M_{\ddot{\epsilon}_n}\ddot{\epsilon}_n \tag{4.12,2}$$

Similar expressions appear for each of the added degrees of freedom, and in each of the aerodynamic force and moment equations. An example of the elastic stability derivatives is given in Sec. 5.10. Alternatively, the aerodynamic forces may be formulated in the form of transfer functions.

THE ADDITIONAL EQUATIONS OF MOTION

The additional equations are most conveniently found by using Newton's laws as expressed by Lagrange's equations of motion (Synge and Griffith, 1942) with the ϵ_n as

[7]The eigenfunctions of the linear vibration problem.

generalized coordinates. The appropriate form of Lagrange's equation for this application is

$$\frac{d}{dt}\frac{\partial T}{\partial \dot{\epsilon}_n} - \frac{\partial T}{\partial \epsilon_n} + \frac{\partial U}{\partial \epsilon_n} = \mathcal{F}_n \tag{4.12,3}$$

where T is the kinetic energy of the elastic motion *relative to* F_B, U is the elastic strain energy, and $\mathcal{F}_n$ is the generalized external force. Since the coordinates are measured in the frame F_B, which is non-Newtonian by virtue of its general motion, an appropriate modification must be made to the external force field acting on the system when calculating the generalized force. This consists of adding to each element of mass δm an inertial body force equal to $-\ \mathbf{a}'\ \delta m$ where $\mathbf{a}'$ is that part of the total acceleration of δm that arises from the acceleration and rotation of F_B (see Appendix A.6).

Since normal modes have been chosen as the degrees of freedom, then the individual equations of motion are independent of one another insofar as elastic and inertia forces are concerned (this is a property of the normal modes), although the equations will be coupled through the aerodynamic contributions to the $\mathcal{F}$'s. The lack of elastic and inertia coupling permits the left-hand side of (4.12,3) to be evaluated by considering only a single elastic degree of freedom to be excited. Let its generalized coordinate be ϵ_n. The kinetic energy is given by

$$T = \tfrac{1}{2}\int (\dot{x}'^2 + \dot{y}'^2 + \dot{z}'^2)\, dm$$

where the integration is over all elements of mass of the body. From (4.12,1) this becomes (with only ϵ_n excited)

$$T = \tfrac{1}{2}\,\dot{\epsilon}_n^2 \int (f_n^2 + g_n^2 + h_n^2)\, dm$$

The integral is the *generalized inertia* in the nth mode, and is denoted by

$$I_n = \int (f_n^2 + g_n^2 + h_n^2)\, dm \tag{4.12,4}$$

so that

$$T = \tfrac{1}{2} I_n \dot{\epsilon}_n^2 \tag{4.12,5}$$

The first term of (4.12,3) is therefore $I_n \ddot{\epsilon}_n$, and the second term is zero.

The strain-energy term is conveniently evaluated in terms of the natural frequency of the nth mode by applying Rayleigh's method. This uses the fact that, when the system vibrates in an undamped normal mode, the maximum strain energy occurs when all elements are simultaneously at the extreme position, and the kinetic energy is zero. This maximum strain energy must be equal to the maximum kinetic energy that occurs when all elements pass simultaneously through their equilibrium position, where the strain energy is zero. Hence, if $\epsilon_n = a \sin \omega_n t$, then the maximum kinetic energy is, from (4.12,5)

$$T_{\max} = \tfrac{1}{2} I_n \omega_n^2 a^2$$

Since the stress-strain relation is assumed to be linear, the strain energy[8] is a quadratic function of ϵ_n; that is, $U = \frac{1}{2}k\epsilon_n^2$. Hence

$$U_{\max} = \tfrac{1}{2}ka^2 = T_{\max} = \tfrac{1}{2}I_n\omega_n^2a^2$$

It follows that $k = I_n\omega_n^2$, and that

$$U = \tfrac{1}{2}I_n\omega_n^2\epsilon_n^2$$

and hence $\partial U/\partial\epsilon_n = I_n\omega_n^2\epsilon_n$. The left side of (4.12,3) is therefore as follows:

$$I_n\ddot{\epsilon}_n + I_n\omega_n^2\epsilon_n = \mathscr{F}_n \tag{4.12,6}$$

When structural damping is present, this simple form of uncoupled equation is not exact but the changes in frequency and mode shape for small damping are not large. Hence damping can be allowed for approximately by adding a damping term to (4.12,6), that is,

$$\ddot{\epsilon}_n + 2\zeta_n\omega_n\dot{\epsilon}_n + \omega_n^2\epsilon_n = \mathscr{F}_n/I_n \tag{4.12,7}$$

without changing ω_n or the mode-shape functions. The value of ζ is ordinarily less than 0.1, and usually must be found by an experimental measurement on the actual structure.

EVALUATION OF $\mathscr{F}_n$

The generalized force is calculated from the work done during a virtual displacement,

$$\mathscr{F}_n = \frac{\partial W}{\partial\epsilon_n} \tag{4.12,8}$$

where W is the work done by all the external forces, including the inertia forces associated with nonuniform motion of the frame of reference. The inertia force field is given by

$$d\mathbf{f}_i = -(\mathbf{a}')\,dm \tag{4.12,9}$$

where the components of the r.h.s. in F_B are given by (A.6,8) without the terms ($\ddot{x}$, $\ddot{y}$, $\ddot{z}$). The work done by these forces in a virtual displacement of the structure is

$$\delta W_i = \int (\delta x\, df_{x_i} + \delta y\, df_{y_i} + \delta z\, df_{z_i})$$

where the integration is over the whole body. Introducing (4.12,1) this becomes

$$\delta W_i = \sum_{n=1}^{\infty} \delta_{\epsilon_n} \int (f_n\, df_{x_i} + g_n\, df_{y_i} + h_n\, df_{z_i})$$

whence

$$\mathscr{F}_{n_i} = \frac{\partial W_i}{\partial\epsilon_n} = \int (f_n\, df_{x_i} + g_n\, df_{y_i} + h_n\, df_{z_i}) \tag{4.12,10}$$

When the inertia-force expressions are linearized to small disturbances, and substituted into (4.12,10), all the remaining first-order terms contain integrals of the following types:

[8]For example, in a spring of stiffness k and stretch x, the strain energy is $U = \frac{1}{2}kx^2$.

$$\int f_n \, dm, \qquad \int (yh_n - zg_n) \, dm$$

The first of these is zero because the origin is the mass center, and the second is zero because the angular momentum associated with the elastic mode vanishes. The net result is that $\mathcal{F}_{n_i} = 0$. This result simply verifies what was stated above; that is, there is no inertial coupling between the elastic and rigid-body degrees of freedom.

The remaining contribution to $\mathcal{F}_n$ is that of the aerodynamic forces. Let the local normal-pressure perturbation at an element dS of the airplane's surface be $p(x_0, y_0, z_0)$, and let the local outward normal be $\mathbf{n}(n_x, n_y, n_z)$. Then the work done by the aerodynamic forces in a virtual displacement is

$$\delta W_a = -\int p\mathbf{n} \cdot (\mathbf{r} - \mathbf{r}_0) \, dS$$

where the integral is over the whole surface of the airplane, and $(\mathbf{r} - \mathbf{r}_0)$ is the vector displacement at dS. It is given by

$$\mathbf{r} - \mathbf{r}_0 = \sum_{n=1}^{\infty} (\mathbf{i}f_n + \mathbf{j}g_n + \mathbf{k}h_n) \, \delta\epsilon_n$$

hence

$$\delta W_a = -\sum_{n=1}^{\infty} \delta\epsilon_n \int p(n_x f_n + n_y g_n + n_z h_n) \, dS$$

and

$$\mathcal{F}_n = \frac{\partial W_a}{\partial \epsilon_n} = -\int p(n_x f_n + n_y g_n + n_z h_n) \, dS \tag{4.12,11}$$

Each of the variables inside the integral is a function of (x_0, y_0, z_0), i.e., of position on the surface, and moreover, p is in the most general case a function of all the generalized coordinates, of their derivatives, and of the control-surface angles. The result is that $\mathcal{F}_n$ is a linear function of all these variables, which may be expressed in terms of a set of generalized aerodynamic derivatives (or alternatively aerodynamic transfer functions), namely,

$$\mathcal{F}_n = A_{nu}\Delta u + A_{n\dot{u}}\dot{u} + \cdots + A_{np}p + \cdots + A_{n\delta_r}\delta_r + \cdots$$

$$+ \sum_{m=1}^{\infty} a_{nm}\epsilon_m + \sum_{m=1}^{\infty} b_{nm}\dot{\epsilon}_m + \sum_{m=1}^{\infty} c_{nm}\ddot{\epsilon}_m \tag{4.12,12}$$

In application, only the important derivatives would be retained in any given case. The values of the derivatives kept would be computed by application of (4.12,11). An example of this computation is given in Sec. 5.10.

RÉSUMÉ

The effects of structural dynamics on the stability and control equations can be incorporated by adding structural degrees of freedom based on free normal modes. For an exact representation, an infinite number of such modes are required; however, in practice only a few of the lowest modes need be employed. The six rigid-body equations are altered only to the extent of additional aerodynamic terms of the type given in (4.12,2). One additional equation is required for each elastic degree of freedom (4.12,7). The generalized forces appearing in the added equations contain only aerodynamic contributions, which are computed from (4.12,11) and expressed as in (4.12,12).

4.13 Exercises

4.1 Prove that the angular momentum vector **h** of an airplane is the same whether or not the wind vector **W** is zero.

4.2 Carry out the expansion of $\tilde{\mathbf{r}}\tilde{\boldsymbol{\omega}}\mathbf{r}$ to derive the result given in (4.3,6).

4.3 Carry out the multiplication of the three rotation matrices indicated in (4.4,2) to obtain the result given in (4.4,3).

4.4 Derive (4.5,11) for the moments and product of inertia in frame F_S when they are given for principal axes.

4.5 Prove that when an airplane has spinning rotors, such as jet engines, the angular momentum is given by (4.6,1). Derive the additional terms in the moment equations given in (4.6,2).

4.6 Two airplanes are geometrically similar and have similar mass distributions. Airplane A has a span of 100 ft (30.48 m) and a weight of 100,000 lb (445,000 N). B has 150-ft (47.72 m) span and weighs 225,000 lb (1,001,250 N). Both fly at speeds low enough to neglect Mach number effects, and high enough to neglect Reynolds number effects.

When flying at 400 knots and 20,000-ft (6,096-m) altitude, airplane B has a spiral divergence (a lateral instability) that has a characteristic time of 20 seconds.

(a) At what speed and altitude will A be dynamically similar to B?

(b) What will be the characteristic time of the spiral divergence of A at that speed and altitude?

(c) What is the ratio of the C_L values for the two flight conditions?

4.7 Substitute the linear expressions for ΔZ and ΔM into the right side of (4.9,7*c*) and (4.9,8*b*) and solve the resulting equations to get the second and third components of (4.9,18).

4.8 Derive the expressions given in Tables 4.4 and 4.5 for the derivatives X_q, M_w, Y_p, N_v.

4.9 Let $[p_E \;\; q_E \;\; r_E]^T$ be the angular velocity of an airplane in frame F_E. Find a set of equations that produces these components, given the body axis components $[p \;\; q \;\; r]^T$.

4.10 A hovercraft in ground effect is acted on by the following aerodynamic forces, expressed as body frame components:

$$X = Y = 0$$
$$Z = -\text{mg} + Z_z z_E$$
$$L = L_\phi \phi; \quad M = M_\theta \theta; \quad N = 0$$

The body axes are principal axes, and the engine/rotor angular momentum is

$$\mathbf{h}'_B = [0 \quad 0 \quad H]^T$$

Derive a set of small-disturbance equations of motion.

4.11 Suppose that the wind vector **W** is not zero but instead is given by

$$\mathbf{W}_B = [u_g \quad v_g \quad w_g]^T$$

where **W** is assumed to be "small" and a known function of time. Derive the additional terms that would have to be added to the small-disturbance equations of motion (4.9,18) and (4.9,19).

*4.12** An aircraft is performing a rolling pullup. At the instant of observation, the vehicle is at the bottom of a vertical circle of 2000 ft (610 m) radius moving at a constant speed of 500 fps (152 m/s) with wings horizontal. (See Fig. 3.1). At the same time the roll rate is constant at $p = 90°\ \text{s}^{-1}$. Given that

$$I_y - I_x = 300 \text{ slug ft}^2 \ (406 \text{ kg m}^2) \quad \text{and} \quad I_z = 500 \text{ slug ft}^2 \ (677 \text{ kg m}^2)$$

determine the moments required at this time to perform this maneuver. Assume that the axes are principal axes, with Cx horizontal. (You may assume constant Euler angle rates and $\psi = 0$.)

4.14 Additional Symbols Introduced in Chapter 4

C_{w_0}	$mg/\frac{1}{2}\rho u_0^2 S$
I_x, I_y, I_z	moments of inertia about (x, y, z) axes
I'_x, I'_z	see (4.9,19)
I_{yz}	product of inertia $\int yz\, dm$
I_{xz}	product of inertia $\int xz\, dm$
I'_{zx}	see (4.9,19)
I_{xy}	product of inertia $\int xy\, dm$
$\mathbf{f}$	resultant external force vector
$\mathcal{F}$	generalized force in Lagrange's equation
$\mathbf{G}$	resultant external moment vector, about the mass center
$\mathbf{h}$	angular momentum vector of the airplane with respect to its mass center
$\mathbf{h}'$	angular momentum vector of spinning rotors with respect to rotor mass center
(h_x, h_y, h_z)	scalar components of $\mathbf{h}$ in F_B
(h'_x, h'_y, h'_z)	scalar components of $\mathbf{h}'$ in F_B
(L, M, N)	scalar components of $\mathbf{G}$ in F_B
(p, q, r)	scalar components of $\boldsymbol{\omega}$, radians/sec in F_B

*Problem courtesy of Prof. F. H. Lutze, Virginia Polytechnic Institute

(u, v, w)	scalar components of $\mathbf{V}$ in F_B
$\mathbf{V}$	airspeed vector of airplane mass center
(X, Y, Z)	components of resultant aerodynamic force acting on the airplane, in F_B
x_E, y_E, z_E	coordinates of airplane mass center relative to fixed axes (see Fig. 4.2)
ϵ_n	generalized coordinate of the nth elastic mode
$\delta_e, \delta_r, \delta_a$	angles of elevator, rudder, and aileron
δ_p	propulsion control
$\boldsymbol{\omega}$	angular velocity vector of the airplane
(ψ, θ, ϕ)	Euler angles, radians (see Sec. 4.4)

See also Tables 4.1 to 4.3.

See also Secs. 2.11 and 3.15.

CHAPTER 5

The Stability Derivatives

5.1 General Remarks

We saw in Chap. 4 how the aerodynamic actions on the airplane can be represented approximately by means of stability derivatives (or more exactly by aerodynamic transfer functions). Indeed, all the aerodynamics involved in airplane dynamics is concentrated in this section of the subject: i.e., in the determination of these derivatives (or transfer functions). Each of the stability derivatives contained in the equations of motion is discussed in the following sections. Wherever possible, formulas for them are given in terms of the more elementary parameters used in static stability and performance. Where this is not feasible, it is shown in a qualitative way how the particular force or moment is related to the relevant perturbation quantity. No data for estimation are contained in this chapter; these are all in Appendix B.

EXPRESSIONS FOR C_x AND C_z

For convenience, we shall want the derivatives of C_x and C_z expressed in terms of lift, drag, and thrust coefficients. The relevant forces are shown in Fig. 5.1. As shown, the thrust line does not necessarily lie on the x axis. However, the angle between them is generally small, and we shall assume it be zero. With this assumption, and for small α_x, we get[1]

$$\begin{aligned} C_x &= C_T + C_L\alpha_x - C_D \\ C_z &= -(C_L + C_D\alpha_x) \end{aligned} \tag{5.1,1}$$

where C_T is the coefficient of thrust, $T/\frac{1}{2}\rho V^2 S$.

5.2 The α Derivatives (C_{x_α}, C_{z_α}, C_{m_α})

The α derivatives describe the changes that take place in the forces and moments when the angle of attack of the airplane is increased. They are normally an increase in the lift, an increase in the drag, and a negative pitching moment. The contents of Chap. 2 are relevant to these derivatives.

[1]Since X and Z are the *aerodynamic* forces acting on the airplane, there are no weight components in (5.1,1).

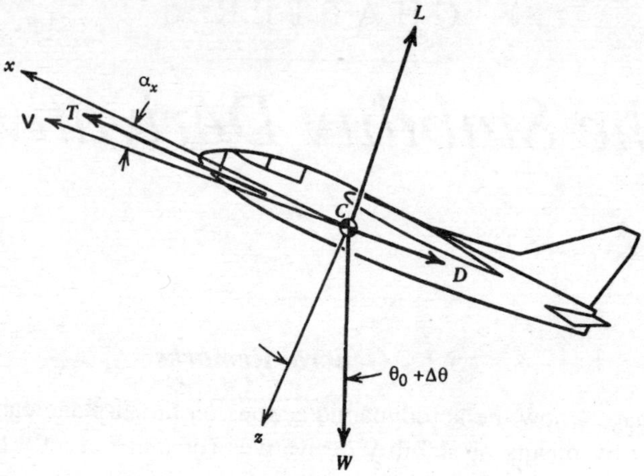

Figure 5.1 Forces in symmetric flight.

THE DERIVATIVE C_{x_α}

By definition, $C_{x_\alpha} = (\partial C_x/\partial \alpha)_0$, where the subscript zero indicates that the derivative is evaluated when the disturbance quantities are zero. From (5.1,1)

$$\frac{\partial C_x}{\partial \alpha} = \frac{\partial C_T}{\partial \alpha} + C_L + \alpha_x \frac{\partial C_L}{\partial \alpha} - \frac{\partial C_D}{\partial \alpha}$$

We may assume that the thrust coefficient is sensibly independent of α_x so that $\partial C_T/\partial \alpha = 0$, and hence

$$C_{x_\alpha} = \left(\frac{\partial C_x}{\partial \alpha}\right)_0 = C_{L_0} - \left(\frac{\partial C_D}{\partial \alpha}\right)_0 \tag{5.2,1}$$

where the subscript zero again indicates the reference flight condition, in which, with stability axes, $\alpha_x = 0$. When the drag is given by a parabolic polar in the form $C_D = C_{D_{\min}} + C_L^2/\pi Ae$, then

$$\left(\frac{\partial C_D}{\partial \alpha}\right)_0 = \frac{2C_{L_0}}{\pi Ae} C_{L_\alpha} \tag{5.2,2}$$

THE DERIVATIVE C_{z_α}

By definition, $C_{z_\alpha} = (\partial C_z/\partial \alpha)_0$. From (5.1,1) we get

$$\frac{\partial C_z}{\partial \alpha} = -\left(C_{L_\alpha} + C_D + \alpha_x \frac{\partial C_D}{\partial \alpha}\right)$$

Therefore

$$C_{z_\alpha} = -(C_{L_\alpha} + C_{D_0}) \tag{5.2,3}$$

C_{D_0} will frequently be negligible compared to C_{L_α}, and consequently $C_{z_\alpha} \doteqdot -C_{L_\alpha}$.

THE DERIVATIVE C_{m_α}

C_{m_α} is the static stability derivative, which was treated at some length in Chap. 2. It is conveniently expressed in terms of the stick-fixed neutral point (2.3,25):

$$C_{m_\alpha} = a(h - h_n) \tag{5.2,4}$$

For airplanes with positive pitch stiffness, $h < h_n$, and C_{m_α} is negative.

5.3 The u Derivatives (C_{x_u}, C_{z_u}, C_{m_u})

The u derivatives give the effect on the forces and moments of an increase in the forward speed, while the angle of attack, the elevator angle, and the throttle position remain fixed. If the *coefficients* of lift and drag did not change, then this would imply an increase in these forces in accordance with the speed-squared law, i.e.,

$$\frac{\text{Force or moment}}{\text{Initial force or moment}} = \frac{(u_0 + \Delta u)^2}{u_0^2} \doteq 1 + 2\Delta\hat{u}$$

Since the pitching moment is initially zero, then, so long as C_m does not change with u, it will remain zero. The situation is actually more complicated than this, for the nondimensional coefficients are in general functions of Mach number and Reynolds number, both of which increase with increasing u. The variation with Reynolds number is usually neglected, but the effect of Mach number must be included.

The thrust effect shows up in two different ways. One stems simply from the derivative of thrust with speed, which depends on the type of propulsive system-jet, propeller, and so forth. The other, related mainly to propeller configurations, derives from the propulsion/airframe interaction, for example, the propeller slipstream impinging on the wing. This is an important effect, and for some STOL airplanes, may be dominant at low speeds.

Finally, the increased loading on the airframe due to the speed increase may induce significant structural distortion. This is a static aeroelastic effect. For example, the tail lift coefficient may be influenced appreciably by the loading (see Sec. 3.5). An appropriate variable to use for aeroelastic effects is the dynamic pressure $p_d = \frac{1}{2}\rho V^2$.

In order to formally include each of these three major effects, compressibility, aeroelasticity, and propulsive, even though they would rarely all be present at the same time, each of the coefficients C_x, C_z, C_m is assumed to be a function of $\mathbf{M}$, p_d, and C_T as well as angle of attack.

We then have

$$C_{x_u} = \left(\frac{\partial C_x}{\partial \hat{u}}\right)_0 = \left(\frac{\partial C_x}{\partial \mathbf{M}}\frac{\partial \mathbf{M}}{\partial \hat{u}}\right)_0 + \left(\frac{\partial C_x}{\partial p_d}\frac{\partial p_d}{\partial \hat{u}}\right)_0 + \left(\frac{\partial C_x}{\partial C_T}\frac{\partial C_T}{\partial \hat{u}}\right)_0 \tag{5.3,1}$$

and similarly for C_{z_u} and C_{m_u}.

CALCULATION OF $\partial\mathbf{M}/\partial\hat{u}$

The Mach number is $\mathbf{M} = V/a$, where a is the speed of sound, so

$$\frac{\partial \mathbf{M}}{\partial \hat{u}} = u_0\frac{\partial \mathbf{M}}{\partial u} = \frac{u_0}{a}\frac{\partial V}{\partial u} = \mathbf{M}_0\frac{\partial V}{\partial u}$$

But

$$V^2 = u^2 + v^2 + w^2$$

so
$$\left(\frac{\partial V}{\partial u}\right)_0 = 1$$

and
$$\left(\frac{\partial \mathbf{M}}{\partial \hat{u}}\right)_0 = \mathbf{M}_0 \tag{5.3,2}$$

CALCULATION OF $\partial p_d/\partial \hat{u}$

$$p_d = \tfrac{1}{2}\rho V^2 = \tfrac{1}{2}\rho(u^2 + v^2 + w^2)$$

so
$$\frac{\partial p_d}{\partial u} = \rho u$$

and
$$\left(\frac{\partial p_d}{\partial \hat{u}}\right)_0 = \left(u_0\frac{\partial p_d}{\partial u}\right)_0 = \rho u_0^2 \tag{5.3,3}$$

The thrust derivative is defined in a manner consistent with Table 4.2 to be

$$C_{T_u} = \frac{\partial C_T}{\partial \hat{u}} \tag{5.3,4}$$

It follows that C_{x_u} is given by

$$C_{x_u} = \mathbf{M}_0\left(\frac{\partial C_x}{\partial \mathbf{M}}\right)_0 + \rho u_0^2\left(\frac{\partial C_x}{\partial p_d}\right)_0 + C_{T_u}\left(\frac{\partial C_x}{\partial C_T}\right)_0 \tag{5.3,5}$$

THE DERIVATIVE C_{T_u}

Since
$$C_T = T/\tfrac{1}{2}\rho V^2 S$$

then
$$\frac{\partial C_T}{\partial u} = \frac{\partial T/\partial u}{\tfrac{1}{2}\rho V^2 S} - \frac{2T}{\tfrac{1}{2}\rho V^3 S}\frac{\partial V}{\partial u}$$

In the reference flight condition $V = u_0$ and $\partial V/\partial u = 1$, so

$$C_{T_u} = \left(\frac{\partial C_T}{\partial \hat{u}}\right)_0 = u_0\left(\frac{\partial C_T}{\partial u}\right)_0 = \frac{(\partial T/\partial u)_0}{\tfrac{1}{2}\rho u_0 S} - 2C_{T_0} \tag{5.3,6}$$

For unpowered gliding flight, $T = 0$ and

$$C_{T_u} = 0 \tag{5.3,7}$$

For constant-thrust propulsion, which is an approximation for jet aircraft in cruising flight, $\partial T/\partial u = 0$, and

$$C_{T_u} = -2C_{T_0} \tag{5.3,8}$$

For constant-power propulsion, which is an approximation for constant-speed propellers in cruising flight, TV is constant, so that

$$(\partial T/\partial u)_0 = -T_0/u_0$$

and
$$C_{T_u} = -3C_{T_0} \tag{5.3,9}$$

The values of C_{T_0} in the preceding expressions can be related to the reference lift and drag coefficients [see Fig. 5.1 and (5.1,1)]. Note that **T**, **V**, x are assumed to be colinear, that is,

$$C_{T_0} = C_{D_0} + C_{W_0} \sin \theta_0 \tag{5.3,10}$$

THE DERIVATIVE C_{x_u}

From (5.1,1) we have

$$\left(\frac{\partial C_x}{\partial \mathbf{M}}\right)_0 = \left(\frac{\partial C_T}{\partial \mathbf{M}}\right)_0 - \left(\frac{\partial C_D}{\partial \mathbf{M}}\right)_0$$

$$\left(\frac{\partial C_x}{\partial p_d}\right)_0 = \left(\frac{\partial C_T}{\partial p_d}\right)_0 - \left(\frac{\partial C_D}{\partial p_d}\right)_0$$

$$\left(\frac{\partial C_x}{\partial C_T}\right)_0 = 1 - \left(\frac{\partial C_D}{\partial C_T}\right)_0$$

Since the direct aeroelastic effect on thrust is likely to be negligible, we neglect $\partial C_T/\partial p_d$, and then (5.3,5) gives

$$C_{x_u} = \mathbf{M}_0\left(\frac{\partial C_T}{\partial \mathbf{M}} - \frac{\partial C_D}{\partial \mathbf{M}}\right)_0 - \rho u_0^2\left(\frac{\partial C_D}{\partial p_d}\right)_0 + C_{T_u}\left(1 - \frac{\partial C_D}{\partial C_T}\right)_0 \tag{5.3,11}$$

When a powered wind-tunnel model is tested, it is common practice to measure the net axial force coefficient C_x and not its component parts C_T and C_D. In that case, the test data can provide the C_{x_u} derivative directly.

THE DERIVATIVE C_{z_u}

From (5.1,1) we have that

$$\left(\frac{\partial C_z}{\partial \mathbf{M}}\right)_0 = -\left(\frac{\partial C_L}{\partial \mathbf{M}}\right)_0$$

$$\left(\frac{\partial C_z}{\partial p_d}\right)_0 = -\left(\frac{\partial C_L}{\partial p_d}\right)_0$$

$$\left(\frac{\partial C_z}{\partial C_T}\right)_0 = -\left(\frac{\partial C_L}{\partial C_T}\right)_0$$

so that

$$C_{z_u} = -\mathbf{M}_0\left(\frac{\partial C_L}{\partial \mathbf{M}}\right)_0 - \rho u_0^2\left(\frac{\partial C_L}{\partial p_d}\right)_0 - C_{T_u}\left(\frac{\partial C_L}{\partial C_T}\right)_0 \tag{5.3,12}$$

The derivative $\mathbf{M}_0(\partial C_L/\partial \mathbf{M}_0)$ tends to be small except at transonic speeds. Theoretical values are easily calculated for high-aspect-ratio swept wings. At subsonic speeds, the Prandtl–Glauert rule combined with simple sweep theory (Kuethe and Chow, 1976) gives the lift coefficient for two-dimensional flow as

$$C_L = \frac{a_i \alpha}{\sqrt{1 - \mathbf{M}^2 \cos^2 \Lambda}} \qquad \mathbf{M} \cos \Lambda < 1$$

where a_i is the lift-curve slope in incompressible flow and Λ is the sweepback angle of the $\frac{1}{4}$ chord line. Upon differentiation with respect to **M**, we get

$$\mathbf{M}\frac{\partial C_L}{\partial \mathbf{M}} = \frac{\mathbf{M}^2\cos^2\Lambda}{1-\mathbf{M}^2\cos^2\Lambda}C_L$$

and hence

$$\mathbf{M}_0\left(\frac{\partial C_L}{\partial \mathbf{M}}\right)_0 = \frac{\mathbf{M}_0^2\cos^2\Lambda}{1-\mathbf{M}_0^2\cos^2\Lambda}C_{L_0} \tag{5.3,13}$$

In level flight with the lift equal to the weight, $\mathbf{M}_0^2 C_{L_0}$ is constant, and hence $\mathbf{M}_0(\partial C_L/\partial \mathbf{M})_0$ is proportional to $1/(1-\mathbf{M}_0^2\cos^2\Lambda)$. At supersonic speeds, the two-dimensional lift is given by Kuethe and Chow (1976)

$$C_L = \frac{4\alpha\cos\Lambda}{\sqrt{\mathbf{M}^2\cos^2\Lambda - 1}}$$

After differentiation with respect to **M**, we get exactly the same result as for subsonic speeds. That is (5.3,13) applies over the whole Mach-number range, except of course near $\mathbf{M} = 1$ where the cited airfoil theories do not apply. Low-aspect-ratio wings are less sensitive to changes in **M**.

THE DERIVATIVE C_{m_u}

From (5.3,5) and (5.3,6) C_{m_u} is given as

$$C_{m_u} = \mathbf{M}\left(\frac{\partial C_m}{\partial \mathbf{M}}\right)_0 + \rho u_0^2\left(\frac{\partial C_m}{\partial p_d}\right)_0 + C_{T_u}\left(\frac{\partial C_m}{\partial C_T}\right)_0 \tag{5.3,14}$$

Values of $\partial C_m/\partial \mathbf{M}$ can be found from wind-tunnel tests on a rigid model. They are largest at transonic speeds and are strongly dependent on the wing planform. The main factor that contributes to this derivative is the backward shift of the wing center of pressure that occurs in the transonic range. On two-dimensional symmetrical wings, for example, the center of pressure moves from approximately $0.25c$ to approximately $0.50c$ as the Mach number increases from subsonic to supersonic values. Thus an increase in **M** in this range produces a diving-moment increment; that is, C_{m_u} is negative. For wings of very low aspect ratio, the center of pressure movement is much less, and the values of C_{m_u} are correspondingly smaller.

To find $\partial C_m/\partial p_d$ requires either an aeroelastic analysis or tests on a flexible model. As an example of this phenomenon, let us consider an airplane with a tail and a flexible fuselage.[2] We found in Sec. 3.5 that the tail lift coefficient is given by

$$C_{L_t} = \frac{a_t}{1 + ka_t p_d S_t}(\alpha_{wb} - \epsilon - i_t) \tag{5.3,15}$$

The pitching moment contributed by the tail is (2.2,9)

$$C_{m_t} = -V_H C_{L_t}$$

Hence

$$\left(\frac{\partial C_m}{\partial p_d}\right)_{\text{tail}} = -V_H\frac{\partial C_{L_t}}{\partial p_d} \tag{5.3,16}$$

[2]It is not meant to imply that fuselage bending is the only important aeroelastic contribution to C_{m_u}. Distortion of the wing and tail may also be important.

When (5.3,15) is differentiated with respect to p_d and simplified, and the resulting expression is substituted into (5.3,16), we obtain the result

$$\left(\frac{\partial C_m}{\partial p_d}\right)_{\text{tail}} = -C_{m_t}\frac{ka_tS_t}{1 + ka_tp_dS_t} \tag{5.3,17}$$

The corresponding contribution to C_{m_u} is [see (5.3,14)]

$$(C_{m_u})_{\text{tail}} = -C_{m_t}\frac{2p_{d_0}ka_tS_t}{1 + ka_tp_{d_0}S_t} \tag{5.3,18}$$

All the factors in this expression are positive, except for C_{m_t}, which may be of either sign. The contribution of the tail to C_{m_u} may therefore be either positive or negative. The tail pitching moment is usually positive at high speeds and negative at low speeds. Therefore its contribution to C_{m_u} is usually negative at high speeds and positive at low speeds. Since the dynamic pressure occurs as a multiplying factor in (5.3,18), then the aeroelastic effect on C_{m_u} goes up with speed and down with altitude.

5.4 The q Derivatives (C_{z_q}, C_{m_q})

These derivatives represent the aerodynamic effects that accompany rotation of the airplane about a spanwise axis through the CG while α_x remains zero. An example of this kind of motion was treated in Sec. 3.1 (i.e., the steady pull-up). Figure 5.2*b* shows the general case in which the flight path is arbitrary. This should be contrasted with the situation illustrated in Fig. 5.2*a*, where $q = 0$ while α_x is changing.

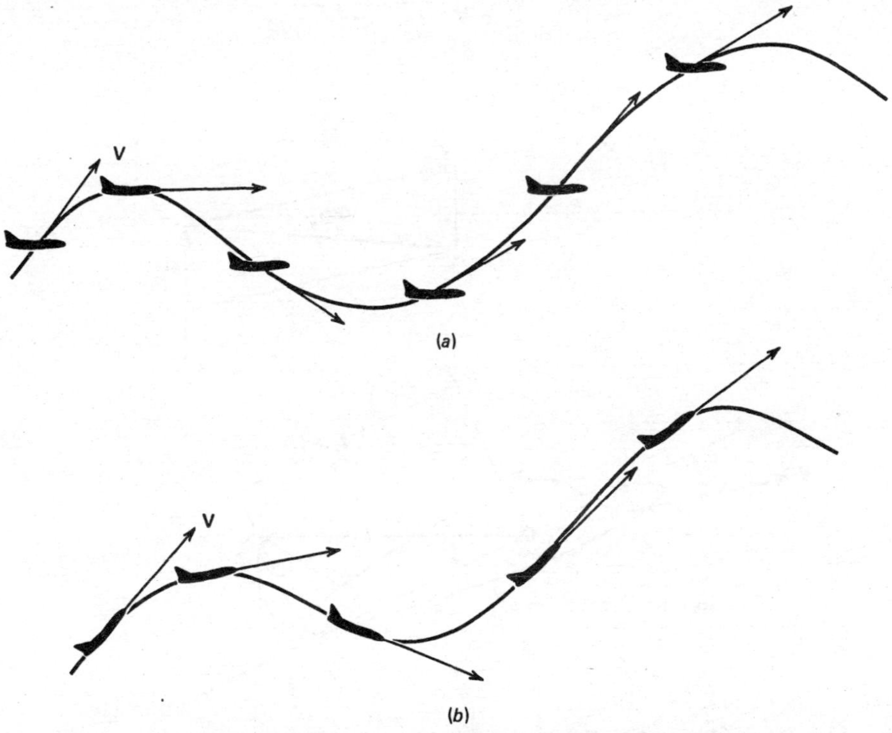

Figure 5.2 (*a*) Motion with zero q, but varying α_x. (*b*) Motion with zero α_x but varying q.

Both the wing and the tail are affected by the rotation, although, when the airplane has a tail, the wing contribution to C_{z_q} and C_{m_q} is often negligible in comparison with that of the tail. In such cases it is common practice to increase the tail effect by an arbitrary amount, of the order of 10%, to allow for the wing and body.

CONTRIBUTIONS OF A TAIL

As illustrated in Fig. 5.3, the main effect of q on the tail is to increase its angle of attack by (ql_t/u_0) radians, where u_0 is the flight speed. It is this change in α_t that accounts for the changed forces on the tail. The assumption is implicit in the following derivations that the instantaneous forces on the tail correspond to its instantaneous angle of attack; i.e., no account is taken of the fact that it takes a finite time for the tail lift to build up to its steady-state value following a sudden change in q. (A method of including this refinement has been given by Tobak, 1954.) The derivatives obtained are therefore *quasistatic.*

C_{z_q} of the Tail

By definition, $C_{z_q} = (\partial C_z/\partial \hat{q})_0 = (2u_0/\bar{c})(\partial C_z/\partial q)_0$, and, from (5.1,1), $(\partial C_z/\partial q)_0 = -(\partial C_L/\partial q)_0$. The change in tail lift coefficient caused by the rotation q is

$$\Delta C_{L_t} = a_t \Delta\alpha_t = a_t \frac{ql_t}{u_0} \tag{5.4,1}$$

and the corresponding change in airplane lift coefficient is

$$\Delta C_L = \frac{S_t}{S} \Delta C_{L_t} = \frac{S_t}{S} a_t \frac{ql_t}{u_0}$$

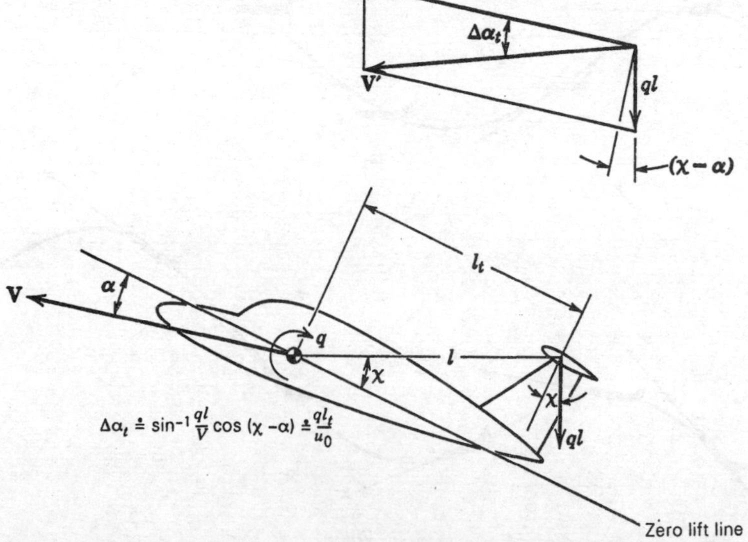

Figure 5.3 Effect of pitch velocity on tail angle of attack.

Therefore

$$\frac{\partial C_L}{\partial q} = a_t \frac{S_t}{S} \frac{l_t}{u_0}$$

and

$$(C_{z_q})_{\text{tail}} = -\frac{2u_0}{\bar{c}} a_t \frac{S_t l_t}{S u_0} = -2a_t V_H \tag{5.4,2}$$

C_{m_q} of the Tail

The increment in pitching moment that corresponds to ΔC_{L_t} is [see (2.2,9)]

$$\Delta C_m = -V_H \Delta C_{L_t} = -a_t V_H \frac{q l_t}{u_0}$$

Hence

$$\frac{\partial \Delta C_m}{\partial q} = -a_t V_H \frac{l_t}{u_0}$$

and

$$(C_{m_q})_{\text{tail}} = \frac{2u_0}{\bar{c}} \left(\frac{\partial \Delta C_m}{\partial q} \right)_0 = -2a_t V_H \frac{l_t}{\bar{c}} \tag{5.4,3}$$

CONTRIBUTIONS OF A WING

As previously remarked, on airplanes with tails the wing contributions to the q derivatives are frequently negligible. However, if the wing is highly swept or of low aspect ratio, it may have significant values of C_{z_q} and C_{m_q}; and of course, on tailless airplanes, the wing supplies the major contribution. The q derivatives of wings alone are therefore of great engineering importance.

Unfortunately, no simple formulas can be given, because of the complicated dependence on the wing planform and the Mach number. However, the following discussion of the physical aspects of the flow indicates how linearized wing theory can be applied to the problem. Consider a plane lifting surface, at zero α_x, with forward speed u_0 and angular velocity q about a spanwise axis (see Fig. 5.4). Each point in the wing has a velocity component, relative to the resting atmosphere, of qx normal to the surface. This velocity distribution is shown in the figure for the central and tip chords. Now there is an equivalent cambered wing that would have the identical distribution of velocities normal to its surface when in rectilinear translation at speed u_0. This is illustrated in Fig. 5.5*a*. The cross section of the curved surface S is shown in (*b*). The normal velocity distribution will be the same as in Fig. 5.4 if

$$u_0 \frac{\partial z}{\partial x} = qx \quad \text{or} \quad \frac{\partial z}{\partial x} = \frac{q}{u_0} x$$

Hence

$$z = \frac{1}{2} \frac{q}{u_0} x^2 \tag{5.4,5}$$

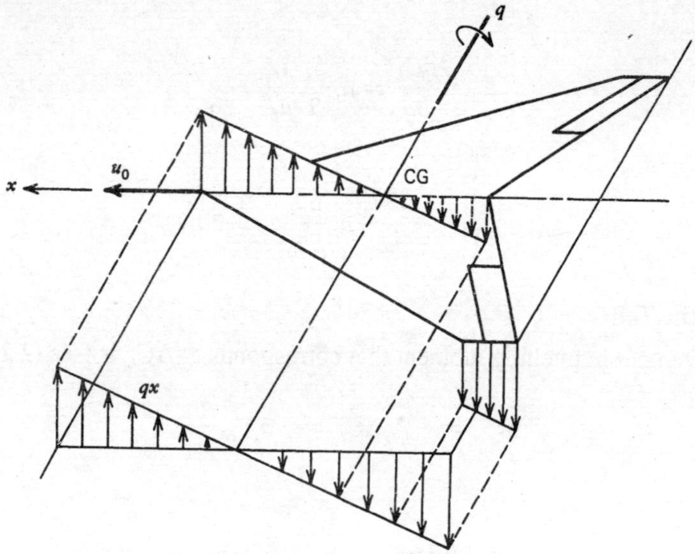

Figure 5.4 Wing velocity distribution due to pitching.

and the cross section of S is a parabolic arc. In linearized wing theory, both subsonic and supersonic, the boundary condition is the same for the original plane wing with rotation q and the equivalent curved wing in rectilinear flight. The problem of finding the q derivatives then is reduced to that of finding the pressure distribution over the equivalent cambered wing. Because of the form of (5.4,5), the pressures are proportional to q/u_0. From the pressure distribution, C_{z_q} and C_{m_q} can be calculated. The derivatives can in principle also be found by experiment, by testing a model of the equivalent wing.

The values obtained by this approach are quasistatic; i.e., they are steady-state values corresponding to $\alpha_x = 0$ and a small constant value of q. This implies that the flight path is a circle (as in Fig. 3.1), and hence that the vortex wake is not rectilinear. Now both the linearized theory and the wind-tunnel measurement apply to a straight wake, and to this extent are approximate. Since the values of the derivatives obtained are in the end applied to arbitrary flight paths, as in Fig. 5.2*b*, there is little point in correcting them for the curvature of the wake.

The error involved in the application of the quasistatic derivatives to unsteady flight is not as great as might be expected. It has been shown that, when the flight path is a sine wave, the quasistatic derivatives apply so long as the reduced frequency is small, that is,

$$k = \frac{\omega \bar{c}}{2u_0} \ll 1 \tag{5.4,6}$$

where ω is the circular frequency of the pitching oscillation. If l is the wavelength of the flight path, then

$$k = \pi \frac{\bar{c}}{l}$$

so that the condition $k \ll 1$ implies that the wavelength must be long compared to the chord, for example, $l > 60\bar{c}$ for $k < 0.05$.

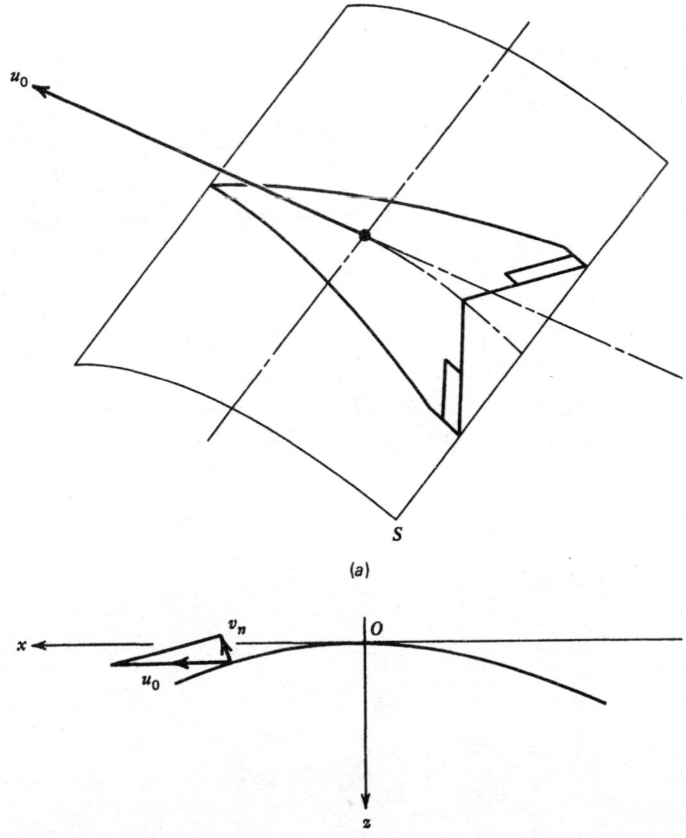

Figure 5.5 The equivalent cambered wing.

DEPENDENCE ON *h*

Because the axis of rotation in Fig. 5.5, passes through the CG, the results obtained are dependent on h. The nature of this variation is found as follows. Let the axis of rotation be at A in Fig. 5.6, and let the associated lift and moment be

$$C_{L_A} = C_{L_{qA}}\hat{q}; \qquad C_{m_A} = C_{m_{qA}}q \tag{5.4,7}$$

Now let the axis of rotation be moved to B, with the change in normal velocity distribution shown on the figure. Since the two normal velocity distributions differ by a constant, (the upward translation $q\bar{c}\,\Delta h$) the difference between the two pressure distributions is that associated with a flat plate at angle of attack

$$\alpha = -\frac{q\bar{c}}{u_0}\Delta h \tag{5.4,8}$$

This angle of attack introduces a lift increment acting at the wing mean aerodynamic center of amount

$$\Delta C_L = C_{L_\alpha}\alpha = -\frac{q\bar{c}}{u_0}\Delta h C_{L_\alpha} = -2C_{L_\alpha}\hat{q}\,\Delta h \tag{5.4,9}$$

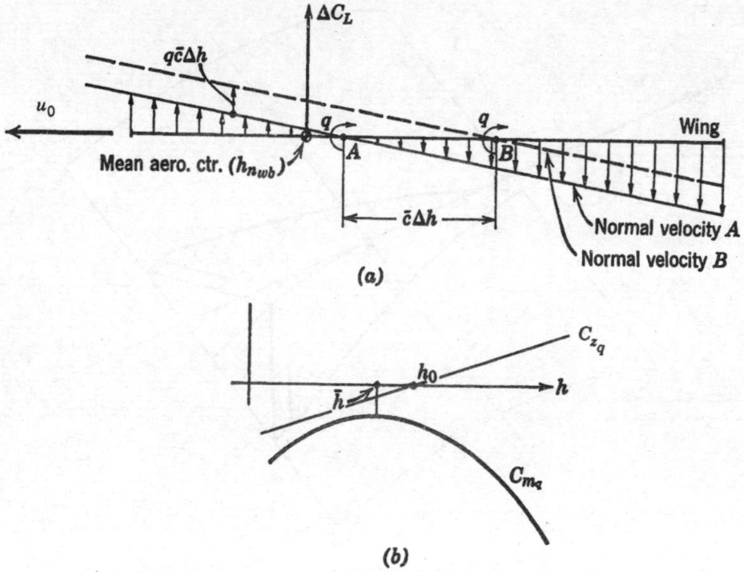

Figure 5.6 Effect of CG location on C_{z_q}, C_{m_q}.

that is,
$$\frac{\partial C_L}{\partial h} = \frac{\Delta C_L}{\Delta h} = -2C_{L_\alpha} q$$

and
$$\frac{\partial^2 C_L}{\partial h \partial \hat{q}} = \frac{\partial}{\partial h} C_{L_q} = -2C_{L_\alpha} \tag{5.4,10}$$

We see that C_{L_q} is linear in h, and can therefore be expressed as

$$C_{L_q} = -2C_{L_\alpha}(h - h_0) \tag{5.4,11}$$

where h_0 is the CG location at which C_{z_q} is zero. By virtue of (5.1,1) we get

$$C_{z_q} = -C_{L_q} = 2C_{L_\alpha}(h - h_0) \tag{5.4,12}$$

The pitching moment about the CG is

$$C_m = C_{m_{ac}} + C_L(h - h_{n_w}) \tag{5.4,13}$$

so that
$$\begin{aligned} C_{m_q} &= \frac{\partial C_{m_{ac}}}{\partial \hat{q}} + C_{L_q}(h - h_{n_w}) \\ &= \frac{\partial C_{m_{ac}}}{\partial \hat{q}} - 2C_{L_\alpha}(h - h_0)(h - h_{n_w}) \end{aligned} \tag{5.4,14}$$

Equation (5.4,14) shows that C_{m_q} is quadratic in h. We can write it without loss of generality as

$$C_{m_q} = \bar{C}_{m_q} - 2C_{L_\alpha}(h - \bar{h})^2 \tag{5.4,15}$$

where $\bar{C}_{m_q}$ is the maximum (least negative) value of C_{m_q} and $\bar{h}$ is the CG location where it occurs (see Fig. 5.6*b*). The value of $\bar{h}$ is found by differentiating (5.4,14). This yields

$$\bar{h} = \tfrac{1}{2}(h_0 + h_{n_w}) \tag{5.4,16}$$

The linear theory of two-dimensional thin wings gives for supersonic flow:

$$h_0 = \bar{h} = \tfrac{1}{2}$$

$$\bar{C}_{m_q} = -\frac{2}{3\sqrt{\mathbf{M}^2 - 1}} \tag{5.4,17}$$

and for subsonic flow:

$$\begin{aligned} h_0 &= \tfrac{3}{4} \\ \bar{h} &= \tfrac{1}{2} \\ \bar{C}_{m_q} &= 0 \end{aligned} \tag{5.4,18}$$

PITCH DAMPING OF PROPULSIVE JETS

When gases flow at high speed inside jet or rocket engines at the same time as the vehicle is rotating in pitch or yaw, they react against the walls of the ducts with a force perpendicular to their velocity vector (the Coriolis force). This reaction can result in a pitching moment proportional to q, that is, in a contribution to C_{m_q}, (and similarly to C_{n_r}). An analysis of this effect is given in Sec. 7.9 of Etkin (1972).

For jet airplanes in cruising flight this contribution to C_{m_q} is usually negligible. Only at high values of C_T, and when the C_{m_q} of the rest of the airplane is small, would it be significant. On the other hand, a rocket booster at lift-off, when the speed is low, has practically zero external aerodynamic damping and the jet damping becomes very important.

5.5 The $\dot{\alpha}$ Derivatives ($C_{L_{\dot{\alpha}}}$, $C_{m_{\dot{\alpha}}}$)

The $\dot{\alpha}$ derivatives owe their existence to the fact that the pressure distribution on a wing or tail does not adjust itself instantaneously to its equilibrium value when the angle of attack is suddenly changed. The calculation of this effect, or its measurement, involves *unsteady flow.* In this respect, the $\dot{\alpha}$ derivatives are very different from those discussed previously, which can all be determined on the basis of steady-state aerodynamics.

CONTRIBUTIONS OF A WING

Consider a wing in horizontal flight at zero α. Let it be subjected to a downward impulse, so that it suddenly acquires a constant downward velocity component. Then, as shown in Fig. 5.7, its angle of attack undergoes a step increase. The lift then responds in a transient manner (the indicial response) the form of which depends on whether **M** is greater or less than 1. In subsonic flight, the vortices which the wing leaves behind it can influence it at all future times, so that the steady state is approached only asymptotically. In supersonic flight, the upstream traveling disturbances move more slowly than the wing, so that it outstrips the disturbance field of the initial impulse in a finite time t_1. From that time on the lift remains constant.

In order to find the lift associated with $\dot{\alpha}$, let us consider the motion of an airfoil with a small constant $\dot{\alpha}$, but with $q = 0$. The motion, and the angle of attack, are

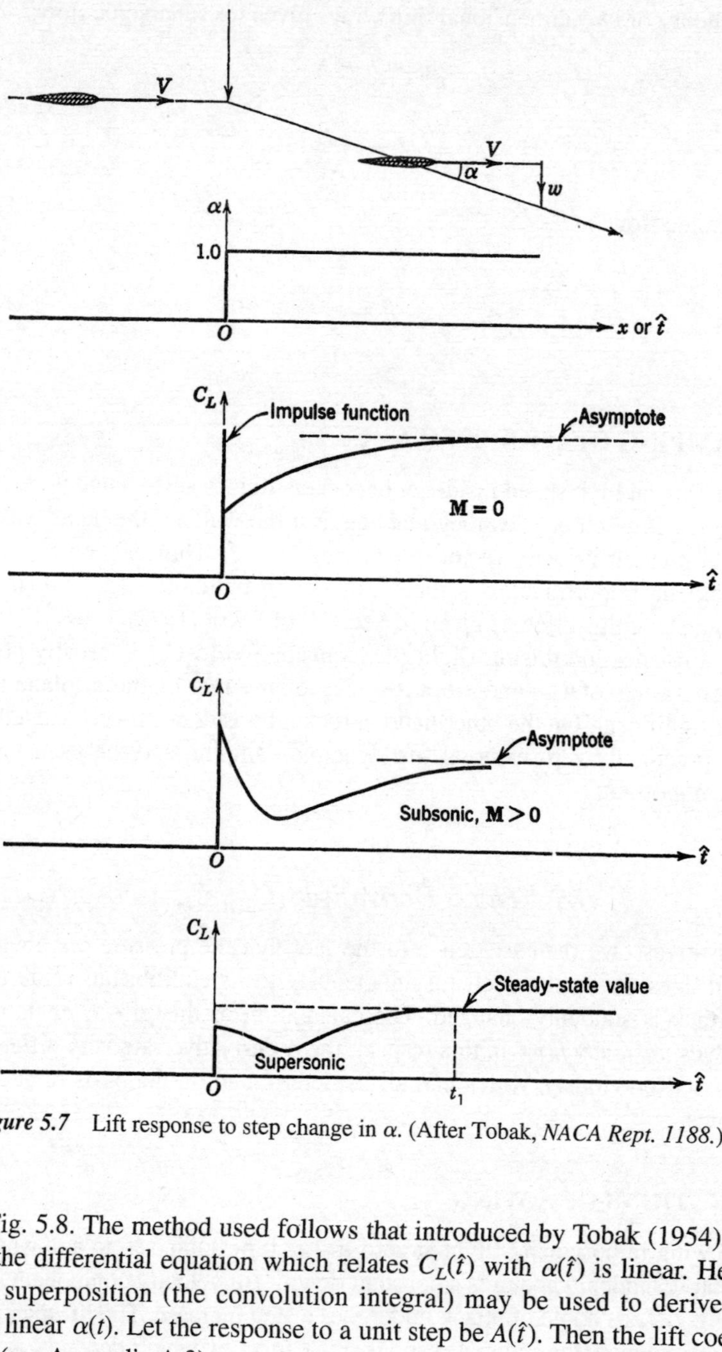

Figure 5.7 Lift response to step change in α. (After Tobak, *NACA Rept. 1188.*)

shown in Fig. 5.8. The method used follows that introduced by Tobak (1954). We assume that the differential equation which relates $C_L(\hat{t})$ with $\alpha(\hat{t})$ is linear. Hence the method of superposition (the convolution integral) may be used to derive the response to a linear $\alpha(t)$. Let the response to a unit step be $A(\hat{t})$. Then the lift coefficient at time $\hat{t}$ is (see Appendix A.3).

$$C_L(\hat{t}) = \int_{\tau=0}^{\hat{t}} A(\hat{t} - \tau)\dot{\alpha}(\tau)\, d\tau$$

Since $\dot{\alpha}(\tau)$ = constant, then

$$C_L(\hat{t}) = \dot{\alpha} \int_{\tau=0}^{\hat{t}} A(\hat{t} - \tau)\, d\tau \tag{5.5,1}$$

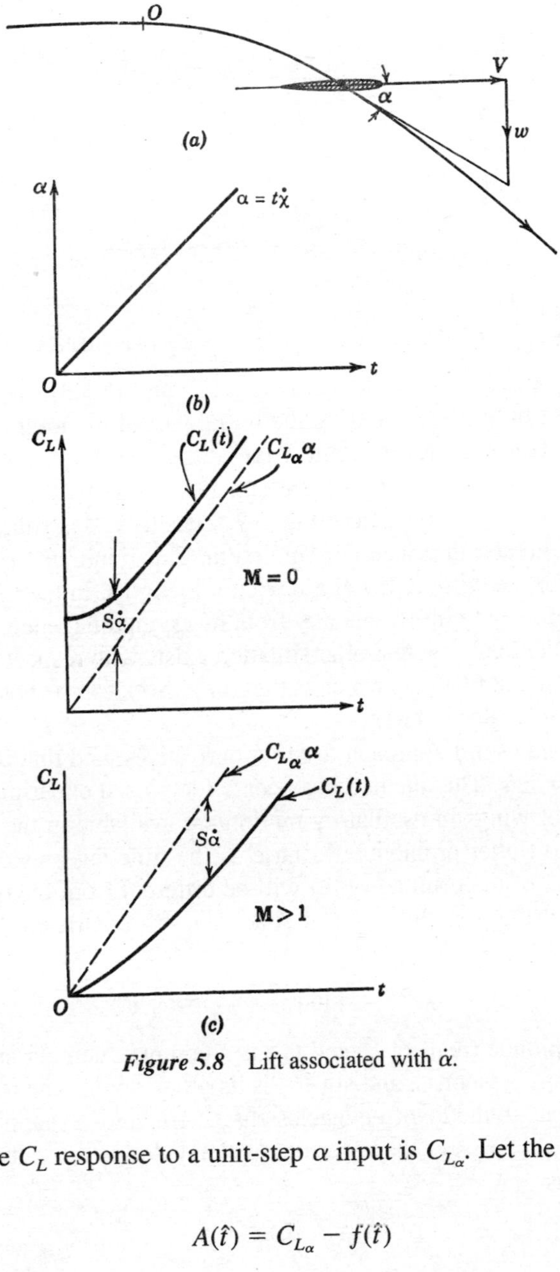

Figure 5.8 Lift associated with $\dot{\alpha}$.

The ultimate C_L response to a unit-step α input is C_{L_α}. Let the lift defect be $f(t)$: that is,

$$A(\hat{t}) = C_{L_\alpha} - f(\hat{t})$$

Then (5.5,1) becomes

$$\begin{aligned} C_L(\hat{t}) &= \dot{\alpha} C_{L_\alpha} \hat{t} - \dot{\alpha} \int_{\tau=0}^{\hat{t}} f(\hat{t} - \tau)\, d\tau \\ &= C_{L_\alpha}\alpha - S\dot{\alpha} \end{aligned} \tag{5.5,2}$$

where $S(\hat{t}) = \int_{\tau=0}^{\hat{t}} f(\hat{t} - \tau)\, d\tau$. The term $S\,\dot{\alpha}$ is shown on Fig. 5.8. Now, if the idea of representing the lift by means of aerodynamic derivatives is to be valid, we must be able to write, for the motion in question,

$$C_L(\hat{t}) = C_{L_\alpha}\alpha(\hat{t}) + C_{L_{\dot{\alpha}}}\dot{\alpha} \tag{5.5,3}$$

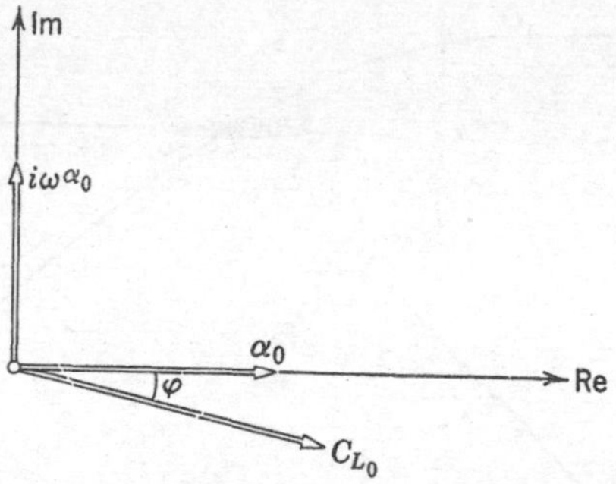

Figure 5.9 Vector diagram of lift response to oscillatory α.

where C_{L_α} and $C_{L_{\dot\alpha}}$ are constants. Comparing (5.5,2) and (5.5,3), we find that $C_{L_{\dot\alpha}} = -S(\hat t)$, a function of time. Hence, during the initial part of the motion, the *derivative concept is invalid.* However, for all finite wings,[3] the area $S(\hat t)$ converges to a finite value as $\hat t$ increases indefinitely. In fact, for supersonic wings, S reaches its limiting value in a finite time, as is evident from Fig. 5.7. Thus (5.5,3) is valid,[4] with constant $C_{L_{\dot\alpha}}$, for values of $\hat t$ greater than a certain minimum. This minimum is not large, being the time required for the wing to travel a few chord lengths. In the time range where S is constant, or differs only infinitesimally from its asymptotic value, the $C_L(\hat t)$ curve of Fig. 5.8*c* is parallel to $C_{L_\alpha}\alpha$. A similar situation exists with respect to C_m.

We see from Fig. 5.8 that $C_{L_{\dot\alpha}}$, which is the $\lim_{\hat t\to\infty} - S(\hat t)$, can be positive for $\mathbf{M} = 0$ and negative for larger values of $\mathbf{M}$.

There is a second useful approach to the $\dot\alpha$ derivatives, and that is via consideration of oscillating wings. This method has been widely used experimentally, and extensive treatments of wings in oscillatory motion are available in the literature,[5] primarily in relation to flutter problems. Because of the time lag previously noted, the amplitude and phase of the oscillatory lift will be different from the quasisteady values. Let us represent the periodic angle of attack and lift coefficient by the complex numbers

$$\alpha = \alpha_0 e^{i\omega t} \qquad \text{and} \qquad C_L = C_{L_0} e^{i\omega t} \tag{5.5,4}$$

where α_0 is the amplitude (real) of α, and C_{L_0} is a complex number such that $|C_{L_0}|$ is the amplitude of the C_L response, and arg C_{L_0} is its phase angle. The relation between C_{L_0} and α_0 appropriate to the low frequencies characteristic of dynamic stability is illustrated in Fig. 5.9. In terms of these vectors, we may derive the value of $C_{L_{\dot\alpha}}$ as follows. The $\dot\alpha$ vector is

$$\dot\alpha = i\omega\alpha_0 e^{i\omega t}$$

Thus C_L may be expressed as

$$\begin{aligned} C_L &= R[C_{L_0}]e^{i\omega t} + iI[C_{L_0}]e^{i\omega t} \\ &= R[C_{L_0}]\frac{\alpha}{\alpha_0} + I[C_{L_0}]\frac{\dot\alpha}{\omega\alpha_0} \end{aligned}$$

[3]For two-dimensional incompressible flow, the area $S(\hat t)$ diverges as $t \to \infty$. That is, the derivative concept is definitely *not* applicable to that case.

[4]Exactly for supersonic wings, and approximately for subsonic wings.

[5]See bibliography.

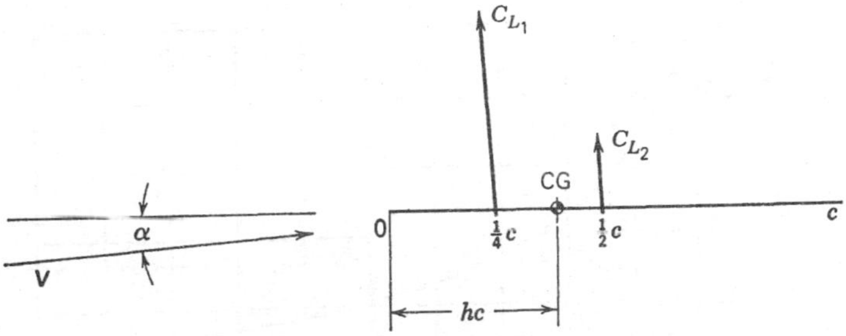

Figure 5.10 Lift on oscillating two-dimensional airfoil.

Hence
$$C_{L_{\dot{\alpha}}} = \frac{\partial C_L}{\partial(\dot{\alpha}\bar{c}/2u_0)} = \frac{I[C_{L_0}]}{k\alpha_0} \tag{5.5,5}$$

or, if the amplitude α_0 is unity, $C_{L_{\dot{\alpha}}} = I[C_{L_0}]/k$, where k is the reduced frequency $\omega\bar{c}/2u_0$.

To assist in forming a physical picture of the behavior of a wing under these conditions, we give here the results for a two-dimensional,[6] airfoil in incompressible flow. The motion of the airfoil is a plunging oscillation; that is, it is like that shown in Fig. 5.2*a*, except that the flight path is a sine wave. The instantaneous lift on the airfoil is given in two parts (see Fig. 5.10):

$$C_L = C_{L_1} + C_{L_2}$$

where
$$C_{L_1} = 2\pi\left[\alpha F(k) + \left(\frac{\dot{\alpha}c}{2u_0}\right)\frac{G(k)}{k}\right] \tag{5.5,6}$$
$$C_{L_2} = \pi\left(\frac{\dot{\alpha}c}{2u_0}\right)$$

and $F(k)$ and $G(k)$ are the real and imaginary parts of the Theodorsen function $C(k)$ plotted in Fig. 5.11 (Theodorsen, 1934). The lift that acts at the midchord is proportional to $\dot{\alpha} = \ddot{z}/u_0$, where z is the translation (vertically downward) of the airfoil. That is, it represents a force opposing the downward acceleration of the airfoil. This force is exactly that which is required to impart an acceleration $\ddot{z}$ to a mass of air contained in a cylinder, the diameter of which equals the chord c. This is known as the "apparent additional mass." It is as though the mass of the airfoil were increased by this amount. Except in cases of very low relative density $\mu = 2m/\rho S\bar{c}$, this added mass is small compared to that of the airplane itself, and hence the force C_{L_2} is relatively unimportant. Physically, the origin of this force is in the reaction of the air which is associated with its downward acceleration. The other component, C_{L_1}, which acts at the $\frac{1}{4}$ chord point, is associated with the circulation around the airfoil, and is a consequence of the imposition of the Kutta–Joukowski condition at the trailing edge. It is seen that it contains one term proportional to α and another proportional to $\dot{\alpha}$. From Fig. 5.10, the pitching-moment coefficient about the CG is obtained as

$$C_m = C_{L_1}(h - \tfrac{1}{4}) + C_{L_2}(h - \tfrac{1}{2}) \tag{5.5,7}$$

[6]Rodden and Giesing (1970) have extended and generalized this method. In particular they give results for finite wings.

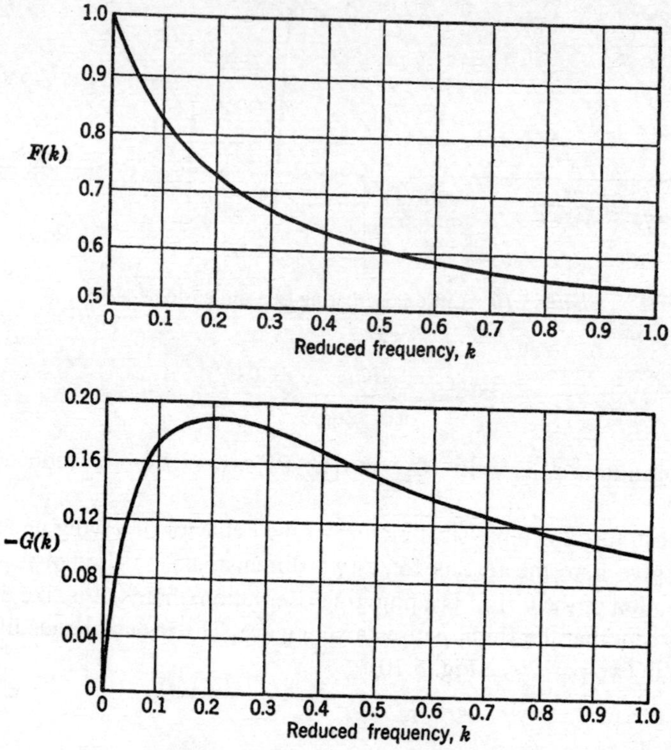

Figure 5.11 The Theodorsen function.

From [(5.5,6) and (5.5,7)], the following derivatives are found for frequency k.

$$C_{L_\alpha} = 2\pi F(k)$$

$$C_{L_{\dot\alpha}} = \pi + 2\pi\frac{G(k)}{k}$$

$$C_{m_\alpha} = 2\pi F(k)(h - \tfrac{1}{2}) \tag{5.5,8}$$

$$C_{m_{\dot\alpha}} = \pi(h - \tfrac{1}{2}) + 2\pi\frac{G(k)}{k}(h - \tfrac{1}{4})$$

The awkward situation is evident, from (5.5.8), that the derivatives are frequency-dependent. That is, in free oscillations one does not know the value of the derivative until the solution to the motion (i.e. the frequency) is known. In cases of forced oscillations at a given frequency, this difficulty is not present.

When dealing with the rigid-body motions of flight vehicles, the characteristic nondimensional frequencies k are usually small, $k \ll 1$. Hence it is reasonable to use the $F(k)$ and $G(k)$ corresponding to $k \to 0$. For the two-dimensional incompressible case described above, $\lim_{k\to 0} F(k) = 1$, so that $C_{L_\alpha} = 2\pi$ and $C_{m_\alpha} = 2\pi(h - \frac{1}{4})$, the theoretical steady-flow values. This conclusion, that C_{L_α} and C_{m_α} are the quasistatic values, also holds for finite wings at all Mach numbers. The results for $C_{L_{\dot\alpha}}$ and $C_{m_{\dot\alpha}}$ are not so clear, however, since $\lim G(k)/k$ given above is infinite. This singularity is marked for the example of two-dimensional flow given above, but is not evident for finite wings at moderate aspect ratio. Miles (1950) indicates that the $k \log k$ term responsible for the singularity is not significant for aspect ratios less than 10, and the numerical calculations of Rodden and Giesing (1970) show no difficulty at values of k as low as 0.001. Filotas' (1971) solutions for finite wings bear out Miles' contention. Thus for finite wings definite values of $C_{L_{\dot{\alpha}}}$ and $C_{m_{\dot{\alpha}}}$ can be associated with small but nonvanishing values of k. The limiting values described above can be obtained from a first-order-in-frequency analysis of an oscillating wing. To summarize, *the $\dot{\alpha}$ derivatives of a wing alone may be computed from the indicial response of lift and pitching moment, or from first-order-in-frequency analysis of harmonically plunging wings.*

CONTRIBUTIONS OF A TAIL

There is an approximate method for evaluating the contributions of a tail surface, which is satisfactory in many cases. This is based on the concept of the *lag of the downwash.* It neglects entirely the nonstationary character of the lift response of the tail to changes in tail angle of attack, and attributes the result entirely to the fact that the downwash at the tail does not respond instantaneously to changes in wing angle of attack. The downwash is assumed to be dependent primarily on the strength of the wing's trailing vortices in the neighborhood of the tail. Since the vorticity is convected with the stream, then a change in the circulation at the wing will not be felt as a change in downwash at the tail until a time $\Delta t = l_t/u_0$ has elapsed, where l_t is the tail length. It is therefore assumed that the instantaneous downwash at the tail, $\epsilon(t)$, corresponds to the wing α at time $(t - \Delta t)$. The corrections to the quasistatic downwash and tail angle of attack are therefore

$$\begin{aligned}\Delta\epsilon &= -\frac{\partial\epsilon}{\partial\alpha}\dot{\alpha}\,\Delta t = -\frac{\partial\epsilon}{\partial\alpha}\dot{\alpha}\frac{l_t}{u_0}\\ &= -\Delta\alpha_t\end{aligned} \tag{5.5,9}$$

$C_{z_{\dot{\alpha}}}$ of a Tail

The correction to the tail lift coefficient for the downwash lag is

$$\Delta C_{L_t} = a_t\Delta\alpha_t = a_t\dot{\alpha}\frac{l_t}{u_0}\frac{\partial\epsilon}{\partial\alpha} \tag{5.5,10}$$

The correction to the airplane lift is therefore

$$\Delta C_L = a_t\dot{\alpha}\frac{l_t}{u_0}\frac{\partial\epsilon}{\partial\alpha}\frac{S_t}{S}$$

Therefore

$$\frac{\partial C_z}{\partial\dot{\alpha}} = -\frac{\partial C_L}{\partial\dot{\alpha}} = -a_t\frac{\partial\epsilon}{\partial\alpha}\frac{l_t}{u_0}\frac{S_t}{S}$$

and

$$(C_{z_{\dot{\alpha}}})_{\text{tail}} = \frac{\partial C_z}{\partial\left(\dfrac{\dot{\alpha}\bar{c}}{2u_0}\right)} = -2a_tV_H\frac{\partial\epsilon}{\partial\alpha} \tag{5.5,11}$$

$C_{m_{\dot{\alpha}}}$ of a Tail

The correction to the pitching moment is obtained from ΔC_{L_t} as

$$\Delta C_m = -V_H \, \Delta C_{L_t} = -a_T \dot{\alpha} \frac{\partial \epsilon}{\partial \alpha} \frac{l_t}{u_0} V_H$$

Therefore

$$\frac{\partial C_m}{\partial \dot{\alpha}} = -a_t V_H \frac{l_t}{u_0} \frac{\partial \epsilon}{\partial \alpha}$$

and

$$(C_{m_{\dot{\alpha}}})_{\text{tail}} = \frac{\partial C_z}{\partial \left(\dfrac{\dot{\alpha} \bar{c}}{2u_0} \right)} = -2a_t V_H \frac{l_t}{\bar{c}} \frac{\partial \epsilon}{\partial \alpha} \tag{5.5,12}$$

5.6 *The β Derivatives (C_{y_β}, C_{l_β}, C_{n_β})*

These derivatives all are obtainable from wind-tunnel tests on yawed models (Campbell and McKinney, 1952). Generally speaking, estimation methods do not give completely reliable results, and testing is a necessity.

THE DERIVATIVE C_{y_β}

This is the side-force derivative, giving the force that acts in the y direction (right) when the airplane has a positive β or v (i.e., a sideslip to the right, see Fig. 3.11). C_{y_β} is usually negative, and frequently small enough to be neglected entirely. The main contributions are those of the body and the vertical tail, although the wing, and wing-body interference, may modify it significantly. Of these, only the tail effect is readily estimated. It may be expressed in terms of the vertical-tail lift-curve slope and the sidewash factor (see Sec. 3.9). (In this and the following sections the fin velocity ratio V_F/V is assumed to be unity.)

$$(C_y)_{\text{tail}} = -a_F(\beta - \sigma) \frac{S_F}{S}$$

or

$$(C_{y_\beta})_{\text{tail}} = -a_F \left(1 - \frac{\partial \sigma}{\partial \beta} \right) \frac{S_F}{S} \tag{5.6,1}$$

The most troublesome component of this equation is the sidewash derivative $\partial\sigma/\partial\beta$, which is difficult to estimate because of its dependence on the wing and fuselage geometry (see Sec. 3.9).

THE DERIVATIVE C_{l_β}

C_{l_β} is the dihedral effect, which was discussed at some length in Sec. 3.12.

THE DERIVATIVE C_{n_β}

C_{n_β} is the weathercock stability derivative, dealt with in Sec. 3.9.

5.7 The p Derivatives (C_{y_p}, C_{l_p}, C_{n_p})

When an airplane rolls with angular velocity p about its x axis (the flight direction), its motion is instantaneously like that of a screw. This motion affects the airflow (local angle of attack) at all stations of the wing and tail surfaces. This is illustrated in Fig. 5.12 for two points: a wing tip and the fin tip. It should be noted that the nondimensional rate of roll, $\hat{p} = pb/2u_0$ is, for small p, the angle (in radians) of the helix traced by the wing tip. These angle of attack changes bring about alterations in the aerodynamic load distribution over the surfaces, and thereby introduce perturbations in the forces and moments. The change in the wing load distribution also causes a modification to the trailing vortex sheet. The vorticity distribution in it is no longer symmetrical about the x axis, and a sidewash (positive, i.e., to the right) is induced at a vertical tail conventionally placed. This further modifies the angle-of-attack distribution on the vertical-tail surface. This sidewash due to rolling is characterized by the derivative $\partial\sigma/\partial\hat{p}$. It has been studied theoretically and experimentally by Michael (1952), who has shown its importance in relation to correct estimation of the tail contributions to the rolling derivatives. Finally, the helical motion of the wing produces a trailing vortex sheet that is not flat, but helical. For the small rates of roll admissible in a linear theory, this effect may be neglected with respect to both wing and tail forces.

THE DERIVATIVE C_{y_p}

The side force due to rolling is often negligible. When it is not, the contributions that need to be considered are those from the wing[7] and from the vertical tail. The vertical-tail effect may be estimated in the light of its angle-of-attack change (see Fig. 5.12) as follows. Let the mean change in α_F (see Fig. 3.12) due to the rolling velocity be

$$\Delta\alpha_F = -\frac{pz_F}{u_0} + p\frac{\partial\sigma}{\partial p}$$

where z_F is an appropriate mean height of the fin. Introducing the nondimensional rate of roll, we may rewrite this as

$$\Delta\alpha_F = -\hat{p}\left(2\frac{z_F}{b} - \frac{\partial\sigma}{\partial\hat{p}}\right) \tag{5.7,1}$$

The incremental side-force coefficient on the fin is obtained from $\Delta\alpha_F$,

$$\Delta C_{y_F} = a_F\,\Delta\alpha_F = -a_F\hat{p}\left(2\frac{z_F}{b} - \frac{\partial\sigma}{\partial\hat{p}}\right) \tag{5.7,2}$$

[7]For the effect of the wing at low speeds, see Campbell and McKinney (1952).

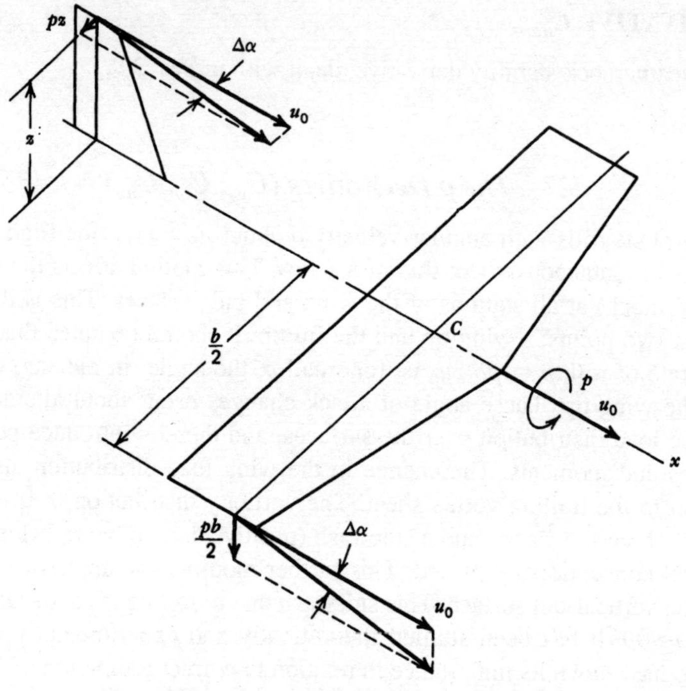

Figure 5.12 Angle of attack changes due to p.

where a_F is the lift-curve slope of the vertical tail. The incremental side force on the airplane is then given by

$$\Delta C_y = \frac{S_F}{S}\Delta C_{y_F} = -a_F\hat{p}\frac{S_F}{S}\left(2\frac{z_F}{b} - \frac{\partial\sigma}{\partial\hat{p}}\right)$$

thus

$$(C_{y_p})_{\text{tail}} = -a_F\frac{S_F}{S}\left(2\frac{z_F}{b} - \frac{\partial\sigma}{\partial\hat{p}}\right) \tag{5.7,3}$$

THE DERIVATIVE C_{l_p}

C_{l_p} is known as the *damping-in-roll* derivative. It expresses the resistance of the airplane to rolling. Except in unusual circumstances, only the wing contributes significantly to this derivative. As can be seen from Fig. 5.12, the angle of attack due to p varies linearly across the span, from the value $pb/2u_0$ at the right wing tip to $-pb/2u_0$ at the left tip. This antisymmetric α distribution produces an anti-symmetric increment in the lift distribution as shown in Fig. 5.13. In the linear range this is superimposed on the symmetric lift distribution associated with the wing angle of attack in undisturbed flight. The large rolling moment L produced by this lift distribution is proportional to the tip angle of attack $\hat{p}$ (see Fig. 5.12), and C_{l_p} is a negative constant, so long as the local angle of attack remains below the local stalling angle.

If the wing angle of attack at the center line, $\alpha_w(0)$, is large, then the incremental value due to p may take some sections of the wing beyond the stalling angle, as

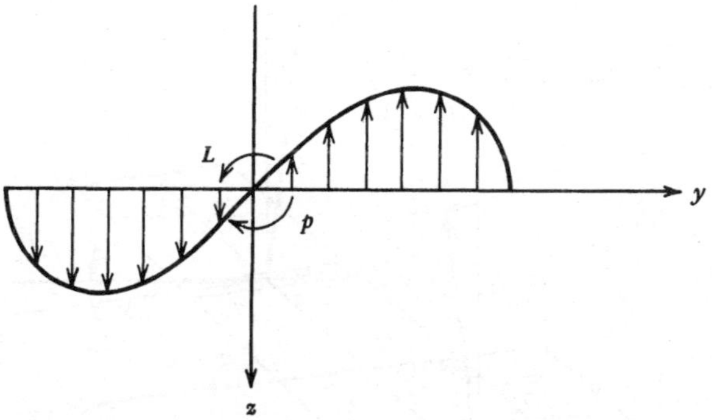

Figure 5.13 Spanwise lift distribution due to rolling.

shown in Fig. 5.14. [Actually, for finite span wings, there is an additional induced angle of attack distribution $\alpha_i(y)$ due to the vortex wake that modifies the net sectional value still further. We neglect this correction here in the interest of making the main point.] When this happens $|C_{l_p}\hat{p}|$ is reduced in magnitude from the linear value and if $\alpha_w(0)$ is large enough, will even change sign. When this happens, the wing will autorotate, the main characteristic of spinning flight.

THE DERIVATIVE C_{n_p}

The yawing moment produced by the rolling motion is one of the so-called *cross derivatives.* It is the existence of these cross derivatives that causes the rolling and yawing motions to be so closely coupled. The wing and tail both contribute to C_{n_p}.

The wing contribution is in two parts. The first comes from the change in profile drag associated with the change in wing angle of attack. The wing α is increased on the right-hand side and decreased on the left-hand side. These changes will normally be accompanied by an increase in profile drag on the right side, and a decrease on the

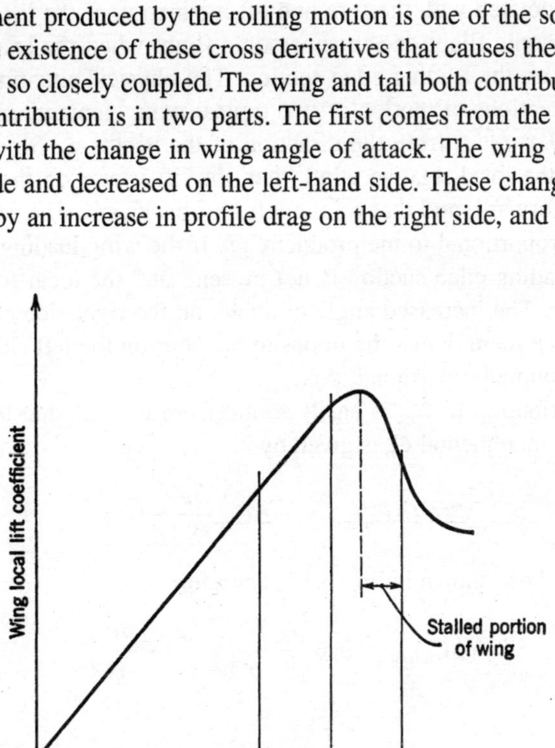

Figure 5.14 Reduction of C_{l_p} due to wing stall.

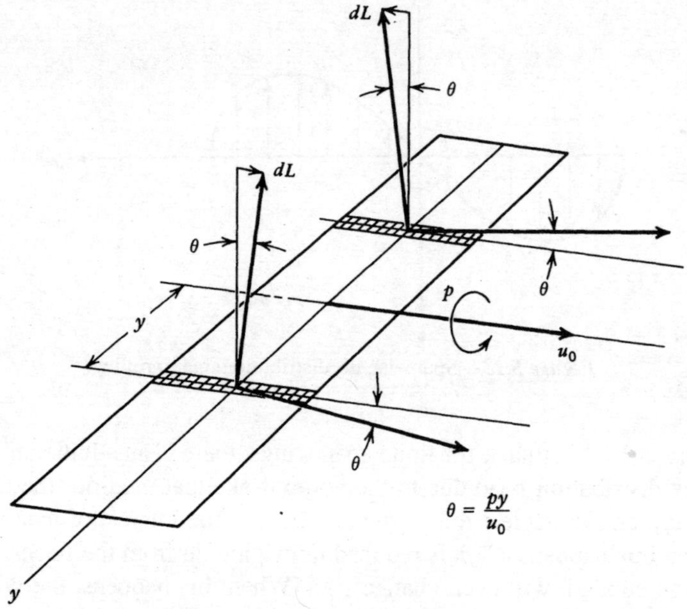

Figure 5.15 Inclination of C_L vector due to rolling.

left side, combining to produce a *positive* (nose-right) yawing moment. The second wing effect is associated with the fore-and-aft inclination of the lift vector caused by the rolling in subsonic flight and in supersonic flight when the leading edge is subsonic. It depends on the leading-edge suction. The physical situation is illustrated in Fig. 5.15. The directions of motion of two typical wing elements are shown inclined by the angles $\pm\theta = py/u_0$ from the direction of the vector u_0. Since the local lift is perpendicular to the local relative wind, then the lift vector on the right half of the wing is inclined forward, and that on the left half backward. The result is a *negative* yawing couple, proportional to the product $C_L\hat{p}$. If the wing leading edges are supersonic, then the leading-edge suction is not present, and the local force remains normal to the surface. The increased angle of attack on the right side causes an increase in this normal force there, while the opposite happens on the left side. The result is a positive yawing couple proportional to $\hat{p}$.

The tail contribution to C_{n_p} is easily found from the tail side force given previously (5.7,2). The incremental C_n is given by

$$(\Delta C_n)_{\text{tail}} = -\Delta C_{y_F}\frac{S_F}{S}\frac{l_F}{b}$$

where l_F is the distance shown in Fig. 3.12. Therefore

$$(\Delta C_n)_{\text{tail}} = a_F\hat{p}\frac{S_F}{S}\frac{l_F}{b}\left(2\frac{z_F}{b} - \frac{\partial\sigma}{\partial\hat{p}}\right)$$

and (5.7,4)

$$(C_{n_p})_{\text{tail}} = a_F V_V\left(2\frac{z_F}{b} - \frac{\partial\sigma}{\partial\hat{p}}\right)$$

where V_V is the vertical-tail volume ratio.

5.8 The r Derivatives (C_{y_r}, C_{l_r}, C_{n_r})

When an airplane has a rate of yaw r superimposed on the foward motion u_0, its velocity field is altered significantly. This is illustrated for the wing and vertical tail in Fig. 5.16. The situation on the wing is clearly very complicated when it has much sweepback. The main feature however, is that the velocity of the $\frac{1}{4}$ chord line normal to itself is increased by the yawing on the left-hand side, and decreased on the right side. The aerodynamic forces at each section (lift, drag, moment) are therefore increased on the left-hand side, and decreased on the right-hand side. As in the case of the rolling wing, the unsymmetrical lift distribution leads to an unsymmetrical trailing vortex sheet, and hence a sidewash at the tail. The incremental tail angle of attack is then

$$\Delta\alpha_F = \frac{rl_F}{u_0} + r\frac{\partial\sigma}{dr}$$

or

$$\Delta\alpha_F = \hat{r}\left(2\frac{l_F}{b} + \frac{\partial\sigma}{\partial\hat{r}}\right) \tag{5.8,1}$$

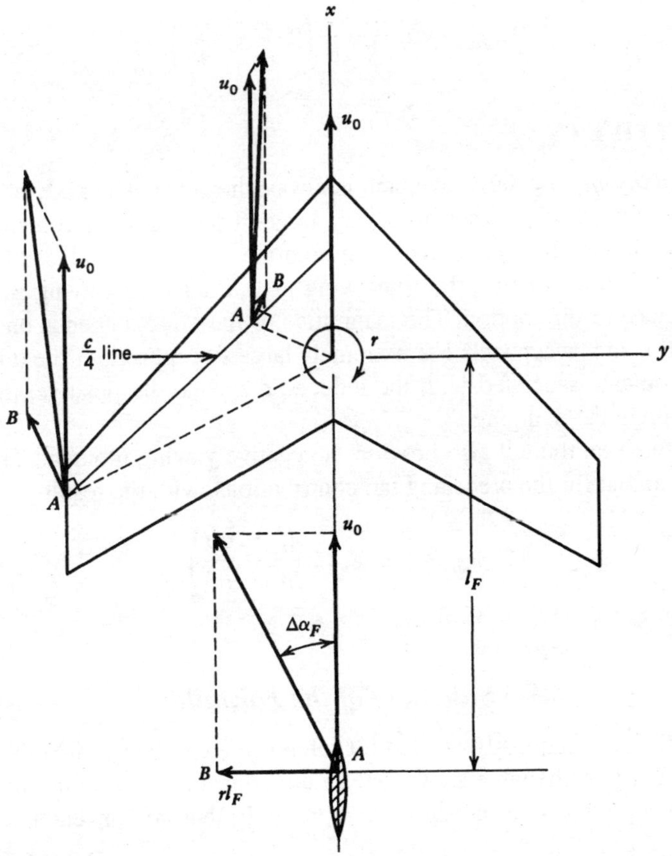

Figure 5.16 Velocity field due to yawing. $\vec{AB}$ = velocity vector due to rate of yaw r.

THE DERIVATIVE C_{y_r}

The only contribution to C_{y_r} that is normally important is that of the tail. From the angle of attack change we find the incremental C_y to be

$$(\Delta C_y)_{\text{tail}} = a_F \hat{r} \frac{S_F}{S}\left(2\frac{l_F}{b} + \frac{\partial\sigma}{\partial\hat{r}}\right)$$

thus

$$(C_{y_r})_{\text{tail}} = a_F \frac{S_F}{S}\left(2\frac{l_F}{b} + \frac{\partial\sigma}{\partial\hat{r}}\right) \tag{5.8,2}$$

THE DERIVATIVE C_{l_r}

This is another important cross derivative; the rolling moment due to yawing. The increase in lift on the left wing, and the decrease on the right wing combine to produce a positive rolling moment proportional to the original lift coefficient C_L. Hence this derivative is largest at low speed. Aspect ratio, taper ratio, and sweepback are all important parameters.

When the vertical tail is large, its contribution may be significant. A formula for it can be derived in the same way as for the previous tail contributions, with the result

$$(C_{l_r})_{\text{tail}} = a_F \frac{S_F}{S}\frac{z_F}{b}\left(2\frac{l_F}{b} + \frac{\partial\sigma}{\partial\hat{r}}\right) \tag{5.8,3}$$

THE DERIVATIVE C_{n_r}

C_{n_r} is the *damping-in-yaw* derivative, and is always negative. The body adds a negligible amount to C_{n_r} except when it is very large. The important contributions are those of the wing and tail. The increases in both the profile and induced drag on the left wing and the decreases on the right wing give a negative yawing moment and hence a resistance to the motion. The magnitude of the effect depends on the aspect ratio, taper ratio, and sweepback. For extremely large sweepback, of the order of 60°, the yawing moment associated with the induced drag may be positive; that is, produce a reduction in the damping.

The side force on the tail also provides a negative yawing moment. The calculation is similar to that for the preceding tail contributions, with the result

$$(C_{n_r})_{\text{tail}} = -a_F V_V\left(2\frac{l_F}{b} + \frac{\partial\sigma}{\partial\hat{r}}\right) \tag{5.8,4}$$

5.9 Summary of the Formulas

The formulas that are frequently wanted for reference are collected in Tables 5.1 and 5.2. Where an entry in the table shows only a tail contribution, it is not implied that the wing and body effects are not important, but only that no convenient formula is available.

Table 5.1
Summary—Longitudinal Derivatives

	C_x	C_z	C_m
$\hat{u}$†	$\mathbf{M}_0\left(\dfrac{\partial C_T}{\partial \mathbf{M}} - \dfrac{\partial C_D}{\partial \mathbf{M}}\right) - \rho u_0^2 \dfrac{\partial C_D}{\partial p_d} + C_{T_u}\left(1 - \dfrac{\partial C_D}{\partial C_T}\right)$	$-\mathbf{M}_0 \dfrac{\partial C_L}{\partial \mathbf{M}} - \rho u_0^2 \dfrac{\partial C_L}{\partial p_d} - C_{T_u} \dfrac{\partial C_L}{\partial C_T}$	$\mathbf{M}_0 \dfrac{\partial C_m}{\partial \mathbf{M}} + \rho u_0^2 \dfrac{\partial C_m}{\partial p_d} + C_{T_u} \dfrac{\partial C_m}{\partial C_T}$
α	$C_{l_0} - C_{D_\alpha}$	$-(C_{L_\alpha} + C_{D_0})$	$-a(h_n - h)$
$\dot{\alpha}$	Neg.	$*-2a_t V_H \dfrac{\partial \epsilon}{\partial \alpha}$	$*-2a_t V_H \dfrac{l_t}{\bar{c}} \dfrac{\partial \epsilon}{\partial \alpha}$
$\hat{q}$	Neg.	$*-2a_t V_H$	$*-2a_t V_H \dfrac{l_t}{\bar{c}}$

Neg. means usually negligible.
*means contribution of the *tail only,* formula for wing-body not available.

$$\dagger C_{T_u} = \frac{(\partial T/\partial u)_0}{\frac{1}{2}\rho u_0 S} - 2C_{T_0};\ C_{T_0} = C_{D_0} + C_{w_0} \sin \theta_0$$

Table 5.2
Summary—Lateral Derivatives

	C_y	C_l	C_n
β	$*-a_F\dfrac{S_F}{S}\left(1-\dfrac{\partial\sigma}{\partial\beta}\right)$	N.A.	$*a_FV_V\left(1-\dfrac{\partial\sigma}{\partial\beta}\right)$
$\hat{p}$	$*-a_F\dfrac{S_F}{S}\left(2\dfrac{z_F}{b}-\dfrac{\partial\sigma}{\partial\hat{p}}\right)$	N.A.	$*a_FV_V\left(2\dfrac{z_F}{b}-\dfrac{\partial\sigma}{\partial\hat{p}}\right)$
$\hat{r}$	$*a_F\dfrac{S_F}{S}\left(2\dfrac{l_F}{b}+\dfrac{\partial\sigma}{\partial\hat{r}}\right)$	$*a_F\dfrac{S_F}{S}\dfrac{z_F}{b}\left(2\dfrac{l_F}{b}+\dfrac{\partial\sigma}{\partial\hat{r}}\right)$	$*-a_FV_V\left(2\dfrac{l_F}{b}+\dfrac{\partial\sigma}{\partial\hat{r}}\right)$

*means contribution of the *tail only,* formula for wing-body not available; $V_F/V = 1$.
N.A. means no formula available.

5.10 Aeroelastic Derivatives

In Sec. 4.12 there were introduced aerodynamic derivatives associated with the deformations of the airplane. These are of two kinds: those that appear in the rigid-body equations and those that appear in the added equations of the elastic degrees of freedom. These are illustrated in this section by consideration of the hypothetical vibration mode shown in Fig. 5.17. In this mode it is assumed that the fuselage and tail are rigid, and have a motion of vertical translation only. The flexibility is all in the wing, and it bends without twisting. The functions describing the mode (4.11,1) are therefore:

$$\begin{aligned} x' &= 0 \\ y' &= 0 \\ z' &= h(y)z_T \end{aligned} \tag{5.10,1}$$

For the generalized coordinate, we have used the wing-tip deflection z_T. $h(y)$ is then a normalized function describing the wing bending mode.

Since the elastic degrees of freedom are only important in relation to stability and control when their frequencies are relatively low, approaching those of the rigid-body modes, then it is reasonable to use the same approximation for the aerodynamic forces as is used in calculating stability derivatives. That is, if quasisteady flow theory is adequate for the aerodynamic forces associated with the rigid-body motions, then we may use the same theory for the elastic motions.

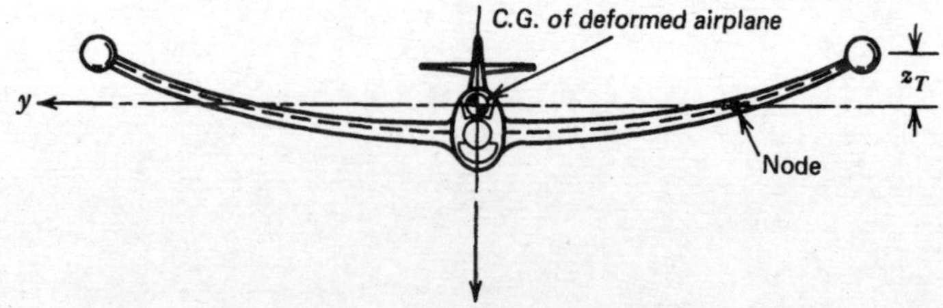

Figure 5.17 Symmetrical wing bending.

In the example chosen, we assume that the only significant forces are those on the wing and tail, and that these are to be computed from quasisteady flow theory. In the light of these assumptions, some of the representative derivatives of both types are discussed below. As a preliminary, the forces induced on the wing and tail by the elastic motion are treated first.

FORCES ON THE WING

The vertical velocity of the wing section distant y from the center line is

$$\dot{z} = h(y)\dot{z}_T \tag{5.10,2}$$

and the corresponding change in wing angle of attack is

$$\Delta\alpha(y) = h(y)\dot{z}_T/u_0 \tag{5.10,3}$$

This angle of attack distribution can be used with any applicable steady-flow wing theory to calculate the incremental local section lift. (It will of course be proportional to $\dot{z}_T/u_0$.) Let it be denoted in coefficient form by $C_l'(y)\dot{z}_T/u_0$, and the corresponding increment in wing total lift coefficient by $C_{L_w}'\dot{z}_T/u_0$. $C_l'(y)$ and C_{L_w}' are thus the values corresponding to unit value of the nondimensional quantity $\dot{z}_T/u_0$.

FORCE ON THE TAIL

The tail experiences a downward velocity $h(0)\dot{z}_T$, and also, because of the altered wing lift distribution, a downwash change $(\partial\epsilon/\partial\dot{z}_T)\dot{z}_T$. Hence the net change in tail angle of attack is

$$\begin{aligned}\Delta\alpha_t &= h(0)\dot{z}_T/u_0 - \frac{\partial\epsilon}{\partial\dot{z}_T}\dot{z}_T \\ &= \left[h(0) - \frac{\partial\epsilon}{\partial(\dot{z}_T/u_0)}\right]\frac{\dot{z}_T}{u_0}.\end{aligned}$$

This produces an increment in the tail lift coefficient of amount

$$\Delta C_{L_t} = a_t\left[h(0) - \frac{\partial\epsilon}{\partial(\dot{z}_T/u_0)}\right]\frac{\dot{z}_T}{u_0} \tag{5.10,4}$$

THE DERIVATIVE $Z_{\dot{z}_T}$

This derivative describes the contribution of wing bending velocity to the Z force acting on the airplane. A suitable nondimensional form is $\partial C_z/\partial(\dot{z}_T/u_0)$. Since $C_z = -C_L$, we have that

$$\Delta C_z = -C_{L_w}'\frac{\dot{z}_T}{u_0} - a_t\left[h(0) - \frac{\partial\epsilon}{\partial(\dot{z}_T/u_0)}\right]\frac{\dot{z}_T}{u_0}$$

and hence

$$\frac{\partial C_z}{\partial(\dot{z}_T/u_0)} = -C_{L_w}' - a_t\left[h(0) - \frac{\partial\epsilon}{\partial(\dot{z}_T/u_0)}\right] \tag{5.10,5}$$

THE DERIVATIVE A_{n_w}

This derivative [see (4.12,12)] represents the contribution to the generalized force in the bending degree of freedom, associated with a change in the w velocity of the airplane. A suitable nondimensional form is obtained by defining

$$C_{\mathscr{F}} = \frac{\mathscr{F}}{\frac{1}{2}\rho u_0^2 S}$$

and using α_x in place of w ($w = u_0\alpha_x$). Then the appropriate nondimensional derivative is $C_{\mathscr{F}_\alpha}$.

Let the wing lift distribution due to a perturbation α in the angle of attack (constant across the span) be given by $C_{l_\alpha}(y)\alpha$. Then in a virtual displacement in the wing bending mode δz_T, the work done by this wing loading is

$$\delta W = -\int_{-b/2}^{b/2} \alpha C_{l_\alpha}(y) h(y)\ \delta z_T \frac{1}{2}\rho u_0^2 c(y)\ dy$$

where $c(y)$ is the local wing chord. The corresponding contribution to $\mathscr{F}$ is

$$\frac{\delta W}{\partial z_T} = -\alpha \frac{1}{2}\rho u_0^2 \int_{-b/2}^{b/2} C_{l_\alpha}(y) h(y) c(y)\ dy$$

and to $C_{\mathscr{F}_\alpha}$ is

$$\frac{1}{\frac{1}{2}\rho u_0^2 S}\frac{\partial^2 W}{\partial z_T \partial\alpha} = -\frac{1}{S}\int_{-b/2}^{b/2} C_{l_\alpha}(y) h(y) c(y)\ dy \tag{5.10,6}$$

The tail also contributes to this derivative, for the tail lift associated with α is

$$a_t\alpha\left(1 - \frac{\partial\epsilon}{\partial\alpha}\right)\frac{1}{2}\rho u_0^2 S_t$$

and the work done by this force during the virtual displacement is

$$-a_t\alpha\left(1 - \frac{\partial\epsilon}{\partial\alpha}\right)\frac{1}{2}\rho u_0^2 S_t h(0)\ \delta z_T$$

Therefore the contribution to $C_{\mathscr{F}}$ is

$$-a_t\alpha\left(1 - \frac{\partial\epsilon}{\partial\alpha}\right)\frac{S_t}{S}h(0)$$

and to $C_{\mathscr{F}_\alpha}$ is

$$-a_t\frac{S_t}{S}h(0)\left(1 - \frac{\partial\epsilon}{\partial\alpha}\right) \tag{5.10,7}$$

The total value of $C_{\mathscr{F}_\alpha}$ is then the sum of 5.10,6 and 5.10,7.

THE DERIVATIVE b_{11} (see 4.12,12)

This derivative identifies the contribution of $\dot{z}_T$ to the generalized aerodynamic force in the distortion degree of freedom. We have defined the associated wing load distribution above by the local lift coefficient $C_l'(y)\dot{z}_T/u_0$. As in the case of the derivative A_{nw} above, the work done by this loading is calculated, with the result that the wing contributes

$$\frac{\partial C_{\mathcal{F}}}{\partial(\dot{z}_T/u_0)} = \frac{1}{\frac{1}{2}\rho u_0^2 S}\frac{\partial^2 W}{\partial z_T \partial(\dot{z}_T/u_0)} = -\frac{1}{S}\int_{-b/2}^{b/2} C_l'(y)h(y)c(y)\,dy \qquad (5.10,8)$$

Likewise, the contribution of the tail is calculated here as for A_{nw}, and is found to be

$$-a_t\frac{S_t}{S}h(0)\left[h(0) - \frac{\partial\epsilon}{\partial(\dot{z}_T/u_0)}\right] \qquad (5.10,9)$$

The total value of $\partial C_{\mathcal{F}}/\partial(\dot{z}_T/u_0)$ is then the sum of 5.10,8 and 5.10,9.

5.11 Exercises

5.1 The derivative C_{z_u} contains the term $\mathbf{M}_0\left(\frac{\partial C_L}{\partial \mathbf{M}}\right)_0$. Estimate the magnitude of this term for an airplane with wing loading 70 psf (3,352 Pa) flying at 20,000 ft (6,096 m) altitude, for Mach numbers between 0.2 and 0.8. The following data pertain to the wing:

Sweep ($\frac{1}{4}$ chord) $\quad \Lambda = 30°$

$\quad S = 5{,}500$ ft^2 (511.0 m^2)

Plot the result vs. $\mathbf{M}_0$. Calculate the contribution this term makes to Z_u and plot this as well. (Compare with Z_u for the B747 from Table 6.2, and comment).

5.2 A wind-tunnel model is mounted with one degree of freedom–pivoted so that it can only rotate about the y-axis of the body frame, which is perpendicular to the relative wind. It is elastically restrained with a pitching moment $M = -k\theta$. Show how the sum $(C_{m_q} + C_{m_{\dot{\alpha}}})$ can be estimated from experiments in which the model is free to oscillate in pitch with wind on and off. Assume M_q can be neglected with the wind off and Z_q and $Z_{\dot{w}}$ can be neglected.

5.3 Consider the wind/fin system of Fig. 5.16, with the following properties:

Wing: $\quad A = 5$; $\lambda = 0.5$; $\Lambda_{1/4} = 30°$; Γ variable

Fin: $\quad a_F = 3.5$ rad^{-1}; $l_F/b = 0.5$; $z_F/b = 0.1$; V_V variable; $\partial\sigma/\partial\beta$ negligible.

Estimate values of the stability derivatives (for $h_{n_w} = h$ and $L/D = 12$)

$$C_{l_\beta}, C_{l_r}, C_{n_\beta}, C_{n_r}$$

at $C_{L_0} = 1.0$. Plot the spiral stability boundary for horizontal flight:

$$E = C_{l_\beta}C_{n_r} - C_{l_r}C_{n_\beta} = 0$$

[see (6.8,6) with $\theta_0 = 0$] in the plane of V_V vs. Γ. (Make any reasonable assumptions you need to supplement the given data).

5.4 A jet airplane has a thrust line that passes above the CG by a distance equal to 10% of the M.A.C. With the assumption $\partial T/\partial u = 0$, estimate the increment thus caused in C_{m_u}.

5.5 Find C_{n_p} due to the tilting of the lift vector for a wing with an elliptic lift distribution $\left(\text{i.e., a wing with lift per unit span } l(y) \text{ which obeys } \frac{l^2}{a^2} + \frac{4y^2}{b^2} = 1\right)$. Assume that the tilt angle is small. Express C_{n_p} in terms of C_L, the lift coefficient of the wing when it is not rolling.

5.6 Assume that Figs. 5.7 and 5.8 are experimental measurements. Select an analytic function $C_{L_{\text{step}}}(t)$ that can represent Fig. 5.7 ($\mathbf{M} = 0$ case). Find the corresponding transfer function relating C_L to α. Use this transfer function to generate a function of time corresponding to Fig. 5.8*b* and demonstrate that it has the desired form.

5.12 Additional Symbols Introduced in Chapter 5

e	efficiency factor for wing-induced drag
k	reduced frequency, $\omega\bar{c}/2u_0$
p_d	dynamic pressure, $\frac{1}{2}\rho V^2$
ω	circular frequency

See also Secs. 2.11, 3.15, and 4.14.

CHAPTER 6

Stability of Uncontrolled Motion

The preceding chapters have been to some extent simply the preparation for what follows in this and the succeeding two chapters, that is, a treatment of the uncontrolled and controlled motions of an airplane. The system model was developed in Chap. 4, and the aerodynamic ingredients were described in Chaps. 2, 3, and 5. In this chapter we tackle first the simplest of these cases, the uncontrolled motion, that is, the motion when all the controls are locked in position. An airplane in steady flight may be subjected to a momentary disturbance by a nonuniform or nonstationary atmosphere, or by movements of passengers, release of stores, and so forth. In this circumstance some of the questions to be answered are, "What is the character of the motion following the cessation of the disturbance? Does it subside or increase? If it subsides what is the final flight path?" The stability of small disturbances from steady flight is an extremely important property of aircraft—first, because steady flight conditions make up most of the flight time of airplanes, and second, because the disturbances in this condition *must be small* for a satisfactory vehicle. If they were not it would be unacceptable for either commercial or military use. The required dynamic behavior is ensured by design—by making the small-disturbance properties of concern (*the natural modes*) such that either human or automatic control can keep the disturbances to an acceptably small level. Finally the small-disturbance model is actually valid for disturbance magnitudes that seem quite violent to human occupants.

6.1 Form of Solution of Small-Disturbance Equations

The small-disturbance equations are (4.9,18 and 4.9,19). They are both of the form

$$\dot{\mathbf{x}} = \mathbf{A}\mathbf{x} + \Delta\mathbf{f}_c \tag{6.1,1}$$

where $\mathbf{x}$ is the $(N \times 1)$ *state vector,* $\mathbf{A}$ is the $(N \times N)$ *system matrix,* a constant, and $\Delta\mathbf{f}_c$ is the $(N \times 1)$ vector of incremental control forces and moments. In this application, the control force vector is zero, so the equation to be studied is

$$\dot{\mathbf{x}} = \mathbf{A}\mathbf{x} \tag{6.1,2}$$

Solutions of this first-order differential equation are well known. They are of the form

$$\mathbf{x}(t) = \mathbf{x}_0 e^{\lambda t} \tag{6.1,3}$$

$\mathbf{x}_0$ is an *eigenvector* and λ is an *eigenvalue* of the system. $\mathbf{x}_0$ is also seen to be the value of the state vector at $t = 0$. Substitution of (6.1,3) into (6.1,2) gives

$$\lambda \mathbf{x}_0 = \mathbf{A}\mathbf{x}_0 \tag{6.1,4}$$

or

$$(\mathbf{A} - \lambda \mathbf{I})\mathbf{x}_0 = \mathbf{0} \tag{6.1,5}$$

where $\mathbf{I}$ is the identity matrix. Since the scalar expansion of (6.1,5) is a system of N homogeneous equations (zeros on the right-hand side) then there is a nonzero solution for $\mathbf{x}_0$ only when the system determinant vanishes, that is, when

$$\det(\mathbf{A} - \lambda \mathbf{I}) = 0 \tag{6.1,6}$$

The determinant in (6.1,6) is the *characteristic determinant* of the system. When expanded, the result is a polynomial in λ of degree N, the *characteristic polynomial*, and the Nth degree algebraic equation (6.1,6) is the *characteristic equation* of the system. Since the equation is of the Nth degree it has in general N roots λ_i, some real and some occurring in conjugate complex pairs. Corresponding to each real eigenvalue λ is a real eigenvector $\mathbf{x}_0$, and to each complex pair λ_i and λ_i^* there corresponds a conjugate complex pair of eigenvectors $\mathbf{x}_0$ and $\mathbf{x}_0^*$. Since any one of the λ's can provide a solution to (6.1,2) and since the equation is linear, the most general solution is a sum of all the corresponding $\mathbf{x}(t)$ of (6.1,3), that is,

$$\mathbf{x}(t) = \sum_i \mathbf{x}_{0_i} e^{\lambda_i t} \tag{6.1,7}$$

Each of the solutions described by (6.1,3) is called a *natural mode*, and the general solution (6.1,7) is a sum of all the modes. A typical variable, say w, would, according to (6.1,7) have the form

$$w(t) = a_1 e^{\lambda_1 t} + a_2 e^{\lambda_2 t} + \cdots \tag{6.1,8}$$

where the a_i would be fixed by the initial conditions. The pair of terms corresponding to a conjugate pair of eigenvalues

$$\lambda = n \pm i\omega \tag{6.1,9}$$

is

$$a_1 e^{(n+i\omega)t} + a_2 e^{(n-i\omega)t} \tag{6.1,10}$$

Upon expanding the exponentials, (6.1,10) becomes

$$e^{nt}(A_1 \cos \omega t + A_2 \sin \omega t) \tag{6.1,11}$$

where $A_1 = (a_1 + a_2)$ and $A_2 = i(a_1 - a_2)$ are always real. That is, (6.1,11) describes an oscillatory mode, of period $T = 2\pi/\omega$, that either grows or decays, depending on the sign of n. The four kinds of mode that can occur, according to whether λ is real or complex, and according to the sign of n are illustrated in Fig. 6.1. The disturbances shown in (a) and (c) increase with time, and hence these are unstable modes. It is conventional to refer to (a) as a static instability or *divergence*, since there is no tendency for the disturbance to diminish. By contrast, (c) is called dynamic instability or a *divergent oscillation*, since the disturbance quantity alternately increases and diminishes, the amplitude growing with time. (b) illustrates a *subsidence* or *convergence*, and (d) a *damped* or *convergent* oscillation. Since in both (b) and (d) the disturbance quantity ultimately vanishes, they represent stable modes.

It is seen that a "yes" or "no" evaluation of the stability is obtained simply from the signs of the real parts of the λs. If there are no positive real parts, there is no instability. This information is not sufficient, however, to evaluate the handling quali-

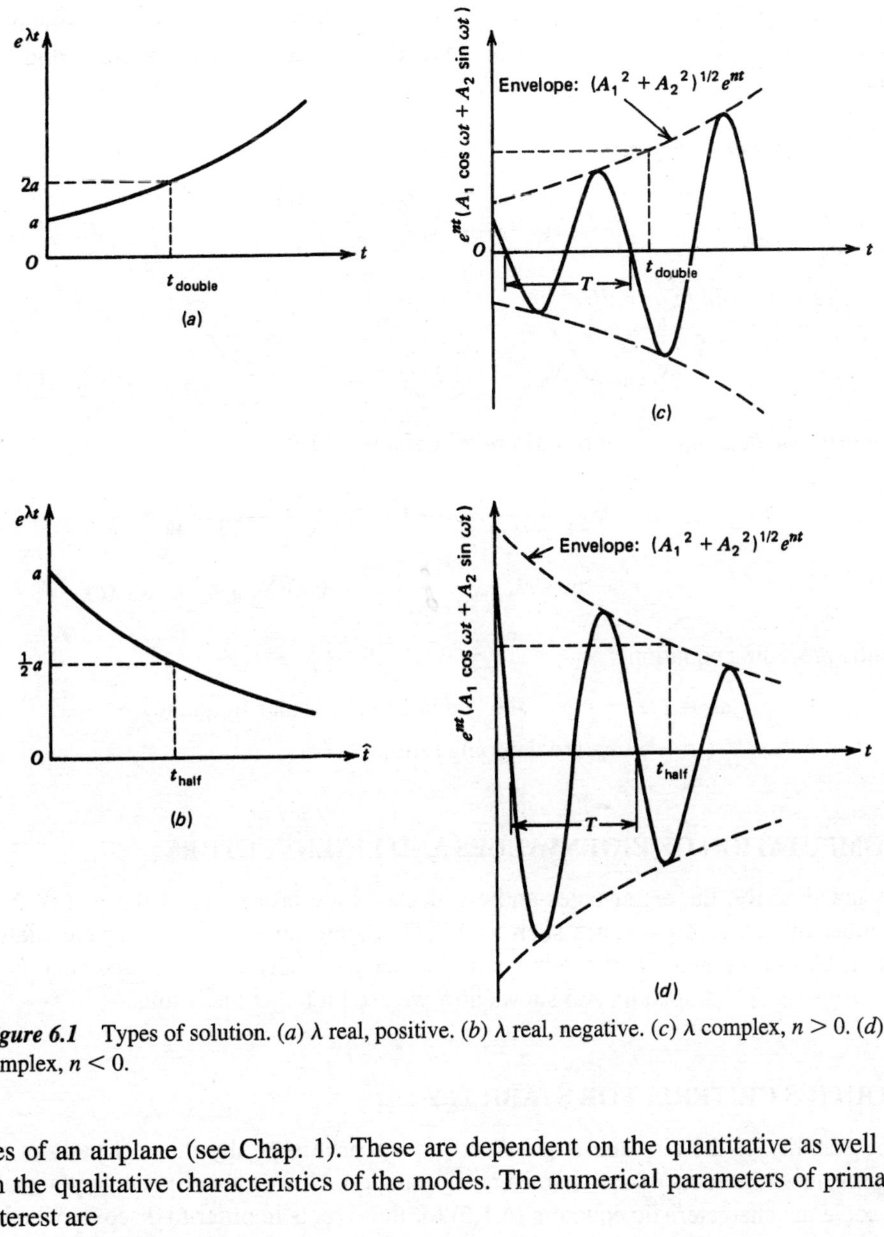

Figure 6.1 Types of solution. (*a*) λ real, positive. (*b*) λ real, negative. (*c*) λ complex, $n > 0$. (*d*) λ complex, $n < 0$.

ties of an airplane (see Chap. 1). These are dependent on the quantitative as well as on the qualitative characteristics of the modes. The numerical parameters of primary interest are

1. Period, $T = \dfrac{2\pi}{\omega}$
2. Time to double or time to half.
3. Cycles to double (N_{double}) or cycles to half (N_{half}).

The first two of these are illustrated in Fig. 6.1. When the roots are real, there is of course no period, and the only parameter is the time to double or half. These are the times that must elapse during which any disturbance quantity will double or halve itself, respectively. When the modes are oscillatory, it is the envelope ordinate that doubles or halves. Since the envelope may be regarded as an amplitude modulation, then we may think of the doubling or halving as applied to the variable amplitude. By noting that $\log_e 2 = -\log_e \frac{1}{2} = 0.693$, the reader will easily verify the following relations:

Time to double or half:

$$t_{\text{double}} \text{ or } t_{\text{half}} = \frac{.693}{|n|} = \frac{.693}{|\zeta|\omega_n} \qquad (a)$$

Cycles to double or half:

$$N_{\text{double}} \text{ or } N_{\text{half}} = .110\frac{\omega}{|n|} = .110\frac{\sqrt{1-\zeta^2}}{|\zeta|} \qquad (b) \quad (6.1,12)$$

Logarithmic decrement (log of ratio of successive peaks):

$$\delta = \log_e \frac{e^{nt}}{e^{n(t+T)}} = -nT = 2\pi\frac{\zeta}{\sqrt{1-\zeta^2}}$$

$$= -.693/N_{\text{double}} \quad \text{or} \quad .693/N_{\text{half}} \qquad (c)$$

In the preceding equations,

$\omega_n = (\omega^2 + n^2)^{1/2}$, the "undamped" circular frequency
$\zeta = -n/\omega_n$, the damping ratio

COMPUTATION OF EIGENVALUES AND EIGENVECTORS

As noted above, the eigenvalues and eigenvectors are properties of the matrix **A**. A number of software packages such as MATLAB are now available for calculating them. For all the numerical examples in this book we have used the *Student Version* of *Program CC*.[1] Appendix A.5 shows how we used it to get the results.

ROUTH'S CRITERIA FOR STABILITY

The stability of the airplane is governed, as we have seen, by the real parts of the eigenvalues, the roots of the characteristic equation. Now it is not necessary actually to solve the characteristic equation (6.1,6) for these roots in order to discover whether there are any unstable ones. E. J. Routh (1905) has derived a criterion that can be applied to the coefficients of the equation to get the desired result. The criterion is that a certain set of test functions shall all be positive (Etkin, 1972). We present below the result for the important case of the quartic equation, which will turn up later in this chapter.

Let the quartic equation be

$$A\lambda^4 + B\lambda^3 + C\lambda^2 + D\lambda + E = 0 \qquad (A > 0) \qquad (6.1,13)$$

Then the test functions are $F_0 = A$, $F_1 = B$, $F_2 = BC - AD$, $F_3 = F_2D - B^2E$, $F_4 =$

[1]Available from Systems Technology Inc. Hawthorne, CA.

F_3BE. The necessary and sufficient conditions for these test functions to be positive are

$$A, B, D, E > 0$$

and

$$R = D(BC - AD) - B^2E > 0 \qquad (6.1,14)$$

It follows that C also must be positive. The quantity on the left-hand side of (6.1,14) is commonly known as *Routh's discriminant.*

Duncan (1952) has shown that the vanishing of E and of R represent significant critical cases. If the airplane is stable, and some design parameter is then varied in such a way as to lead to instability, then the following conditions hold:

1. If only E changes from + to −, then one real root changes from negative to positive; that is, one divergence appears in the solution (see Fig. 6.1*a*).
2. If only R changes from + to −, then the real part of one complex pair of roots changes from negative to positive; that is, one divergent oscillation appears in the solution (see Fig. 6.1*c*).

Thus the conditions $E = 0$ and $R = 0$ define *boundaries* between stability and instability. The former is the boundary between stability and static instability, and the latter is the boundary between stability and a divergent oscillation. These stability boundaries are very useful for certain analytical purposes. We shall in particular make use of the $E = 0$ boundary.

6.2 Longitudinal Modes of a Jet Transport

The foregoing theory is now illustrated by applying it to the Boeing 747 transport. The needed geometrical and aerodynamic data for this airplane are given in Appendix E. The flight condition for this example is cruising in horizontal flight at approximately 40,000 ft at Mach number 0.8. Relevant data are as follows:

$W = 636{,}636$ lb (2.83176×10^6 N)	$S = 5500$ ft^2 (511.0 m^2)
$\bar{c} = 27.31$ ft (8.324 m)	$b = 195.7$ ft (59.64 m)
$I_x = 0.183 \times 10^8$ slug ft^2 (0.247×10^8 kg m^2)	$I_y = 0.331 \times 10^8$ slug ft^2 (0.449×10^8 kg m^2)
$I_z = 0.497 \times 10^8$ slug ft^2 (0.673×10^8 kg m^2)	$I_{zx} = -.156 \times 10^7$ slug ft^2 ($-.212 \times 10^7$ kg m^2)
$u_0 = 774$ fps (235.9 m/s) $\theta_0 = 0$	$\rho = 0.0005909$ slug/ft^3 (0.3045 kg/m^3)
$C_{L_0} = 0.654$	$C_{D_0} = 0.0430$

The preceding four inertias are for stability axes at the stated flight condition. In the numerical examples of this and the following two chapters, the system matrices and the solutions are all given in English units. The nondimensional stability derivatives

Table 6.1
Nondimensional Derivatives—B747 Airplane

	C_x	C_z	C_m
$\hat{u}$	−0.1080	−0.1060	0.1043
α	0.2193	−4.920	−1.023
$\hat{q}$	0	−5.921	−23.92
$\hat{\dot{\alpha}}$	0	5.896	−6.314

are given in Table 6.1, and the dimensional derivatives in Table 6.2. With the above data we calculate the system matrix **A** for this case. (Recall that the state vector is $[\Delta u \ w \ q \ \Delta\theta]^T$).

$$\mathbf{A} = \begin{bmatrix} -0.006868 & 0.01395 & 0 & -32.2 \\ -0.09055 & -0.3151 & 773.98 & 0 \\ 0.0001187 & -0.001026 & -0.4285 & 0 \\ 0 & 0 & 1 & 0 \end{bmatrix} \tag{6.2,1}$$

The characteristic equation (6.1,6) is next calculated to be:

$$\lambda^4 + 0.750468\lambda^3 + 0.935494\lambda^2 + 0.0094630\lambda + 0.0041959 = 0 \tag{6.2,2}$$

The two stability criteria are

$$E = 0.0041959 > 0$$

and

$$R = 0.004191 > 0$$

so that there are no unstable modes.

EIGENVALUES

The roots of the characteristic equation (6.2,2), the eigenvalues, are

$$\begin{aligned} &\text{Mode 1 (Phugoid mode):} && \lambda_{1,2} = -0.003289 \pm 0.06723i \\ &\text{Mode 2 (Short-period mode):} && \lambda_{3,4} = -0.3719 \pm 0.8875i \end{aligned} \tag{6.2,3}$$

We see that the natural modes are two damped oscillations, one of long period and lightly damped, the other of short period and heavily damped. This result is quite typ-

Table 6.2
Dimensional Derivatives—B747 Airplane

	X (lb)	*Z (lb)*	*M (ft·lb)*
u(ft/s)	-1.358×10^2	-1.778×10^3	3.581×10^3
w(ft/s)	2.758×10^2	-6.188×10^3	-3.515×10^4
q(rad/s)	0	-1.017×10^5	-1.122×10^7
$\dot{w}$(ft/s^2)	0	1.308×10^2	-3.826×10^3
	X(N)	*Z(N)*	*M(m·N)*
u(m/s)	-1.982×10^3	-2.595×10^4	1.593×10^4
w(m/s)	4.025×10^3	-9.030×10^4	-1.563×10^4
q(rad/s)	0	-4.524×10^5	-1.521×10^7
$\dot{w}$(m/s^2)	0	1.909×10^3	-1.702×10^4

Table 6.3

Mode	Name	Period (s)	t_{half} (s)	N_{half} (cycles)
1	Phugoid[a]	93.4	211	22.5
2	Short period	7.08	1.86	0.26

[a]The phugoid mode was first described by Lanchester (1908), who also named it. The name comes from the Greek root for *flee* as in *fugitive*. Actually Lanchester wanted the root for *fly*. Appropriate or not, the word phugoid has become established in aeronautical jargon.

ical. The modes are conventionally named as in Table 6.3, which also gives their periods and damping. The transient behavior of the state variables in these two modes is displayed in Fig. 6.2.

EIGENVECTORS

The eigenvectors corresponding to the above modes are given in Table 6.4. They are for the nondimensionable variables, in polar form, the values given corresponding to $n + i\omega$. Eigenvectors are arbitrary to within a complex factor, so it is only the *relative* values of the state variables that are significant. We have therefore factored them to

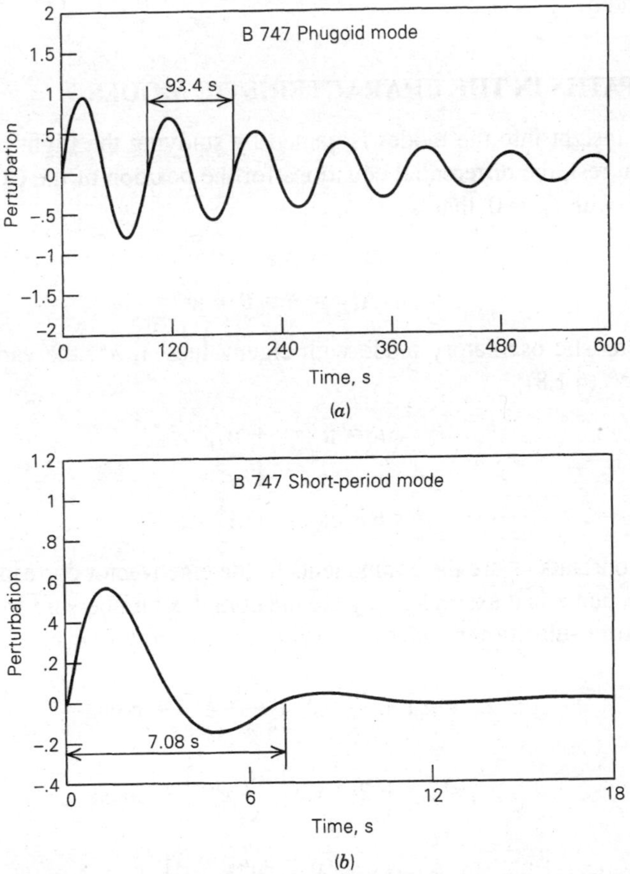

Figure 6.2 Characteristic transients. (*a*) Phugoid mode. (*b*) Short-period (pitching) mode.

Table 6.4
Eigenvectors (polar form)

	Phugoid		Short-Period	
	Magnitude	*Phase*	*Magnitude*	*Phase*
$\Delta\hat{u}$	0.62	92.4°	0.029	57.4°
$\alpha = \hat{w}$	0.036	82.8°	1.08	19.2°
$\hat{q}$	0.0012	92.8°	0.017	112.7°
$\Delta\theta$	1.0	0°	1.0	0°

make $\Delta\theta$ equal to unity, as displayed in the Argand diagram of Fig. 6.3. As we have chosen the positive value of ω, the diagrams can be imagined to be rotating counterclockwise and shrinking, with their projections on the real axis being the real values of the variables.

The phugoid is seen to be a motion in which the pitch rate $\hat{q}$ and the angle of attack change α are very small, but $\Delta\hat{u}$ and $\Delta\theta$ are present with significant magnitude. The speed leads $\Delta\theta$ by about 90° in phase.

The short-period mode, by contrast, is one in which there is negligible speed variation, while the angle of attack oscillates with an amplitude and phase not much different from that of $\Delta\theta$. This mode behaves like one with only two degrees of freedom, $\Delta\theta$ and α.

FLIGHT PATHS IN THE CHARACTERISTIC MODES

Additional insight into the modes is gained by studying the flight path. With the atmosphere at rest, the differential equations for the position of the CG in F_E are given by (4.9,10), with $\theta_0 = 0$, that is,

$$\begin{aligned} \Delta\dot{x}_E &= \Delta u \\ \Delta\dot{z}_E &= -u_0\theta + w \end{aligned} \tag{6.2,4}$$

In a characteristic oscillatory mode with eigenvalues λ, λ^*, the variations of Δu, θ, and w are [cf. (6.1,8)]

$$\begin{aligned} \Delta u &= u_{1j}e^{\lambda t} + u_{1j}^*e^{\lambda^* t} \\ w &= u_{2j}e^{\lambda t} + u_{2j}^*e^{\lambda^* t} \\ \theta &= u_{4j}e^{\lambda t} + u_{4j}^*e^{\lambda^* t} \end{aligned} \tag{6.2,5}$$

where the constants u_{ij} are the components of the eigenvector corresponding to λ. For the previous numerical example, they are the complex numbers given in polar form in Table 6.4. After substituting (6.2,5) in (6.2,4) and integrating from $t = 0$ to t we get

$$\begin{aligned} x_E &= u_0 t + \frac{u_{1j}}{\lambda}e^{\lambda t} + \frac{u_{1j}^*}{\lambda^*}e^{\lambda^* t} + \text{const} \\ &= u_0 t + 2e^{nt}\,\text{Re}\left[\frac{u_{1j}}{\lambda}e^{i\omega t}\right] + \text{const} \\ z_E &= 2e^{nt}\,\text{Re}\left[\frac{u_{2j} - u_0 u_{4j}}{\lambda}e^{i\omega t}\right] + \text{const} \end{aligned} \tag{6.2,6}$$

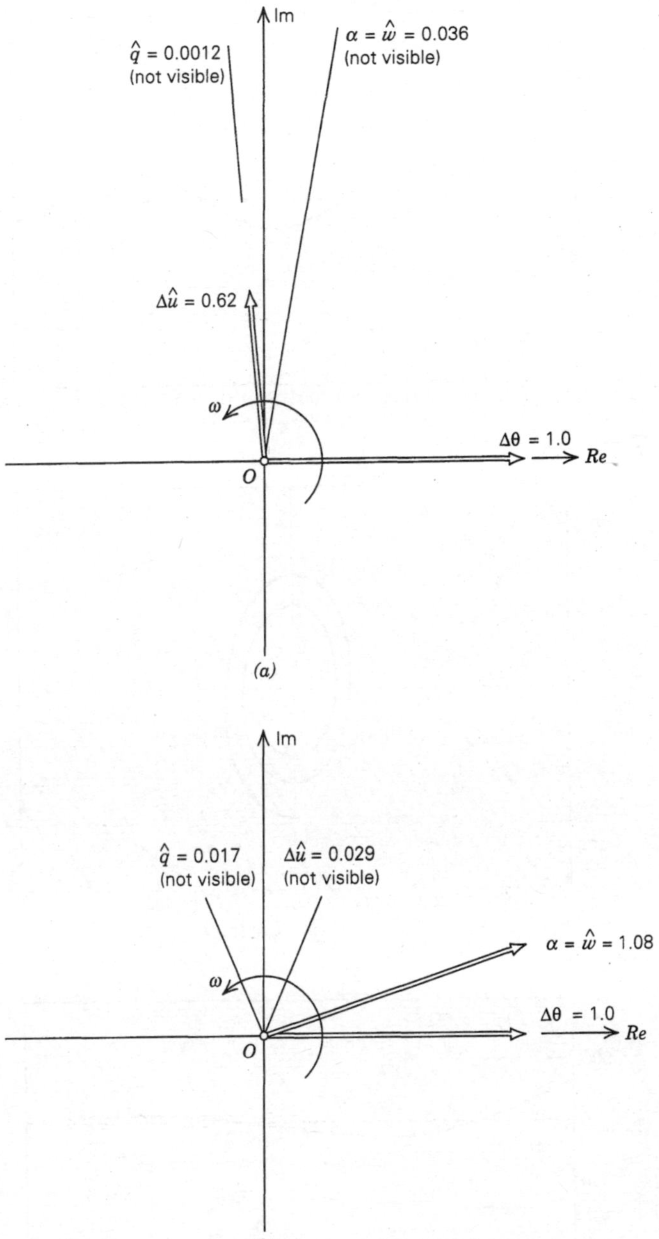

Figure 6.3 (*a*) Vector diagram of phugoid mode. (*b*) Vector diagram of short-period mode.

where Re denotes the real part of the complex number in the square brackets. For the numerical data of the above example (6.2,6) has been used to calculate the flight paths in the two modes, plotted in Fig. 6.4. The nonzero initial conditions are arbitrary, and the trajectories for both modes asymptote to the steady reference flight path. Figure 6.4 shows that the phugoid is an undulating flight of very long wavelength. The mode diagram, Fig. 6.3*a* shows that the speed leads the pitch angle by about 90°, from which we can infer that u is largest near the bottom of the wave and least near the top. This variation in speed results in different distances being traversed

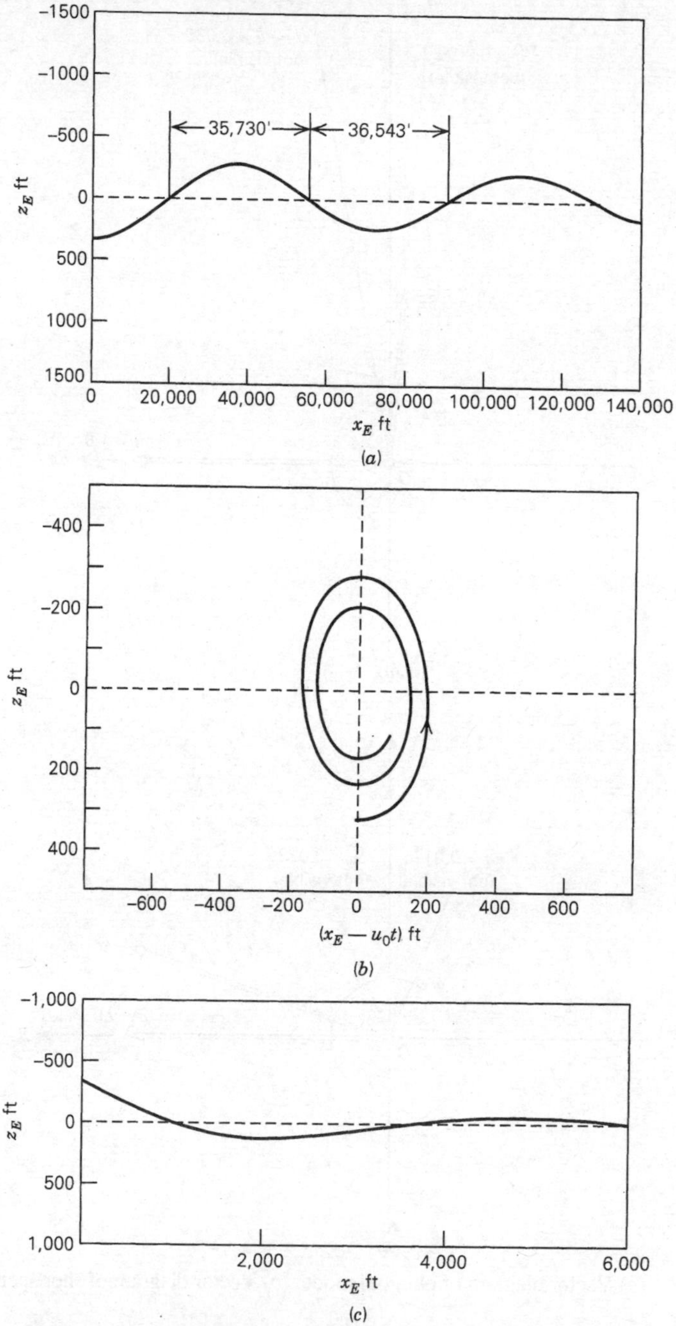

Figure 6.4 (*a*) Phugoid flight path (fixed reference frame). (*b*) Phugoid flight path (moving reference frame). (*c*) Short-period flight path.

during the upper and lower halves of the cycle, as shown in Fig. 6.4*a*. For larger amplitude oscillations, this lack of symmetry in the oscillation becomes much more pronounced (although the linear theory then fails to describe it accurately) until ultimately the upper part becomes first a cusp and then a loop (see Miele, 1962, p. 273). The motion (see Sec. 6.3) is approximately one of constant total energy, the rising and falling corresponding to an exchange between kinetic and potential energy. Figure 6.4*b* shows the phugoid motion relative to axes moving at the reference speed u_0. This is the relative path that would be seen by an observer flying alongside at speed u_0.

Figure 6.4*c* shows the path for the short-period mode. The disturbance is rapidly damped. The transient has virtually disappeared within 3000 ft of flight, even though the initial $\Delta\alpha$ and $\Delta\theta$ were very large. The deviation of the path from a straight line is small, the principal feature of the motion being the rapid rotation in pitch.

6.3 Approximate Equations for the Longitudinal Modes

The numerical solutions for the modes, although they certainly show their properties, do not give much physical insight into their genesis. Now each oscillatory mode is equivalent to some second-order mass-spring-damper system, and each nonoscillatory mode is equivalent to some mass-damper system. To understand the modes, and the influence on them of the main flight and vehicle parameters, it is helpful to know what contributes to the equivalent masses, springs, and dampers. To achieve this requires *analytical* solutions, which are simply not available for the full system of equations. Hence we are interested in getting *approximate* analytical solutions, if they can reasonably represent the modes. Additionally, approximate models of the individual modes are frequently useful in the design of automatic flight control systems (McRuer et al., 1973). In the following we present some such approximations and the methods of arriving at them.

There are two approaches generally used to arrive at these approximations. One is to write out a literal expression for the characteristic equation and, by studying the order of magnitude of the terms in it, to arrive at approximate linear or quadratic factors. For example, if the characteristic equation (6.1,13) is known to have a "small" real root, an approximation to it may be obtained by neglecting all the higher powers of λ, that is,

$$D\lambda + E \doteq 0$$

Or if there is a "large" complex root, it may be approximated by keeping only the first three terms, that is,

$$A\lambda^2 + B\lambda + C \doteq 0$$

This method is frequently useful, and is sometimes the only reasonable way to get an approximation.

The second method, which has the advantage of providing more physical insight, proceeds from a foreknowledge of the modal characteristics to arrive at approximate system equations of lower order than the exact ones. For the longitudinal modes we use the second method (see below), and for the lateral modes (see Sec. 6.8) both methods are needed.

It should be noted that no simple analytical approximations can be relied on to give accurate results under all circumstances. Machine solutions of the exact matrix is the only certain way. The value of the approximations is indicated by examples in the following.

To proceed now to the phugoid and short-period modes, we saw in Fig. 6.3 that some state variables are negligibly small in each of the two modes. This fact suggests certain approximations to them based on reduced sets of equations of motion arrived at by physical reasoning. These approximations, which are quite useful, are developed below.

PHUGOID MODE

Lanchester's original solution (Lanchester, 1908) for the phugoid used the assumptions that $\alpha_T = 0$, $\Delta\alpha \equiv 0$ and $T - D \equiv 0$ (see Fig. 2.1). It follows that there is no net aerodynamic force tangent to the flight path, and hence no work done on the vehicle except by gravity. The motion is then one of constant total energy, as suggested previously. This simplification makes it possible to treat the most general case with large disturbances in speed and flight-path angle (see Miele, 1962, p. 271 et seq.) Here we content ourselves with a treatment of only the corresponding small-disturbance case, for comparison with the exact numerical result given earlier. The energy condition is

$$E = \tfrac{1}{2}mV^2 - mgz_E = \text{const}$$

or

$$V^2 = u_0^2 + 2gz_E \tag{6.3,1}$$

where the origin of F_E is so chosen that $V = u_0$ when $z_E = 0$. With α constant, and in addition neglecting the effect of q on C_L, then C_L is constant at the value for steady horizontal flight, that is, $C_L = C_{L_0} = C_{W_0}$, and $L = C_{W_0}\frac{1}{2}\rho V^2 S$ or, in view of (6.3,1),

$$L = C_{W_0}\tfrac{1}{2}\rho u_0^2 S + (C_{W_0}\rho g S)z_E = W + kz_E \tag{6.3,2}$$

Thus the lift is seen to vary linearly with the height in such a manner as always to drive the vehicle back to its reference height, the "spring constant" being

$$k = C_{W_0}\rho g S \tag{6.3,3}$$

The equation of motion in the vertical direction is clearly, when $T - D = 0$, and $\gamma =$ angle of climb (see Fig. 2.1)

$$W - L\cos\gamma = m\ddot{z}_E$$

or for small γ,

$$W - L = m\ddot{z}_E \tag{6.3,4}$$

On combining (6.3,2) and (6.3,4) we get

$$m\ddot{z}_E + kz_E = 0$$

which identifies a simple harmonic motion of period

$$T = 2\pi\sqrt{\frac{m}{k}} = 2\pi\sqrt{\frac{m}{C_{W_0}\rho g S}}$$

Since $C_{W_0} = mg/\frac{1}{2}\rho u_0^2 S$, this becomes

$$T = \pi\sqrt{2}\frac{u_0}{g} = 0.138u_0 \tag{6.3,5}$$

when u_0 is in fps, a beautifully simple result, suggesting that the phugoid period depends only on the *speed* of flight, and not at all on the airplane or the altitude! This elegant result is not only of historical interest—it actually gives a reasonable approximation to the phugoid period of rigid airplanes at speeds below the onset of significant compressibility effects. Thus for the B747 example, the Lanchester approximation gives $T = 107$s, a value not very far from the true 93s. It is possible to get an even better approximation, one that gives an estimate of the damping as well. Because q is approximately zero in this mode, we can infer that the pitching moment is approximately zero—that the airplane is in quasistatic pitch equilibrium during the motion. Moreover the pitching moment can reasonably be simplified in these circumstances by keeping only the first two terms on the right side of (4.9,17e). Because q and $\dot{w}$ are both relatively small we further neglect Z_q and $Z_{\dot{w}}$ as well. On making these simplifications to (4.9,18) and setting $\Delta\mathbf{f}_c = 0$ and $\theta_0 = 0$ we get the reduced system of equations:

$$\begin{bmatrix} \Delta\dot{u} \\ \dot{w} \\ 0 \\ \Delta\dot{\theta} \end{bmatrix} = \begin{bmatrix} \dfrac{X_u}{m} & \dfrac{X_w}{m} & 0 & -g \\ \dfrac{Z_u}{m} & \dfrac{Z_w}{m} & u_0 & 0 \\ M_u & M_w & 0 & 0 \\ 0 & 0 & 1 & 0 \end{bmatrix} \begin{bmatrix} \Delta u \\ w \\ q \\ \Delta\theta \end{bmatrix} \tag{6.3,6}$$

These equations are not in the canonical form $\dot{\mathbf{x}} = \mathbf{A}\mathbf{x}$, but we can still get the characteristic equation by substituting $\mathbf{x} = \mathbf{x}_0 e^{\lambda t}$ and factoring out the exponential. The result is

$$\begin{vmatrix} (X_u/m - \lambda) & X_w/m & 0 & -g \\ Z_u/m & (Z_w/m - \lambda) & u_0 & 0 \\ M_u & M_w & 0 & 0 \\ 0 & 0 & 1 & -\lambda \end{vmatrix} = 0 \tag{6.3,7}$$

Equation (6.3,7) expands to

$$A\lambda^2 + B\lambda + C = 0$$

or

$$\lambda^2 + 2\zeta\omega_n\lambda + \omega_n^2 = 0 \tag{6.3,8}$$

which is a convenient way to write the characteristic equation of a second-order system. The constants are

$$\begin{aligned} A &= -u_0 M_w \\ B &= gM_u + \frac{u_0}{m}(X_u M_w - M_u X_w) \\ C &= \frac{g}{m}(Z_u M_w - M_u Z_w) \end{aligned} \tag{6.3,9}$$

from which we derive the radian frequency and damping to be

$$\begin{aligned} \omega_n^2 &= -\frac{g}{mu_0}\left(Z_u - \frac{M_u Z_w}{M_w}\right) \\ \zeta &= -\frac{1}{2}\left[\frac{g}{mu_0}\left(\frac{M_u}{M_w}Z_w - Z_u\right)\right]^{-1/2}\left[\frac{g}{u_0}\frac{M_u}{M_w} + \frac{1}{m}\left(X_u - \frac{M_u}{M_w}X_w\right)\right] \end{aligned} \tag{6.3,10}$$

When $M_u = 0$, these reduce to:

$$A = -u_0 M_w$$

$$B = \frac{u_0}{m} X_u M_w \qquad (6.3,11)$$

$$C = \frac{g}{m} Z_u M_w$$

from which

$$\omega_n^2 = \frac{C}{A} = -\frac{gZ_u}{mu_0} \qquad (a)$$

$$\zeta = -\frac{X_u}{2}\sqrt{\frac{u_0}{-mgZ_u}} \qquad (b) \quad (6.3,12)$$

When Z_u from Table 4.4 with $C_{z_u} = 0$ is substituted into (6.3,12) we find (see Exercise 6.1) that $T_n = 2\pi/\omega_n$ is exactly the Lanchester period (6.3,5). Moreover if we make the further assumption that the airplane is a jet with constant thrust ($\partial T/\partial u_0 = 0$) we find the damping ratio to be

$$\zeta = \frac{1}{\sqrt{2}}\frac{C_{D_0}}{C_{L_0}}$$

that is, it is proportional to the inverse of the L/D ratio of the airplane. In fact the approximation for the period is good over the whole range of C_{m_u}, whereas that for the damping is poor for large positive C_{m_u}. For the example airplane the above approximation gives $\zeta = 0.046$, compared with the exact value 0.049.

SHORT-PERIOD MODE

Figure 6.3*b* shows that the short-period mode is essentially one with two degrees of freedom, the speed being substantially constant while the airplane pitches relatively rapidly. We can therefore arrive at approximate system equations by neglecting the X-force equation entirely and putting $\Delta u = 0$. Examination of the magnitudes of the terms in the numerical example shows that $Z_{\dot{w}}$ is small compared to m and Z_q is small compared to mu_0. The result after simplifying (4.9,18) with $\theta_0 = 0$ is a pair of equations for w and q.

$$\begin{bmatrix}\dot{w}\\ \dot{q}\end{bmatrix} = \begin{bmatrix}\dfrac{Z_w}{m} & u_0\\ \dfrac{1}{I_y}\left[M_w + \dfrac{M_{\dot{w}}Z_w}{m}\right] & \dfrac{1}{I_y}[M_q + M_{\dot{w}}u_0]\end{bmatrix}\begin{bmatrix}w\\ q\end{bmatrix} \qquad (6.3,13)$$

The characteristic equation of (6.3,13) is found to be

$$\lambda^2 - \lambda\left[\frac{Z_w}{m} + \frac{1}{I_y}(M_q + M_{\dot{w}}u_0)\right] - \frac{1}{I_y}\left(u_0 M_w - \frac{M_q Z_w}{m}\right) = 0 \qquad (6.3,14)$$

When converted with the aid of Tables 4.1 and 4.4, (6.3,14) becomes

$$\lambda^2 + B\lambda + C = 0 \qquad (a)$$

where

$$B = -\frac{1}{t^*}\left[\frac{C_{z_\alpha}}{2\mu} + \frac{1}{\hat{I}_y}(C_{m_q} + C_{m_{\dot\alpha}})\right] \qquad (b) \qquad (6.3,15)$$

$$C = -\frac{1}{t^{*2}\hat{I}_y}\left(C_{m_\alpha} - \frac{C_{m_q}C_{z_\alpha}}{2\mu}\right) \qquad (c)$$

When the data for the B747 is substituted into (6.3,15) the result obtained is

$$\lambda^2 + 0.741\lambda + 0.9281 = 0$$

with roots

$$\lambda = -0.371 \pm 0.889i$$

which are seen to be almost the same as those in (6.2,3) obtained from the complete matrix equation. The short-period approximation is actually very good for a wide range of vehicle characteristics and flight conditions.

6.4 General Theory of Static Longitudinal Stability

In Chap. 2 we used positive pitch stiffness (negative C_{m_α}) as an approximate criterion for static longitudinal stability. Now static instability really means the presence of a real positive root of the characteristic equation, and we saw in Sec. 6.1 that the condition for no such root to occur is that the coefficient E of the stability quartic must be positive. Thus the boundary between static stability and instability is defined by $E = 0$. We get E by putting $\lambda = 0$ in the characteristic determinant. Thus

$$E = \det \mathbf{A} \qquad (6.4,1)$$

In evaluating (6.4,1) for the matrix of (4.9,18) we put $\theta_0 = 0$, and as in Sec. 6.3, we neglect the two derivatives $Z_{\dot w}$ and Z_q. The result is

$$E = \frac{g}{mI_y}(Z_u M_w - M_u Z_w) \qquad (6.4,2)$$

Since g, m and I_y are all positive, the criterion for static stability is

$$Z_u M_w - M_u Z_w > 0. \qquad (6.4,3)$$

When converted to nondimensional form, this becomes

$$C_{m_\alpha}(C_{z_u} - 2C_{W_0}) - C_{m_u}C_{z_\alpha} > 0 \qquad (6.4,4)$$

When there are no speed effects, that is, C_{z_u} and C_{m_u} are both zero, then the criterion does indeed reduce to the simple $C_{m_\alpha} < 0$.

We now compare the above criterion for stability with the trim slope (2.4,24). In making this comparison, we must take note of a minor difference in basic assumptions. In the preceding development, it was specifically assumed that the thrust vector rotates with the vehicle when α is changed [see (5.1,1)]. In the development leading to (2.4,24) by contrast, there is an implicit assumption that the thrust provides *no* component of force perpendicular to **V** [see (2.4,18)]. It is this difference that leads to the presence of C_{z_α} in (6.4,4) instead of C_{L_α} in (2.4,24). Had the assumptions been the same, the expressions would be strictly compatible. In any case, C_{D_0} is usually small compared to C_{L_α}, so that the difference is not important, see Table 5.1. We see that the justification for the statement made in Sec. 2.4, that the slope of the elevator trim curve $(d\delta_{e_{\text{trim}}}/d\hat{V})_{\delta_p}$ is a criterion of static stability, is provided by (6.4,4). [Note that $C_{W_0} = C_{L_0}$ in (6.4,4).]

Another stability criterion referred to in Chap. 2 is the derivative dC_m/dC_L (2.3,8). It was pointed out there that this derivative can only be said to exist if enough constraints are imposed on the independent variables α, V, δ_e, q, etc., on which C_m and C_L separately depend. Such a situation results if we postulate that the vehicle is in rectilinear motion ($q \equiv 0$) at constant elevator angle and throttle setting, with $L = W$, but with varying speed and angle of attack. Such a condition cannot, of course, actually occur in flight because the pitching moment could be zero at only one speed, but it can readily be simulated in a wind tunnel where the model is restrained by a balance. With the above stipulations, C_m and C_L reduce to functions of the two variables $\hat{u}$ and α, and incremental changes from a reference state $(\;)_0$ are given by

$$\begin{aligned} dC_L &= C_{L_\alpha}\, d\alpha + C_{L_u}\, d\hat{u} \\ dC_m &= C_{m_\alpha}\, d\alpha + C_{m_u}\, d\hat{u} \end{aligned} \tag{6.4,5}$$

The required derivative is then

$$\frac{dC_m}{dC_L} = \frac{C_{m_\alpha} + C_{m_u}\dfrac{d\hat{u}}{d\alpha}}{C_{L_\alpha} + C_{L_u}\dfrac{d\hat{u}}{d\alpha}} \tag{6.4,6}$$

provided $d\hat{u}/d\alpha$ exists. This is guaranteed by the remaining condition imposed, that is, $L = W$ (implying $\alpha_T \equiv 0$). For then we have

$$W = C_L(\alpha, \hat{u})\tfrac{1}{2}\rho V^2 S = \text{const}$$

from which we readily derive

$$(C_{L_\alpha}\, d\alpha + C_{L_u}\, d\hat{u})\tfrac{1}{2}\rho u_0^2 S + C_{L_0}\rho u_0 S\, du = 0 \tag{6.4,7}$$

From (6.4,7)

$$(C_{L_u} + 2C_{L_0})\, d\hat{u} + C_{L_\alpha}\, d\alpha = 0$$

or

$$\frac{d\hat{u}}{d\alpha} = -\frac{C_{L_\alpha}}{C_{L_u} + 2C_{L_0}} \tag{6.4,8}$$

After substituting (6.4,8) into (6.4,6) and simplifying we get

$$\left.\frac{dC_m}{dC_L}\right|_{L=W} = \frac{1}{2C_{L_0}C_{L_\alpha}}[C_{m_\alpha}(C_{L_u} + 2C_{L_0}) - C_{L_\alpha}C_{m_u}] \tag{6.4,9}$$

On comparing (6.4,9) with (6.4,4), with the same caveats as for the trim slope, we see that the static stability criterion is

$$\left.\frac{dC_m}{dC_L}\right|_{L=W} < 0 \tag{6.4,10}$$

provided that dC_m/dC_L is calculated with the constraints $\Delta\delta_e = \Delta\delta_p = q = 0$ and $L \equiv W$. [The quantity on the left side of (6.4,9) is sometimes referred to as *speed stability* in the USA, by contrast with "angle of attack" stability. In Great Britain, this term usually has a different meaning, as in Sec. 8.5.]

On using the definition of h_s given in (2.4,26) we find from (6.4,9) that

$$\left.\frac{dC_m}{dC_L}\right|_{L=W} = \left(1 + \frac{C_{L_u}}{2C_{L_0}}\right)(h - h_s) \tag{6.4,11}$$

that is, that it is proportional to the "stability margin," and when $C_{L_u} \ll 2C_{L_0}$, is equal to it.

6.5 Effect of Flight Condition on the Longitudinal Modes of a Subsonic Jet Transport

In Sec. 6.2 we gave the representative characteristic modes of a subsonic jet airplane for a single set of parameters. It is of considerable interest to enquire into how these characteristics are affected by changes in the major flight variables—speed, altitude, angle of climb, and stability margin. It is also of interest to look into the effect of the vertical density gradient in the atmosphere. In this section and by means of exercises we examine some of these effects.

EFFECT OF SPEED AND ALTITUDE

The data in Heffley and Jewel (1972) for the example airplane include several comparable cases, all having the same geometry, static margin, and gross weight. There are two speeds at sea-level, and three each at 20,000- and 40,000-ft altitudes. The modal periods and damping for these eight cases are displayed in Fig. 6.5. (Since there are so few points, the shapes of the curves are conjectural!) It is an understatement to say that there is no simple pattern to these data. It can be said, however, that the phugoid period increases with speed, as predicted by the Lanchester theory, and decreases with altitude at fixed Mach number. The short-period does the opposite, decreasing with speed and increasing with altitude.

The most striking feature of the data is the sudden and large increase in the phugoid period at high Mach number at the two higher altitudes. This phenomenon is a result of a loss of true static stability at these Mach numbers brought about by a negative value of C_{m_u}, which has the effect of reducing E in (6.4,2). This happens because this large aircraft is necessarily quite flexible, and because at these Mach numbers it is entering the transonic regime, where air compressibility leads to substantial alterations in the aerodynamic pressure distribution. To show that C_{m_u} is the reason for the behavior of the graphs, we vary it over a large range for the flight condition $\mathbf{M} = 0.8$ and 20,000-ft altitude. Figure 6.6 shows the result and substantiates the important role of this derivative. In fact, from (6.4,4) we calculate that $E = 0$ when $C_{m_u} = -0.0968$, a value only 4% more negative than that of the example at the given flight

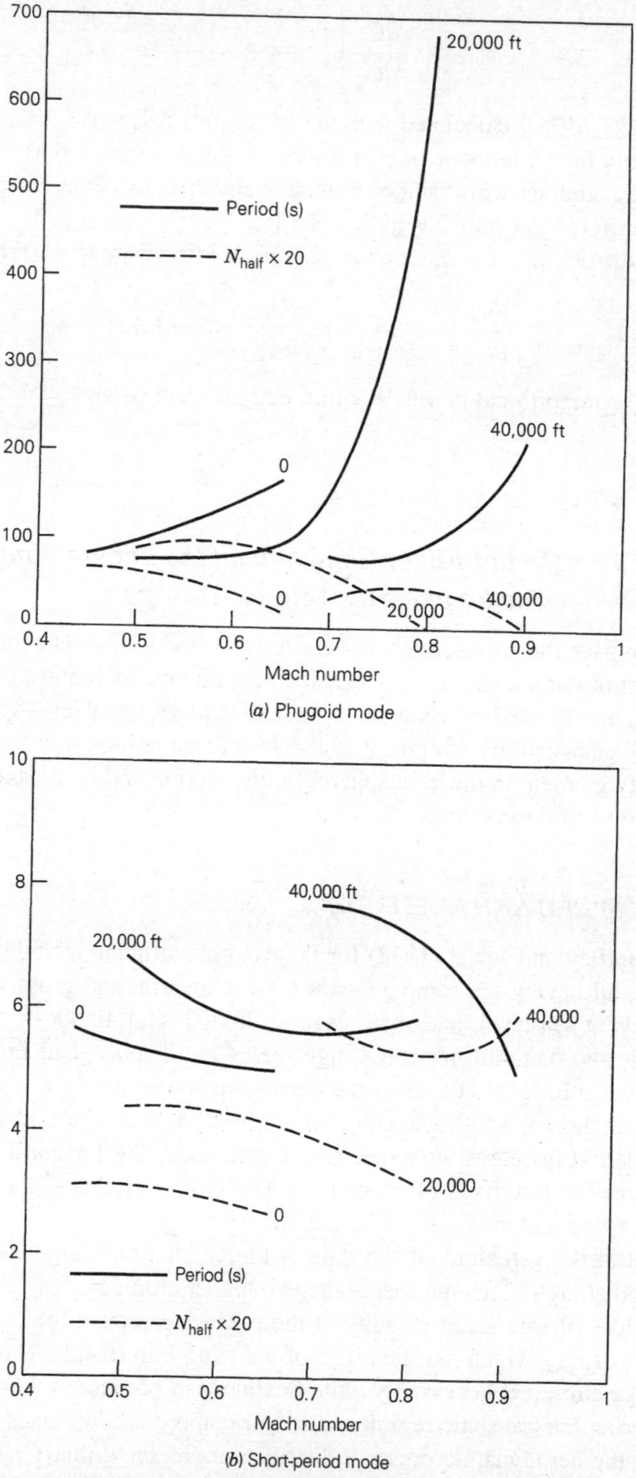

Figure 6.5 Variation of longitudinal modes with speed and altitude. (*a*) Phugoid mode. (*b*) Short-period mode.

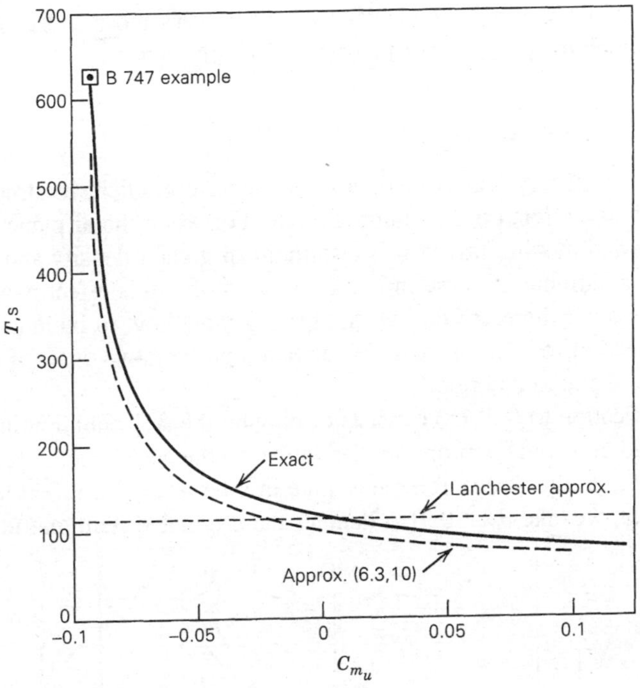

(a) Period

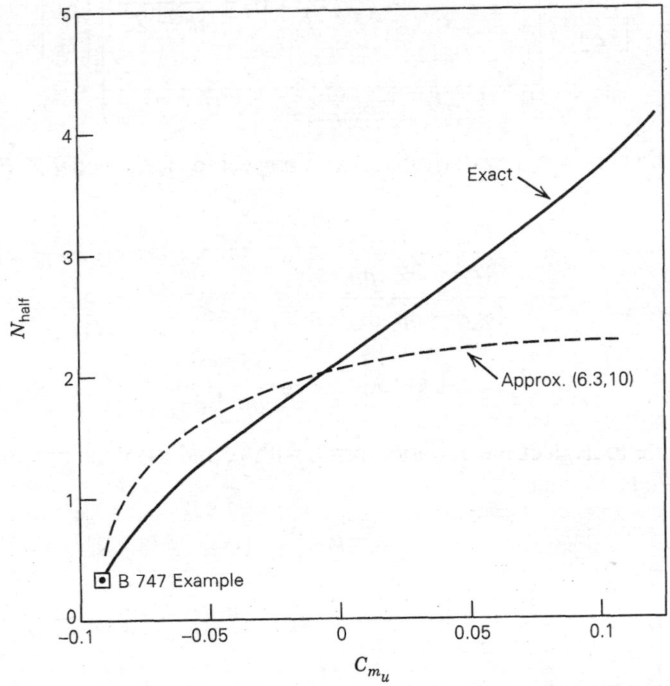

(b) Damping

Figure 6.6 Effect of C_{m_u} on the phugoid mode. (*a*) Period. (*b*) Damping.

condition. Thus at this point the airplane would be very close to the static stability boundary, at which the period would go asymptotically to ∞.

EFFECT OF VERTICAL DENSITY GRADIENT

We might expect on physical grounds that the vertical gradient in atmospheric density would have an effect on the phugoid mode. For when the airplane is at the bottom of a cycle and moving fastest it is also in air of greater density and hence would experience an additional increase in lift. It turns out that this effect is appreciable in magnitude. We shall therefore do two things: (1) show how to include this effect in the general equations of motion, and (2) derive a representative order of magnitude of the change in the phugoid period.

The modification to (4.9,18) consists of moving the $\Delta\dot{z}_E$ equation into the matrix equation and adding some appropriate derivatives to the aerodynamic forces (4.9,17). Since the only possible steady reference state in a vertically stratified atmosphere is horizontal flight, we take $\theta_0 = 0$. If $\mathbf{A}$ denotes the original system matrix, the result is

$$\begin{bmatrix} \Delta\dot{u} \\ \dot{w} \\ \dot{q} \\ \Delta\dot{\theta} \\ \dot{z}_E \end{bmatrix} = \left[\begin{array}{cccc:c} & & & & \dfrac{X_z}{m} \\ & \mathbf{A} & & & \dfrac{Z_z}{m} \\ & & & & \dfrac{1}{I_y}\left[M_z + \dfrac{M_{\dot{w}}Z_z}{m - Z_{\dot{w}}}\right] \\ & & & & 0 \\ \hdashline 0 & 1 & 0 & -u_0 & 0 \end{array}\right] \begin{bmatrix} \Delta u \\ w \\ q \\ \Delta\theta \\ z_E \end{bmatrix} \tag{6.5,1}$$

In (6.5,1) there are three new derivatives with respect to z_E. Consider Z_z first:

$$Z = C_z \tfrac{1}{2}\rho V^2 S$$

$$\frac{\partial Z}{\partial z_E} = \frac{\partial Z}{\partial \rho}\frac{\partial \rho}{\partial z_E} \tag{6.5,2}$$

$$= \tfrac{1}{2}V^2 S\left[C_z + \rho\frac{\partial C_z}{\partial \rho}\right]\frac{\partial \rho}{\partial z_E}$$

It is reasonable to neglect the variation of C_z with ρ, and the density varies exponentially with height,[2] so that

$$\rho = \rho_0 e^{\kappa z_E}$$

and

$$\frac{\partial \rho}{\partial z_E} = \kappa\rho \tag{6.5,3}$$

where κ is constant over a sufficient range of altitude for a linear analysis. It follows that

[2]Exactly, in an isothermal atmosphere of uniform composition; approximately, in the real atmosphere.

$$\frac{\partial Z}{\partial z_E} = \tfrac{1}{2}\kappa\rho V^2 SC_z$$

$$Z_z = \left.\frac{\partial Z}{\partial z_E}\right|_0 = \kappa Z_0 \tag{6.5,4}$$

Similarly,

$$X_z = \kappa X_0$$
$$M_z = \kappa M_0$$

From (4.9,6) we get the reference values, leading to

$$\begin{aligned} Z_z &= -mg\kappa \\ X_z &= M_z = 0 \end{aligned} \tag{6.5,5}$$

The result is a rather simple elaboration of the original matrix equation. To get an estimate of the order of magnitude of the density gradient effect, it is convenient to return to the Lanchester approximation to the phugoid, and modify it to suit.

In Sec. 6.3 we saw that with this approximation, there is a vertical "spring stiffness" k given by (6.3,3) that governs the period. When the density varies there is a second "stiffness" k' resulting from the fact that the increased density when the vehicle is below its reference altitude increases the lift, and vice versa. This incremental lift associated with a density change is

$$\Delta L = C_L \tfrac{1}{2} V^2 S\, \Delta\rho$$

so that

$$k' = \frac{\partial \Delta L}{\partial z_E} = C_{L_0} \tfrac{1}{2} u_0^2 S \frac{\partial \rho}{\partial z_E}$$

Using (6.5,3) we get

$$k' = \kappa W \tag{6.5,6}$$

Thus we find that k' is approximately constant, whereas k from (6.3,3) depends on $C_{W_0}\rho$, which varies as V^{-2} for constant weight. The density gradient therefore has its greatest relative effect at high speed. The correction factor for the period, which varies inversely as the square root of the stiffness, is

$$F = \left(\frac{k}{k + k'}\right)^{1/2} = \frac{1}{(1 + k'/k)^{1/2}} \tag{6.5,7}$$

so that the period, when there is a density gradient, is $T' = FT$. With the given values of k and k' this becomes

$$F = \frac{1}{\left(1 + \dfrac{\kappa u_0^2}{2g}\right)^{1/2}} \tag{6.5,8}$$

in which the principal variable is seen to be the speed. Using a representative value for κ of 4.2×10^{-5} (6.5,8) gives a reduction in the phugoid period of 18% for the example airplane at 774 fps. This is seen to be a very substantial effect. If the full sys-

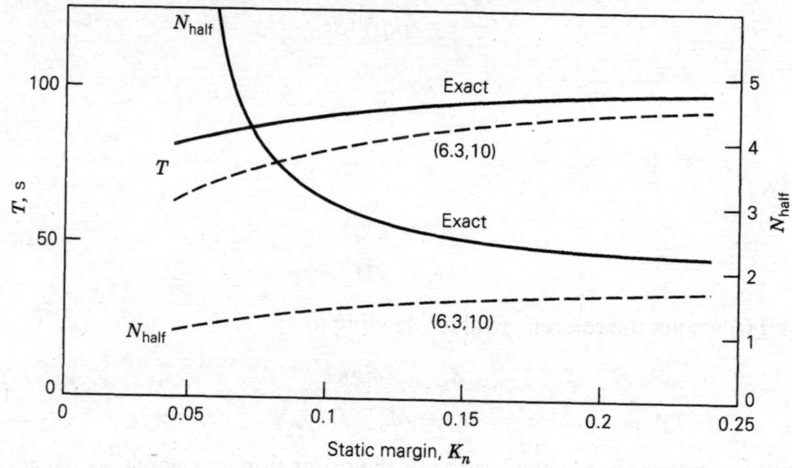

Figure 6.7 Variation of period and damping of phugoid mode with static margin.

tem model (6.5,1) is used, comparable effects can be found on the damping of the phugoid as well.

EFFECT OF CG LOCATION

It was indicated in Chap. 2 that the single most important aerodynamic characteristic for longitudinal stability is the pitch stiffness C_{m_α}, and that it varies strongly with the CG position, that is,

$$C_{m_\alpha} = C_{L_\alpha}(h - h_n)$$

where the static margin is $K_n = h_n - h$. The effect of this parameter is demonstrated by using (4.9,18) with variable K_n. The results, with all the other numerical data identical with that in Sec. 6.2, are shown in Figs. 6.7–6.9. Figure 6.7 shows that the phugoid period and damping vary rapidly at low static margin and that the approximation (6.3,10) is useful mainly at large K_n. Figure 6.8 shows the variation of the

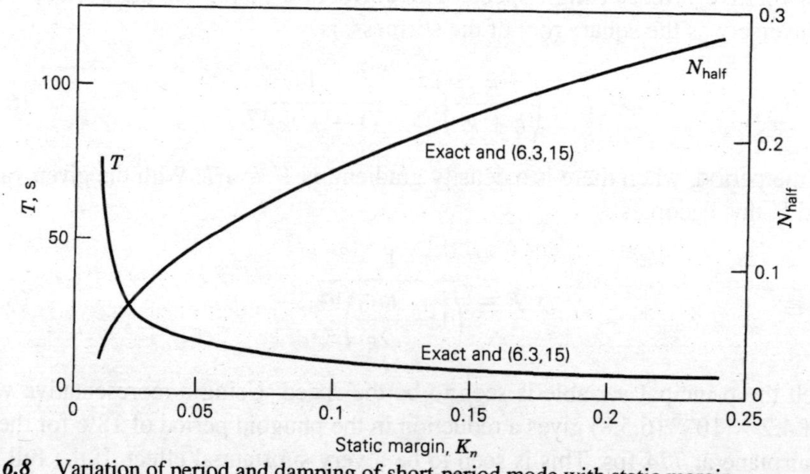

Figure 6.8 Variation of period and damping of short-period mode with static margin.

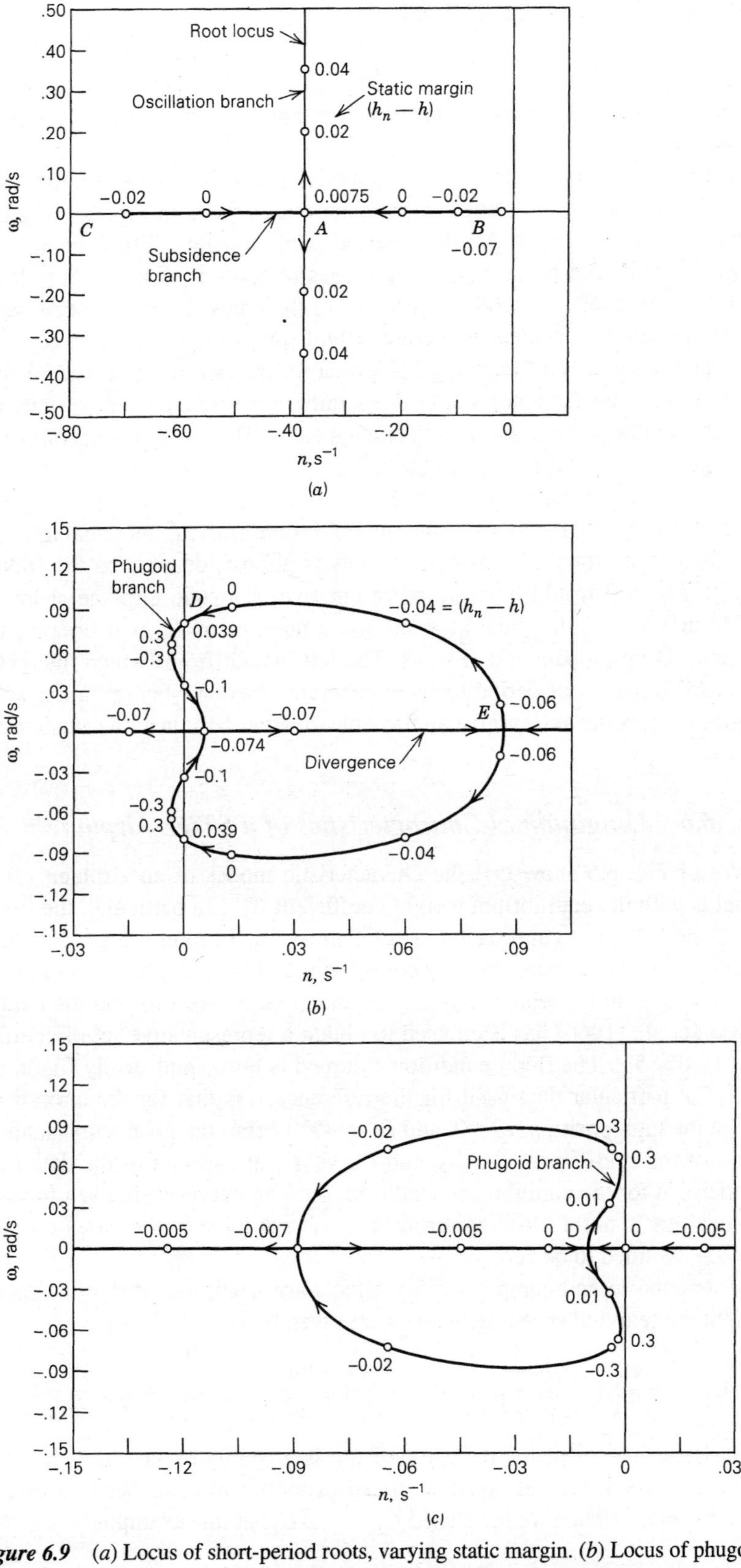

Figure 6.9 (*a*) Locus of short-period roots, varying static margin. (*b*) Locus of phugoid roots, varying static margin. (*c*) Locus of phugoid roots, varying static margin, $M_u = 0$.

short-period roots. These, too, vary strongly with pitch stiffness, the mode becoming nonoscillatory at a static margin near zero. The approximation (6.3,15) is excellent over the whole oscillatory range.

Important additional insight into these modes is obtained by examining the root loci obtained by varying the static margin. These are shown in Fig. 6.9. Figure 6.9*a* shows that the damping, n, of the short-period mode remains virtually constant with decreasing K_n while the frequency, ω, decreases to zero at point A where the locus splits into two real roots, branches AB and AC of the locus. These of course represent nonperiodic modes or subsidences. Figure 6.9*b* displays the much more complex behavior of the phugoid. With reducing static margin (rearward movement of the CG) this mode becomes unstable at point D. At (totally unrealistic!) negative static margin beyond -0.1, a new stable oscillation has reappeared. However, it is accompanied by a catastrophic positive real root far to the right.

The importance of the M_u derivative was shown earlier. It is again displayed in Fig. 6.9*c*, which repeats the locus of the phugoid roots with M_u set equal to zero. The corresponding locus for the short-period roots is almost identical to Fig. 6.9*a*. The pattern with $M_u = 0$ would be more representative of a rigid airplane at low Mach number. It shows a stable phugoid at all static margins as K_n is decreased until it splits at point D into a pair of real roots. The left branch from D then interacts with the branch AB of the short-period locus to generate a new stable oscillation while the right branch crosses the axis to give an unstable divergence at negative static margin.

6.6 Longitudinal Characteristics of a STOL Airplane

The curves of Fig. 6.5 show that the characteristic modes of an airplane vary with speed, that is with the equilibrium weight coefficient C_{W_0}. In particular, the two characteristic periods begin to approach one another as C_{W_0} becomes large. It is of interest to explore this range more fully by considering a STOL airplane, operating in the "powered-lift" region for which C_{W_0} may be much larger. To this end the data given in Margason et al., (1966) has been used to obtain a representative set of coefficients for $2.0 \le C_{W_0} \le 5.0$. The flight condition assumed is horizontal steady flight, so that $C_{x0} = 0$. (The particular data used for the reference was that for the aircraft with a large tail in the high position, $i_t = 0$, and $\delta_f = 45°$.) From the given curves, and from cross-plots of the coefficients C_L, C_D, and C_m vs. C_T at constant α, the data in Table 6.5 were derived for the equilibrium condition. Smooth curves were used for interpolation. Since this is not a tilt-wing airplane, α_T is not large in the cases considered, and has been assumed to be zero.

Since aeroelastic and compressibility effects are negligible at the low speeds of STOL flight the required speed derivatives are given by (see Table 5.1)

$$C_{x_u} = C_{T_u}\left(1 - \frac{\partial C_D}{\partial C_T}\right); \qquad C_{z_u} = -C_{T_u}\frac{\partial C_L}{\partial C_T}; \qquad C_{m_u} = C_{T_u}\frac{\partial C_m}{\partial C_T}$$

For a propeller-driven airplane, the value of C_{T_u} is given by (5.3,6), and an examination of the data for a typical constant-speed propeller at low speed[3] showed that $\partial T/\partial u$ is very small. Hence we have used $C_{T_u} = -2C_{T_0}$ in this example.

[3]The De Havilland Buffalo airplane.

Table 6.5
Basic Data for STOL Airplane

C_{W_0}	C_{T_0}	C_{L_α}	C_{D_α}†	h_n‡	$\dfrac{\partial C_L}{\partial C_T}$	$\dfrac{\partial C_D}{\partial C_T}$	$\dfrac{\partial C_m}{\partial C_T}$
2.0	0.53	5.75	1.19	0.500	0.705	0.285	−0.090
2.5	0.72	6.20	1.80	0.475	0.790	0.328	−0.070
3.0	0.90	6.65	2.41	0.450	0.875	0.370	−0.050
3.5	1.09	7.10	3.02	0.424	0.955	0.411	−0.030
4.0	1.28	7.55	3.63	0.398	1.025	0.450	−0.010
4.5	1.46	8.00	4.24	0.371	1.097	0.488	+0.010
5.0	1.65	8.45	4.85	0.346	1.165	0.525	+0.030

†$C_{D_\alpha} = \dfrac{\partial C_D}{\partial C_L} C_{L_\alpha}$

‡$h_n = 0.30 - C_{m_\alpha}/C_{L_\alpha}$

Using the formulae of Table 5.1, the following estimates were made of the q and $\dot{\alpha}$ derivatives:

$$C_{z_q} = -14, \qquad C_{m_q} = -17.9, \qquad C_{z_{\dot{\alpha}}} = -5.5, \qquad C_{m_{\dot{\alpha}}} = -13$$

Finally the following inertial and geometric characteristics were assumed:

$$W = 40{,}000 \text{ lb } (177{,}920 \text{ N}), \qquad S = 1000 \text{ ft}^2 \ (92.9 \text{ m}^2),$$
$$A = 5.42, \qquad \bar{c} = 13.60 \text{ ft } (4.145\text{m}),$$
$$\mu = 76.8, \qquad \hat{I}_y = 385, \qquad h = 0.30$$

With the above data, the coefficients of the system matrix were evaluated, and its eigenvalues and eigenvectors calculated. The main results are shown on Figs. 6.10–6.13. Figures 6.10 and 6.11 show the loci of the roots as C_{W_0} varies between 2 and 5. The effect of C_{W_0} is seen to be large on both modes, the short-period mode becoming nonoscillatory at a value of C_{W_0} somewhat greater than 3.5, and the damping

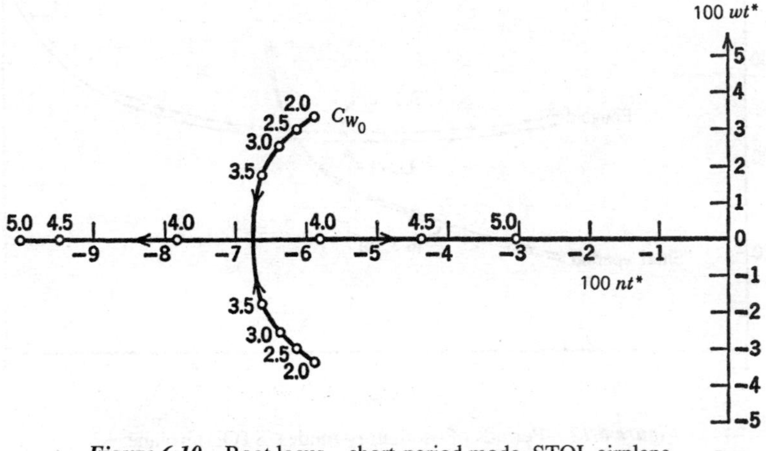

Figure 6.10 Root locus—short-period mode, STOL airplane.

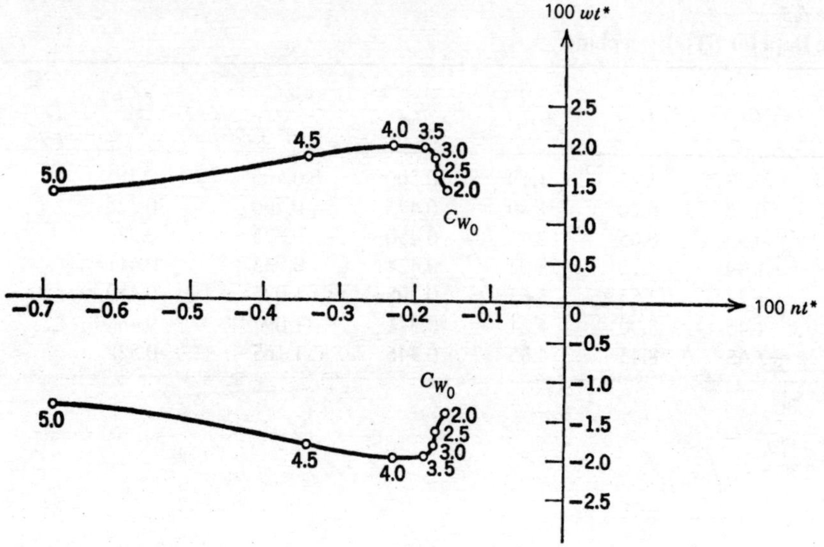

Figure 6.11 Root locus—phugoid mode, STOL airplane.

of the phugoid increasing rapidly at the same time. Figure 6.12 shows the two periods, and that they actually cross over at $C_{W_0} \doteq 3.4$. The concept of the phugoid as a "long" period oscillation is evidently not applicable in this situation! The approximations to the phugoid and the pitching mode are also shown for comparison. It is seen that they give the two periods quite well, and that (6.3,15) also depicts quite accurately the damping of the pitching oscillation and of the two nonperiodic modes into which it degenerates at high C_{W_0}. The phugoid damping, however, is not at all well predicted by the approximate solution. Figure 6.13 shows that the modes are all heavily damped over the whole range of C_{W_0}.

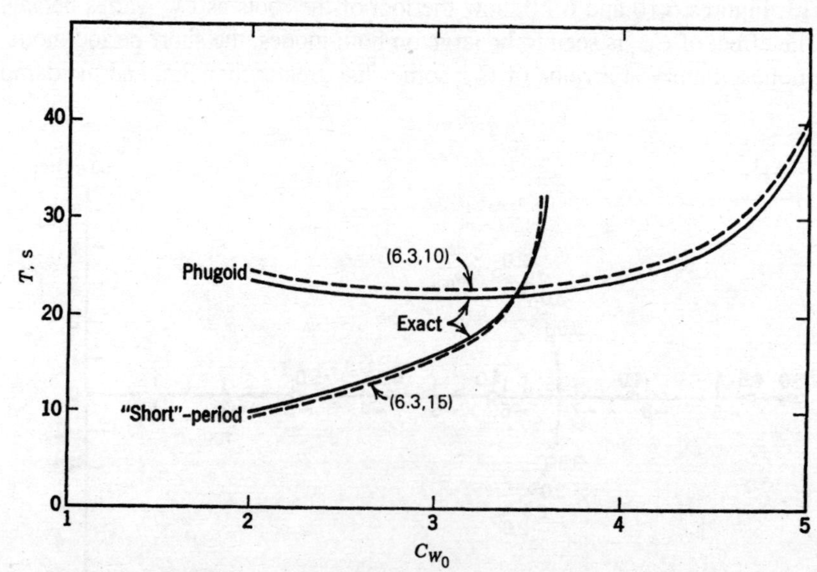

Figure 6.12 Periods of oscillatory modes, STOL airplane.

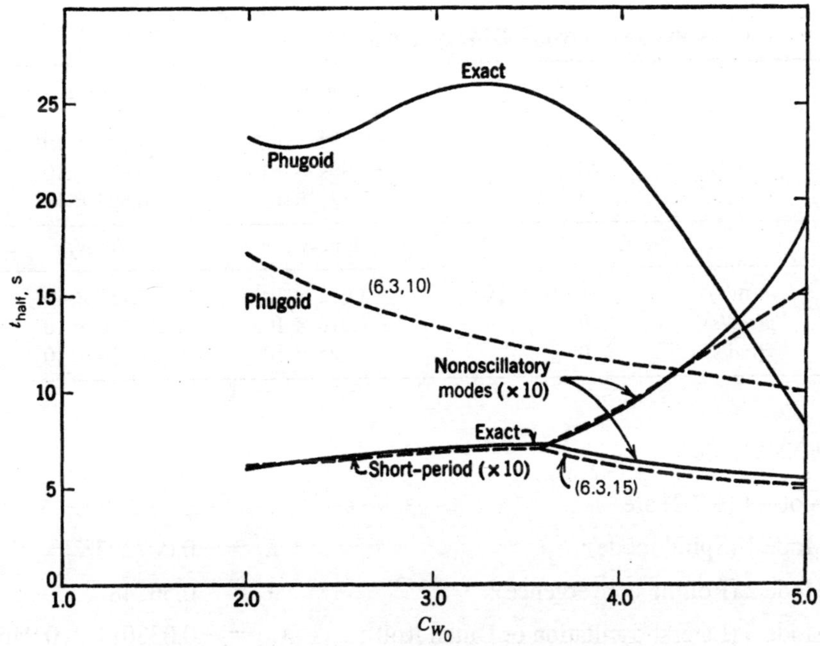

Figure 6.13 Time to damp of modes, STOL airplane.

6.7 Lateral Modes of a Jet Transport

We use the same airplane and flight condition as for the longitudinal modes in Sec. 6.2, and calculate the lateral modes. The nondimensional and dimensional derivatives are given in Tables 6.6 and 6.7. Using these, the system matrix of (4.9,19) is found to be (note that the state vector is $[v\ \ p\ \ r\ \ \phi]^T$.):

$$\mathbf{A} = \begin{bmatrix} -0.0558 & 0 & -774 & 32.2 \\ -0.003865 & -0.4342 & 0.4136 & 0 \\ 0.001086 & -0.006112 & -0.1458 & 0 \\ 0 & 1 & 0 & 0 \end{bmatrix}$$

This yields the characteristic equation

$$\lambda^4 + 0.6358\lambda^3 + 0.9388\lambda^2 + 0.5114\lambda + 0.003682 = 0 \qquad (6.7,2)$$

The stability criteria are

$$E = 0.003682 > 0$$
$$R = 0.04223 > 0$$

so there are no unstable modes.

Table 6.6
Nondimensional Derivatives—B747 Airplane

	C_y	C_l	C_n
β	−0.8771	−0.2797	0.1946
$\hat{p}$	0	−0.3295	−0.04073
$\hat{r}$	0	0.304	−0.2737

Table 6.7
Dimensional Derivatives—B747 Airplane

	$Y(lb)$	$L(ft\cdot lb)$	$N(ft\cdot lb)$
v(ft/s)	-1.103×10^3	-6.885×10^4	4.790×10^4
p(rad/s)	0	-7.934×10^6	-9.809×10^5
r(rad/s)	0	7.321×10^6	-6.590×10^6
	$Y(N)$	$L(m\cdot N)$	$N(m\cdot N)$
v(m/s)	-1.610×10^4	-3.062×10^5	2.131×10^5
p(rad/s)	0	-1.076×10^7	-1.330×10^6
r(rad/s)	0	9.925×10^6	-8.934×10^6

EIGENVALUES

The roots of (6.7,2) are

Mode 1 (Spiral mode): $\lambda_1 = -0.0072973$

Mode 2 (Rolling convergence): $\lambda_2 = -0.56248$

Mode 3 (Lateral oscillation or Dutch Roll): $\lambda_{3,4} = -0.033011 \pm 0.94655i$

Table 6.8 shows the characteristic times of these modes. We see that two of them are convergences, one very rapid, one very slow, and that one is a lightly damped oscillation with a period similar to that of the longitudinal short-period mode.

EIGENVECTORS

The eigenvectors corresponding to the above eigenvalues are given in Table 6.9. In addition to the basic 4 state variables, Table 6.9 contains two extra rows that show the values of the two state variables ψ and y_E (see Exercise 6.2).

MODE 1: THE SPIRAL MODE

From Table 6.9, we find the ratios of the angle variables in the spiral mode to be

$$\beta:\phi:\psi = -0.00119:-0.177:1$$

so that the motion is seen to consist mainly of yawing at nearly zero sideslip with some rolling. This is, of course, the condition for a truly banked turn, and this mode

Table 6.8
Characteristic Times—Lateral Modes

Mode	*Name*	*Period (s)*	$t_{half}(s)$	N_{half} *(cycles)*
1	Spiral	—	95	—
2	Rolling convergence	—	1.23	—
3	Lateral oscillation (Dutch Roll)	6.64	21	3.16

Table 6.9
Eigenvectors (polar form)

	Spiral		*Rolling convergence*		*Dutch Roll*	
	Magnitude	*Phase*	*Magnitude*	*Phase*	*Magnitude*	*Phase*
$\beta = \hat{v}$	0.00119	180°	0.0198	180°	0.33	−28.1°
$\hat{p}$	1.63×10^{-4}	0°	0.0712	180°	0.12	92.0°
$\hat{r}$	9.20×10^{-4}	180°	0.0040	0°	0.037	−112.3°
ϕ	0.177	180°	1.0	0°	1.0	0°
ψ	1.0	0°	0.0562	180°	0.31	155.7°
$\dfrac{y_E}{u_0 t^*}$	7.772×10^3	180°	7.65	0°	1.69	−165.8°

can be thought of as a variable-radius turn. The aerodynamically important variables are

$$\beta{:}\hat{p}{:}\hat{r} = 1{:}-0.137{:}0.773$$

and the largest of these, β, has already been seen to be negligibly small for moderate values of ϕ and ψ. The aerodynamic forces in this mode are therefore very small, and it may be termed a "weak" mode. This is consistent with its long time constant.

The flight path in the spiral mode can readily be constructed for any given initial yaw angle from the eigenvector. For example, with an initial ψ of 20° (0.35 rad), we have from Table 6.9

$$\psi = 0.35e^{\lambda_1 t}$$
$$v = 774(-0.00119)0.35e^{\lambda_1 t} \text{ fps}$$

where

$$\lambda_1 = -0.0072973 \text{ s}^{-1}$$

From (4.9,19) and the above it follows that (for $\theta_0 = 0$)

$$y_E = -37079e^{-0.0072973t} \text{ ft}$$
$$x_E = 774\, t \text{ ft}$$

Figure 6.14 shows the path—it is seen to be a long, smooth return to the reference flight path, corresponding to $y_E = 0$. When the spiral mode is unstable as is frequently the case, ψ, ϕ, and y_E are all of the same sign, and of course all increase with time instead of decreasing, as shown in the figure.

MODE 2: THE ROLLING CONVERGENCE

The ratios of the angle variables in this mode are, from Table 6.9,

$$\beta{:}\phi{:}\psi = -0.0198{:}1{:}-0.0562$$

The mode is evidently one of almost pure rotation around the x axis, and hence its name. The variables that are significant for aerodynamic forces are $(\beta, \hat{p}, \hat{r})$ and they are in the ratios

$$\beta{:}\hat{p}{:}\hat{r} = 0.278{:}1{:}-0.0561$$

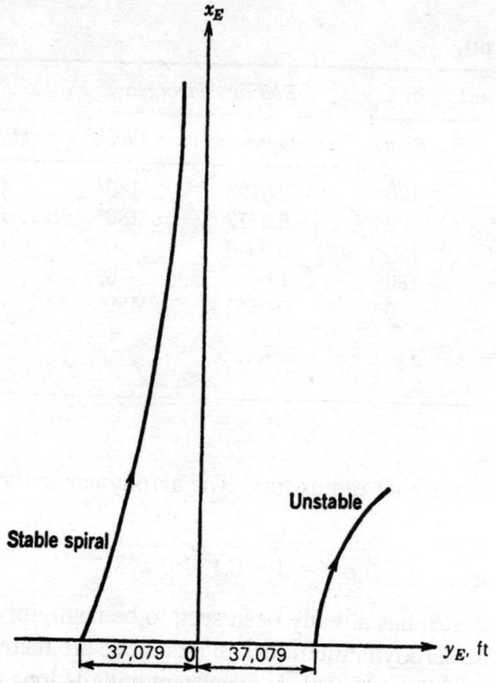

Figure 6.14 Flight path in spiral mode.

so that the largest rolling moment in this mode of motion is $C_{l_p}\hat{p}$, and the $\hat{r}$ contributions are negligible by comparison.

MODE 3: THE LATERAL OSCILLATION (DUTCH ROLL)

The vector diagram for this mode is shown in Fig. 6.15. It is seen that the three angle variables β, ϕ, ψ are of the same order of magnitude, that $\hat{r}$ is an order smaller, and

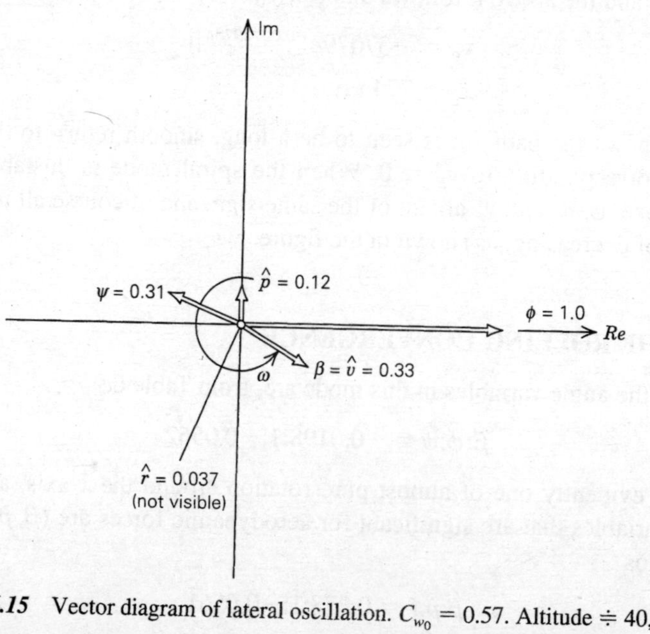

Figure 6.15 Vector diagram of lateral oscillation. $C_{w_0} = 0.57$. Altitude $\doteqdot$ 40,000 ft.

that β and ψ are almost equal and opposite. It follows from (4.9,19) that $\dot{y}_E$ is nearly zero. In dimensional terms, when $|\phi| = 20°$, $|y_E| \doteq 8$ ft, whereas the wavelength of the oscillation is about 5000 ft. The vehicle mass center is seen to follow a nearly rectilinear path in this mode, the motion consisting mainly of yawing and rolling, the latter lagging the former by about 160° in phase.

EFFECT OF SPEED AND ALTITUDE

Even for the "basic" case of a rigid airplane at low Mach number, the variation of the lateral modes with speed and altitude may not be simple. This is because some of the lateral stability derivatives are dependent on the lift coefficient in complex ways. That is especially true of airplanes with swept wings and low aspect ratio for which C_{l_β} increases markedly with C_L. These effects will appear most strongly at low speed and high altitude, both of which require high C_L. (Note that in the B747 example at **M** = 0.8 and 40,000 ft, $C_L = 0.654$, which is quite large for cruising flight.) For a rigid swept-wing airplane at low Mach number the period of the Dutch Roll mode would be expected first to increase and then to decrease as the airplane speed increases. The damping of this mode would be expected to be weak at low speed and to increase at higher speeds. The rolling convergence is well damped at all speeds, but the damping would normally increase with speed. The spiral mode is frequently unstable over some portion of the speed/altitude flight envelope, depending on the interplay of the derivatives that appear in (6.8,6). The characteristic times of this mode are, however, usually so long that the instability does not degrade the handling qualities unduly.

The effect of increasing altitude at fixed C_L is primarily an increase in the damping time constants of all the modes. The period of the Dutch Roll is not much affected.

When substantial aeroelastic and compressibility effects are added to the already complex behavior of the lateral modes, the result is an even more irregular pattern of modal characteristics. The data of (Heffley and Jewel, 1972) for the B747, reproduced in Table 6.10 show this. (Note that a negative t_{half} implies an unstable mode.) At the two lower altitudes, with relatively low values of C_L, the modes are seen to behave in a fairly regular way (see Fig. 6.16). However, at 40,000 ft and high Mach

Table 6.10
Variation of Lateral Modes with Speed and Altitude

		Spiral mode	*Rolling convergence*	*Lateral oscillation (Dutch Roll)*	
Altitude, ft	*Mach No.*	t_{half} *(s)*	t_{half} *(s)*	*Period (s)*	N_{half} *(cycles)*
0	0.45	35.7	0.56	5.98	0.87
0	0.65	34.1	0.44	4.54	0.71
20,000	0.5	76.7	0.93	7.3	1.58
20,000	0.65	64.2	0.76	5.89	1.33
20,000	0.8	67.3	0.85	4.82	1.12
40,000	0.7	−296	1.5	7.99	1.93
40,000	0.8	94.9	1.23	6.64	3.15
40,000	0.9	−89.2	1.45	6.19	1.18

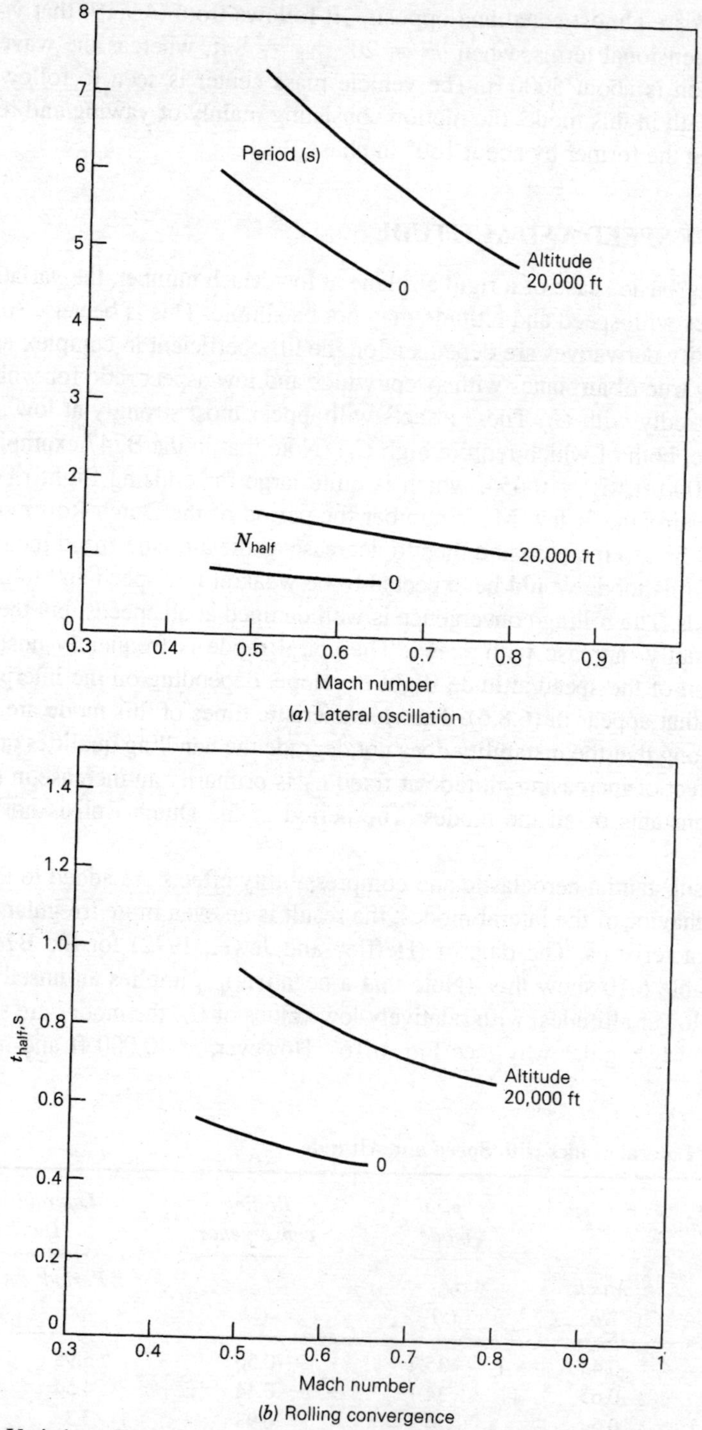

Figure 6.16 Variation of lateral modes with speed and altitude. (*a*) Lateral oscillation. (*b*) Rolling convergence. (*c*) Spiral mode.

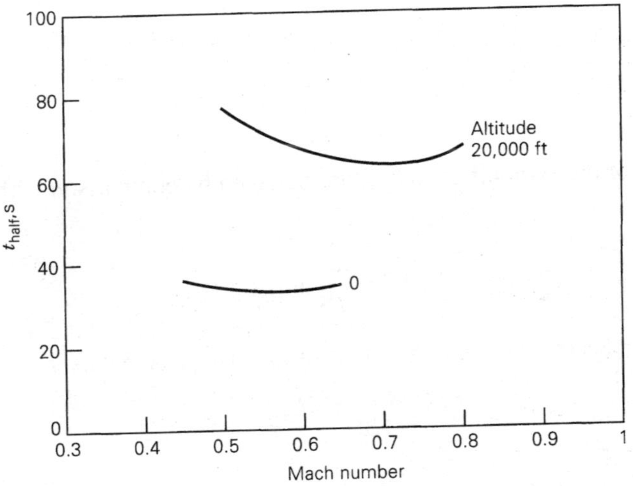

(c) Spiral mode

Figure 6.16 (*Continued*)

number the lateral behavior is quite irregular, especially the variation of the damping of all the modes with **M**. The spiral mode is seen to be unstable (albeit with long time constant) at both **M** = 0.7 and 0.9 but stable at **M** = 0.8. This behavior is primarily the result of the complex variation of C_{l_β} with C_L and **M** in this region.

6.8 Approximate Equations for the Lateral Modes

As with the longitudinal modes we should like if possible to have useful analytical approximations to the lateral characteristics. We find that there are reasonable approximations to all three modes, but the application of all such approximations must be made with caution. Their accuracy can really be verified only a posteriori, by comparison with exact solutions. They can only be used with confidence in situations similar to those in which they have previously been found to work well.

SPIRAL MODE

Comparison of the eigenvalues in Sec. 6.7 shows that λ for the spiral mode is two orders of magnitude smaller than the next larger one. This suggests that a good approximation to this root may be obtained by keeping only the two lowest-order terms in the characteristic equation, that is,

$$D\lambda + E \doteq 0 \qquad (6.8,1)$$

or

$$\lambda_S \doteq -E/D$$

where λ_S denotes the real root for the spiral mode. Before deriving expressions for D and E, we rewrite the matrix of (4.9,19) in a more compact notation for convenience, including the approximation $Y_p = 0$.

$$\mathbf{A} = \begin{bmatrix} \mathcal{Y}_v & 0 & \mathcal{Y}_r & g\cos\theta_0 \\ \mathcal{L}_v & \mathcal{L}_p & \mathcal{L}_r & 0 \\ \mathcal{N}_v & \mathcal{N}_p & \mathcal{N}_r & 0 \\ 0 & 1 & \tan\theta_0 & 0 \end{bmatrix} \tag{6.8,2}$$

The meanings of the symbols in (6.8,2) are obtained by comparison with (4.9,19), for example

$$\mathcal{L}_v = \frac{L_v}{I'_x} + I'_{zx} N_v$$

and in the special case when the stability axes are also principal axes, $I_{zx} = 0$ and

$$\mathcal{L}_v = \frac{L_v}{I_x}$$

With the notation of (6.8,2), expanding det $(\mathbf{A} - \lambda\mathbf{I})$ yields

$$\begin{aligned} E &= g[(\mathcal{L}_v\mathcal{N}_r - \mathcal{L}_r\mathcal{N}_v)\cos\theta_0 + (\mathcal{L}_p\mathcal{N}_v - \mathcal{L}_v\mathcal{N}_p)\sin\theta_0] \quad (a) \\ D &= -g(\mathcal{L}_v\cos\theta_0 + \mathcal{N}_v\sin\theta_0) + \mathcal{Y}_v(\mathcal{L}_r\mathcal{N}_p - \mathcal{L}_p\mathcal{N}_r) \\ &\quad + \mathcal{Y}_r(\mathcal{L}_p\mathcal{N}_v - \mathcal{L}_v\mathcal{N}_p) \quad (b) \end{aligned} \tag{6.8,3}$$

When the orders of the various terms in D are compared, it is found that the second term can be neglected entirely and Y_r can be neglected in $\mathcal{Y}_r$. The approximation that then results is

$$D = -g(\mathcal{L}_v\cos\theta_0 + \mathcal{N}_v\sin\theta_0) + u_0(\mathcal{L}_v\mathcal{N}_p - \mathcal{L}_p\mathcal{N}_v) \tag{6.8,4}$$

The result obtained from (6.8,1), (6.8,3*a*), and (6.8,4) for the jet transport example of Sec. 6.7 is $\lambda_s = -0.00725$, less than 1% different from the correct value. Equation (6.8,1) is seen to give a good approximation in this case.

It will be recalled that the coefficient E has special significance with respect to static stability (see Sec. 6.1). We note here that in consequence of (6.8,1) the spiral mode may exhibit exponential growth, and that the criterion for static lateral stability is

$$(\mathcal{L}_v\mathcal{N}_r - \mathcal{L}_r\mathcal{N}_v)\cos\theta_0 + (\mathcal{L}_p\mathcal{N}_v - \mathcal{L}_v\mathcal{N}_p)\sin\theta_0 > 0 \tag{6.8,5}$$

On substituting the expanded expressions for $\mathcal{L}_v$ and so forth, (6.8,5) reduces to

$$(C_{l_\beta}C_{n_r} - C_{l_r}C_{n_\beta})\cos\theta_0 + (C_{l_p}C_{n_\beta} - C_{l_\beta}C_{n_p})\sin\theta_0 > 0 \tag{6.8,6}$$

Since some of the derivatives in (6.8,6) depend on C_{L_0}, the static stability will vary with flight speed. It is not at all unusual for the spiral mode to be unstable over some portion of the flight envelope (see Table 6.10 and Exercise 6.3).

ROLLING MODE

It was observed in Sec. 6.7 that the rolling convergence is a motion of almost a single degree of freedom, rotation about the x-axis. This suggests that it can be approximated with the equation obtained from (4.9,19) by putting $v = r = 0$, and considering only the second row, that is,

$$\dot{p} = \mathcal{L}_p p \tag{6.8,7}$$

which gives the approximate eigenvalue

$$\lambda_R \doteq \mathcal{L}_p = L_p/I'_x + I'_{zx}N_p \tag{6.8,8}$$

The result obtained from (6.8,8) for the B747 example is $\lambda_R = -0.434$, 23% smaller than the true value -0.562. This approximation is quite rough.

An alternative approximation has been given by McRuer et al. (1973). This approximation leads to a second-order system, the two roots of which are approximations to the roll and spiral modes. In some cases the roots may be complex, corresponding to a "lateral phugoid"—a long-period lateral oscillation. The approximation corresponds to the physical assumption that the side-force due to gravity produces the same yaw rate r that would exist with $\beta = 0$. Additionally Y_p and Y_r are neglected. With no approximation to the rolling and yawing moment equations the system that results for horizontal flight is

$$\begin{aligned}
0 &= -u_0 r + g\phi && (a)\\
\dot{p} &= \mathcal{L}_v v + \mathcal{L}_p p + \mathcal{L}_r r && (b)\\
\dot{r} &= \mathcal{N}_v v + \mathcal{N}_p p + \mathcal{N}_r r && (c)\\
\dot{\phi} &= p && (d)
\end{aligned} \tag{6.8,9}$$

The procedure used to get the characteristic equation of (6.8,9) is the same as that used previously for the phugoid approximation in Sec. 6.3. The result is

$$\begin{vmatrix} 0 & 0 & -u_0 & g \\ \mathcal{L}_v & (\mathcal{L}_p - \lambda) & \mathcal{L}_r & 0 \\ \mathcal{N}_v & \mathcal{N}_p & (\mathcal{N}_r - \lambda) & 0 \\ 0 & 1 & 0 & -\lambda \end{vmatrix} = 0 \tag{6.8,10}$$

which expands to

$$C\lambda^2 + D\lambda + E = 0 \qquad (a)$$

where

$$\begin{aligned}
C &= u_0\mathcal{N}_v && (b)\\
D &= u_0(\mathcal{L}_v\mathcal{N}_p - \mathcal{L}_p\mathcal{N}_v) - g\mathcal{L}_v && (c)\\
E &= g(\mathcal{L}_v\mathcal{N}_r - \mathcal{L}_r\mathcal{N}_v) && (d)
\end{aligned} \tag{6.8,11}$$

The result of applying (6.8,11) to the B747 example is

$$\lambda_s = -0.00734 \quad \text{and} \quad \lambda_R = -0.597$$

These are within about 1% and 6% of the true values, respectively, so this is seen to be a good approximation for both modes, certainly much better than (6.8,7) for the rolling mode.

DUTCH ROLL MODE

A physical model that gives an approximation to the lateral oscillation is a "flat" yawing/sideslipping motion in which rolling is suppressed. The corresponding equations are obtained from (4.9,19) by setting $p \equiv \phi \equiv 0$ and dropping the second (rolling moment) equation. The term in Y_r is also neglected in the first equation. The result is

$$\dot{v} = \mathcal{Y}_v v - u_0 r$$
$$\dot{r} = \mathcal{N}_v v + \mathcal{N}_r r$$

The corresponding characteristic equation is readily found to be

$$\lambda^2 - (\mathcal{Y}_v + \mathcal{N}_r)\lambda + (\mathcal{Y}_v \mathcal{N}_r + u_0 \mathcal{N}_v) = 0 \tag{6.8,12}$$

The result obtained from (6.8,12) for our example is $\lambda_{DR} = -0.1008 \pm 0.9157i$, or

$$T = 6.86 \text{ sec}$$
$$N_{\text{half}} = 1.0$$

The approximation for the period is seen to be useful (an error of about 3%) but the damping is very much overestimated.

There is another approximation available for the damping in this mode that may give a better answer. It follows from the fact that the coefficient of the next-to-highest power of λ in the characteristic equation is the "sum of the dampings" (see Exercise 6.4). Thus it follows from the complete system matrix of (6.8,2) that

$$2n_{DR} + \lambda_R + \lambda_S = \mathcal{Y}_v + \mathcal{L}_p + \mathcal{N}_r$$

or

$$n_{DR} = \tfrac{1}{2}\{\mathcal{Y}_v + \mathcal{L}_p + \mathcal{N}_r - (\lambda_R + \lambda_S)\} \tag{6.8,13}$$

But the approximation (6.8,11) for the roll and spiral modes gives precisely

$$\lambda_R + \lambda_S = -\frac{D}{C}$$

On using (6.8,11) we get the expression

$$n_{DR} = \frac{1}{2}\left[\mathcal{Y}_v + \mathcal{N}_r + \frac{\mathcal{L}_v}{\mathcal{N}_v}\left(\mathcal{N}_p - \frac{g}{u_0}\right)\right] \tag{6.8,14}$$

which is to be compared with $\frac{1}{2}(\mathcal{Y}_v + \mathcal{N}_r)$ given by (6.8,12). The damping obtained from (6.8,14) is $n_{DR} = -0.0159$, better than that obtained from (6.8,12) but still quite far from the true value of -0.0330. The simple average of the two preceeding approximations for the Dutch Roll damping has also been used. In this instance it gives $n_{DR} = -0.0584$, which although better is still 77% off the true value.

This example of an attempt to get an approximation to the Dutch Roll damping illustrates the difficulty of doing so. Although the approximation tends to be better at low values of C_L, nevertheless it is clear that it must be used with caution, and that only the full system matrix can be relied on to give the correct answer.

6.9 Effects of Wind

In all the preceding examples, the atmosphere has been assumed to be at rest or to have a velocity uniform in space and constant in time. Since this is the exceptional rather than the usual case, it is necessary to examine the effects of nonuniform and unsteady motion of the atmosphere on the behavior of flight vehicles. The principal effects are those associated with atmospheric turbulence, and these are treated at some length in Etkin (1972, 1981). However, quite apart from turbulence, the wind may have a mean structure which is not uniform in space, that is, there can be *spatial* gradients in the *time-averaged* velocity. The examples of most concern are downbursts and the boundary layer next to the ground produced by the wind blowing over it. Downbursts are vertical outflows from low level clouds that impinge on the ground, somewhat in the manner of a circular jet, and spread horizontally. The resulting wind field has strong gradients, both horizontal and vertical. A number of airplane accidents have been attributed to this phenomenon.

In order to introduce wind into the analytical model, we must make any alterations that may be needed, because of the presence of the wind, to the aerodynamic forces and moments. In the trivial case when the wind is uniform and steady, no change is necessary to the representation of aerodynamic forces from that used before. However, turbulence and wind gradients may require such changes.

Since the linear model that was developed in Chap. 4 is based on small disturbances from a steady reference condition, and since there is no such steady state when the aircraft is landing or taking off through a boundary layer or downburst, the linear model is of limited use in these situations. For this kind of analysis, one must use the nonlinear equations (4.7,1)–(4.7,5) and introduce a model for the aerodynamic forces that embraces the whole range of speeds and attitudes that will occur throughout the transient (Etkin, B. and Etkin, D. A. (1990)). Such an analysis is beyond the scope of this volume.

There is one relevant steady state, however, that can be investigated with the linear model, and that is horizontal flight in the boundary layer. The planetary boundary layer has characteristics quite similar to the classical flat-plate turbulent boundary layer of aerodynamics. The vertical extent of this layer in strong winds depends mainly on the roughness of the underlying terrain, but is usually many hundreds of feet. Figure 6.17 shows the power-law profiles associated with different roughnesses. These are all of the form

$$W = kh^n \tag{6.9,1}$$

where, as indicated in the figure, h is height above the ground. The vertical gradient is then given by

$$\frac{dW}{dh} = nkh^{n-1} \tag{6.9,2}$$

For example, for smooth terrain ($n = 0.16$), and for a wind of 50 fps at 50 ft altitude, the gradient would be $dW/dh = 0.16$ fps/ft.

By way of example, we shall analyze what effect the vertical wind gradient has on the longitudinal modes of the STOL airplane of Sec. 6.6 in low speed flight, when wind effects can be expected to be largest. In order to generate the analytical model, we go back to the exact equations (4.7,1) et seq. The longitudinal equations, when linearized for small perturbations around a reference state of horizontal flight at speed u_0 and $\theta_0 = 0$ are

$$\begin{aligned}
\Delta X - mg\theta &= m\Delta \dot{u}^E \\
\Delta Z &= m(\dot{w}^E - u_0^E q) \\
\Delta M &= I_y \dot{q} \\
\dot{\theta} &= q \\
\dot{z}_E &= -u_0^E\theta + w^E
\end{aligned} \tag{6.9,3}$$

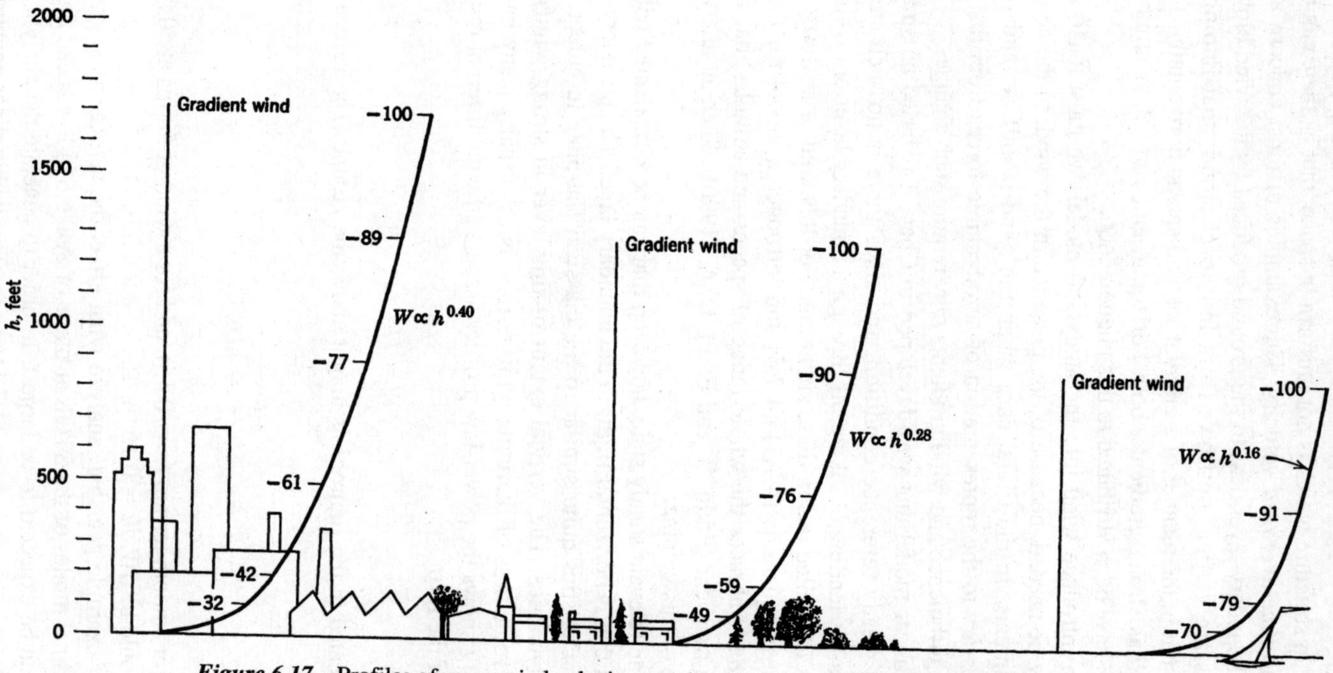

Figure 6.17 Profiles of mean wind velocity over level terrains of differing roughness. (After A. G. Davenport, 1960)

The wind is specified to vary linearly with altitude with gradient $\Gamma = dW/dz_E$ and to be parallel to the xz plane. The wind vector is prescribed in frame F_E as

$$\mathbf{W}_E = \begin{bmatrix} 1 \\ 0 \\ 0 \end{bmatrix} (W_0 + \Gamma z_E) \tag{6.9,4}$$

We will assume that this implies a tailwind of strength W_0 at the reference height. For a wind that increases with altitude, $\Gamma = -\text{sgn}(W_0)|\Gamma|$. To convert $\mathbf{W}_E$ to body axes, we use the transformation matrix $\mathbf{L}_{BE}$ [the transpose of (4.4,3)] for small angles and $\phi = \psi = 0$ to get (see Exercise 6.6)

$$\mathbf{W}_B = \begin{bmatrix} W_0 + \Gamma z_E \\ 0 \\ (W_0 + \Gamma z_E)\theta \end{bmatrix} \approx \begin{bmatrix} W_0 + \Gamma z_E \\ 0 \\ W_0\theta \end{bmatrix} \tag{6.9,5}$$

In the third component of (6.9,5) we have neglected the second-order product $z_E\theta$. We can now use (6.9,5) together with $\mathbf{V}^E = \mathbf{V} + \mathbf{W}$ to eliminate u^E and w^E from (6.9,3) with the result

$$\begin{aligned} \Delta X - mg\theta &= m(\Delta\dot{u} + \Gamma\dot{z}_E) \\ \Delta Z &= m(\dot{w} + W_0\dot{\theta} - u_0^E q) \\ \Delta M &= I_y\dot{q} \\ \dot{\theta} &= q \\ \dot{z}_E &= -u_0^E\theta + w + W_0\theta \end{aligned} \tag{6.9,6}$$

We note that $u_0^E = u_0 + W_0$ and eliminate $\dot{\theta}$ and $\dot{z}_E$ from the right side of (6.9,6) with the result

$$\begin{aligned} \Delta X - mg\theta &= m(\Delta\dot{u} + \Gamma[-u_0\theta + w]) \\ \Delta Z &= m(\dot{w} - u_0 q) \\ \Delta M &= I_y\dot{q} \\ \dot{\theta} &= q \\ \dot{z}_E &= -u_0\theta + w \end{aligned} \tag{6.9,7}$$

It is observed that the only *explicit* effect of wind on the system equations in this case is the term containing Γ on the right side of the first equation. Since a uniform wind (i.e., $\Gamma = 0$) does not affect the airplane dynamics the system must be independent of W_0, and we see that W_0 does indeed not appear anywhere in the equations. In addition to what we see in (6.9,7), however, there are some *implicit* effects of the wind gradient on the aerodynamic derivatives.

It is clear that the changes in pressure distribution over the surfaces of a vehicle, and hence its basic aerodynamic derivatives, are not the same when the incident flow has a gradient Γ as when it is spatially uniform. Two simple examples suffice to make this clear. (1) When there is a perturbation $\Delta\alpha$ from the reference state, the tail moves downward into a region of lower air velocity and on this account one would expect $|C_{m_\alpha}|$ to be smaller than normal. (2) When the wing rolls through an angle ϕ, the right tip moves into a low-wind region, and the left tip into a high-wind region. The gradient in velocity across the span is like that associated with yaw rate r, and hence we should expect values of C_{l_ϕ} and C_{n_ϕ} proportional to the wing contributions to C_{l_r} and

C_{n_r}. Note that for upwind flight this leads to an *unstable* roll "stiffness" $C_{l_\phi} > 0$ where none existed before (C_{l_ϕ} is negative for downwind flight).

Reasonable estimates of the major changes in the basic derivatives associated with Γ can be made from available aerodynamic theories, but a complete account of these is not currently available, and to develop them here would take us too far afield. Instead we simply incorporate the additional terms given by (6.9,7) into the longitudinal equations of motion and note the extent of the changes they make in the characteristic modes previously calculated. The appropriate matrix for this case is obtained from (4.9,18) by adding the two terms containing Γ, and is given by (6.9,8).

$$\mathbf{A} = \begin{bmatrix} \dfrac{X_u}{m} & \left(\dfrac{X_w}{m} - \Gamma\right) & 0 & (-g + \Gamma u_0) \\ \hline & \mathbf{A}' & & \end{bmatrix} \tag{6.9,8}$$

In (6.9,8), $\mathbf{A}'$ is the 3×4 matrix consisting of the last three rows of the matrix of (4.9,18), with $\theta_0 = 0$.

An example of the results for the STOL airplane is shown in Figs. 6.18 and 6.19. The numerical data used was the same as in Sec. 6.6, with wind gradient variable from -0.30 fps/ft (the headwind case) to $+0.30$ fps/ft (the tailwind case). The effects on both the phugoid and pitching modes are seen to be large. A strong headwind decreases both the frequency and damping of the phugoid, and a strong tailwind changes the real pair of pitching roots into a complex pair representing a pitching oscillation of long period and heavy damping.

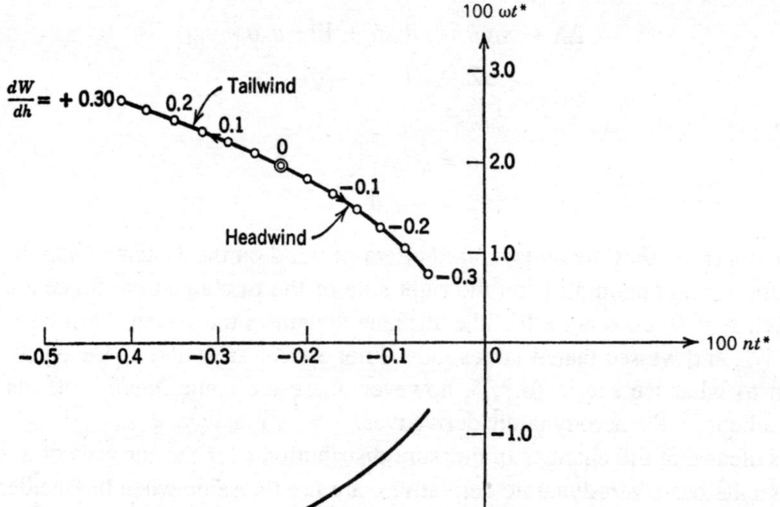

Figure 6.18 Effect of wind gradient on phugoid roots—STOL airplane. $C_{w_0} = 4.0$.

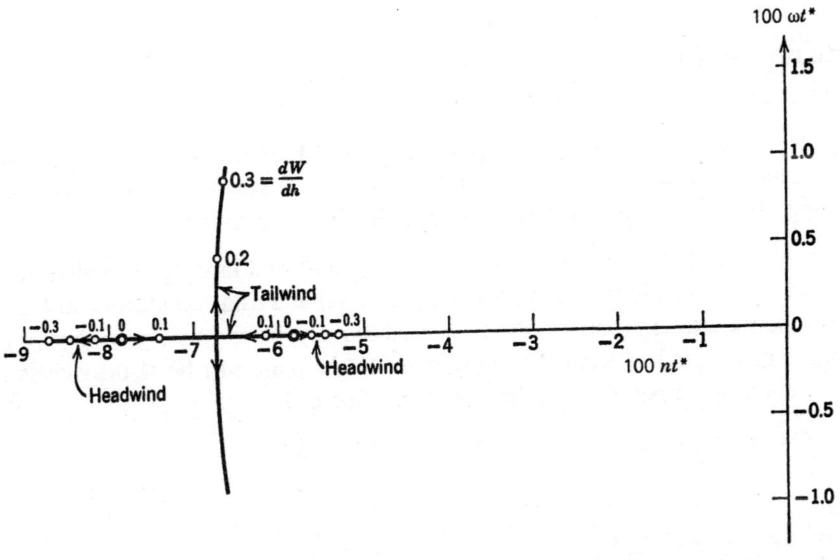

Figure 6.19 Effect of wind gradient on short-period roots—STOL airplane. $C_{w_0} = 4.0$.

6.10 Exercises

6.1 Use (6.3,10) to calculate the approximate period $T_n = 2\pi/\omega_n$ of the phugoid oscillation. Assume $M_u = 0$.

6.2 Derive the entries for ψ and (y_E/u_0t^*) in Table 6.9 from those for $\hat{v}$ *and* $\hat{r}$.

6.3 The stability derivatives of a general aviation airplane are given in Table 7.2. The airplane weighs 2400 lb (10,675 N) and has a wing area of 160 ft² (14.9 m²). The flight altitude is sea level. Calculate and plot the spiral stability criterion E as a function of speed $(0.15 < C_L < 1.7)$ for values of $\theta_0 = -10°, 0°, 10°$.

6.4 The characteristic equation is of order N. Prove that the coefficient of λ^{N-1} is the negative of the sum of the real parts of all the roots, and hence is aptly termed "the sum of the dampings."

6.5 Find the critical climb angle for spiral stability of the jet transport of Sec. 6.7. [Hint: start with (6.8,6)]. Having regard to its expected influence on the stability derivatives, state the effect on spiral stability in horizontal flight of increasing the wing dihedral angle.

6.6 Carry out the transformation $\mathbf{W}_B = \mathbf{L}_{BE}\mathbf{W}_E$ to get (6.9,5).

6.7 Using the stability derivatives given in Table 7.2 for a general aviation airplane, calculate the lateral modes *in the absence of gravity.* The relevant data are:

$W = 2400$ lb (10,675 N)	$I_x = 170$ slug·ft² (230 kg·m²)
$S = 160$ ft² (14.9 m²)	$I_z = 1{,}312$ slug·ft² (1,778 kg·m²)
$b = 30$ ft (9.14 m)	$I_{zx} = 0$

$V = 150$ knots (77.3 m/s) $\theta_0 = 0$

altitude = sea level

Compare the results with those for gravity present.

6.8 Find the characteristic equation of the hovercraft of Exercise 4.10. Show that when it is statically unstable with both M_θ and L_ϕ positive it can be gyrostabilized (like a spinning top, i.e., solutions remain bounded) if H is large enough.

6.9 A conventional stable aircraft is on a steady descent to a landing on a shallow glide slope when the headwind suddenly vanishes. What initial condition problem describes the subsequent motion? Describe qualitatively, from your knowledge of longitudinal natural modes, what the subsequent flight path will be if the elevator and throttle controls remain fixed at their prior positions.

6.10 Show that if we neglect all the Y force derivatives, then the small-disturbance equations (for $\theta_0 = 0$) yield the following approximation for the lateral displacement:

$$\Delta y_E(t) = g \int_0^t \int_0^t \phi(\tau)\, d\tau\, dt$$

6.11 Assume that in the lateral oscillation of an airplane, the modal diagram shows that $|\psi| > |\phi|$ and the $\boldsymbol{\psi}$ vector leads the $\boldsymbol{\phi}$ vector by an angle between 90° and 180°. This is reasonably representative of flight at low Mach number. Describe the relative motion of the two wing tips. To simplify the situation, neglect damping, the motion associated with sideslip and with the forward motion of the airplane, assume the angles are "small," and interpret "wing tip" to mean a point on the y-axis.

Hold a model airplane by its wing tips and practise until you can execute a motion of the type you have deduced. Observe it in a mirror. (The name Dutch Roll was given to this motion because of a perceived resemblance to that of an ice skater, Holland being noted for this sport.)

6.12 The theory for a stable airplane shows that if the controls are neutral it will fly in a steady state on a straight line at constant speed. However, the development in the text assumed that the airplane produced no lateral aerodynamic force or moments when the controls were set to neutral and the lateral state variables were zero. Real airplanes cannot achieve this owing to design and construction factors. You are asked in this exercise to examine the steady states that are possible for such an airplane. The steady state here is defined to be one in which the linear and angular velocity components (u, v, w, p, q, r), the aerodynamic forces and moments (X, Y, Z, L, M, N) and the gravity force components (X_g, Y_g, Z_g) are all constant.

(a) Show that in the steady state the two Euler angles θ and ϕ are also constant.

(b) Since θ is constant, it can be set to zero by a suitable choice of body axes. For this case show that $p = 0$ and the angular velocity vector $\boldsymbol{\omega}$ is vertical in F_E.

(c) Starting with (4.5,8) and (4.5,9) develop the lateral steady state equations. Assume that in the steady state q, r, ϕ and v are small quantities. Assume that the controls are neutral and thus (since $p = 0$ in the steady state)

$$Y = Y_a + Y_v v + Y_r r$$

$$L = L_a + L_v v + L_r r$$

$$N = N_a + N_v v + N_r r$$

where ( )a stands for effects due to lateral asymmetries in the aircraft. What condition must be met in order for a unique steady state to exist? How is this condition related to E of (6.8,3a) when the body axes are principal axes?

(d) Relate the results of (c) to the flight of a statically stable hand-launched glider.

6.11 Additional Symbols Introduced in Chapter 6

h	altitude
$\mathcal{L}_v$	$L_v/I'_x + I'_{zx}N_v$
$\mathcal{L}_p$	$L_p/I'_x + I'_{zx}N_p$
$\mathcal{L}_r$	$L_r/I'_x + I'_{zx}N_r$
n	real part of eigenvalue
$\mathcal{N}_v$	$N_v/I'_z + I'_{zx}L_v$
$\mathcal{N}_p$	$N_p/I'_z + I'_{zx}L_p$
$\mathcal{N}_r$	$N_r/I'_z + I'_{zx}L_r$
T	period of oscillation
$\mathcal{Y}_v$	Y_v/m
$\mathcal{Y}_r$	$Y_r/m - u_0$
Γ	vertical wind gradient, $-\mathrm{d}W/\mathrm{d}h$
κ	density gradient parameter (see 6.5,3)
λ	eigenvalue
λ_{DR}	eigenvalue of lateral oscillation
λ_R	eigenvalue of rolling convergence
λ_S	eigenvalue of spiral mode
ζ	damping ratio
ω	imaginary part of eigenvalue, circular frequency
ω_n	undamped circular frequency

CHAPTER 7

Response to Actuation of the Controls-Open Loop

7.1 General Remarks

In this chapter we study how an airplane responds to actuation of the primary controls—elevator, aileron/spoiler, rudder, and throttle. These are of course not the only controls that can be incorporated in the design of an airplane. Also used, less frequently, are vectored thrust and direct lift control. Closely related to the control-response problem is the response of the airplane to an in-flight change of configuration such as flap deflection, lowering the undercarriage, releasing stores or armaments, deploying dive brakes, or changing wing sweep. The analysis of the response of the airplane to any of these uses methods generally similar to those that are described in the following. In the remainder of this chapter, it is assumed that there is no wind.

LONGITUDINAL CONTROL

The two principal quantities that need to be controlled in symmetric flight are the speed and the flight-path angle, that is to say, the vehicle's velocity vector. To achieve this obviously entails the ability to apply control forces both parallel and perpendicular to the flight path. The former is provided by thrust or drag control, and the latter by lift control via elevator deflection or wing flaps. It is evident from simple physical reasoning (or from the equations of motion) that the main *initial* response to opening the throttle (increasing the thrust) is a forward acceleration, i.e. control of speed. The main *initial* response to elevator deflection is a rotation in pitch, with subsequent change in angle of attack and lift, and hence a rate of change of flight-path direction. When the transients that follow such control actions have ultimately died away, the new steady state that results can be found in the conventional way used in performance analysis. Figure 7.1 shows the basic relations. The steady speed V at which the airplane flies is governed by the lift coefficient, which is in turn fixed by the elevator angle—see Fig. 2.19. Hence a constant δ_e implies a fixed V. The flight-path angle $\gamma = \theta - \alpha_x$ at any given speed is determined, as shown in Fig. 7.1, by the thrust. Thus the *ultimate* result of moving the throttle at fixed elevator angle (when the thrust line passes through the CG) is a change in γ without change in speed. But we saw above that the initial response to throttle is a change in speed—hence the short-term and long-term effects of this control are quite contrary. Likewise we saw that the main initial effect of moving the elevator is to rotate the vehicle and influence γ, whereas the ultimate effect at fixed throttle is to change both speed and γ. The short-term and long-term effects of elevator motion are therefore also quite different. The

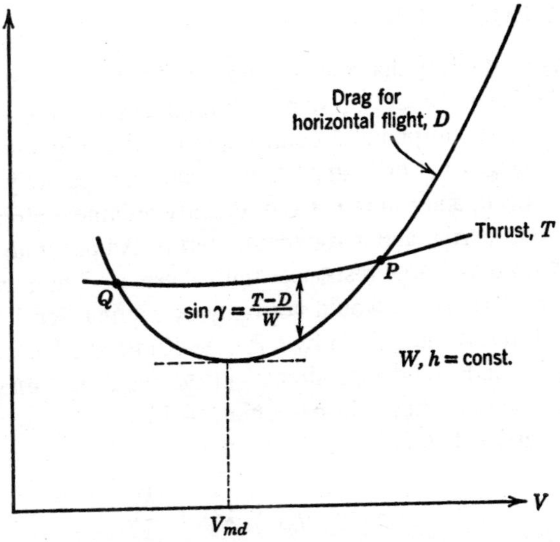

Figure 7.1 Basic performance graph.

total picture of longitudinal control is clearly far from simple, and the transients that connect the initial and final responses require investigation. We shall see in the following that these are dominated by the long-period, lightly damped phugoid oscillation, and that the final steady state with step inputs is reached only after a long time. These matters are explored more fully in the following sections.

LATERAL CONTROL

The lateral controls (the aileron and rudder) on a conventional airplane have three principal functions.

1. To provide trim in the presence of asymmetric thrust associated with power plant failure.
2. To provide corrections for unwanted motions associated with atmospheric turbulence or other random events.
3. To provide for turning maneuvers—that is, rotation of the velocity vector in a horizontal plane.

The first two of these purposes are served by having the controls generate aerodynamic moments about the x and z axes—rolling and yawing moments. For the third a force must be provided that has a component normal to $\mathbf{V}$ and in the horizontal plane. This is, of course, the component $L \sin \phi$ of the lift when the airplane is banked at angle ϕ. Thus the lateral controls (principally the aileron) produce turns as a secondary result of controlling ϕ.

Ordinarily, the long-term responses to deflection of the aileron and rudder are very complicated, with all the lateral degrees of freedom being excited by each. Solution of the complete nonlinear equations of motion is the only way to appreciate these fully. Certain useful approximations of lower order are however available.

THE CONTROL EQUATIONS

Whereas the study of stability that was the subject of Chap. 6 is generally sufficiently well served by the linear model of small disturbances from a condition of steady flight, the response of an airplane to control action or configuration change can involve very large changes in some important variables, especially bank angle, pitch angle, load factor, speed, and roll rate. Consequently nonlinear effects may be present in any of the gravity, inertia, and aerodynamic terms. An accurate system model capable of dealing with these large responses must therefore begin with the more exact equations (4.7,1)–(4.7,4). These would normally be reorganized into first-order state space form for subsequent integration by a Runge-Kutta or other integration scheme. Equations (4.7,3*d*–*f*) and (4.7,4) are already in the required form. However (4.7,1) and (4.7,2) need to be rearranged. In particular, (4.7,2*a* and *c*) need to be solved simultaneously for $\dot{p}$ and $\dot{r}$. The functional form that results is as follows:

$$\begin{aligned}
\dot{u}^E &= f(v^E, w^E, q, r\,\theta, X') + \frac{X_c}{m} && (a)\\
\dot{v}^E &= f(u^E, w^E, p, r, \theta, \phi, Y') + \frac{Y_c}{m} && (b) \quad (7.1,1)\\
\dot{w}^E &= f(u^E, v^E, p, q, \theta, \phi, Z') + \frac{Z_c}{m} && (c)
\end{aligned}$$

$$\begin{aligned}
\dot{p} &= f(p, q, r, L', N') + \frac{L_c}{I'_x} + I'_{zx}N_c && (a)\\
\dot{q} &= f(p, r, M') + \frac{M_c}{I_y} && (b) \quad (7.1,2)\\
\dot{r} &= f(p, q, r, L', N') + \frac{N_c}{I'_z} + I'_{zx}L_c && (c)
\end{aligned}$$

$$\begin{aligned}
\dot{\theta} &= f(q, r, \phi) && (a)\\
\dot{\phi} &= f(p, q, r, \theta, \phi) && (b) \quad (7.1,3)
\end{aligned}$$

In the preceding equations, the subscript c denotes the control forces and moments, and the prime on force and moment symbols denotes the remainder of the aerodynamic forces and moments. The solution of these equations would require that an aerodynamic submodel be constructed for each case to calculate the forces and moments at each computing step from a knowledge of the state vector, the control vector, the current configuration, and the wind field. To follow this course *in extenso* would take us beyond the scope of this text, so for the most part the treatments that follow are restricted to the responses of *linear invariant* systems, that is, ones described by (4.9,20) with **A** and **B** constant viz

$$\dot{\mathbf{x}} = \mathbf{A}\mathbf{x} + \mathbf{B}\mathbf{c} \qquad (7.1,4)$$

Although we are thereby restricted to relatively small departures from the steady state, these responses are nevertheless extremely useful and informative. Not only do they reveal important dynamic features, but when used in the design and analysis of automatic flight control systems that are designed to maintain small disturbances they are in fact quite appropriate.

Table 7.1
Dimensional Control Derivatives

	X	Z	M
δ_e	$C_{x_{\delta_e}}\frac{1}{2}\rho u_0^2 S$	$C_{z_{\delta_e}}\frac{1}{2}\rho u_0^2 S$	$C_{m_{\delta_e}}\frac{1}{2}\rho u_0^2 S\bar{c}$
δ_p	$C_{x_{\delta_p}}\frac{1}{2}\rho u_0^2 S$	$C_{z_{\delta_p}}\frac{1}{2}\rho u_0^2 S$	$C_{m_{\delta_p}}\frac{1}{2}\rho u_0^2 S\bar{c}$

	Y	L	N
δ_a	$C_{y_{\delta_a}}\frac{1}{2}\rho u_0^2 S$	$C_{l_{\delta_a}}\frac{1}{2}\rho u_0^2 Sb$	$C_{n_{\delta_a}}\frac{1}{2}\rho u_0^2 Sb$
δ_r	$C_{y_{\delta_r}}\frac{1}{2}\rho u_0^2 S$	$C_{l_{\delta_r}}\frac{1}{2}\rho u_0^2 Sb$	$C_{n_{\delta_r}}\frac{1}{2}\rho u_0^2 Sb$

In the examples that follow, we use $\{\delta_e, \delta_p\}$ for the longitudinal controls, elevator and throttle; and $\{\delta_a, \delta_r\}$ for the lateral controls, aileron, and rudder. The aerodynamic forces and moments are expressed just like the stability derivatives in terms of sets of nondimensional and dimensional derivatives. The nondimensional set is the partial derivatives of the six force and moment coefficients $\{C_x, C_y, C_z, C_l, C_m, C_n\}$ with respect to the above control variables, such as $C_{z_{\delta_e}} = \partial C_z/\partial \delta_e$ or $C_{l_{\delta_a}} = \partial C_l/\partial \delta_a$, and so on. The dimensional derivatives are displayed in Table 7.1.

The powerful and well-developed methods of modern control theory are directly applicable to this restricted class of airplane control responses. Before proceeding to specific applications, however, we first present a review of some of the highlights of the general theory. Readers who are well versed in this material may skip directly to Sec. 7.6.

7.2 Response of Linear/Invariant Systems

For linear/invariant systems there are four basic single-input, single-response cases, illustrated in Fig. 7.2. They are characterized by the inputs, which are, respectively:

1. a unit impulse at $t = 0$
2. a unit step at $t = 0$
3. a sinusoid of unit amplitude and frequency f
4. white noise

In the first two the system is specified to be quiescent for $t < 0$ and to be subjected to a control or disturbance input at $t = 0$. In the last two cases the input is presumed to have been present for a very long time. In these two the system is assumed to be stable, so that any initial transients have died out. Thus in case 3 the response is also a steady sinusoid and in case 4 it is a statistically steady state. We discuss the first three of these cases in the following, but the fourth, involving the theory of random processes, is outside the scope of this text. The interested reader will find a full account of that topic in Etkin, 1972.

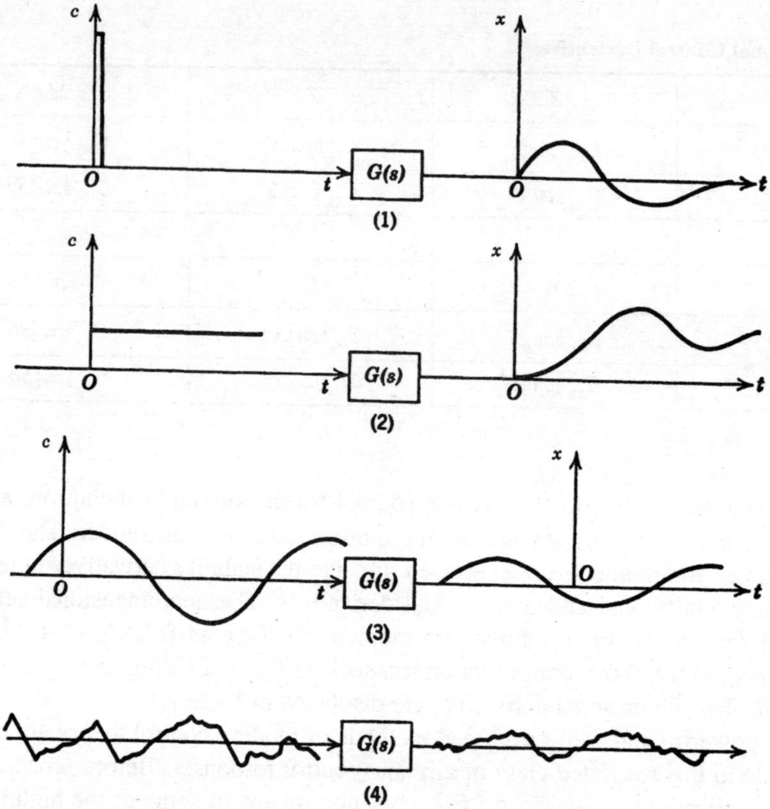

Figure 7.2 The four basic response problems. (1) Impulse response. (2) Step response. (3) Frequency response. (4) Response to white noise.

TRANSFER FUNCTIONS

A central and indispensable concept for response analysis is the *transfer function* that relates a particular input to a particular response. The transfer function, almost universally denoted $G(s)$, is the ratio of the Laplace transform of the response to that of the input for the special case when the system is quiescent for $t < 0$. A system with n state variables x_i and m controls c_j would therefore have a matrix of nm transfer functions $G_{ij}(s)$.

The Laplace transform of (7.1,4) is

$$s\bar{\mathbf{x}} = \mathbf{A}\bar{\mathbf{x}} + \mathbf{B}\bar{\mathbf{c}} \tag{7.2,1}$$

hence

$$(s\mathbf{I} - \mathbf{A})\bar{\mathbf{x}} = \mathbf{B}\bar{\mathbf{c}}$$

and

$$\bar{\mathbf{x}} = \mathbf{G}\bar{\mathbf{c}} \qquad (a)$$

where

$$\mathbf{G}(s) = (s\mathbf{I} - \mathbf{A})^{-1}\mathbf{B} \qquad (b) \tag{7.2,2}$$

is the matrix of transfer functions. The response of the ith state variable is then given by

$$\bar{x}_i(s) = \sum_j G_{ij}(s)\bar{c}_j(s) \tag{7.2,3}$$

For a single-input single-response system with transfer function $G(s)$ we have simply

$$\bar{x}(s) = G(s)\bar{c}(s) \tag{7.2,4}$$

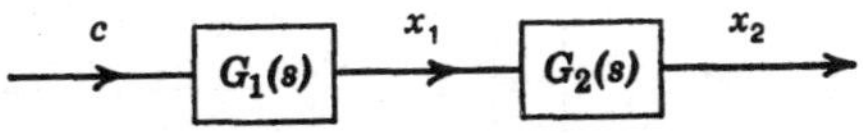

Figure 7.3 Systems in series.

SYSTEMS IN SERIES

When two systems are in series, so that the response of the first is the input to the second, as in Fig. 7.3, the overall transfer function is seen to be the product of the two. That is,

$$\bar{x}_1(s) = G_1(s)\bar{c}(s)$$

and

$$\bar{x}_2(s) = G_2(s)\bar{x}_1(s) = G_2(s)G_1(s)\bar{c}(s)$$

Thus the overall transfer function is

$$G(s) = \bar{x}_2(s)/\bar{c}(s) = G_1(s)G_2(s) \tag{7.2,5}$$

Similarly for n systems in series the overall transfer function is

$$G(s) = G_1(s)G_2(s) \cdots G_n(s) \tag{7.2,6}$$

HIGH-ORDER SYSTEMS

High-order linear/invariant systems, such as those that occur in aerospace practise, can always be represented by a chain of subsystems like (7.2,6). This is important, because the elemental building blocks that make up the chain are each of a simple kind—either first-order or second-order. To prove this we note from the definition of an inverse matrix (Appendix A.1) that

$$(s\mathbf{I} - \mathbf{A})^{-1} = \frac{\text{adj}\,(s\mathbf{I} - \mathbf{A})}{\det\,(s\mathbf{I} - \mathbf{A})} \tag{7.2,7}$$

We saw in Sec. 6.1 that det $(\mathbf{A} - s\mathbf{I})$ is the characteristic polynomial of the system. We also have, from the definition of the adjoint matrix as the transpose of the matrix of cofactors, that each element of the numerator of the right side of (7.2,7) is also a polynomial in s. (See Exercise 7.1.) Thus it follows from (7.2,2*b*), on noting that **B** is a matrix of constants, that each element of **G** is a ratio of two polynomials, which can be written as

$$G_{ij} = \frac{N_{ij}(s)}{f(s)} \tag{7.2,8}$$

in which $f(s)$ is the characteristic polynomial. It is seen that all the transfer functions of the system have the same denominator and differ from one another only in the different numerators. Since $f(s)$ has the roots $\lambda_1 \ldots \lambda_n$, the denominator can be factored to give

$$G_{ij} = \frac{N_{ij}(s)}{(s - \lambda_1)(s - \lambda_2) \cdots (s - \lambda_n)} \tag{7.2,9}$$

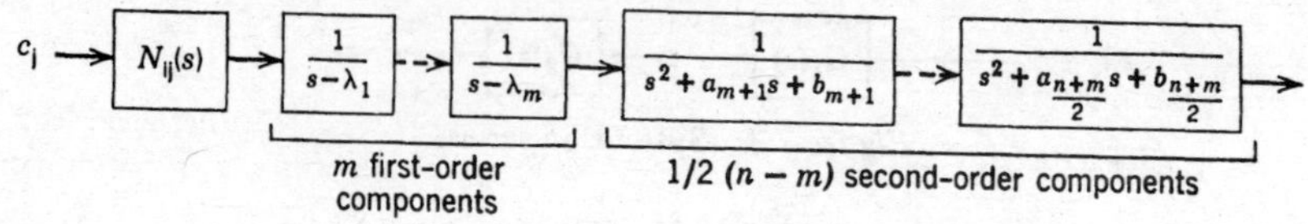

Figure 7.4 High-order systems as a "chain."

Now some of the eigenvalues λ_r are real, but others occur in complex pairs, so to obtain a product of factors containing only real numbers we rewrite the denominator thus

$$f(s) = \prod_{r=1}^{m} (s - \lambda_r) \prod_{r=m+1}^{1/2(n+m)} (s^2 + a_r s + b_r) \tag{7.2,10}$$

Here λ_r are the m real roots of $f(s)$ and the quadratic factors with real coefficients a_r and b_r produce the $(n - m)$ complex roots. It is then clearly evident that the transfer function (7.2,9) is also the overall transfer function of the fictitious system made up of the series of elements shown in Fig. 7.4. The leading component $N_{ij}(s)$ is of course particular to the system, but all the remaining ones are of one or other of two simple kinds. These two, first-order components and second-order components, may therefore be regarded as the basic building blocks of linear/invariant systems. It is for this reason that it is important to understand their characteristics well—the properties of all higher-order systems can be inferred directly from those of these two basic elements.

7.3 Impulse Response

The system is specified to be initially quiescent and at time zero is subjected to a single impulsive input

$$c_j(t) = \delta(t) \tag{7.3,1}$$

The Laplace transform of the ith component of the output is then

$$\bar{x}_i(s) = G_{ij}(s)\bar{\delta}(s)$$

which, from Table (A.1), item 1, becomes

$$\bar{x}_i(s) = G_{ij}(s)$$

This response to the unit impulse is called the *impulse response* or *impulsive admittance* and is denoted $h_{ij}(t)$. It follows that

$$\bar{h}_{ij}(s) = G_{ij}(s) \qquad (a)$$

that is, $G(s)$ is the Laplace transform of $h(t)$

$$G_{ij}(s) = \int_0^\infty h_{ij}(t)e^{-st}\, dt \qquad (b) \quad (7.3,2)$$

From the inversion theorem, (A.2,11) $h_{ij}(t)$ is then given by

$$h_{ij}(t) = \frac{1}{2\pi i} \int_c G_{ij}(s)e^{st}\, ds \tag{7.3,3}$$

Now if the system is stable, all the eigenvalues, which are the poles of $G_{ij}(s)$ lie in the left half of the s plane, and this is the usual case of interest. The line integral of (7.3,3) can then be taken on the imaginary axis, $s = i\omega$, so that (7.3,3) leads to

$$h_{ij}(t) = \frac{1}{2\pi} \int_{-\infty}^{\infty} e^{i\omega t} G_{ij}(i\omega)\, d\omega \tag{7.3,4}$$

that is, it is the inverse Fourier transform of $G_{ij}(i\omega)$. The significance of $G_{ij}(i\omega)$ will be seen later.

For a first-order component of Fig. 7.4 with eigenvalue λ the differential equation is

$$\dot{x} - \lambda x = c \tag{7.3,5}$$

for which we easily get

$$G(s) = \bar{h}(s) = \frac{1}{s - \lambda} \tag{7.3,6}$$

The inverse is found directly from item 8 of Table A.1 as

$$h(t) = e^{\lambda t}$$

For convenience in interpretation, λ is frequently written as $\lambda = -1/T$, where T is termed the *time constant* of the system. Then

$$h(t) = e^{-t/T} \tag{7.3,7}$$

A graph of $h(t)$ is presented in Fig. 7.5*a*, and shows clearly the significance of the time constant T.

For a second-order component of Fig. 7.4 the differential equation is

$$\ddot{y} + 2\zeta\omega_n\dot{y} + \omega_n^2 = c \tag{7.3,8}$$

where $\mathbf{x} = [y \ \dot{y}]^T$ is the state vector. It easily follows that

$$G(s) = \bar{h}(s) = \frac{1}{s^2 + 2\zeta\omega_n s + \omega_n^2} \tag{7.3,9}$$

Let the eigenvalues be $\lambda = n \pm i\omega$, where

$$n = -\zeta\omega_n$$
$$\omega = \omega_n(1 - \zeta^2)^{1/2}$$

then $\bar{h}(s)$ becomes

$$\begin{aligned} \bar{h}(s) &= \frac{1}{(s - n - i\omega)(s - n + i\omega)} \\ &= \frac{1}{(s - n)^2 + \omega^2} \end{aligned} \tag{7.3,10}$$

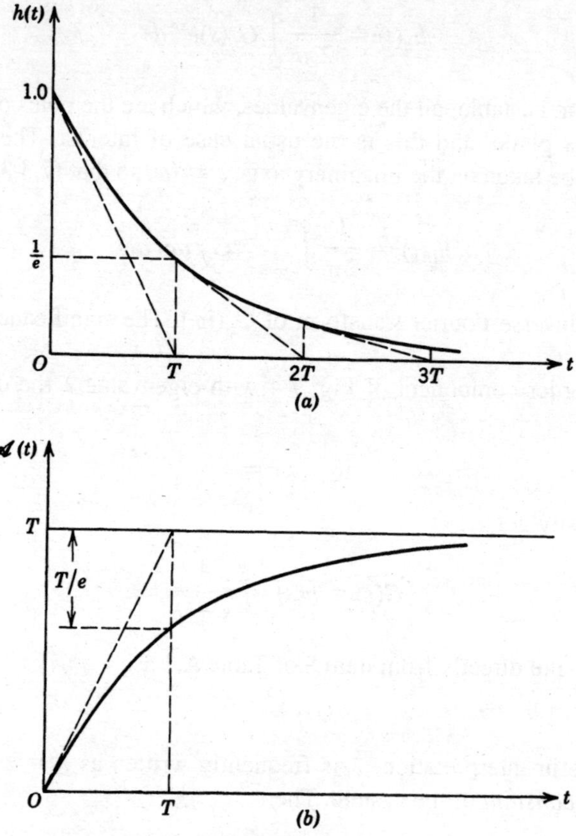

Figure 7.5 Admittances of a first-order system.

and the inverse is found from item 13, Table A.1 to be

$$h(t) = \frac{1}{\omega} e^{nt} \sin \omega t \qquad (7.3,11)$$

For a stable system n is negative and (7.3,11) describes a damped sinusoid of frequency ω. This is plotted for various ζ in Fig. 7.6. Note that the coordinates are so chosen as to lead to a one-parameter family of curves. Actually the above result only applies for $\zeta \leq 1$. The corresponding expression for $\zeta \geq 1$ is easily found by the same method and is

$$h(t) = \frac{1}{\omega'} e^{nt} \sinh \omega' t \qquad (7.3,12)$$

where

$$\omega' = \omega_n(\zeta^2 - 1)^{1/2}$$

Graphs of (7.3,12) are also included in Fig. 7.6, although in this case the second-order representation could be replaced by two first-order elements in series.

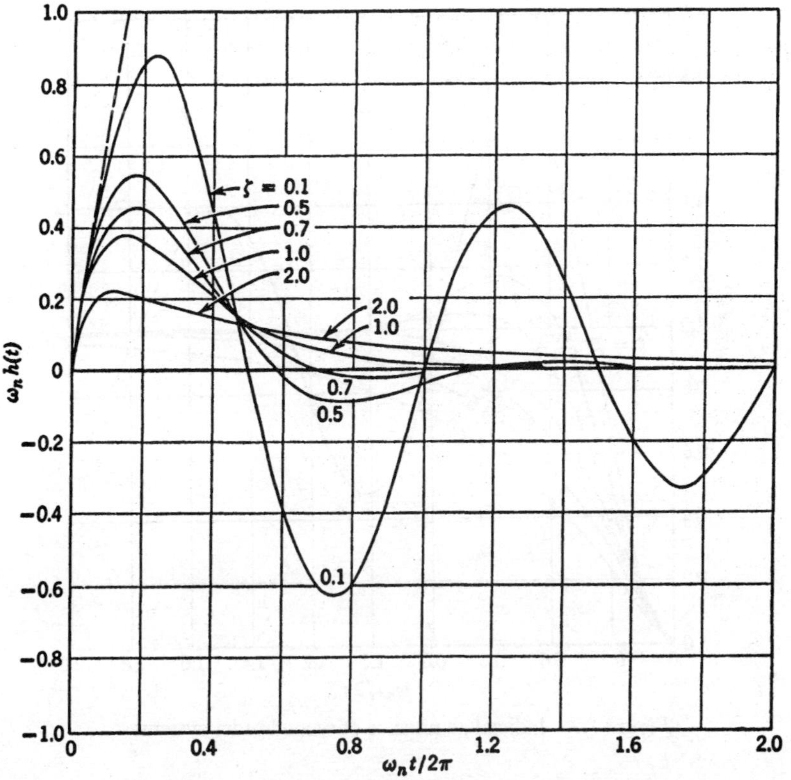

Figure 7.6 Impulsive admittance of second-order systems.

7.4 Step-Function Response

This is like the impulse response treated above except that the input is the unit step function $\mathbf{I}(t)$, with transform $1/s$ (Table A.1). The response in this case is called the *step response* or *indicial admittance,* and is denoted $\mathscr{A}_{ij}(t)$. It follows then that

$$\bar{\mathscr{A}}_{ij}(s) = G_{ij}(s)\bar{\mathbf{I}}(s) = \frac{G_{ij}(s)}{s} \qquad (a)$$

$$(7.4,1)$$

or

$$\bar{\mathscr{A}}_{ij}(s) = \frac{\bar{h}_{ij}(s)}{s} \qquad (b)$$

Since the initial values (at $t = 0^-$) of $h_{ij}(t)$ and $\mathscr{A}_{ij}(t)$ are both zero, the theorem (A.2,4) shows that

$$\mathscr{A}_{ij}(t) = \int_0^t h_{ij}(\tau)\, d\tau \qquad (a)$$

$$(7.4,2)$$

or

$$h_{ij}(t) = \frac{d\mathscr{A}_{ij}(t)}{dt} \qquad (b)$$

Thus $\mathscr{A}_{ij}(t)$ can be found either by direct inversion of (7.4,1*b*) or by integration of $h_{ij}(t)$. By either method the results for first- and second-order systems are readily obtained, and are as follows (for a single input/response pair the subscript is dropped):

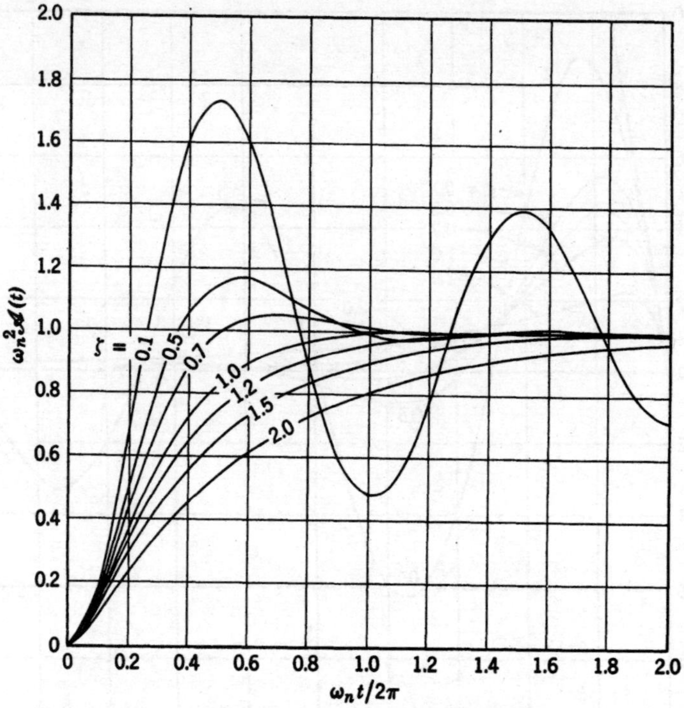

Figure 7.7 Indicial admittance of second-order systems.

First-order system:

$$\mathscr{A}(t) = T(1 - e^{-t/T}) \tag{7.4,3}$$

Second-order system:

$$\mathscr{A}(t) = \frac{1}{\omega_n^2}\left[1 - e^{nt}\left(\cos \omega t - \frac{n}{\omega} \sin \omega t\right)\right], \qquad \zeta < 1 \tag{7.4,4}$$

For $\zeta > 1$, see Appendix A.2.

Graphs of the indicial responses are given in Figs. 7.5*b* and 7.7.

The asymptotic value of $\mathscr{A}(t)$ as $t \to \infty$ is called the static gain K. Applying the final value theorem (A.2,12) (7.4,1) yields

$$\lim_{t\to\infty} \mathscr{A}(t) = \lim_{s\to 0} s\bar{\mathscr{A}}(s) = \lim_{s\to 0} G(s)$$

Thus

$$K = \lim_{s\to 0} G(s) \tag{7.4,5}$$

7.5 Frequency Response

When a stable linear/invariant system has a sinusoidal input, then after some time the transients associated with the starting conditions die out, and there remains only a steady-state sinusoidal response at the same frequency as that of the input. Its amplitude and phase are generally different from those of the input, however, and the expression of these differences is embodied in the frequency-response function.

Consider a single input/response pair, and let the input be the sinusoid $a_1 \cos \omega t$. We find it convenient to replace this by the complex expression $c = A_1 e^{i\omega t}$, of which $a_1 \cos \omega t$ is the real part. A_1 is known as the *complex amplitude* of the wave. The response sinusoid can be represented by a similar expression, $x = A_2 e^{i\omega t}$, the real part of which is the physical response. As usual, x and c are interpreted as rotating vectors whose projections on the real axis give the relevant physical variables (see Fig. 7.8*a*).

From Table A.1, item 8, the transform of c is

$$\bar{c} = \frac{A_1}{s - i\omega}$$

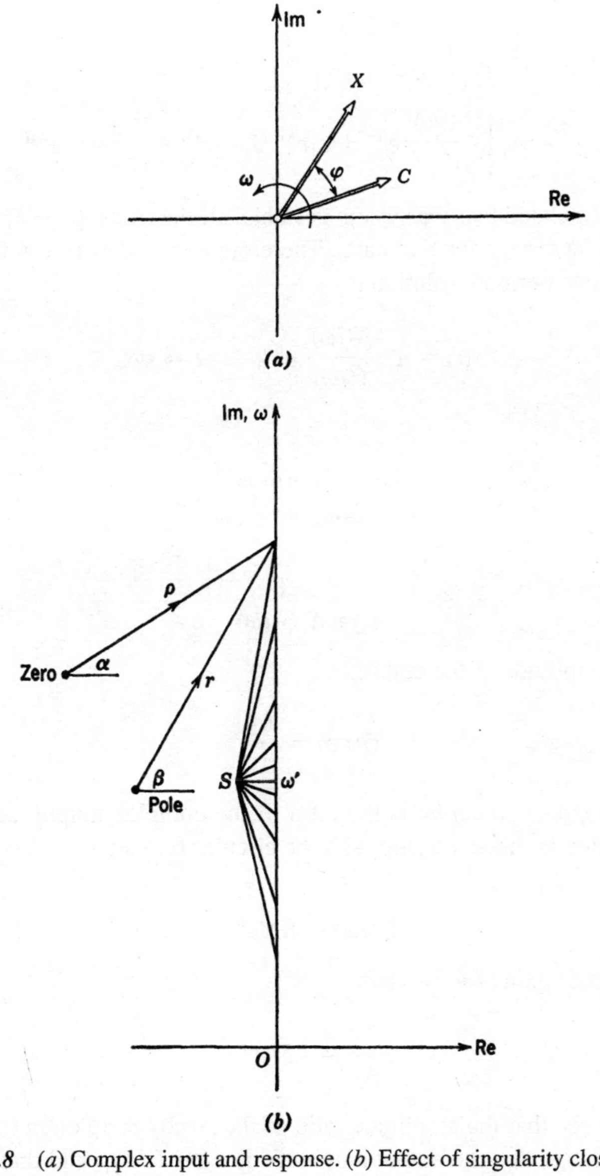

Figure 7.8 (*a*) Complex input and response. (*b*) Effect of singularity close to axis.

Therefore

$$\bar{x} = A_1 \frac{G(s)}{s - i\omega}$$

The function $G(s)$ is given by (7.2,9) so that

$$\bar{x} = A_1 \frac{N(s)}{(s - i\omega)f(s)} \tag{7.5,1}$$

The roots of the denominator of the r.h.s. are

$$\lambda_1 \cdots \lambda_n, i\omega$$

so that the application of the expansion theorem (A.2,10) yields the complex output

$$\begin{aligned} x(t) &= A_1 \sum_{r=1}^{n+1} \left[\frac{(s - \lambda_r)N(s)}{(s - i\omega)f(s)} \right]_{s=\lambda_r} e^{\lambda_r t} \\ &= A_1 \left[\frac{N(i\omega)}{f(i\omega)} e^{i\omega t} + c_1 e^{\lambda_1 t} + c_2 e^{\lambda_2 t} + \cdots + c_n e^{\lambda_n t} \right] \end{aligned} \tag{7.5,2}$$

Since we have stipulated that the system is stable, all the roots $\lambda_1 \cdots \lambda_n$ of the characteristic equation have negative real parts. Therefore $e^{\lambda_r t} \rightarrow 0$ as $t \rightarrow \infty$ for $r = 1 \cdots n$, and the steady-state periodic solution is

$$x(t) = A_1 \frac{N(i\omega)}{f(i\omega)} e^{i\omega t}, \qquad t \rightarrow \infty$$

or

$$\begin{aligned} x(t) &= A_1 G(i\omega) e^{i\omega t} \\ &= A_2 e^{i\omega t} \end{aligned}$$

Thus

$$A_2 = A_1 G(i\omega) \tag{7.5,3}$$

is the complex amplitude of the output, or

$$G(i\omega) = \frac{A_2}{A_1} \tag{7.5,4}$$

the *frequency response function,* is the ratio of the complex amplitudes. In general, $G(i\omega)$ is a complex number, varying with the circular frequency ω. Let it be given in polar form by

$$G(i\omega) = KMe^{i\varphi} \tag{7.5,6}$$

where K is the static gain (7.4,5). Then

$$\frac{A_2}{A_1} = KMe^{i\varphi} \tag{7.5,7}$$

From (7.5,7) we see that the amplitude ratio of the steady-state output to the input is $|A_2/A_1| = KM$: that is, that the output amplitude is $a_2 = KMa_1$, and that the phase re-

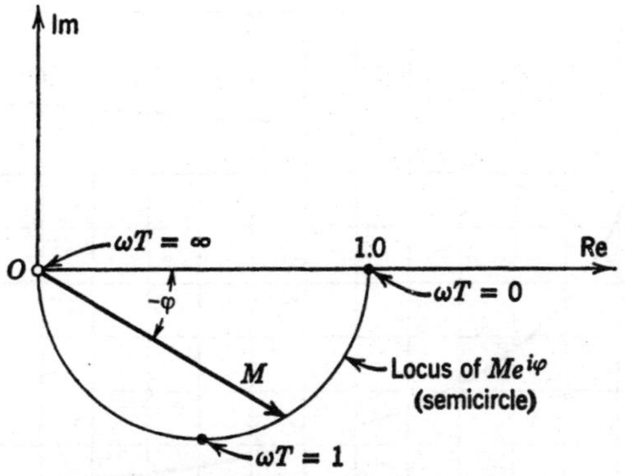

Figure 7.9 Vector plot of $Me^{i\varphi}$ for first-order systems.

lation is as shown on Fig. 7.8*a*. The output leads the input by the angle φ. The quantity M, which is the modulus of $G(i\omega)$ divided by K, we call the magnification factor, or *dynamic gain,* and the product KM we call the *total gain.* It is important to note that M and φ are frequency-dependent.

Graphical representations of the frequency response commonly take the form of either vector plots of $Me^{i\varphi}$ (Nyquist diagram) or plots of M and φ as functions of frequency (Bode diagram). Examples of these are shown in Figs. 7.9 to 7.13.

EFFECT OF POLES AND ZEROS ON FREQUENCY RESPONSE

We have seen (7.2,9) that the transfer function of a linear/invariant system is a ratio of two polynomials in s, the denominator being the characteristic polynomial. The roots of the characteristic equation are the *poles* of the transfer function, and the roots of the numerator polynomial are its *zeros.* Whenever a pair of complex poles or zeros lies close to the imaginary axis, a characteristic peak or valley occurs in the amplitude of the frequency-response curve together with a rapid change of phase angle at the corresponding value of ω. Several examples of this phenomenon are to be seen in the frequency response curves in Figs. 7.14 to 7.18. The reason for this behavior is readily appreciated by putting (7.2,9) in the following form:

$$G(s) = \frac{(s - z_1) \cdot (s - z_2) \cdots (s - z_m)}{(s - \lambda_1) \cdot (s - \lambda_2) \cdots (s - \lambda_n)}$$

where the λ_i are the characteristic roots (poles) and the z_i are the zeros of $G(s)$. Let

$$(s - z_k) = \rho_k e^{i\alpha_k}$$
$$(s - \lambda_k) = r_k e^{i\beta_k}$$

where ρ, r, α, β are the distances and angles shown in Fig. 7.8*b* for a point $s = i\omega$ on the imaginary axis. Then

$$|G| = \prod_{k=1}^{m} \rho_k \Big/ \prod_{k=1}^{n} r_k$$

$$\varphi = \sum_{1}^{m} \alpha_k - \sum_{1}^{n} \beta_k$$

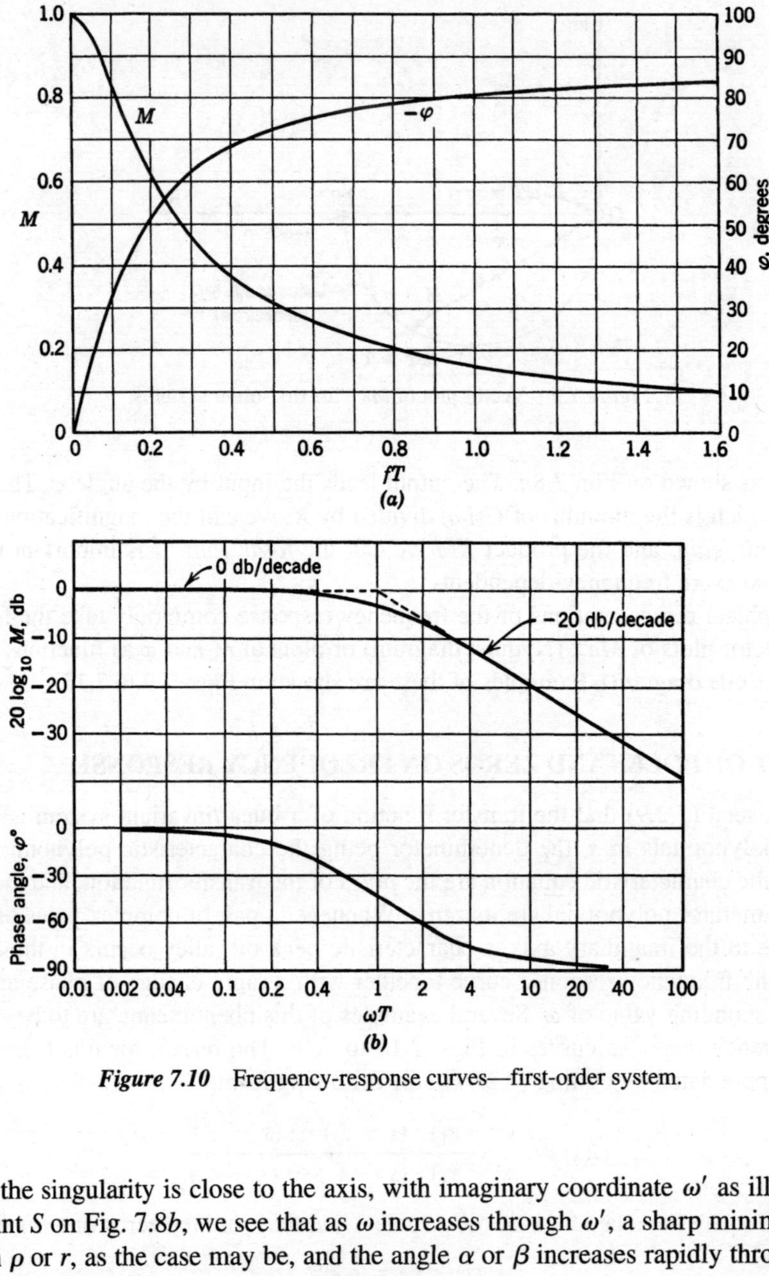

Figure 7.10 Frequency-response curves—first-order system.

When the singularity is close to the axis, with imaginary coordinate ω' as illustrated for point S on Fig. 7.8b, we see that as ω increases through ω', a sharp minimum occurs in ρ or r, as the case may be, and the angle α or β increases rapidly through approximately 180°. Thus we have the following cases:

1. For a pole, in the left half-plane, there results a peak in $|G|$ and a reduction in φ of about 180°.
2. For a zero in the left half-plane, there is a valley in $|G|$ and an increase in φ of about 180°.
3. For a zero in the right half-plane, there is a valley in $|G|$ and a decrease in φ of about 180°.

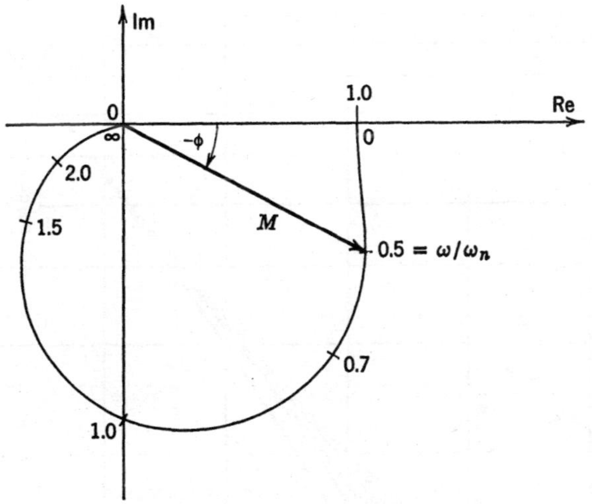

Figure 7.11 Vector plot of $Me^{i\varphi}$ for second-order system. Damping ratio $\zeta = 0.4$.

FREQUENCY RESPONSE OF FIRST-ORDER SYSTEM

The first-order transfer function, written in terms of the time constant T is

$$G(s) = \frac{1}{s + 1/T} \tag{7.5,8}$$

whence

$$K = \lim_{s \to 0} G(s) = T$$

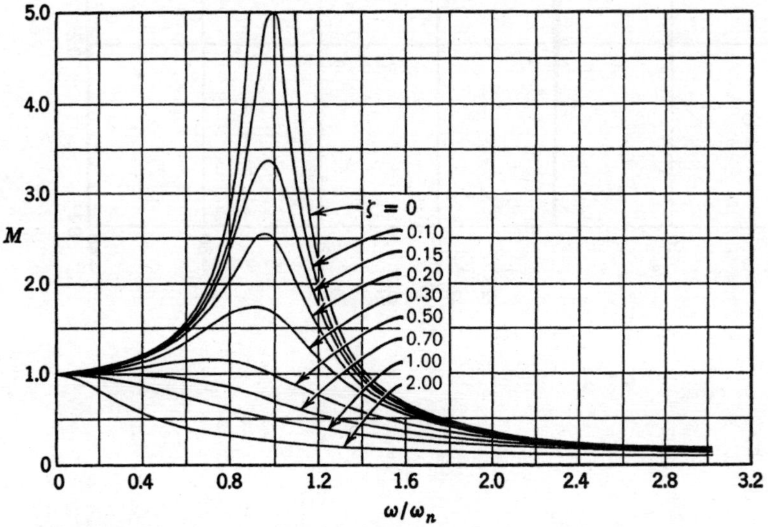

Figure 7.12 Frequency-response curves—second-order system.

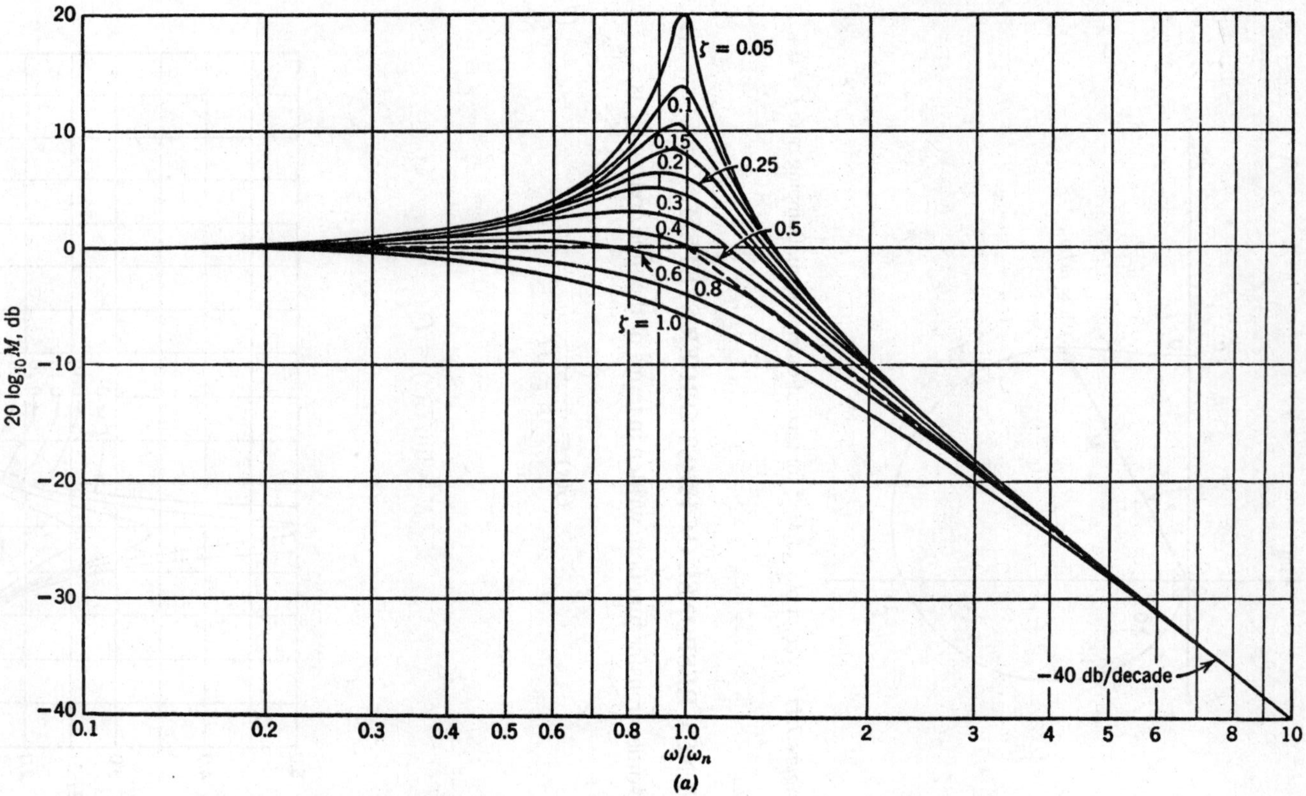

Figure 7.13 Frequency-response curves—second-order system.

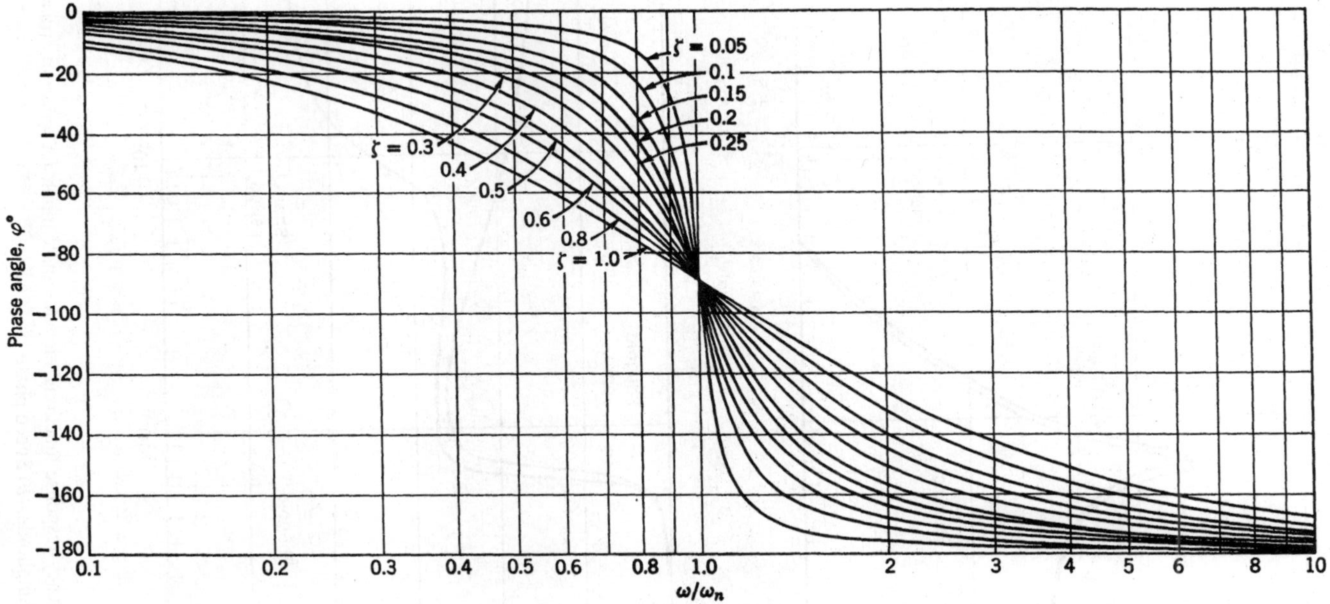

(b)

Figure 7.13 *(Continued)*

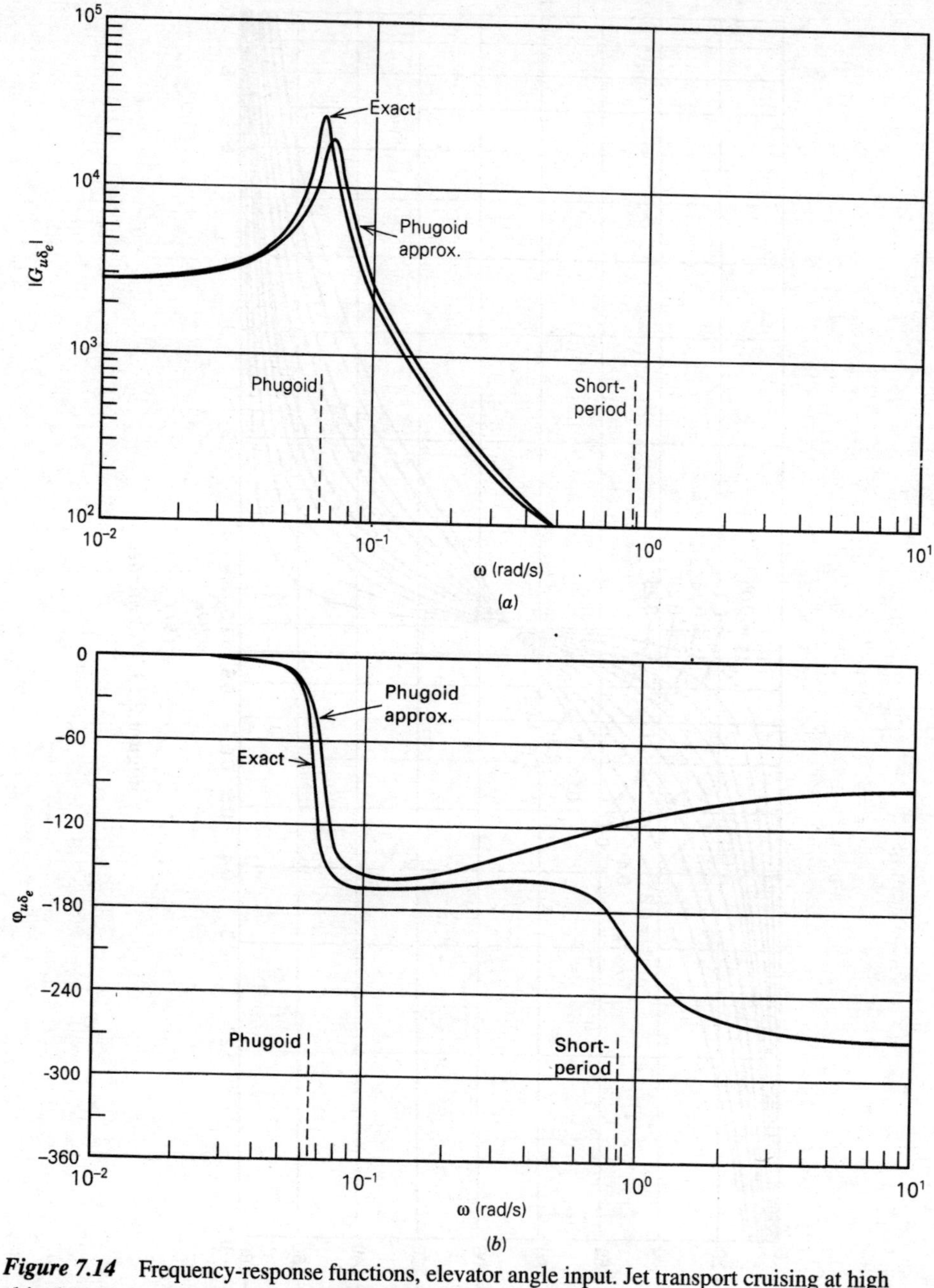

Figure 7.14 Frequency-response functions, elevator angle input. Jet transport cruising at high altitudes. (*a*) Speed amplitude. (*b*) Speed phase.

The frequency response is determined by the vector $G(i\omega)$

$$G(i\omega) = KMe^{i\varphi} = \frac{T}{1 + i\omega T}$$

whence

$$Me^{i\varphi} = \frac{1 - i\omega T}{1 + \omega^2 T^2} \tag{7.5,9}$$

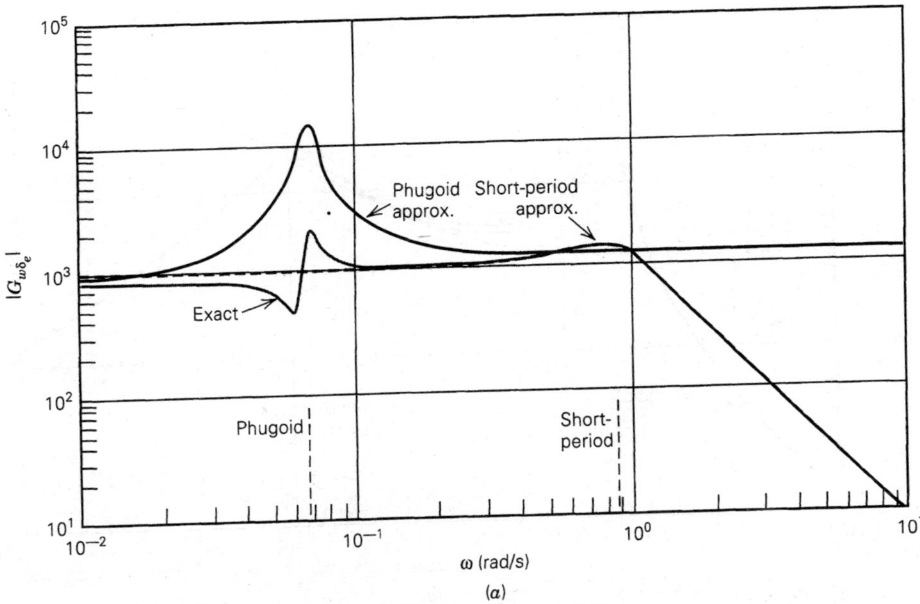

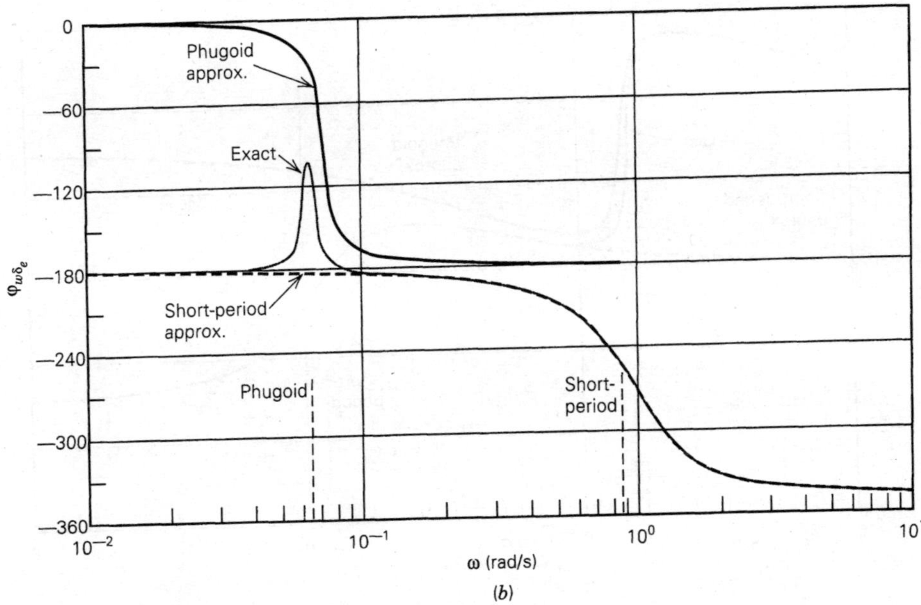

Figure 7.15 Frequency-response functions, elevator angle input. Jet transport cruising at high altitude. (*a*) Angle of attack amplitude. (*b*) Angle of attack phase.

From (7.5,9), M and φ are found to be

$$M = \frac{1}{(1 + \omega^2 T^2)^{1/2}} \tag{7.5,10}$$

$$\varphi = -\tan^{-1} \omega T$$

A vector plot of $Me^{i\varphi}$ is shown in Fig. 7.9. This kind of diagram is sometimes called the transfer-function locus. Plots of M and φ are given in Figs. 7.10*a* and *b*. The abscissa is fT or $\log \omega T$ where $f = \omega/2\pi$, the input frequency. This is the only parame-

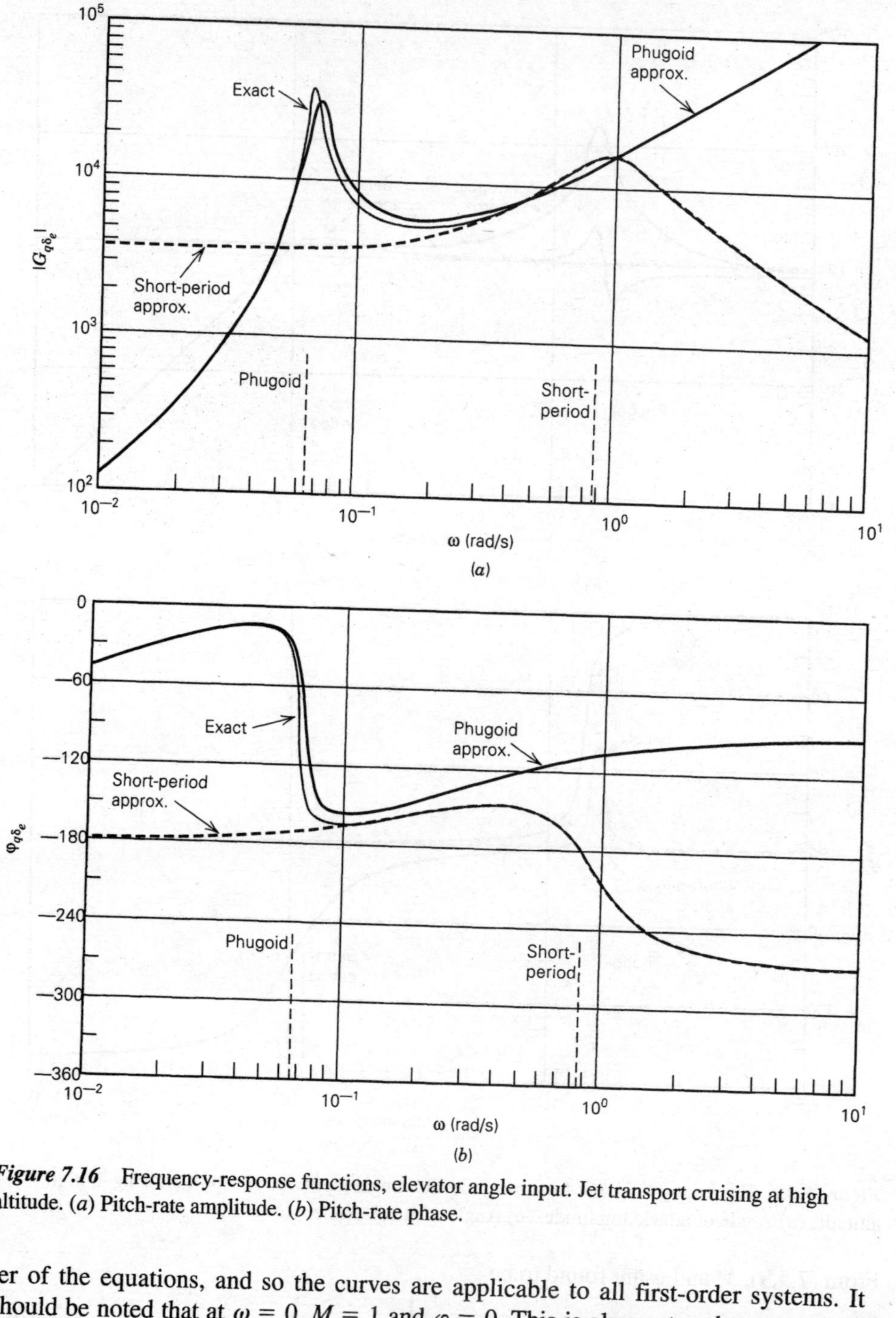

Figure 7.16 Frequency-response functions, elevator angle input. Jet transport cruising at high altitude. (*a*) Pitch-rate amplitude. (*b*) Pitch-rate phase.

ter of the equations, and so the curves are applicable to all first-order systems. It should be noted that at $\omega = 0$, $M = 1$ and $\varphi = 0$. This is always true because of the definitions of K and $G(s)$—it can be seen from (7.4,5) that $G(0) = K$.

FREQUENCY RESPONSE OF A SECOND-ORDER SYSTEM

The transfer function of a second-order system is given in (7.3,9). The frequency-response vector is therefore

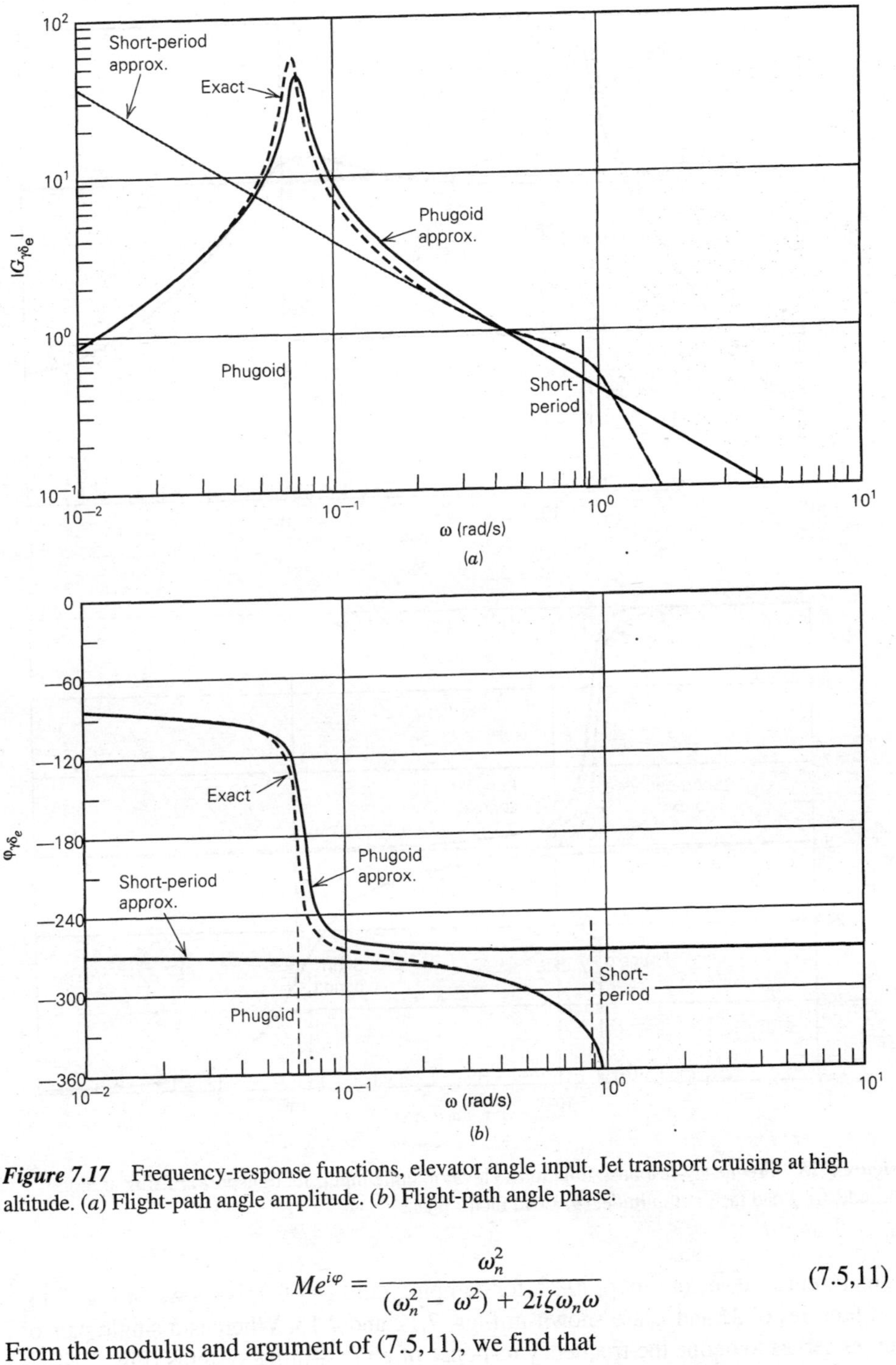

Figure 7.17 Frequency-response functions, elevator angle input. Jet transport cruising at high altitude. (*a*) Flight-path angle amplitude. (*b*) Flight-path angle phase.

$$Me^{i\varphi} = \frac{\omega_n^2}{(\omega_n^2 - \omega^2) + 2i\zeta\omega_n\omega} \tag{7.5,11}$$

From the modulus and argument of (7.5,11), we find that

$$\begin{aligned} M &= \frac{1}{\{[1 - (\omega/\omega_n)^2]^2 + 4\zeta^2(\omega/\omega_n)^2\}^{1/2}} \\ \varphi &= -\tan^{-1}\frac{2\zeta\omega/\omega_n}{1 - (\omega/\omega_n)^2} \end{aligned} \tag{7.5,12}$$

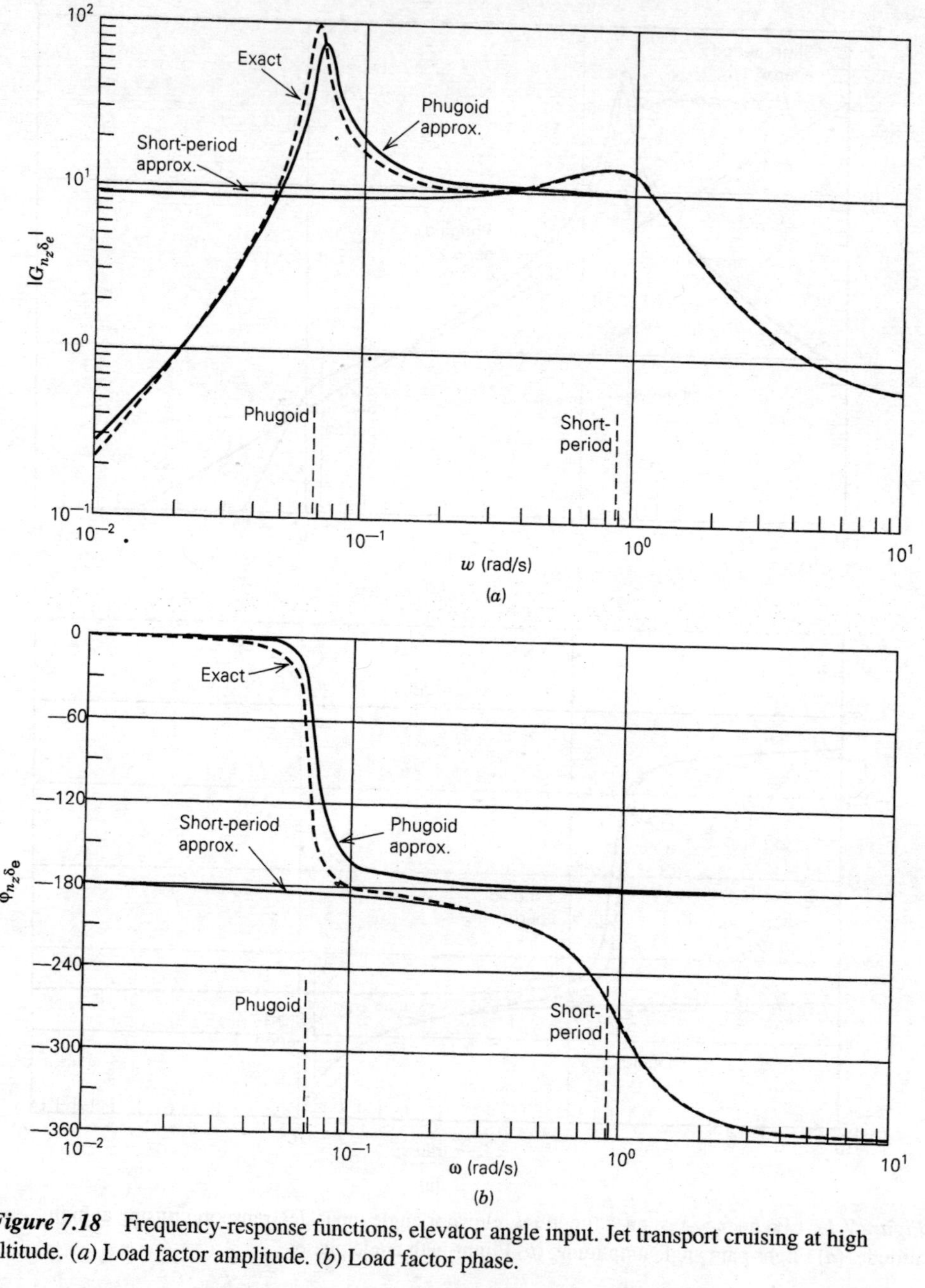

Figure 7.18 Frequency-response functions, elevator angle input. Jet transport cruising at high altitude. (*a*) Load factor amplitude. (*b*) Load factor phase.

A representative vector plot of $Me^{i\varphi}$, for damping ratio $\zeta = 0.4$, is shown in Fig. 7.11, and families of M and φ are shown in Figs. 7.12 and 7.13. Whereas a single pair of curves serves to define the frequency response of all first-order systems (Fig. 7.10), it takes two families of curves, with the damping ratio as parameter, to display the characteristics of all second-order systems. The importance of the damping as a parameter should be noted. It is especially powerful in controlling the magnitude of the resonance peak which occurs near unity frequency ratio. At this frequency the phase lag is by contrast independent of ζ, as all the curves pass through $\varphi = -90°$ there. For all values of ζ, $M \rightarrow 1$ and $\varphi \rightarrow 0$ as $\omega/\omega_n \rightarrow 0$. This shows that, whenever a system is driven by an oscillatory input whose frequency is low compared to the undamped natural frequency, the response will be *quasistatic*. That is, at each instant, the output will be the same as though the instantaneous value of the input were applied statically.

The behavior of the output when ζ is near 0.7 is interesting. For this value of ζ, it is seen that φ is very nearly linear with ω/ω_n up to 1.0. Now the phase lag can be interpreted as a time lag, $\tau = (\varphi/2\pi)T = \varphi/\omega$ where T is the period. The output wave form will have its peaks retarded by τ sec relative to the input. For the value of ζ under consideration, $\varphi/(\omega/\omega_n) \doteqdot \pi/2$ or $\varphi/\omega = \pi/2\omega_n = \frac{1}{4}T_n$, where $T_n = 2\pi/\omega_n$, the undamped natural period. Hence we find that, for $\zeta \doteqdot 0.7$, there is a nearly constant time lag $\tau \doteqdot \frac{1}{4}T_n$, independent of the input frequency, for frequencies below resonance.

The "chain" concept of higher-order systems is especially helpful in relation to frequency response. It is evident that the phase changes through the individual elements are simply additive, so that higher-order systems tend to be characterized by greater phase lags than low-order ones. Also the individual amplitude ratios of the elements are multiplied to form the overall ratio. More explicitly, let

$$G(s) = G_1(s) \cdot G_2(s) \cdots G_n(s)$$

be the overall transfer function of n elements. Then

$$\begin{aligned} G(i\omega) &= G_1(i\omega) \cdot G_2(i\omega) \cdots G_n(i\omega) \\ &= (K_1M_1 \cdot K_2M_2 \cdots K_nM_n)e^{i(\varphi_1+\varphi_2+\cdots\varphi_n)} \\ &= KMe^{i\varphi} \end{aligned}$$

so that

$$KM = \prod_{r=1}^{n} K_rM_r \qquad (a)$$

$$\varphi = \sum_{r=1}^{n} \varphi_r \qquad (b) \qquad (7.5,13)$$

On logarithmic plots (Bode diagrams) we note that

$$\log KM = \sum_{r=1}^{n} \log K_rM_r \qquad (7.5,14)$$

Thus the log of the overall gain is obtained as a sum of the logs of the component gains, and this fact, together with the companion result for phase angle (7.5,13) greatly facilitates graphical methods of analysis and system design.

RELATION BETWEEN IMPULSE RESPONSE AND FREQUENCY RESPONSE

We saw earlier (7.3,4), that $h(t)$ is the inverse Fourier transform of $G(i\omega)$, which we can now identify as the frequency response vector. The reciprocal Fourier transform relation then gives

$$G(i\omega) = \int_{-\infty}^{\infty} h(t)e^{-i\omega t}\,dt \qquad (7.5,15)$$

that is, *the frequency response and impulsive admittance are a Fourier transform pair.*

7.6 Longitudinal Response

To treat this case we need the matrices **A** and **B** of (7.1,4). **A** is given in (4.9,18), but we have not yet given **B** explicitly. On the right side of (4.9,18) we have the product $\mathbf{B}\Delta\mathbf{c}$ given by

$$\mathbf{B}\Delta\mathbf{c} = \begin{bmatrix} \dfrac{\Delta X_c}{m} \\ \dfrac{\Delta Z_c}{(m - Z_{\dot{w}})} \\ \dfrac{\Delta M_c}{I_y} + \dfrac{M_{\dot{w}}}{I_y}\dfrac{\Delta Z_c}{(m - Z_{\dot{w}})} \\ 0 \end{bmatrix} \tag{7.6,1}$$

We now have to specify the control vector **c** and the corresponding aerodynamic forces and moment. For longitudinal control, we assume here that the available controls are well enough represented by

$$\mathbf{c} = [\delta_e \quad \delta_p]^T \tag{7.6,2}$$

and that the incremental aerodynamic forces and moment that result from their actuation are given by a set of *control derivatives* X_{δ_e} and so on, in the form

$$\begin{bmatrix} \Delta X_c \\ \Delta Z_c \\ \Delta M_c \end{bmatrix} = \begin{bmatrix} X_{\delta_e} & X_{\delta_p} \\ Z_{\delta_e} & Z_{\delta_p} \\ M_{\delta_e} & M_{\delta_p} \end{bmatrix} \begin{bmatrix} \Delta\delta_e \\ \Delta\delta_p \end{bmatrix} \tag{7.6,3}$$

Additional elements can be added to **c** and to (7.6,3) if the situation requires it.

The use of constant derivatives, as in (7.6,3), to describe the force output of the propulsion system in response to throttle input does not allow for any time lag in the buildup of engine thrust since it implies that the thrust is instantaneously proportional to the throttle position. This is not unreasonable for propeller airplanes, but it is not a good model for jets in situations when the short-term response is important, as for example in a balked landing. To allow for this effect when the system is modeled in the Laplace domain, one can use control transfer functions instead of control derivatives. That is, one can replace, for example, X_{δ_p} by $G_{x\delta_p}(s)$. If the system model is in the time domain, the same result can be obtained by adding an additional differential equation and an additional variable. This latter method is illustrated in the example of Sec. 8.5.

By substituting (7.6,3) into (7.6,1) we derive the matrix **B** to be

$$\mathbf{B} = \begin{bmatrix} \dfrac{X_{\delta_e}}{m} & \dfrac{X_{\delta_p}}{m} \\ \dfrac{Z_{\delta_e}}{(m - Z_{\dot{w}})} & \dfrac{Z_{\delta_p}}{(m - Z_{\dot{w}})} \\ \dfrac{M_{\delta_e}}{I_y} + \dfrac{M_{\dot{w}} Z_{\delta_e}}{I_y(m - Z_{\dot{w}})} & \dfrac{M_{\delta_p}}{I_y} + \dfrac{M_{\dot{w}} Z_{\delta_p}}{I_y(m - Z_{\dot{w}})} \\ 0 & 0 \end{bmatrix} \tag{7.6,4}$$

With **A** and **B** known, we can compute the desired transfer functions and responses. (In this example the elevator angle is in radians, and English units are used for all other quantities). We calculate responses for the same jet transport as was used previously in Sec. 6.2, with **A** given by (6.2,1). The nondimensional elevator derivatives are:

$$\begin{aligned} C_{x_{\delta_e}} &= -3.818 \times 10^{-6} \\ C_{z_{\delta_e}} &= -0.3648 \\ C_{m_{\delta_e}} &= -1.444 \end{aligned}$$

from which the dimensional derivatives are calculated as

$$\begin{aligned} X_{\delta_e} &= C_{x_{\delta_e}} \tfrac{1}{2}\rho u_0^2 S = -3.717 \\ Z_{\delta_e} &= C_{z_{\delta_e}} \tfrac{1}{2}\rho u_0^2 S = -3.551 \times 10^5 \\ M_{\delta_e} &= C_{m_{\delta_e}} \tfrac{1}{2}\rho u_0^2 S\bar{c} = -3.839 \times 10^7 \end{aligned}$$

For the throttle, we arbitrarily choose a value of $X_{\delta_p}/m = 0.3\ g$ when $\delta_p = 1$, and Z_{δ_p} and $M_{\delta_p} = 0$. With these values we get for the matrix **B**:

$$\mathbf{B} = \begin{bmatrix} -0.000187 & 9.66 \\ -17.85 & 0 \\ -1.158 & 0 \\ 0 & 0 \end{bmatrix} \tag{7.6,5}$$

7.7 Responses to Elevator and Throttle

RESPONSE TO ELEVATOR

When the only input is the elevator angle $\Delta\delta_e$ the system reduces to

$$(s\mathbf{I} - \mathbf{A}) \begin{bmatrix} \Delta\bar{u} \\ \bar{w} \\ \bar{q} \\ \Delta\bar{\theta} \end{bmatrix} = \begin{bmatrix} b_{11} \\ b_{21} \\ b_{31} \\ 0 \end{bmatrix} \Delta\bar{\delta}_e \tag{7.7,1}$$

where b_{ij} are the elements of **B**. Solving for the ratios $G_{u\delta_e}(s) = \Delta\bar{u}/\Delta\bar{\delta}_e$ and so on yields the four transfer functions.[1] Each is of the form (7.2,8). For this case the char-

[1]In the subscripts for the transfer function symbols, the symbol Δ is omitted in the interest of simplicity.

acteristic polynomial is the left side of (6.2,2) (with s replacing λ), and the numerator polynomials are

$$\begin{aligned} N_{u\delta_e} &= -0.000188s^3 - 0.2491s^2 + 24.68s + 11.16 \\ N_{w\delta_e} &= -17.85s^3 - 904.0s^2 - 6.208s - 3.445 \\ N_{q\delta_e} &= -1.158s^3 - 0.3545s^2 - 0.003873s \\ N_{\theta\delta_e} &= -1.158s^2 - 0.3545s - 0.003873 \end{aligned} \tag{7.7,2}$$

There are two other response quantities of interest, the flight path angle γ and the load factor n_z. Since $\theta_0 = 0$, $\Delta\theta = \theta$, and $\Delta\gamma = \Delta\theta - \Delta\alpha$, it readily follows that

$$G_{\gamma\delta_e} = G_{\theta\delta_e} - G_{\alpha\delta_e} \tag{7.7,3}$$

We define n_z to be (see also Sec. 3.1)

$$n_z = -Z/W \tag{7.7,4}$$

It is equal to unity in horizontal steady flight, and its incremental value during the response to elevator input is

$$\Delta n_z = -\Delta Z/W = -(Z_u\Delta u + Z_w w + Z_q q + Z_{\dot{w}}\dot{w} + Z_{\delta_e}\Delta\delta_e)/W$$

After taking the Laplace transform of the preceding equation and dividing by $\Delta\delta_e$, we get the transfer function for load factor to be

$$G_{n_z\delta_e} = -(Z_u G_{u\delta_e} + Z_w G_{w\delta_e} + Z_q G_{q\delta_e} + sZ_{\dot{w}} G_{w\delta_e} + Z_{\delta_e})/W \tag{7.7,5}$$

The total gain and phase of the frequency responses calculated by (7.5,6) from five of the above six transfer functions (for Δu, w, q, $\Delta\gamma$, and Δn_z) are shown in Figs. 7.14 to 7.18.

The exact solutions show that the responses in the "trajectory" variables u and γ are dominated entirely by the large peak at the low-frequency Phugoid mode. Because of the light damping in this mode, the resonant gains are very large. The peak $|G_{u\delta_e}|$ of nearly 3×10^4 means that a speed amplitude of 100 fps would result from an elevator angle amplitude of about $100/(3 \times 10^4)$ rad, or about 0.2°. Similarly, at resonance an amplitude of 10° in γ would be produced by an elevator amplitude of about $\frac{1}{6}$°. For both of these variables the response diminishes rapidly with increasing frequency, becoming negligibly small above the short-period frequency. The phase angle for u, Fig. 7.14*b*, is zero at low frequency, decreases rapidly to near −180° at the phugoid frequency (very much like the lightly damped second-order systems of Fig. 7.13) and subsequently at the short-period frequency undergoes a second drop characteristic of a heavily damped second-order system. The "chain" concept of high-order systems (Sec. 7.2) is well exemplified by this graph.

By contrast, the attitude variables w and q show important effects at both low and high frequencies. The complicated behavior of w near the phugoid frequency indicates the sort of thing that can happen with high-order systems. It is associated with a pole/zero pair of the transfer function being close to one another. Again, above the short-period frequency, the amplitudes of both w and q fall off rapidly.

The amplitude of the load factor Δn_z has a very large resonant peak at the phugoid frequency, almost 100/rad. It would not take a very large elevator amplitude at this frequency to cause structural failure of the wing!

Step-Function Response

The response of the airplane to a sudden movement of the elevator is shown by the step response. This requires a solution in the *time domain* as distinct from the preceding solution, which was in the *frequency domain.* Time domain solutions are commonly obtained simply by integrating (7.1,4) by a Runge-Kutta, Euler, or other integration scheme, the choice being dependent on the order of the system, accuracy required, computer available, and so on. The software used[2] for the example to follow does not integrate the equations, but instead uses an alternative method. It inverts the transfer function using the Heavyside expansion theorem (A.2,10). For the same jet airplane and flight condition as in the preceding example, the control vector for elevator input is $\mathbf{c} = [\Delta\delta_e \;\; 0]^T$ and $\mathbf{A}$ and $\mathbf{B}$ are as before. (Note that only the first column of $\mathbf{B}$ is needed.) Time traces of speed, angle of attack, and flight path angle are shown in Figs. 7.19 and 7.20 for two time ranges when the elevator displacement is one degree positive, that is, down.

It is seen from Fig. 7.19, which shows the response during the first 10 sec, that only the angle of attack responds quickly to the elevator motion, and that its variation is dominated by the rapid, well-damped short-period mode. By contrast, the trajectory variables, speed, and flight path angle, respond much more slowly. Figure 7.20, which displays a 10 min time span, shows that the dynamic response persists for a very long time, and that after the first few seconds it is primarily the phugoid mode that is evident.

The steady state that is approached so slowly has a slightly higher speed and a slightly smaller angle of attack than the original flight condition—both changes that would be expected from a down movement of the elevator. The flight path angle is seen to be almost unchanged—it increased by about one-tenth of a degree. The reason for an increase instead of the decrease that would be expected in normal cruising flight is that at this flight condition the airplane is flying below its minimum-drag speed.

If the reason for moving the elevator is to establish a new steady-state flight condition, then this control action can hardly be viewed as successful. The long lightly damped oscillation has seriously interfered with it. Clearly, longitudinal control, whether by a human or an automatic pilot, demands a more sophisticated control activity than simply moving it to its new position. We return to this topic in Chap. 8.

Phugoid Approximation

We can get an approximation to the transfer functions by using the phugoid approximation of Sec. 6.3. The differential equation is (6.3,6) with control terms added, that is,

$$\begin{bmatrix} \Delta\dot{u} \\ \dot{w} \\ 0 \\ \Delta\dot{\theta} \end{bmatrix} = \mathbf{A} \begin{bmatrix} \Delta u \\ w \\ q \\ \Delta\theta \end{bmatrix} + \begin{bmatrix} \dfrac{X_{\delta_e}}{m} \\ \dfrac{Z_{\delta_e}}{m} \\ M_{\delta_e} \\ 0 \end{bmatrix} \Delta\delta_e \tag{7.7,6}$$

[2]Program CC

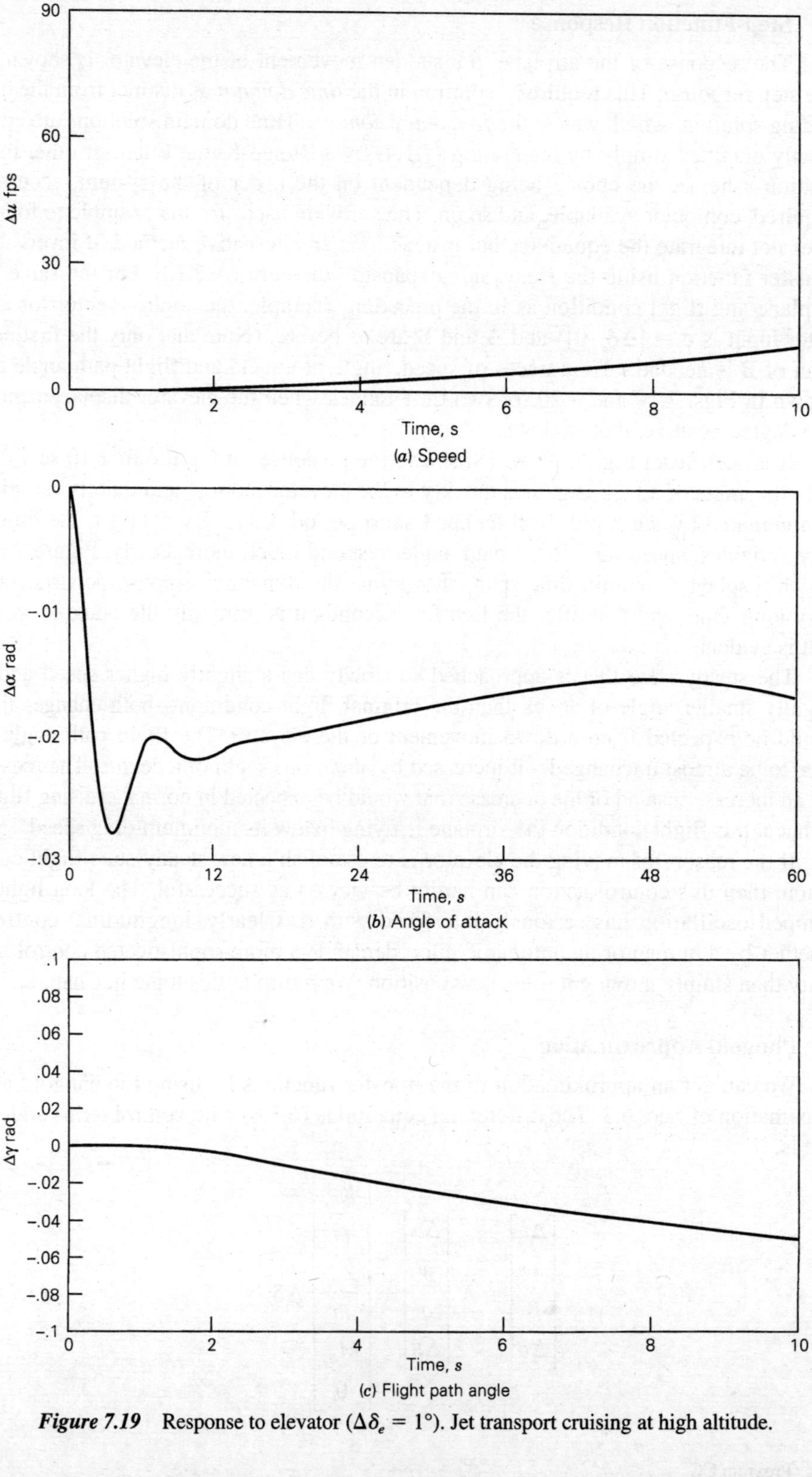

Figure 7.19 Response to elevator ($\Delta\delta_e = 1°$). Jet transport cruising at high altitude.

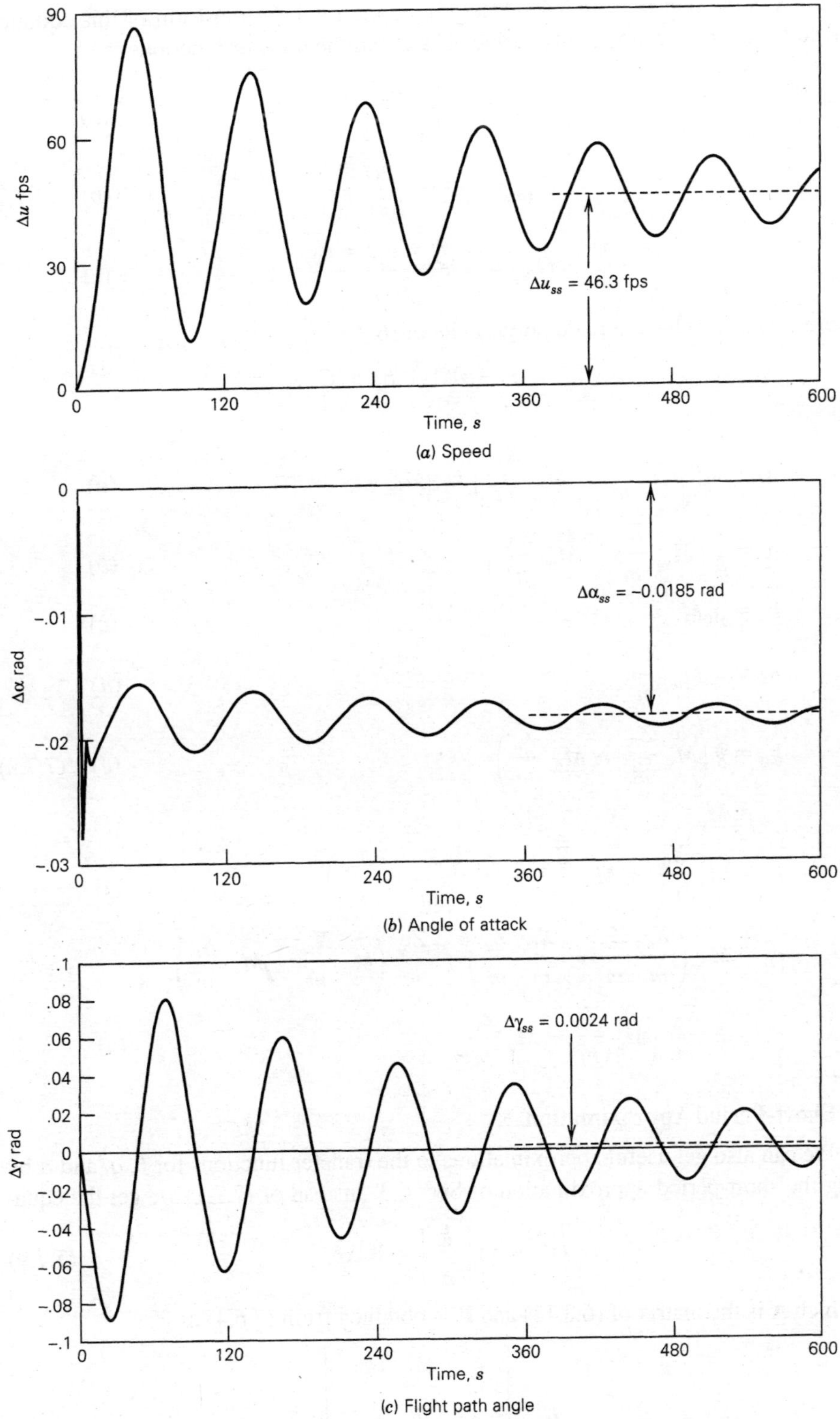

Figure 7.20 Response to elevator ($\Delta\delta_e = 1°$). Jet transport cruising at high altitude.

where **A** is the matrix of (6.3,6). After taking the Laplace transform of this equation and solving for the ratios of the variables, we find the transfer functions to be

$$G_{u\delta_e} = \frac{a_1 s + a_0}{f(s)} \qquad (a)$$

$$G_{w\delta_e} = \frac{b_2 s^2 + b_1 s + b_0}{f(s)} \qquad (b) \qquad (7.7,7)$$

$$G_{\theta\delta_e} = \frac{c_2 s^2 + c_1 s + c_0}{f(s)} \qquad (c)$$

where $f(s)$ is the characteristic polynomial of (6.3,8),

$$f(s) = As^2 + Bs + C$$

and

$$a_1 = u_0\left(M_{\delta_e}\frac{X_w}{m} - M_w\frac{X_{\delta_e}}{m}\right) - gM_{\delta_e} \qquad (a)$$

$$a_0 = g\left(M_{\delta_e}\frac{Z_w}{m} - M_w\frac{Z_{\delta_e}}{m}\right) \qquad (b)$$

$$b_2 = u_0 M_{\delta_e} \qquad (c)$$

$$b_1 = u_0 M_u \frac{X_{\delta_e}}{m} \qquad (d)$$

$$b_0 = g\left(M_u\frac{Z_{\delta_e}}{m} - M_{\delta_e}\frac{Z_u}{m}\right) \qquad (e) \qquad (7.7,8)$$

$$c_2 = M_{\delta_e} \qquad (f)$$

$$c_1 = M_u\frac{X_{\delta_e}}{m} + M_w\frac{Z_{\delta_e}}{m} - M_{\delta_e}\left(\frac{X_u}{m} + \frac{Z_w}{m}\right) \qquad (g)$$

$$c_0 = M_{\delta_e}\left(\frac{X_u}{m}\frac{Z_w}{m} - \frac{X_w}{m}\frac{Z_u}{m}\right) + \frac{Z_{\delta_e}}{m}\left(M_u\frac{X_w}{m} - M_w\frac{X_u}{m}\right) + \frac{X_{\delta_e}}{m}\left(M_w\frac{Z_u}{m} - M_u\frac{Z_w}{m}\right) \qquad (h)$$

Short-Period Approximation

We can also get useful approximations to the transfer functions for θ, q, and α by using the short-period approximation of Sec. 6.3. Instead of (7.7,1) we get the equation

$$(s\mathbf{I} - \mathbf{A})\begin{bmatrix} \bar{w} \\ \bar{q} \end{bmatrix} = \mathbf{B}\,\Delta\delta_e \qquad (7.7,9)$$

in which **A** is the matrix of (6.3,13) and **B** is obtained from (7.6,4) as

$$\mathbf{B} = \begin{bmatrix} \dfrac{Z_{\delta_e}}{m} \\ \dfrac{M_{\delta_e}}{I_y} + \dfrac{M_{\dot{w}}}{I_y}\dfrac{Z_{\delta_e}}{m} \end{bmatrix} \qquad (7.7,10)$$

When calculating the transfer functions we take note of the fact that $\bar{q} = s\Delta\bar{\theta}$. After solving (7.7,9) for the appropriate ratios we get the results

$$G_{w\delta_e} = \frac{a_1 s + a_0}{f(s)} \qquad (a)$$

$$G_{\theta\delta_e} = \frac{b_1 s + b_0}{sf(s)} = \frac{G_{q\delta_e}}{s} \qquad (b) \quad (7.7,11)$$

$$f(s) = s^2 + c_1 s + c_0 \qquad (c)$$

where

$$a_1 = \frac{Z_{\delta_e}}{mu_0} \qquad (a)$$

$$a_0 = u_0 \frac{M_{\delta_e}}{I_y} - \frac{M_q}{I_y}\frac{Z_{\delta_e}}{m} \qquad (b)$$

$$b_1 = \frac{M_{\delta_e}}{I_y} + \frac{M_{\dot{w}}}{I_y}\frac{Z_{\delta_e}}{m} \qquad (c)$$

$$b_0 = \frac{Z_{\delta_e}}{m}\frac{M_w}{I_y} - \frac{Z_w}{m}\frac{M_{\delta_e}}{I_y} \qquad (d) \quad (7.7,12)$$

$$c_1 = -\left(\frac{Z_w}{m} + \frac{M_q}{I_y} + \frac{M_{\dot{w}}}{I_y} u_0\right) \qquad (e)$$

$$c_0 = -u_0 \frac{M_w}{I_y} + \frac{M_q}{I_y}\frac{Z_w}{m} \qquad (f)$$

The frequency responses calculated with the above approximate transfer functions are shown on Figs. 7.15 to 7.18. It is seen that the phugoid approximation is exact at very low frequencies and the short-period approximation is exact in the high-frequency limit. For frequencies between those of the phugoid and short-period modes, one approximation or the other can give reasonable results.

RESPONSE TO THE THROTTLE

For the same jet airplane and flight condition as in previous numerical examples, we calculate the response to a step input in the throttle of $\Delta\delta_p = \frac{1}{6}$, which corresponds to a thrust increment of 0.05W. The matrix **B** is given by (7.6,5) and $\Delta\mathbf{c}$ is

$$\Delta\mathbf{c} = [0 \quad 1/6]^T \qquad (7.7,13)$$

The numerical results are shown in Fig. 7.21. Because the model has not included any engine dynamics, the results are not valid for the first few seconds. However, this region is not of much interest in this case. The motion is clearly seen to be dominated by the lightly damped phugoid. The speed begins to increase immediately, before the other variables have time to change. It then undergoes a slow damped oscillation, ultimately returning to its initial value. The angle of attack varies only slightly, and γ makes an oscillatory approach to its final positive value γ_{ss}. The new steady state is a climb with $\Delta u = \Delta\alpha = 0$. When the thrust line does not pass through the CG, the response is different in several details. Principally, the moment of the thrust causes a

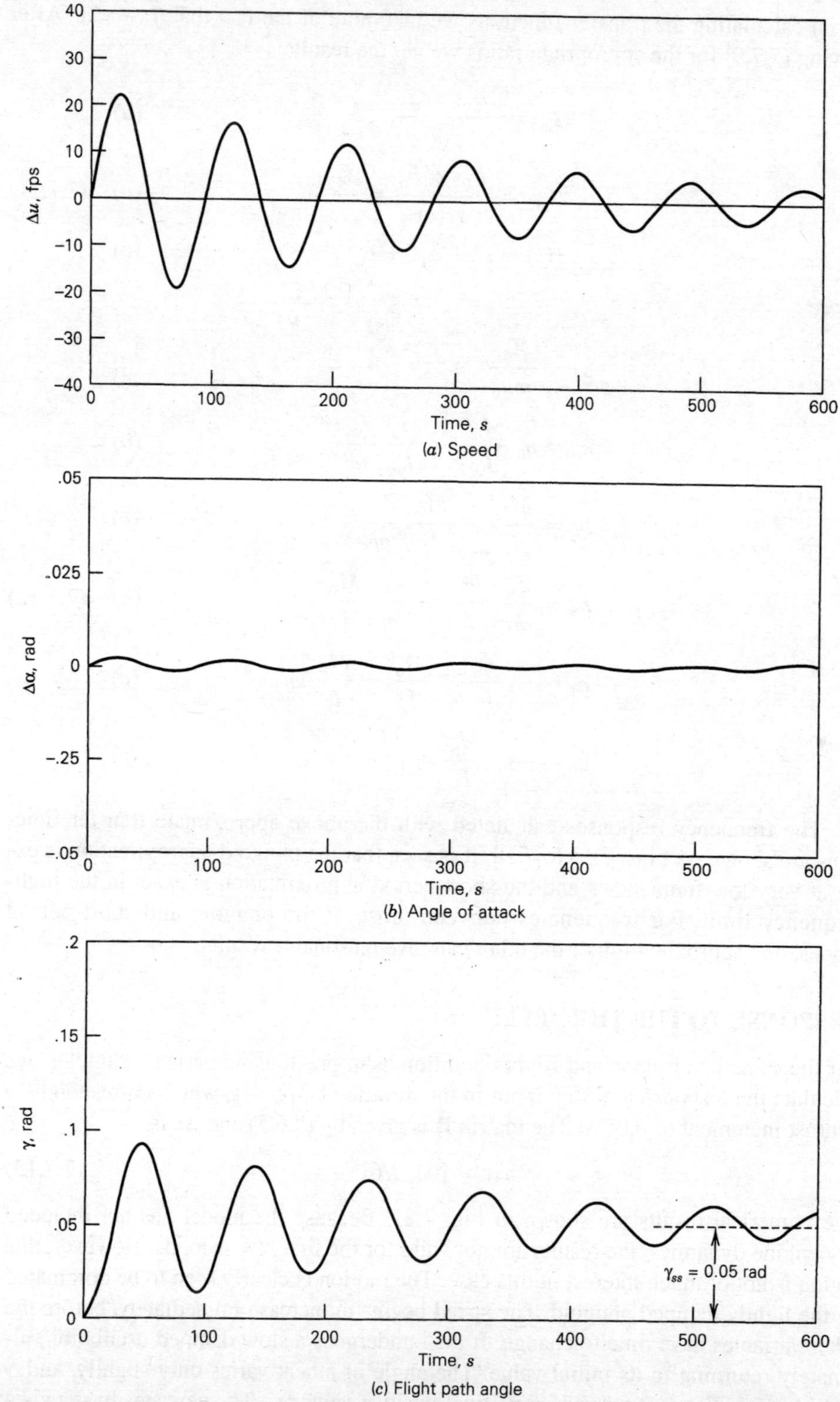

Figure 7.21 Response to throttle. Jet transport cruising at high altitude. Thrust line passing through CG.

rapid change in α, followed by an oscillatory decay to a new $\Delta\alpha_{ss} \neq 0$, and the speed converges to a new value $\Delta u_{ss} \neq 0$.

7.8 *Lateral Steady States*

The basic flight condition is steady symmetric flight, in which all the lateral variables β, p, r, ϕ are identically zero. Unlike the elevator and the throttle, the lateral controls, the aileron and rudder, are not used *individually* to produce changes in the steady state. This is because the steady state values of β, p, r, ϕ that result from a constant δ_a or δ_r are not generally of interest as a useful flight condition. There are two lateral steady states that are of interest, however, each of which requires the joint application of aileron and rudder. These are the *steady sideslip,* in which the flight path is rectilinear, and the *steady turn,* in which the angular velocity vector is vertical. We look into these below before proceeding to the study of dynamic response to the lateral controls.

THE STEADY SIDESLIP

The steady sideslip is a condition of nonsymmetric rectilinear translation. It is sometimes used, particularly with light airplanes, to correct for cross-wind on landing approaches. Glider pilots also use this maneuver to steepen the glide path, since the *L/D* ratio decreases due to increased drag at large β. In this flight condition all the time derivatives in the equations of motion (except $\dot{x}_E$) and the three rotation rates p, q, r are zero. It is simplest in this case to go back to (4.9,2)–(4.9,6), from which we derive the following:

$$\begin{aligned} \Delta Y + mg\phi \cos \theta_0 &= 0 \\ \Delta L &= 0 \\ \Delta N &= 0 \end{aligned} \tag{7.8,1}$$

We use (4.9,17) for aerodynamic forces, and for the control forces use the following as a reasonable representation:

$$\begin{bmatrix} \Delta Y_c \\ \Delta L_c \\ \Delta N_c \end{bmatrix} = \begin{bmatrix} Y_{\delta_r} & 0 \\ L_{\delta_r} & L_{\delta_a} \\ N_{\delta_r} & N_{\delta_a} \end{bmatrix} \begin{bmatrix} \delta_r \\ \delta_a \end{bmatrix} \tag{7.8,2}$$

We now add the assumption that θ_0 is a small angle and get the resulting equation

$$\begin{bmatrix} Y_{\delta_r} & 0 & mg \\ L_{\delta_r} & L_{\delta_a} & 0 \\ N_{\delta_r} & N_{\delta_a} & 0 \end{bmatrix} \begin{bmatrix} \delta_r \\ \delta_a \\ \phi \end{bmatrix} = -\begin{bmatrix} Y_v \\ L_v \\ N_v \end{bmatrix} v \tag{7.8,3}$$

In this form, v is treated as an arbitrary input, and $(\delta_r, \delta_a, \phi)$ as outputs. (See Exercise 7.6.) Clearly, there is an infinity of possible sideslips, since v can be chosen arbitrarily. Note that the other three variables are all proportional to v. We illustrate the steady sideslip with a small general aviation airplane[3] of 30-ft (9.14 m) span and a

[3]Based on the Piper Cherokee. The control derivatives were taken from McCormick (1979). We estimated the stability derivatives. The numerical values used may not truly represent this airplane.

Table 7.2
Nondimensional Derivatives—General Aviation Airplane (expressed in rad^{-1} and (rad/s)$^{-1}$)

	C_y	C_l	C_n
β	-0.14	$-0.0689 - 0.0917C_L$	$0.01326 + 0.017C_L^2$
$\hat{p}$	-0.039	-0.441	$-0.00109 - 0.0966C_L$
$\hat{r}$	0.165	$-0.0144 + 0.271C_L$	$-0.048 - 0.0238C_L^2$
δ_a	0	-0.0531	0.005
δ_r	0.117	0.0105	-0.0509

gross weight of 2400 lb (10,675 N). The altitude is sea level and $C_L = 1.0$, corresponding to a speed of 112.3 fps (34.23 m/s), and the wing area is 160 ft^2 (14.9 m^2). The nondimensional derivatives are given in Table 7.2, from which the numerical system equation is found from (7.8,3) to be (see Exercise 7.5)

$$\begin{bmatrix} 280.7 & 0 & 2400 \\ 755.7 & -3821.9 & 0 \\ -3663.5 & 359 & 0 \end{bmatrix} \begin{bmatrix} \delta_r \\ \delta_a \\ \phi \end{bmatrix} = \begin{bmatrix} 2.991 \\ 102.93 \\ -19.394 \end{bmatrix} v \tag{7.8,4}$$

It is convenient to express the sideslip as an angle instead of a velocity. To do so we recall that $\beta \doteq v/u_0$, with u_0 given above as 112.3 fps. The solution of (7.8,4) is found to be

$$\delta_r/\beta = .303$$
$$\delta_a/\beta = -2.96$$
$$\phi/\beta = .104$$

We see that a positive sideslip (to the right) of say 10° would entail left rudder of 3° and right aileron of 29.6°. Clearly the main control action is the aileron displacement, without which the airplane would, as a result of the sideslip to the right, roll to the left. The bank angle is seen to be only 1° to the right so the sideslip is almost flat.

THE STEADY TURN

We define a "truly banked" turn to be one in which (1) the vehicle angular velocity vector $\boldsymbol{\omega}$ is constant and vertical (see Fig. 7.22) and (2) the resultant of gravity and centrifugal force at the mass center lies in the plane of symmetry (see Fig. 7.23). This corresponds to flying the turn on the turn-and-bank indicator.[4] It is quite common for turns to be made at bank angles that are too large for linearization of sin ϕ and cos ϕ to be acceptable, although all the state variables other than ϕ and V are small. Thus we turn to the basic nonlinear equations in Sec. 4.7 for this analysis. The large bank angle has the consequence that coupling of the lateral and longitudinal equations occurs, since more lift is needed to balance gravity than in level flight. Thus not only the aileron and rudder but the elevator as well must be used for turning at large ϕ.

[4]Neglecting the fact that the pilot and indicator are not right at the CG.

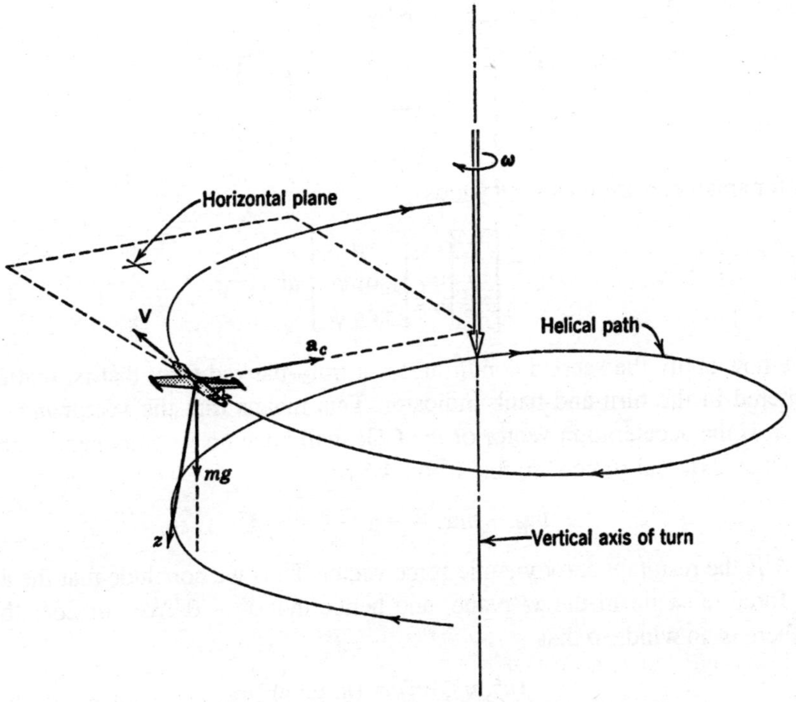

Figure 7.22 Steady climbing turn.

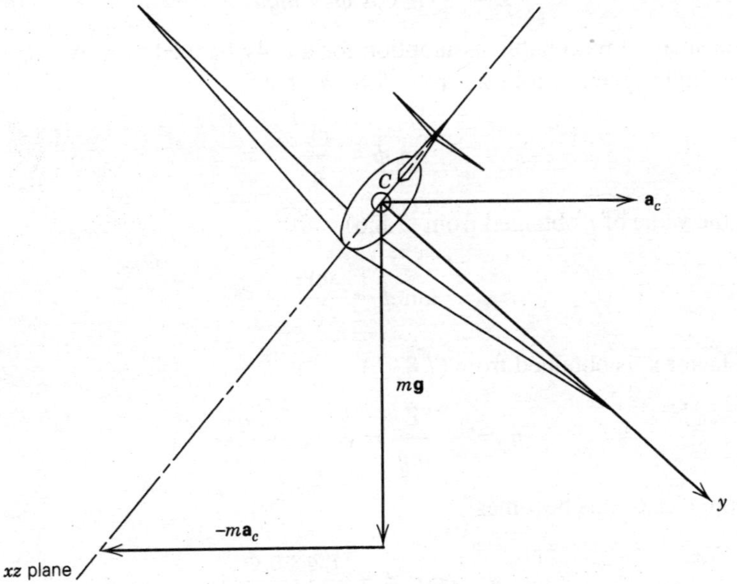

Figure 7.23 Gravity and acceleration in turn.

The body-axis angular rates are given by

$$\begin{bmatrix} p \\ q \\ r \end{bmatrix} = \mathbf{L}_{BE}\boldsymbol{\omega}_E = \mathbf{L}_{BE}\begin{bmatrix} 0 \\ 0 \\ \omega \end{bmatrix} \tag{7.8,5}$$

which for small elevation angle θ yields

$$\begin{bmatrix} p \\ q \\ r \end{bmatrix} = \begin{bmatrix} -\theta \\ \sin\phi \\ \cos\phi \end{bmatrix}\omega \tag{7.8,6}$$

We now apply the second condition for a truly-banked turn, that is, that the ball be centered in the turn-and-bank indicator. This means that the vector $m\mathbf{g} - m\mathbf{a}_c$, where $\mathbf{a}_c$ is the acceleration vector of the CG, shall have no y component. But $m\mathbf{a}_c$ is the resultant external force $\mathbf{f}$, so that from (4.5,6)

$$m\mathbf{g} - m\mathbf{a}_c = m\mathbf{g} - \mathbf{f} = -\mathbf{A}$$

where $\mathbf{A}$ is the resultant aerodynamic force vector. Thus we conclude that the aerodynamic force must lie in the xz plane, and hence that $Y = 0$. We consider the case when there is no wind, so that

$$(u^E, v^E, w^E) = (u, v, w)$$

and choose the body axes so that $\alpha_x = w = 0$. We now use (4.7,1) with all the variables constant and only u and ϕ not small to get:

$$\begin{aligned} Y &= -mg\sin\phi + mru = 0 & (a) \\ Z &= -mg\cos\phi - mqu & (b) \end{aligned} \tag{7.8,7}$$

When v is small, a reasonable assumption for a truly banked turn, we also have that $u \cong V$, the flight speed. It follows from (7.8,7a) that

$$\sin\phi = \frac{rV}{g}$$

and with the value of r obtained from (7.8,6)

$$\tan\phi = \frac{\omega V}{g} \tag{7.8,8}$$

The load factor n_z is obtained from (7.8,7b):

$$n_z = -\frac{Z}{mg} = \cos\phi + \frac{qV}{g}$$

With q from (7.8,6) this becomes

$$n_z = \cos\phi + \frac{V\omega\sin\phi}{g}$$

By using (7.8,8) to eliminate $V\omega$ we get

$$n_z = \sec\phi \tag{7.8,9}$$

We note from (5.1,1) that $Z = -L$ in this case, so that $n = L/W = n_z$. The incremental lift coefficient, as compared with straight flight at the same speed and height, is

$$\Delta C_L = \frac{L - mg}{\frac{1}{2}\rho V^2 S} = (n - 1)C_W \tag{7.8,10}$$

We can now write down the equations governing the control angles. From (4.7,2), to first order, $L = M = N = 0$, so we have the five aerodynamic conditions

$$C_l = C_m = C_n = C_y = 0$$

and

$$\Delta C_L = (n - 1)C_W$$

On expanding these with the usual aerodynamic derivatives, we get

$$\begin{aligned}
C_{l_\beta}\beta + C_{l_p}\hat{p} + C_{l_r}\hat{r} + C_{l_{\delta_r}}\delta_r + C_{l_{\delta_a}}\delta_a &= 0\\
C_{m_\alpha}\Delta\alpha + C_{m_q}\hat{q} + C_{m_{\delta_e}}\Delta\delta_e &= 0\\
C_{n_\beta}\beta + C_{n_p}\hat{p} + C_{n_r}\hat{r} + C_{n_{\delta_r}}\delta_r + C_{n_{\delta_a}}\delta_a &= 0\\
C_{y_\beta}\beta + C_{y_p}\hat{p} + C_{y_r}\hat{r} + C_{y_{\delta_r}}\delta_r &= 0\\
C_{L_\alpha}\Delta\alpha + C_{L_q}\hat{q} + C_{L_{\delta_e}}\Delta\delta_e &= (n-1)C_W
\end{aligned} \tag{7.8,11}$$

In these relations $\hat{p}$, $\hat{q}$, $\hat{r}$ are known from (7.8,6), that is,

$$\begin{bmatrix}\hat{p}\\ \hat{q}\\ \hat{r}\end{bmatrix} = \begin{bmatrix}-\theta\dfrac{b}{2V}\\ \sin\phi\dfrac{\bar{c}}{2V}\\ \cos\phi\dfrac{b}{2V}\end{bmatrix}\omega \tag{7.8,12}$$

The five equations (7.8,11) for the five unknowns $[\beta, \delta_r, \delta_a]$ and $[\Delta\alpha, \Delta\delta_e]$ uncouple into two independent sets:

$$\begin{bmatrix}C_{y_\beta} & C_{y_{\delta_r}} & 0\\ C_{l_\beta} & C_{l_{\delta_r}} & C_{l_{\delta_a}}\\ C_{n_\beta} & C_{n_{\delta_r}} & C_{n_{\delta_a}}\end{bmatrix}\begin{bmatrix}\beta\\ \delta_r\\ \delta_a\end{bmatrix} = \begin{bmatrix}C_{y_p} & C_{y_r}\\ C_{l_p} & C_{l_r}\\ C_{n_p} & C_{n_r}\end{bmatrix}\begin{bmatrix}\theta\\ -\cos\phi\end{bmatrix}\frac{\omega b}{2V} \tag{7.8,13}$$

and

$$\begin{bmatrix}C_{m_\alpha} & C_{m_{\delta_e}}\\ C_{L_\alpha} & C_{L_{\delta_e}}\end{bmatrix}\begin{bmatrix}\Delta\alpha\\ \Delta\delta_e\end{bmatrix} = -\begin{bmatrix}C_{m_q}\\ C_{L_q}\end{bmatrix}\frac{\omega\bar{c}}{2V}\sin\phi + \begin{bmatrix}0\\ (n-1)C_W\end{bmatrix} \tag{7.8,14}$$

When (7.8,9) is used to eliminate ϕ from (7.8,14), and after some routine algebra, the solution for $\Delta\delta_e$ is found to be (see Exercise 7.6)

$$\Delta\delta_e = (n-1)C_W\,\frac{C_{m_\alpha} - \dfrac{n+1}{2\mu n}(C_{L_\alpha}C_{m_q} - C_{L_q}C_{m_\alpha})}{C_{L_{\delta_e}}C_{m_\alpha} - C_{L_\alpha}C_{m_{\delta_e}}} \tag{7.8,15}$$

Except for far forward CG positions and low speeds, the angles given by (7.8,15) are moderate. The similarity of this expression to that for elevator angle per g in a pull-up (3.1,6a) should be noted. They are in fact the same in the limit $n \to \infty$. The elevator angle per g in a turn is therefore not very different from that in a vertical pull-up.

Finally, the lateral control angles are obtained from the solution of (7.8,13).

NUMERICAL EXAMPLE

The rudder and aileron angles in a steady truly-banked turn are calculated by way of example for the same general aviation airplane as was used above for the sideslip. The altitude is sea level, the speed is 125 fps (38.1 m/s) and the stability and control derivatives are as in Table 7.2. The solution of (7.8,13) for climb angles between $-10°$ and $+10°$ shows that the sideslip angle β remains less than 1.5° and the rudder and aileron angles are as shown in Figs. 7.24 and 7.25. The value of C_L varies over the range of bank angles used from 0.8 to 1.6, so several of the stability derivatives are significantly affected. It is seen that the aileron angle is always positive for a right turn—that is, the right aileron is down (stick to the left), and that the rudder is usually negative (right rudder) although its sign may reverse in a steep climb. The strong effect of the climb angle derives from the fact that the roll rate p is proportional to θ and changes sign with it. Thus the terms $C_{l_p}\hat{p}$ and $C_{n_p}\hat{p}$ in the moment equations change sign between climbing and descending and affect the control angles required to produce zero moment.

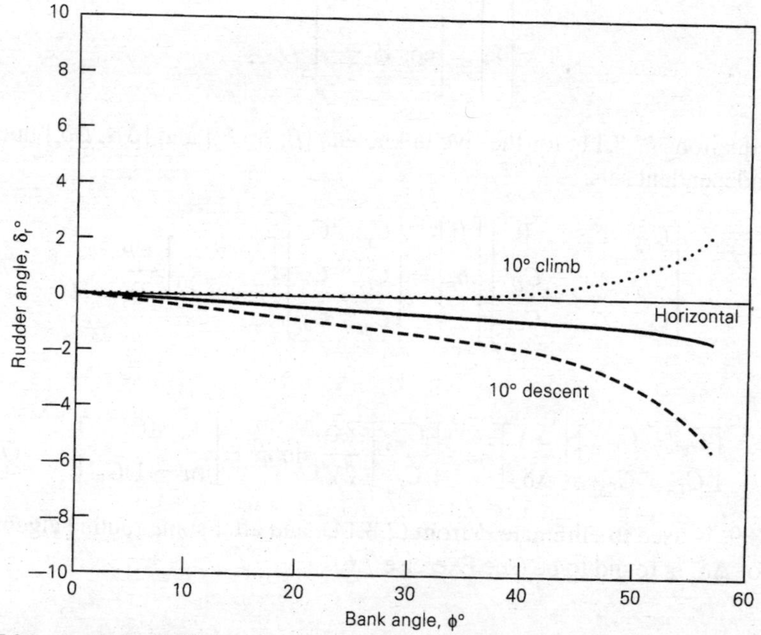

Figure 7.24 Rudder angle in turn.

- General aviation airplane
- Speed 125 fps
- Sea level

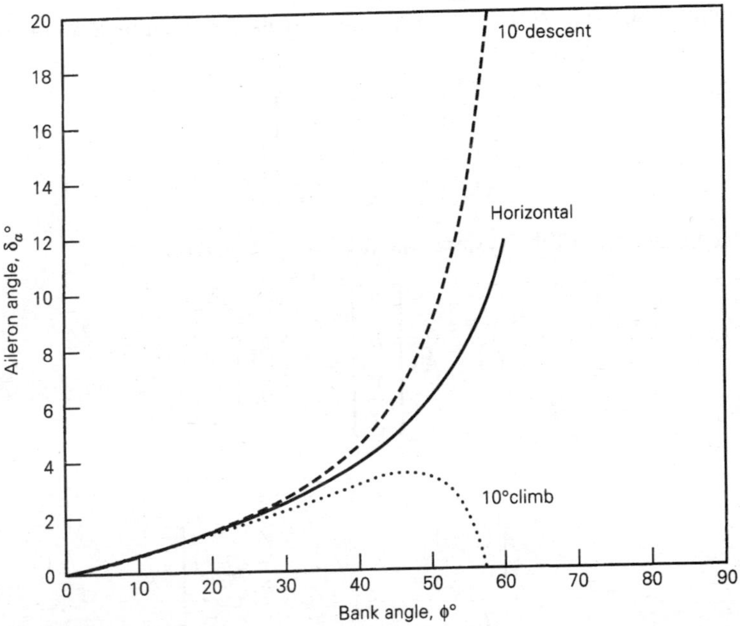

Figure 7.25 Aileron angle in turn.

- General aviation airplane
- Speed 125 fps
- Sea level

Finally, it may be remarked that the control angles obtained would have been substantially different had it been stipulated that β, not C_y, should be zero in the turn. It would not then be possible, however, to satisfy the requirement that the ball be centered in the turn-and-back indicator.

7.9 *Lateral Frequency Response*

The procedure for calculating the response of the airplane to sinusoidal movement of the rudder or aileron is similar to that used for longitudinal response in Sec. 7.6. The

Table 7.3
Control Derivatives—B747 Jet Transport (expressed in rad^{-1})

	C_y	C_l	C_n
δ_a	0	-1.368×10^{-2}	-1.973×10^{-4}
δ_r	0.1146	6.976×10^{-3}	−0.1257

	C_x	C_z	C_m
δ_e	-3.818×10^{-6}	−0.3648	−1.444

aerodynamics associated with the two lateral controls are given by a set of control derivatives:

$$\begin{bmatrix} \Delta Y_c \\ \Delta L_c \\ \Delta N_c \end{bmatrix} = \begin{bmatrix} Y_{\delta_a} & Y_{\delta_r} \\ L_{\delta_a} & L_{\delta_r} \\ N_{\delta_a} & N_{\delta_r} \end{bmatrix} \begin{bmatrix} \delta_a \\ \delta_r \end{bmatrix} \tag{7.9,1}$$

The Laplace transform of the system equation (4.9,19) is then

$$(s\mathbf{I} - \mathbf{A}) \begin{bmatrix} \bar{v} \\ \bar{p} \\ \bar{r} \\ \bar{\phi} \end{bmatrix} = \mathbf{B} \begin{bmatrix} \bar{\delta}_a \\ \bar{\delta}_r \end{bmatrix} \tag{7.9,2}$$

where **B** is

$$\mathbf{B} = \begin{bmatrix} \dfrac{Y_{\delta_a}}{m} & \dfrac{Y_{\delta_r}}{m} \\ \dfrac{L_{\delta_a}}{I'_x} + I'_{zx} N_{\delta_a} & \dfrac{L_{\delta_r}}{I'_x} + I'_{zx} N_{\delta_r} \\ I'_{zx} L_{\delta_a} + \dfrac{N_{\delta_a}}{I'_z} & I'_{zx} L_{\delta_r} + \dfrac{N_{\delta_r}}{I'_z} \\ 0 & 0 \end{bmatrix} \tag{7.9,3}$$

NUMERICAL EXAMPLE

For our numerical example we use the same jet transport and flight condition as in Sec. 7.6. **A** is given by (6.7,1) and the control derivatives are given in Table 7.3, from which, with the definitions of Table 7.1, the elements of **B** are calculated to be

$$\mathbf{B} = \begin{bmatrix} 0 & 5.642 \\ -0.1431 & 0.1144 \\ 0.003741 & -0.4859 \\ 0 & 0 \end{bmatrix} \tag{7.9,4}$$

The eight transfer functions are then as in (7.2,8), where $f(s)$ is the characteristic polynomial of (6.7,2) (with s instead of λ) and with the numerators as follows:

$$\begin{aligned}
N_{v\delta_a} &= 2.896s^2 + 6.542s + 0.6220 && (a) \\
N_{v\delta_r} &= 5.642s^3 + 379.4s^2 + 167.9s - 5.934 && (b) \\
N_{p\delta_a} &= 0.1431s^3 + 0.02730s^2 + 0.1102s && (c) \\
N_{p\delta_r} &= 0.1144s^3 - 0.1997s^2 - 1.368s && (d) \\
N_{r\delta_a} &= -0.003741s^3 - 0.002708s^2 - 0.0001394s + 0.004539 && (e) \quad (7.9,5) \\
N_{r\delta_r} &= -0.4859s^3 - 0.2327s^2 - 0.009018s - 0.05647 && (f) \\
N_{\phi\delta_a} &= .1431s^2 + 0.02730s + 0.1102 && (g) \\
N_{\phi\delta_r} &= 0.1144s^2 - 0.1997s - 1.368 && (h)
\end{aligned}$$

From the transfer functions $G_{ij}(s) = N_{ij}(s)/f(s)$, the frequency response functions $G_{ij}(i\omega)$ were calculated for both aileron and rudder inputs. The results for v, ϕ and r are shown on Figs. 7.26 and 7.27. The most significant feature in all of these responses is the peak in the amplitude at the Dutch Roll frequency, and the associated sharp drop in phase angle.

At zero frequency we see from (7.9,5*c* and *d*) that the roll rate amplitude is zero for both inputs. All the other variables have finite values at $\omega = 0$. Even for moderate

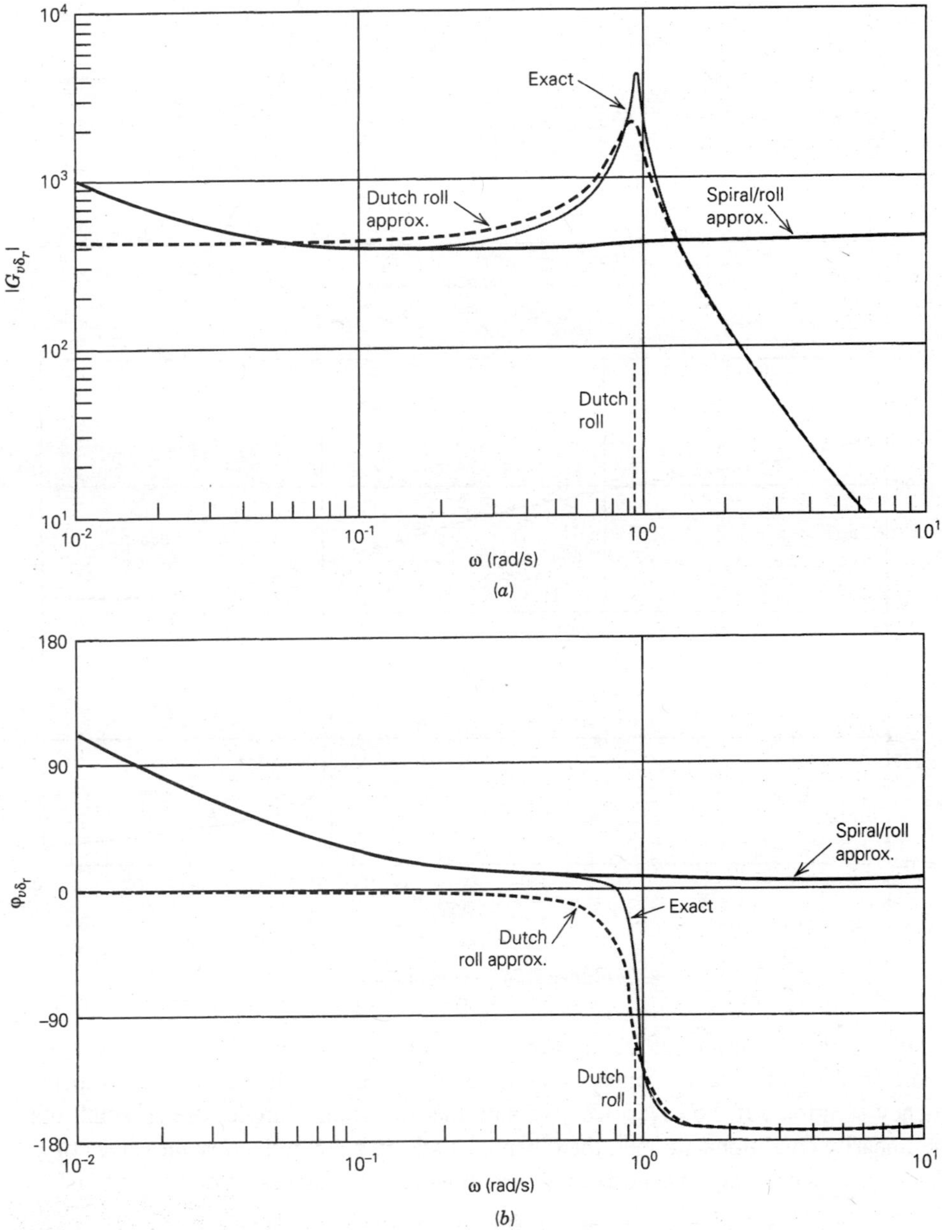

Figure 7.26 Frequency-response functions, rudder angle input. Jet transport cruising at high altitude. (*a*) Sideslip amplitude. (*b*) Sideslip phase. (*c*) Roll amplitude. (*d*) Roll phase. (*e*) Yaw-rate amplitude. (*f*) Yaw-rate phase.

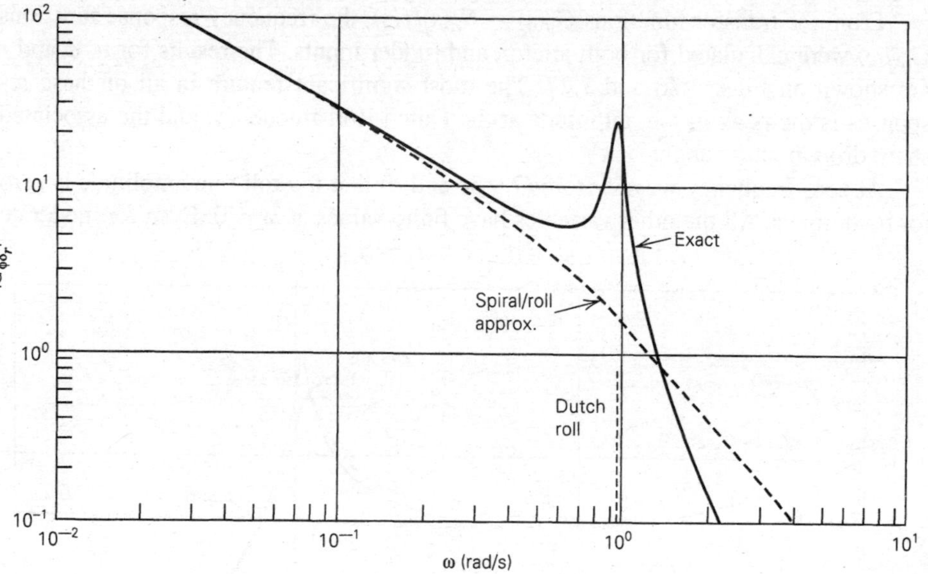

(c)

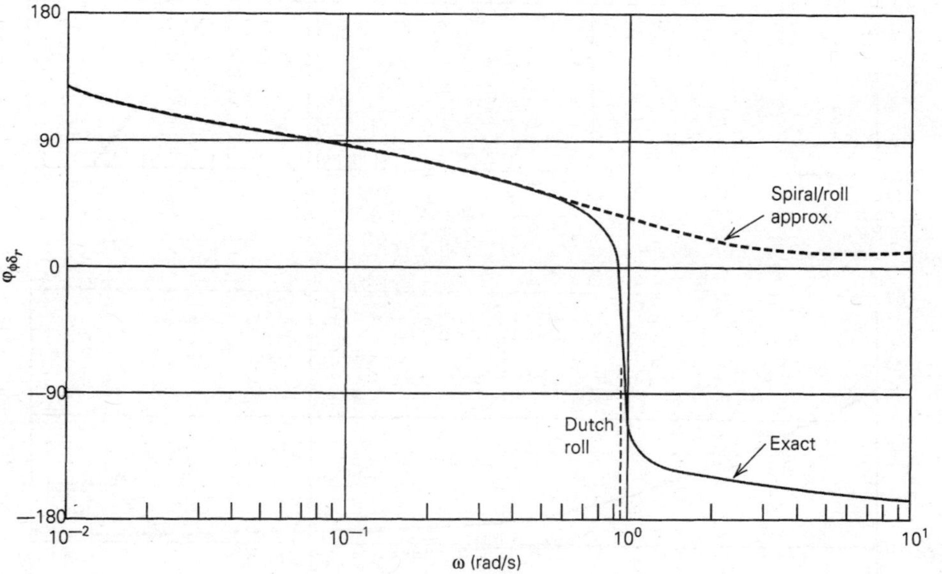

(d)

Figure 7.26 *(Continued)*

control angles, however, the steady-state values of $\beta = v/u_0$ and ϕ are very large (see Exercise 7.10). Hence the linearity assumption severely constrains these zero frequency solutions. If, however, we postulate that the control angles are so small that the linearity conditions are met, then there is a steady state with constant values of ϕ, β, and r. This can only be a horizontal turn in which the angular velocity vector is

$$\mathbf{\Omega}_B = [0 \quad q_{ss} \quad r_{ss}]^T$$

where

$$q_{ss} = \Omega \sin \phi$$

$$r_{ss} = \Omega \cos \phi$$

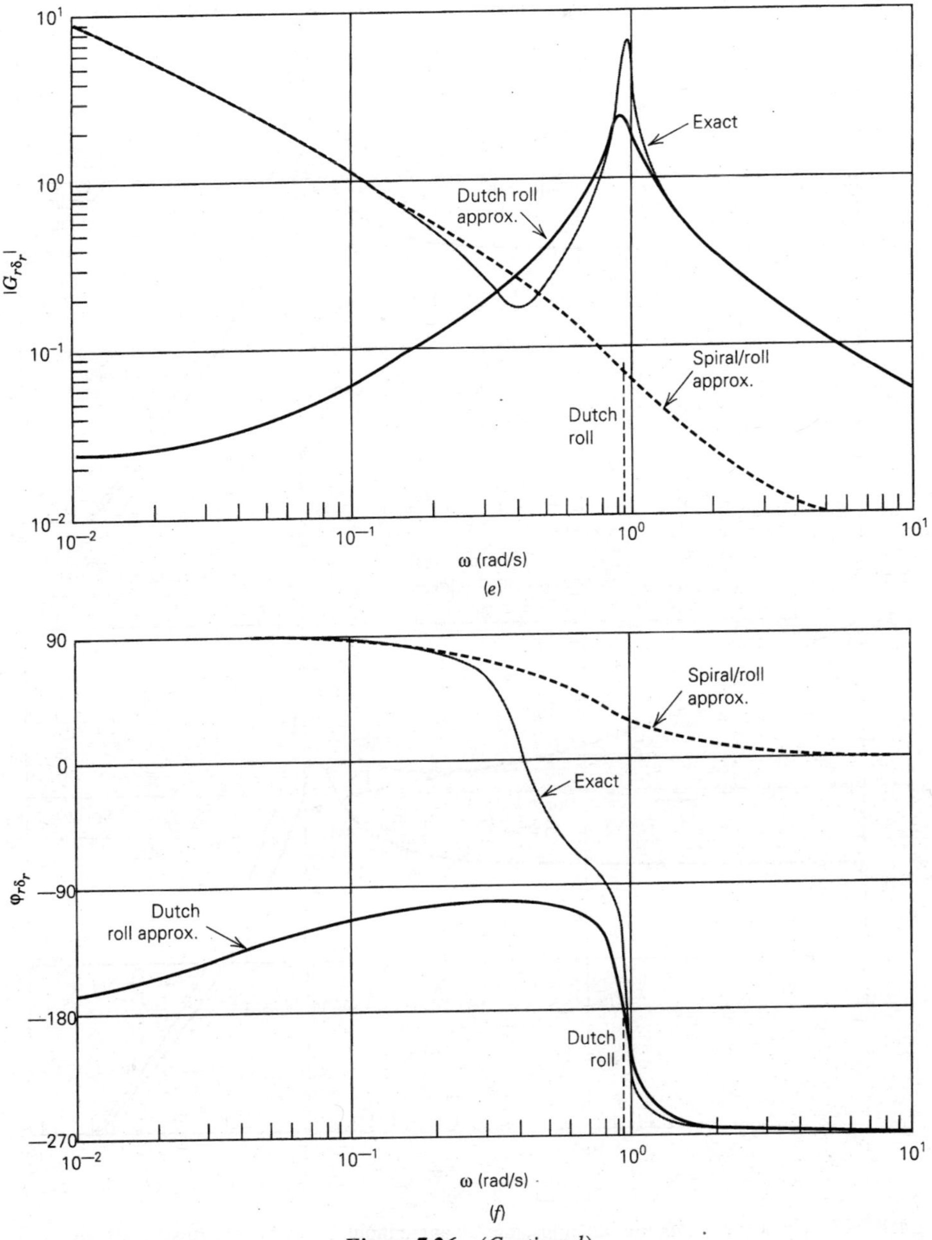

Figure 7.26 (*Continued*)

7.10 Approximate Lateral Transfer Functions

Approximate transfer functions that can be written out explicitly, and that reveal the main aerodynamic influences in a particular frequency range, can be very useful in designing control systems. In Sec. 6.8 we presented two approximate second-order systems that simulate the complete fourth-order system insofar as the characteristic modes are concerned. These same approximations can be used to get approximate transfer functions for control response.

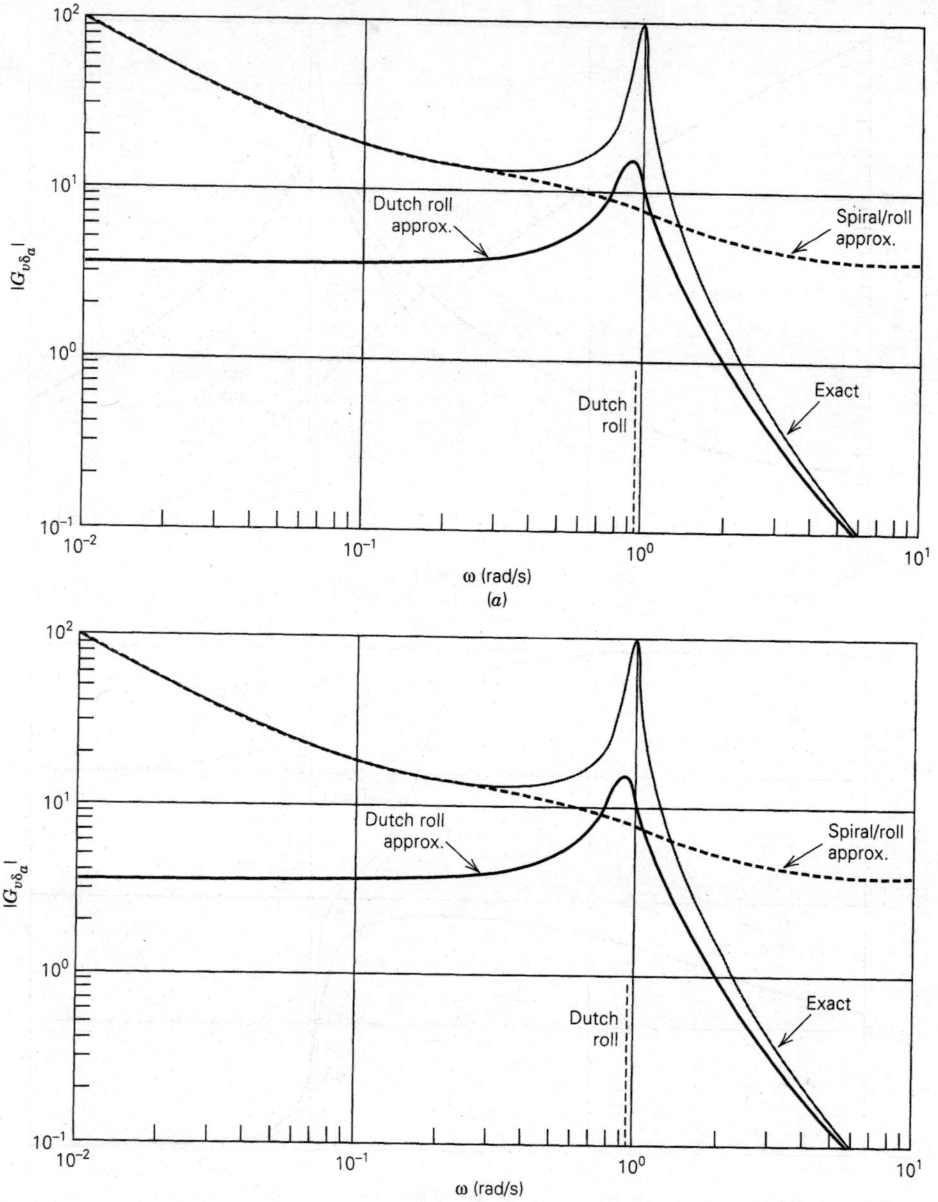

Figure 7.27 Frequency-response functions, aileron angle input. Jet transport cruising at high altitude. (*a*) Sideslip amplitude. (*b*) Sideslip phase. (*c*) Roll amplitude. (*d*) Roll phase. (*e*) Yaw-rate amplitude. (*f*) Yaw-rate phase.

SPIRAL/ROLL APPROXIMATION

When aerodynamic control terms are added to (6.8,9) and the Laplace transform is taken, the result is

$$\begin{bmatrix} 0 & 0 & u_0 & -g \\ -\mathcal{L}_v & (s-\mathcal{L}_p) & -\mathcal{L}_r & 0 \\ -\mathcal{N}_v & -\mathcal{N}_p & (s-\mathcal{N}_r) & 0 \\ 0 & -1 & 0 & s \end{bmatrix} \begin{bmatrix} \bar{v} \\ \bar{p} \\ \bar{r} \\ \bar{\phi} \end{bmatrix} = \begin{bmatrix} \mathcal{Y}_{\delta_a} & \mathcal{Y}_{\delta_r} \\ \mathcal{L}_{\delta_a} & \mathcal{L}_{\delta_r} \\ \mathcal{N}_{\delta_a} & \mathcal{N}_{\delta_r} \\ 0 & 0 \end{bmatrix} \begin{bmatrix} \delta_a \\ \delta_r \end{bmatrix} \quad (7.10,1)$$

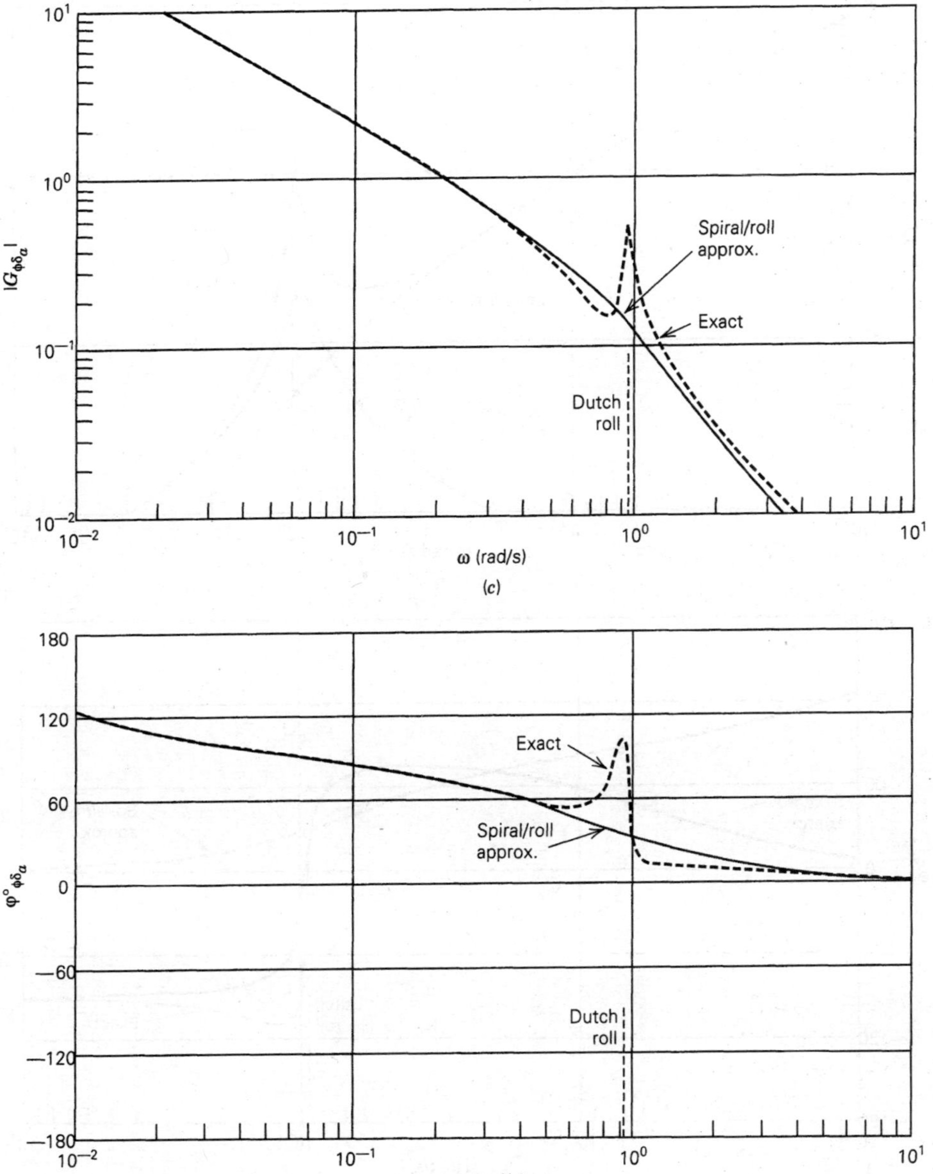

Figure 7.27 *(Continued)*

In (7.10,1) the $\mathcal{Y}$ and $\mathcal{N}$ derivatives are as defined in Sec. 6.8, and the $\mathcal{L}_\delta$ derivatives are

$$\mathcal{L}_\delta = \frac{L_\delta}{I'_x} + I'_{zx}N_\delta \tag{7.10,2}$$

where δ is either δ_a or δ_r. From (7.10,1) we get the desired transfer functions. The denominators are all the same, obtained from (6.8,11) as

$$Cs^2 + Ds + E \tag{7.10,3}$$

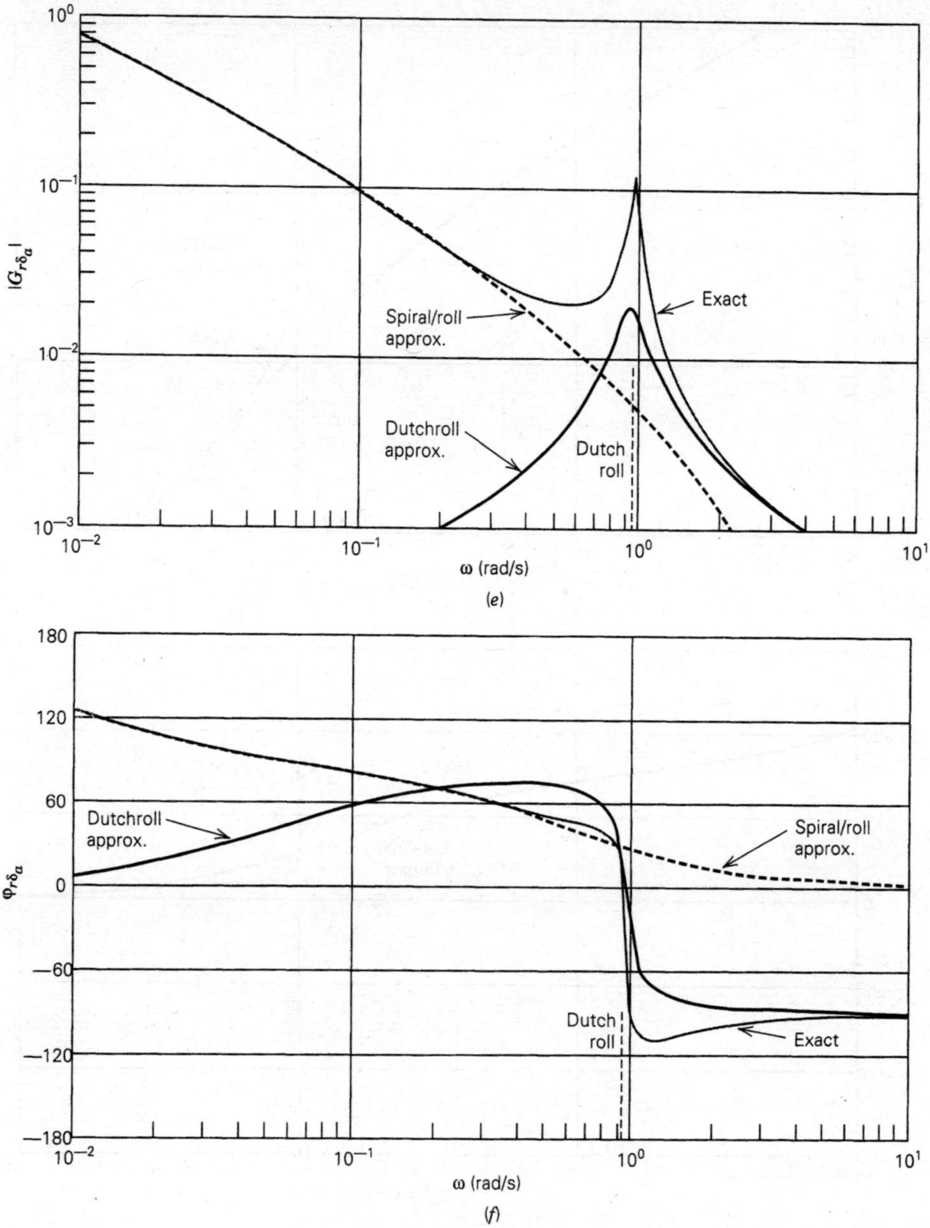

Figure 7.27 *(Continued)*

and the numerators are (again using δ for either δ_a or δ_r):

$$
\begin{aligned}
N_{v\delta} &= a_3s^3 + a_2s^2 + a_1s + a_0 \\
N_{\phi\delta} &= b_1s + b_0 \\
N_{r\delta} &= d_2s^2 + d_1s + d_0 \\
N_{p\delta} &= sN_{\phi\delta}
\end{aligned}
\tag{7.10,4}
$$

The coefficients in these relations are:

$$
\begin{aligned}
&a_3 = \mathcal{Y}_\delta; \qquad a_2 = -\mathcal{Y}_\delta(\mathcal{L}_p + \mathcal{N}_r) - u_0\mathcal{N}_\delta \\
&a_1 = \mathcal{Y}_\delta(\mathcal{L}_p\mathcal{N}_r - \mathcal{L}_r\mathcal{N}_p) - u_0(\mathcal{L}_\delta\mathcal{N}_p - \mathcal{L}_p\mathcal{N}_\delta) + \mathcal{L}_\delta g \\
&a_0 = g(\mathcal{L}_r\mathcal{N}_\delta - \mathcal{L}_\delta\mathcal{N}_r) \\
&b_1 = \mathcal{Y}_\delta\mathcal{L}_v; \qquad b_0 = u_0(\mathcal{L}_\delta\mathcal{N}_v - \mathcal{L}_v\mathcal{N}_\delta) + \mathcal{Y}_\delta(\mathcal{L}_v\mathcal{N}_r - \mathcal{L}_r\mathcal{N}_v) \\
&d_2 = \mathcal{Y}_\delta\mathcal{N}_v; \qquad d_1 = \mathcal{Y}_\delta(\mathcal{L}_v\mathcal{N}_p - \mathcal{L}_p\mathcal{N}_v); \qquad d_0 = g(\mathcal{L}_\delta\mathcal{N}_v - \mathcal{L}_v\mathcal{N}_\delta)
\end{aligned}
\tag{7.10,5}
$$

DUTCH ROLL APPROXIMATION

Following the analysis of Sec. 6.8 and adding control terms to the aerodynamics the reduced system equations are

$$
\begin{aligned}
\dot{v} &= \mathcal{Y}_v v - u_0 r + \Delta\mathcal{Y}_c \\
\dot{r} &= \mathcal{N}_v v + \mathcal{N}_r r + \Delta\mathcal{N}_c
\end{aligned}
\tag{7.10,6}
$$

From (7.10,6) we derive the canonical equation

$$\dot{\mathbf{x}} = \mathbf{A}\mathbf{x} + \mathbf{B}\mathbf{c}$$

where

$$\mathbf{x} = [v \quad r]^T; \qquad \mathbf{c} = [\delta_a \quad \delta_r]^T$$

$$
\mathbf{A} = \begin{bmatrix} \mathcal{Y}_v & -u_0 \\ \mathcal{N}_v & \mathcal{N}_r \end{bmatrix}; \qquad \mathbf{B} = \begin{bmatrix} 0 & \mathcal{Y}_{\delta_r} \\ \mathcal{N}_{\delta_a} & \mathcal{N}_{\delta_r} \end{bmatrix}
\tag{7.10,7}
$$

and where

$$
\mathcal{Y}_{\delta_r} = \frac{Y_{\delta_r}}{m}; \qquad \mathcal{N}_{\delta_a} = I'_{zx}L_{\delta_a} + \frac{N_{\delta_a}}{I'_z}; \qquad \mathcal{N}_{\delta_r} = I'_{zx}L_{\delta_r} + \frac{N_{\delta_r}}{I'_z}
\tag{7.10,8}
$$

With the system matrices given by (7.10,7) the approximate transfer functions are found in the form of (7.2,8) (see Exercise 7.7) with

$$
f(s) = s^2 - (\mathcal{Y}_v + \mathcal{N}_r)s + (\mathcal{Y}_v\mathcal{N}_r + u_0\mathcal{N}_v)
\tag{7.10,9}
$$

and

$$
\begin{aligned}
N_{v\delta_a} &= -u_0\mathcal{N}_{\delta_a} \\
N_{r\delta_a} &= \mathcal{N}_{\delta_a}s - \mathcal{Y}_v\mathcal{N}_{\delta_a} \\
N_{v\delta_r} &= \mathcal{Y}_{\delta_r}s - (\mathcal{Y}_{\delta_r}\mathcal{N}_r + u_0\mathcal{N}_{\delta_r}) \\
N_{r\delta_r} &= \mathcal{N}_{\delta_r}s - (\mathcal{Y}_v\mathcal{N}_{\delta_r} - \mathcal{Y}_{\delta_r}\mathcal{N}_v)
\end{aligned}
\tag{7.10,10}
$$

The accuracy of the preceding approximations is illustrated for the example jet transport on Figs. 7.26 and 7.27. Two general observations can be made: (1) The Dutch Roll approximation is exact in the limit of high frequency, and (2) the spiral/roll approximation is exact as $\omega \rightarrow 0$. In this respect the situation is entirely analogous to that of the longitudinal case, with the spiral/roll corresponding to the phugoid, and the Dutch Roll to the short-period mode. There are ranges of frequency in the middle where neither approximation is good. We repeat that lateral approximations must be used with caution, and that only the exact equations can be relied on to give accurate results.

7.11 Transient Response to Aileron and Rudder

We have seen that useful lateral steady states are produced only by certain definite combinations of the control deflections. It is evident then that our interest in the response to a single lateral control should be focused primarily on the initial behavior. The equations of motion provide some insight on this question directly. Following a step input of one of the two controls the state variables at $t = 0^+$ are all still zero, and from (4.9,19) we can deduce that their initial rates of change are related to the control angles by

$$\begin{aligned}\dot{v} &= \mathcal{Y}_{\delta_r}\delta_r \\ \dot{p} &= \mathcal{L}_{\delta_a}\delta_a + \mathcal{L}_{\delta_r}\delta_r \\ \dot{r} &= \mathcal{N}_{\delta_a}\delta_a + \mathcal{N}_{\delta_r}\delta_r\end{aligned} \tag{7.11,1}$$

The initial sideslip rate $\dot{v}$ is thus seen to be governed solely by the rudder and, since $\mathcal{Y}_{\delta_r} > 0$, is seen to be positive (slip to the right) when δ_r is positive (left rudder). Of somewhat more interest is the rotation generated. The initial angular acceleration is the vector

$$\dot{\boldsymbol{\omega}} = \mathbf{i}\dot{p} + \mathbf{k}\dot{r} \tag{7.11,2}$$

The direction of this vector is the initial axis of rotation, and this is of interest. It lies in the xz plane, the plane of symmetry of the airplane, as illustrated in Fig. 7.28*a*. The angle ξ it makes with the x axis is, of course,

$$\xi = \tan^{-1}\frac{\dot{r}}{\dot{p}} \tag{7.11,3}$$

Let us consider the case of "pure" controls, that is, those with no aerodynamic cross-coupling, so that $L_{\delta_p} = N_{\delta_a} = 0$. The ailerons then produce pure rolling moment and

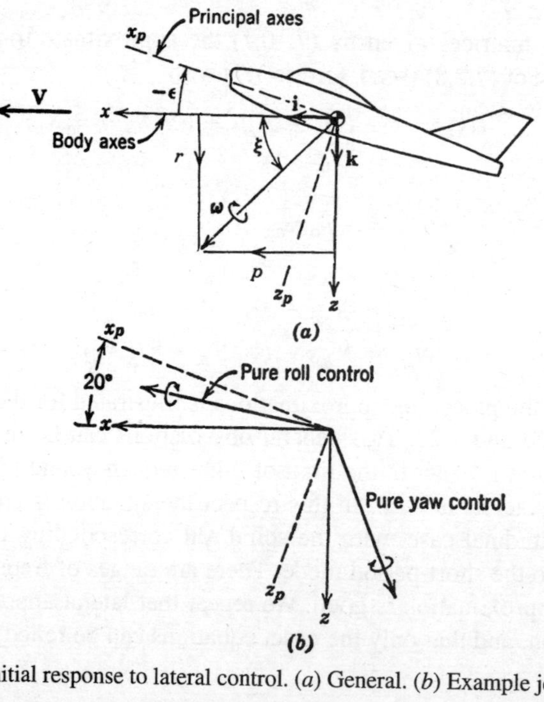

Figurc 7.28 Initial response to lateral control. (*a*) General. (*b*) Example jet transport.

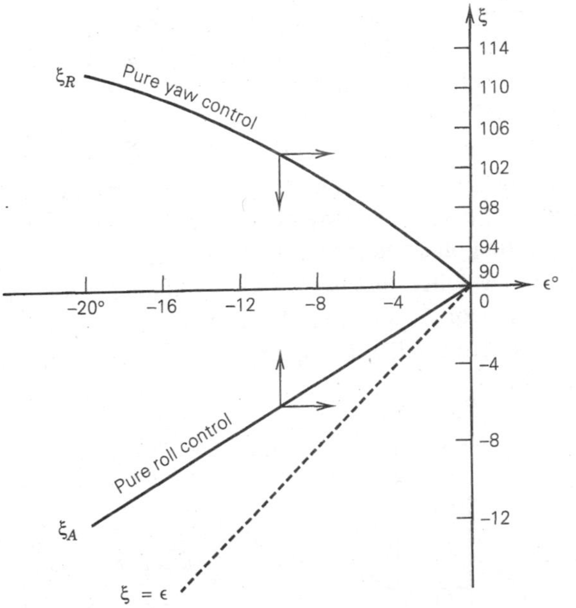

Figure 7.29 Angle of axis of rotation.

the rudder produces pure yawing moment. In that case we get for $\delta_n = 0$ the angle ξ_R for response to rudder from

$$\tan \xi_R = \frac{r}{p} = \frac{\dot{r}}{\dot{p}} = \frac{\mathcal{N}_{\delta_r}}{\mathcal{L}_{\delta_r}} = \frac{N_{\delta_r}/I'_z}{I'_{zx}N_{\delta_r}} = \frac{1}{I'_zI'_{zx}} = I_x/I_{zx} \tag{7.11,4}$$

and similarly for response to aileron:

$$\tan \xi_A = I_{zx}/I_z \tag{7.11,5}$$

The angles ξ_A, ξ_R are seen to depend very much on the product of inertia I_{zx}. When it is zero, the result is as intuitively expected, the rotation that develops is about either the x axis (aileron deflected) or the z axis (rudder deflected). For a vehicle such as the jet transport of previous examples, with $I_{x_p} = 0.4I_{z_p}$, the values of I_x, I_z, I_{zx} given by (4.5,11) yield the results shown in Fig. 7.29. The relations are also shown to scale in Fig. 7.28*b* for $\epsilon = 20°$ (high angle of attack). It can be seen that there is a tendency for the vehicle to rotate about the principal x axis, rather than about the axis of the aerodynamic moment. This is simply because I_x/I_z is appreciably less than unity. Now the jet transport of our example is by no means "slender," in that it is of large span and has wing-mounted engines. For an SST or a slender missile, the trend shown is much accentuated, until in the limit as aspect ratio → 0, both tan ξ_R and tan ξ_A tend to tan ϵ, and the vehicle rotates initially about the x_p axis no matter what control is used!

SOLUTION FOR LARGE ANGLES

The preceding analysis shows how a lateral response starts, but not how it continues. For that we need solutions to the governing differential equations. As remarked earlier, at the beginning of this chapter, control responses can rapidly build up large values of some variables, invalidating the linear equations that we have used so far. There is a compromise available that includes only some nonlinear effects that is useful for transport and general aviation airplanes, which are not subjected to violent maneuvers. The compromise is to retain a linear representation of the inertia and aerodynamic effects, but to put in an exact representation of the gravity forces. This allows the angles ϕ, θ, and ψ to take on any values. As we shall see in the following example, the solution obtained is then limited by the airplane speed growing beyond the range of linear validity, that is, it is an aerodynamic nonlinearity that then controls the useful range of the solution. When the procedure that led to (4.9,18 and 19) is repeated without the small angle approximations we get the following for $\theta_0 = 0$ (see Exercise 7.8):

$$\begin{bmatrix} \dot{u} \\ \dot{w} \\ \dot{q} \\ \dot{\theta} \end{bmatrix} = \begin{bmatrix} \dfrac{\Delta X}{m} - g \sin\theta \\ \dfrac{\Delta Z}{m} - g(1 - \cos\theta\cos\phi) + u_0 q \\ \dfrac{\Delta M}{I_y} \\ q\cos\phi - r\sin\phi \end{bmatrix} \tag{7.11,6}$$

$$\begin{bmatrix} \dot{v} \\ \dot{p} \\ \dot{r} \\ \dot{\phi} \\ \dot{\psi} \end{bmatrix} = \begin{bmatrix} \dfrac{\Delta Y}{m} + g\cos\theta\sin\phi - u_0 r \\ \dfrac{\Delta L}{I'_x} + I'_{zx}\Delta N \\ \dfrac{\Delta N}{I'_z} + I'_{zx}\Delta L \\ p + (q\sin\phi + r\cos\phi)\tan\theta \\ (q\sin\phi + r\cos\phi)\sec\theta \end{bmatrix} \tag{7.11,7}$$

The data for the B747 jet transport previously used was incorporated into the preceding equations. A step aileron input of $-15°$ was applied at time zero, the other controls being kept fixed, and the solution was calculated using a fourth-order Runge-Kutta algorithm. The results are shown in Fig. 7.30. The main feature is the rapid acquisition of roll rate, shown in Fig. 7.30*b*, and its integration into a steadily growing angle of bank (Fig. 7.30*c*) that reaches almost 90° in half a minute. Sideslip, yaw rate, and yaw angle all remain small throughout the time span shown. As the airplane rolls, with its lift remaining approximately equal to its weight, the vertical component of aerodynamic force rapidly diminishes, and a downward net force leads to negative θ and an increase in speed. After 30 seconds, the speed has increased by about 10% of u_0, and the linear aerodynamics becomes increasingly inaccurate. The maximum rotation rate is $p = .05$ rad/s, which corresponds to $\hat{p} = 0.01$. This is small enough that the neglect of the nonlinear inertia terms in the equations of motion is justified.

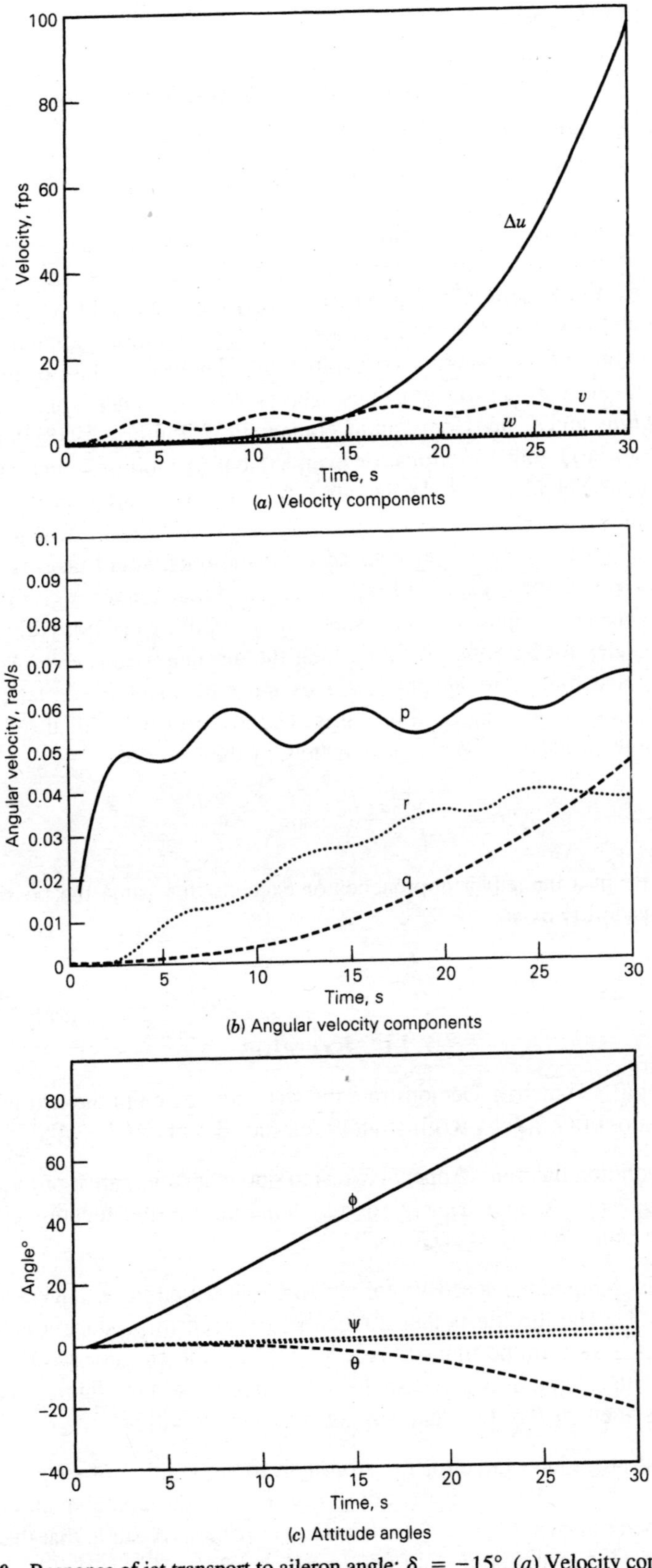

Figure 7.30 Response of jet transport to aileron angle; $\delta_a = -15°$. (*a*) Velocity components. (*b*) Angular velocity components. (*c*) Attitude angles.

7.12 Inertial Coupling in Rapid Maneuvers

We saw in the last section how to include nonlinear gravity effects in control response and how such effects manifest themselves in the response of a relatively sedate vehicle. Of the other two categories of nonlinearity—aerodynamic and inertial—little in a general way can be said about the first. Aerodynamic characteristics, especially for flexible vehicles at high subsonic Mach numbers, are too varied and complex to admit of useful generalizations. A very elaborate (and very costly!) aerodynamic model is required for full and accurate simulation or computation. Not so, however, for the second category of nonlinearity. There is a class of problems, all generically connected, known by names such as *roll resonance, spin-yaw coupling, inertia coupling,* and so on (Heppe and Celinker, 1957; Phillips, 1948; Pinsker, 1958) that pertain to large-angle motions, or even violent instabilities, that can occur on missiles, launch vehicles, and slender aircraft performing rapid rolling maneuvers. These have their source in the pq and pr terms that occur in the pitching and yawing moment equations. A detailed analysis of these motions would take us beyond the scope of this text. Some is given in Etkin (1972), and much more is given in the cited references. One very important conclusion, due to Phillips (1948), is that there is a band of roll rates for airplanes within which the airplane is unstable. At lower roll rates, the usual stability criteria apply. At rates above the band the airplane is gyrostabilized in the way a spinning shell or top is. The lower of the critical roll rates for a normally stable airplane is given approximately by the lesser of

$$p^2 = -\frac{M_\alpha}{I_z - I_x} \quad \text{or} \quad \frac{N_\beta}{I_y - I_x}$$

If the roll rate in a maneuver approaches or exceeds this value the possibility of a dangerous instability exists.

7.13 Exercises

7.1 $\mathbf{A} = [a_{ij}]$ is a (3×3) matrix. Demonstrate the statement made in the text with respect to the numerator of (7.2,7) by writing out in full the adjoint of $(s\mathbf{I} - \mathbf{A})$.

7.2 Use the convolution theorem (Appendix A.3) to obtain an alternative proof of the theorem for frequency response. That is, for a system with transfer function $G(s)$ and input $e^{i\omega t}$ the response for $t \to \infty$ is $G(i\omega)e^{i\omega t}$.

7.3 An airplane is flying at the speed V^* for which the thrust curve is tangent to the drag curve (Fig. 7.1). The throttle is then suddenly advanced to produce a higher thrust curve, such as is seen in the figure. The pilot controls the elevator so as to maintain exactly horizontal flight, in which case the drag curve is as in the figure. The ultimate steady state is at either P or Q. What will govern which it will be?

7.4 In a test flight procedure, the airplane is brought to a condition of steady horizontal flight in quiet air. The elevator is then displaced rapidly through a small angle, held briefly, and then returned as rapidly to its original position. Assume that the resulting input can be treated as an impulse at $t = 0$ (see Sec. 7.3).

(a) Use the short period approximation (7.7,11*b*) to the transfer function for θ to derive a time domain solution for $\theta(t)$. Express the solution in terms of n, ω, b_0, and b_1.

(b) Assuming that θ and t can be determined very accurately from the flight test data, and hence that n and ω can be determined precisely, suggest how the experimental data could be used to determine b_0, b_1, c_0, and c_1. Note that if a_0 and a_1 could likewise be determined accurately, then the six equations (7.7,12) could in principle be used to solve for the six aerodynamic derivatives on the right side of the equations.

7.5 Use the nondimensional derivatives of Table 7.2 to calculate the coefficients of the matrix equation (7.8,4).

7.6 (a) Reformulate the equations for the steady sideslip (7.8,3) to use ϕ as input and (v, δ_a, δ_r) as outputs.

(b) Derive (7.8,15)

7.7 Derive (7.10,10)

7.8 Derive (7.11,6) and (7.11,7).

7.9 An additional vertical control surface (δ_s) is added above the fuselage of an airplane, near the CG. It is capable of providing a side force, accompanied by a rolling moment, given by

$$\Delta Y = Y_{\delta_s}\delta_s; \qquad \Delta L = L_{\delta_s}\delta_s$$

What condition must be satisfied if (δ_a, δ_r, δ_s) are to generate specified (Y, L, N)?

7.10 (a) Using the numerical data for the B747 example (Sec. 7.9), calculate the static gains for each of the eight responses that correspond to (7.9,5)—that is, the values of $|G(i\omega)_{ij}|$ for $\omega = 0$.

(b) Calculate the slopes of the high-frequency asymptotes for each of the eight frequency response amplitudes (express result in decades/decade).

(c) Assume that $\omega = 0$, that is, that a steady state exists in response to one of the controls being deflected, such that $\phi = 15°$. For each of the two controls—aileron and rudder—calculate the control angle, the sideslip angle, and the yaw rate r.

7.11 The elevator of the B747 airplane is oscillated at a frequency a little below that of the short-period mode.

(a) Use the results given in Fig. 7.18 to estimate the amplitude of the load factor if the elevator amplitude is 2°.

(b) What elevator amplitude would lift a passenger seated near the CG from the seat?

(c) What elevator amplitude would cause the load factor to reach the FAR Part 25 limit maneuvering value of 2.5?

7.14 Additional Symbols Introduced in Chapter 7

$\mathcal{A}(t)$	response to a unit step input
$h(t)$	response to a unit pulse input
K	static gain
$\mathcal{L}_\delta$	$L_\delta/I'_x + I'_{zx}N_\delta$
M	magnification factor
$\mathcal{N}_\delta$	$N_\delta/I'_z + I'_{zx}L_\delta$
$\mathcal{Y}_\delta$	Y_δ/m
$\delta(t)$	Dirac's delta function
φ	phase angle

CHAPTER 8

Closed Loop Control

8.1 General Remarks

The development of closed loop control has been one of the major technological achievements of the twentieth century. This technology is a vital ingredient in countless industrial, commercial, and even domestic products. It is a central feature of aircraft, spacecraft, and all robotics. Perhaps the earliest known example of this kind of control is the fly-ball governor that James Watt used in his steam engine in 1784 to regulate the speed of the engine. This was followed by automatic control of torpedoes in the nineteenth century (Bollay, 1951), and later by the dramatic demonstration of the gyroscopic autopilot by Sperry in 1910, highly relevant in the present context. Still later, and the precursor to the development of a general theoretical approach, was the application of negative feedback to improve radio amplifiers in the 1930s. The art of automatic control was quite advanced by the time of the landmark fourteenth Wright Brothers lecture (Bollay, 1951)[1]. Most of what is now known as "classical" control theory—the work of Routh, Nyquist, Bode, Evans, and others was described in that lecture. From that time on the marriage of control concepts with analogue and digital computation led to explosive growth in the sophistication of the technology and the ubiquity of its applications.

Although open-loop responses of aircraft, of the kind studied in some depth in Chap. 7, are very revealing in bringing out inherent vehicle dynamics, they do not in themselves usually represent real operating conditions. Every phase of the flight of an airplane can be regarded as the accomplishment of a set task—that is, flight on a specified trajectory. That trajectory may simply be a straight horizontal line traversed at constant speed, or it may be a turn, a transition from one symmetric flight path to another, a landing flare, following an ILS or navigation radio beacon, homing on a moving target, etc. All of these situations are characterized by a common feature, namely, the presence of a *desired state,* steady or transient, and of departures from it that are designated as errors. These errors are of course a consequence of the unsteady nature of the real environment and of the imperfect nature of the physical system comprising the vehicle, its instruments, its controls, and its guidance system (whether human or automatic). The correction of errors implies a knowledge of them, that is, of error-measuring (or state-measuring) devices, and the consequent actuation of the controls in such a manner as to reduce them. This is the case whether control is by human or by automatic pilot. In the former case—the human pilot—the state information sensed is a complicated blend of visual and motion cues, and instrument readings. The logic by which this information is converted into control action is only

[1]In 1951 most aeronautical engineers were using slide rules and had not heard of a transfer function!

imperfectly understood, but our knowledge of the physiological "mechanism" that intervenes between logical output and control actuation is somewhat better. In the latter case—the automatic control—the sensed information, the control logic, and the dynamics of the control components are usually well known, so that system performance is in principle quite predictable. The process of using state information to govern the control inputs is known as *closing the loop,* and the resulting system as a *closed-loop control* or *feedback control.* The terms *regulator* and *servomechanism* describe particular applications of the feedback principle. Figure 8.1 shows a general block diagram describing the feedback situation in a flight control system. This diagram models a linear invariant system, which is of course an approximation to real nonlinear time-varying systems. The approximation is a very useful one, however, and is used extensively in the design and analysis of flight control systems. In the diagram the arrows show the direction of information flow; the lowercase symbols are vectors (i.e. column matrices), all functions of time; and the uppercase symbols are matrices (in general rectangular). The vectors have the following meanings:

r: reference, input or command signal, dimensions $(p \times 1)$

z: feedback signal, dimensions $(p \times 1)$

e: error, or actuating, signal, dimensions $(p \times 1)$

c: control signal, dimensions $(m \times 1)$

g: gust vector (describing atmospheric disturbances), dimensions $(l \times 1)$

x: airplane state vector, dimensions $(n \times 1)$

y: output vector, dimensions $(q \times 1)$

n: sensor noise vector, dimensions $(q \times 1)$

Of the above, **x** and **c** are the same state and control vectors used in previous chapters. **r** is the system input, which might come from the pilot's controller, from an external navigation or fire control system, or from some other source. It is the command that the airplane is required to follow. The signal **e** drives the system to make **z**

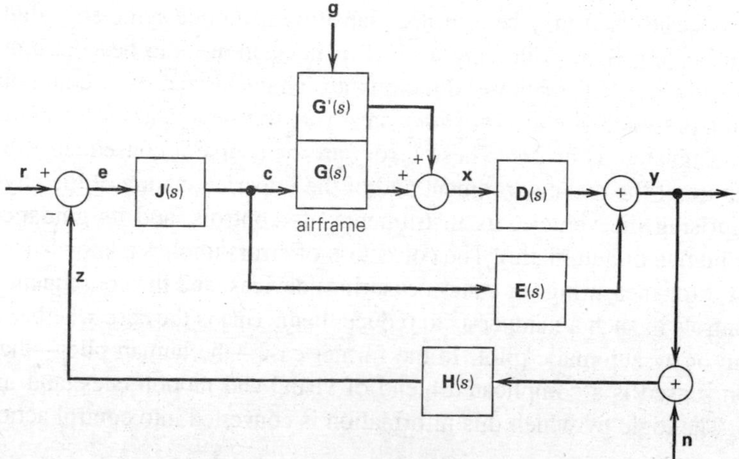

Figure 8.1 A general linear invariant flight control system.

follow $\mathbf{r}$. It will be zero when $\mathbf{z} = \mathbf{r}$. The makeup of the output vector $\mathbf{y}$ is arbitrary, constructed to suit the requirements of the particular control objective. It could be as simple as just one element of $\mathbf{x}$, for example. The feedback signal $\mathbf{z}$ is also at the discretion of the designer via the choice of feedback transfer function $\mathbf{H}(s)$. The choices made for $\mathbf{D}(s)$, $\mathbf{E}(s)$ and $\mathbf{H}(s)$ collectively determine how much the feedback signal differs from the state. With certain choices $\mathbf{z}$ can be made to be simply a subset of $\mathbf{x}$, and it is then the state that is commanded to follow $\mathbf{r}$.

The vector $\mathbf{g}$ describes the local state of motion of the atmosphere. This state may consist of either or both discrete gusts and random turbulence. It is three-dimensional and varies both in space and time. Its description is inevitably complex, and to go into it in depth here would take us beyond the scope of this text. For a more complete discussion of $\mathbf{g}$ and its closely coupled companion $\mathbf{G}'$ the student should consult Etkin (1972) and Etkin (1981).

In real physical systems the state has to be measured by devices (sensors) such as, for example, gyroscopes and Pitot tubes, which are inevitably imperfect. This imperfection is commonly modeled by the noise vector $\mathbf{n}$, usually treated as a random function of time.

The equations that correspond to the diagram are (recall that overbars represent Laplace transforms):

$$\begin{aligned}
\bar{\mathbf{e}} &= \bar{\mathbf{r}} - \bar{\mathbf{z}} && (a)\\
\bar{\mathbf{c}} &= \mathbf{J}(s)\bar{\mathbf{e}} && (b)\\
\bar{\mathbf{x}} &= \mathbf{G}(s)\bar{\mathbf{c}} + \mathbf{G}'(s)\bar{\mathbf{g}} && (c) \quad (8.1,1)\\
\bar{\mathbf{y}} &= \mathbf{D}\bar{\mathbf{x}} + \mathbf{E}\bar{\mathbf{c}} && (d)\\
\bar{\mathbf{z}} &= \mathbf{H}(s)(\bar{\mathbf{y}} + \bar{\mathbf{n}}) && (e)
\end{aligned}$$

In the time domain (8.1,1c) appears as

$$\dot{\mathbf{x}} = \mathbf{A}\mathbf{x} + \mathbf{B}\mathbf{c} + \mathbf{T}\mathbf{g} \tag{8.1,2}$$

It follows that

$$\mathbf{G}(s) = (s\mathbf{I} - \mathbf{A})^{-1}\mathbf{B} \quad \text{and} \quad \mathbf{G}'(s) = (s\mathbf{I} - \mathbf{A})^{-1}\mathbf{T} \tag{8.1,3}$$

The feedback matrix $\mathbf{H}(s)$ represents any analytical operations performed on the output signal. The transfer function matrix $\mathbf{J}(s)$ represents any operations performed on the error signal $\mathbf{e}$, as well as the servo actuators that drive the aerodynamic control surfaces, including the inertial and aërodynamic forces (hinge moments) that act on them. The servo actuators might be hydraulic jacks, electric motors, or other devices. This matrix will be a significant element of the system whenever there are power-assisted controls or when the aircraft has a fly-by-wire or fly-by-light AFCS.

From (8.1,1) we can derive expressions for the three main transfer function matrices. By eliminating $\mathbf{x}$, $\mathbf{e}$, $\mathbf{c}$, and $\mathbf{z}$ we get

$$[\mathbf{I} + (\mathbf{DG} + \mathbf{E})\mathbf{JH}]\mathbf{y} = (\mathbf{DG} + \mathbf{E})\mathbf{J}\mathbf{r} - (\mathbf{DG} + \mathbf{E})\mathbf{JH}\mathbf{n} + \mathbf{DG}'\mathbf{g} \tag{8.1,4}$$

from which the desired transfer functions are

$$\begin{aligned}
\mathbf{G}_{yr} &= [\mathbf{I} + (\mathbf{DG} + \mathbf{E})\mathbf{JH}]^{-1}(\mathbf{DG} + \mathbf{E})\mathbf{J} && (a)\\
\mathbf{G}_{yn} &= -[\mathbf{I} + (\mathbf{DG} + \mathbf{E})\mathbf{JH}]^{-1}(\mathbf{DG} + \mathbf{E})\mathbf{JH} && (b) \quad (8.1,5)\\
\mathbf{G}_{yg} &= [\mathbf{I} + (\mathbf{DG} + \mathbf{E})\mathbf{JH}]^{-1}\mathbf{DG}' && (c)
\end{aligned}$$

The matrices that appear in (8.1,5) have the following dimensions:

$$\mathbf{D}(q \times n);\ \mathbf{G}(n \times m);\ \mathbf{E}(q \times m);\ \mathbf{J}(m \times p);\ \mathbf{H}(p \times q);\ \mathbf{G}'(n \times l)$$

The forward-path transfer function, from **e** to **y**, is

$$\mathbf{F}(s) = (\mathbf{DG} + \mathbf{E})\mathbf{J};\ \text{dimensions } (q \times p)$$

so the preceding transfer functions can be rewritten as

$$\begin{aligned} \mathbf{G}_{yr} &= (\mathbf{I} + \mathbf{FH})^{-1}\mathbf{F} && (a) \\ \mathbf{G}_{yn} &= -(\mathbf{I} + \mathbf{FH})^{-1}\mathbf{FH} && (b) \quad (8.1,6) \\ \mathbf{G}_{yr} &= (\mathbf{I} + \mathbf{FH})^{-1}\mathbf{DG}' && (c) \end{aligned}$$

Note that **F** and **H** are both scalars for a single-input, single-output system.

When the linear system model is being formulated in state space, instead of in Laplace transforms, then one procedure that can be used (see Sec. 8.8) is to generate an *augmented* form of (8.1,2). In general this is done by writing time domain equations for **J** and **H**, adding new variables to **x**, and augmenting the matrices **A** and **B** accordingly. An alternative technique for using differential equations is illustrated in Sec. 8.5. There is a major advantage to formulating the system model as a set of differential equations. Not only can they be used to determine transfer functions, but when they are integrated numerically it is possible, indeed frequently easy, to add a wide variety of nonlinearities. These include second degree inertia terms, dead bands and control limits (see Sec. 8.5), Coulomb friction, and nonlinear aerodynamics given as analytic functions or as lookup tables.

AN AERODYNAMICS VIEWPOINT

It is frequently helpful to view a feedback loop as simply a method of altering one of the airplane's inherent stability derivatives. When one of the main damping derivatives, L_p, M_q, or N_r, is too small, or when one of the two main stiffnesses M_α or N_β is not of the magnitude desired, they can be synthetically altered by feedback of the appropriate control. Specifically let x be any nondimensional state variable, and let a control surface be displaced in response to this variable according to the law

$$\Delta\delta = k\Delta x; \qquad k = \text{const}$$

(Here k is a simplified representation of all the sensor and control system dynamics!) Then a typical aerodynamic force or moment coefficient C_a will be incremented by

$$\begin{aligned} \Delta C_a &= C_{a_\delta}\Delta\delta \\ &= C_{a_\delta}k\Delta x \end{aligned} \tag{8.1,7}$$

This is the same as adding a synthetic increment

$$\Delta C_{a_x} = kC_{a_\delta} \tag{8.1,8}$$

to the aerodynamic derivative C_{a_x}. Thus if x be yaw rate and δ be rudder angle, then the synthetic increment in the yaw-damping derivative is

$$\Delta C_{n_r} = kC_{n_{\delta r}} \tag{8.1,9}$$

which might be the kind of change required to correct a lateral dynamics problem. This example is in fact the basis of the often-applied "yaw damper," a stability-augmentation feature. Again, if x be the roll angle and δ the aileron, we get the *entirely new* derivative

$$C_{l_\phi} = kC_{l_{\delta_a}} \tag{8.1,10}$$

the presence of which can profoundly change the lateral characteristics (see Exercise 8.1).

SENSORS

We have already alluded to the general nature of feedback control, and the need to provide *sensors* that ascertain the *state* of the vehicle. When human pilots are in control, their eyes and kinesthetic senses, aided by the standard flight information displayed by their instruments, provide this information. (In addition, of course, their brains supply the logical and computational operations needed, and their neuro-muscular systems all or part of the actuation.) In the absence of human control, when the vehicle is under the command of an autopilot, the sensors must, of course, be physical devices. As already mentioned, some of the state information needed is measured by the standard flight instruments—air speed, altitude, rate of climb, heading, etc. This information may or may not be of a quality and in a form suitable for incorporation into an automatic control system. In any event it is not generally enough. When both guidance and attitude-stabilization needs are considered, the state information needed may include:

Position and velocity vectors relative to a suitable reference frame.

Vehicle attitude (θ, ϕ, ψ).

Rotation rates (p, q, r).

Aerodynamic angles (α, β).

Acceleration components of a reference point in the vehicle.

The above is not an exhaustive list. A wide variety of devices are in use to measure these variables, from Pitot-static tubes to sophisticated inertial-guidance platforms. Gyroscopes, accelerometers, magnetic and gyro compasses, angle of attack and sideslip vanes, and other devices all find applications as sensors. The most common form of sensor output is an electrical signal, but fluidic devices have also been used. Although in the following examples we tend to assume that the desired variable can be measured independently, linearly, and without time lag, this is of course an idealization that is only approached but never reached in practice. Every sensing device, together with its associated transducer and amplifier, is itself a dynamic system with characteristic frequency response, noise, nonlinearity, and cross-coupling. These attributes cannot finally be ignored in the design of real systems, although one can usefully do so in preliminary work. As an example of cross-coupling effects, consider the sideslip sensor assumed to be available in the gust alleviation system of Sec. 8.9. Assume, as might well be the case, that it consists of a sideslip vane mounted on a boom projecting forward from the nose. Such a device would in general respond not only to β but also to atmospheric turbulence (side gusts), to roll and yaw rates, and to lateral acceleration a_y at the vane hinge. Thus the output signal would in fact be a complicated mathematical function of several state variables, representing several feedback loops. The objective in sensor design is, of course, to minimize all the unwanted extraneous effects, and to provide sufficiently high frequency response and low noise in the sensing system.

This brief discussion serves only to draw attention to the important design and analytical problems related to sensors, and to point out that their real characteristics, as opposed to their idealizations, need finally to be taken into account in design.

8.2 Stability of Closed Loop Systems

For linear invariant systems such as we have discussed above, the methods available for assessing stability include those used with open loop systems. One such method is to formulate the governing differential equations, find the characteristic equation of the system, and solve for its roots. Another is to find the transfer function from input to output and determine its poles. With the powerful computing methods available, it is feasible to plot loci of the roots (or poles) as one or more of the significant design parameters are varied, as we shall see in examples to follow. For the multivariable high-order systems that commonly occur in aerospace practise this is a very useful technique.

Let us now consider the stability of the loop associated with one particular input/output pair in the light of (8.1,6*a*). Since $p = 1$ and **F** and **H** are scalars, the transfer function is

$$G_{yr}(s) = \frac{F(s)}{1 + F(s)H(s)} \tag{8.2,1}$$

The transfer functions $F(s)$ and $H(s)$ are ratios of polynomials in s, that is, $F(s) = N_1/D_1$, and $H(s) = N_2/D_2$. Equation (8.2,1) then leads to

$$G_{yr}(s) = \frac{N_1D_2}{D_1D_2 + N_1N_2} \tag{8.2,2}$$

The characteristic equation is evidently

$$D_1(s)D_2(s) + N_1(s)N_2(s) = 0 \tag{8.2,3}$$

This should be contrasted with the characteristic equation for the airframe alone, which is $D(s) = 0$, where D is the denominator of $G(s)$. The block diagram corresponding to (8.2,1) is shown in Fig. 8.2. FH is the *open loop transfer function,* that is, the ratio of feedback to error, $\bar{z}/\bar{e}$. Its absolute value $|FH|$ is the *open loop gain.*

The stability of the system can be assessed from the frequency response $F(i\omega)H(i\omega)$. It is clear that if there is a frequency and open loop gain for which $FH = -1$ then un-

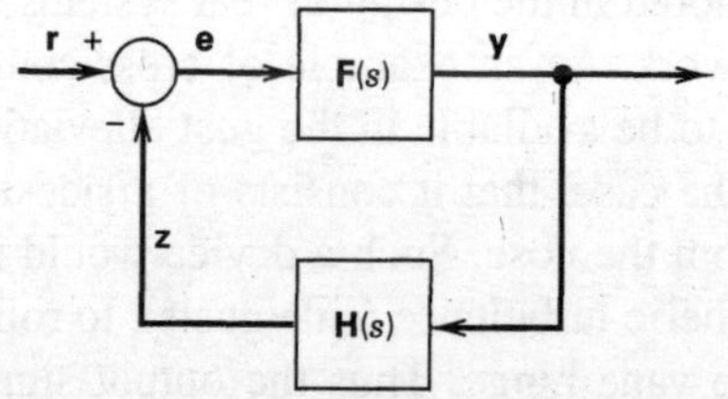

Figure 8.2 Consolidated block diagram of feedback controller.

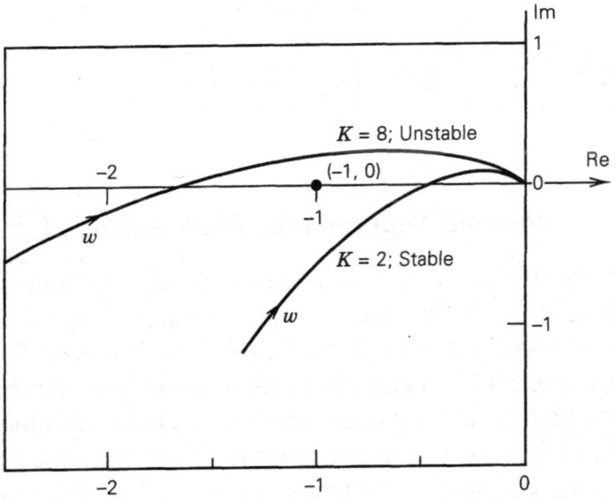

Figure 8.3 Nyquist diagram.

der those conditions the denominator of (8.2,1) is zero and $G_{yr}(i\omega)$ is infinite. When these conditions hold, the feedback signal that is returned to the junction point is precisely the negative of the error signal that generated it. This means that the system can oscillate at this frequency *without any input.* This is exactly the situation with the whistling public address system. For then the acoustic signal that returns to the microphone from the loudspeakers, in response to an input pulse, is equal in strength to the originating pulse. Clearly the point $(-1,0)$ of the complex plane has special significance. Nyquist (1932) has shown how the relationship of the frequency response curve (the Nyquist diagram) of FH to this special point indicates stability (see Fig. 8.3). In brief, if the loop gain is <1 when the phase angle is 180°, or if the phase is $<180°$ when the gain is unity, then the system is stable. The amounts by which the curve misses the critical point define two measures of stability, the *gain margin* and *phase margin,* illustrated on the Nichols diagram (see McLean, 1990) of Fig. 8.4. The

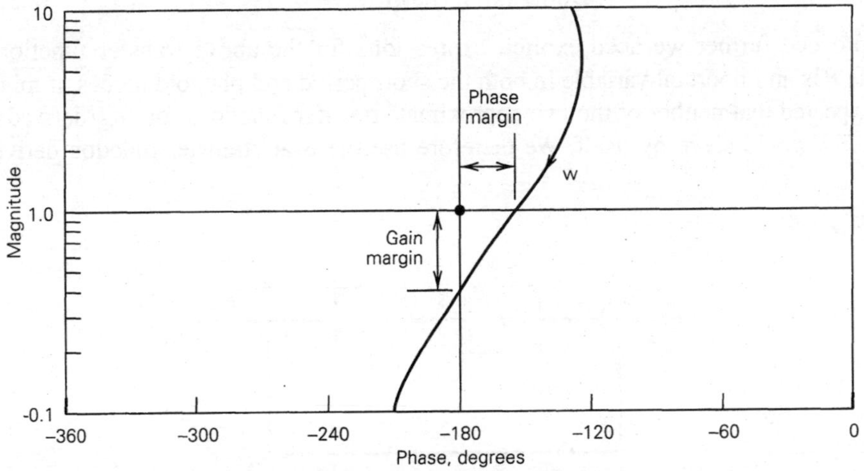

Figure 8.4 Nichols diagram, $K = 2$.

examples of Figs. 8.3 and 8.4 are for the open loop transfer function

$$FH = \frac{K(s + 0.125)}{s^2(0.15s^2 + 0.8s + 1)}$$

8.3 Phugoid Suppression: Pitch Attitude Controller

The characteristic lightly damped, low-frequency oscillation in speed, pitch attitude, and altitude that was identified in Chap. 6, was seen in Chap. 7 to lead to large peaks in the frequency-response curves (Figs. 7.14 to 7.18) and long transients (Fig. 7.20). Similarly, in the control-fixed case, there are large undamped responses in this mode to disturbances such as atmospheric turbulence. These variations in speed, height, and attitude are in fact not in evidence in actual flight; the pilot (human or automatic) effectively suppresses them, maintaining flight at more or less constant speed and height. The logic by which this process of suppression takes place is not unique. In principle it can be achieved by using feedback signals derived from any one or a combination of pitch attitude θ, altitude h, speed v, and their derivatives. In practice, the availability and accuracy of the state information determines what feedback is used.

Since the phugoid oscillation cannot occur if the pitch angle θ is not allowed to change (except when commanded to), a pitch-attitude-hold feature in the autopilot would be expected to suppress the phugoid. This feature is commonly present in airplane autopilots. We shall therefore look at the design of an attitude hold system for the jet transport of our previous examples. Pitch attitude is readily available from either the real horizon (human pilot) or the vertical gyro (autopilot). Consider the controller illustrated in Fig. 8.5. From (8.2,1) we see that the overall transfer function is

$$\frac{\bar{\theta}}{\bar{\theta}_c} = \frac{G_{\theta\delta_e}(s)J(s)}{1 + G_{\theta\delta_e}(s)J(s)} \tag{8.3,1}$$

If we write $G_{\theta\delta_e}(s) = N(s)/D(s)$, and $J(s) = N'(s)/D'(s)$, then the characteristic equation is

$$D(s)D'(s) + N(s)N'(s) = 0 \tag{8.3,2}$$

To proceed further we need explicit expressions for the above transfer functions. Since θ is an important variable in both the short period and phugoid modes, it might be expected that neither of the two approximate transfer functions for $G_{\theta\delta_e}$ derived in Sec. 7.7 would serve by itself. We therefore use the exact transfer function derived

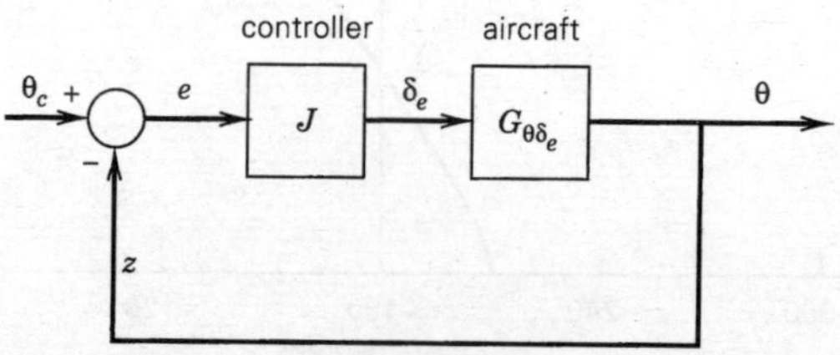

Figure 8.5 Pitch attitude controller.

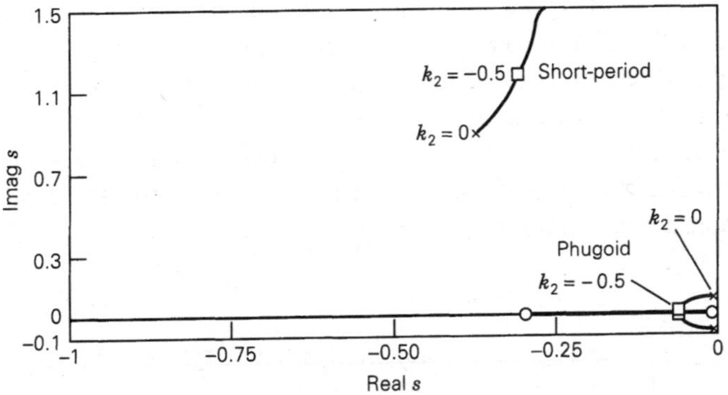

Figure 8.6 Root locus of pitch controller with proportional control.

from the full system of linearized longitudinal equations of motion. Then $N(s)$ is given by (7.7,2) and $D(s)$ by (6.2,2). The result is

$$G_{\theta\delta_e} = \frac{-(1.158s^2 + 0.3545s + 0.003873)}{s^4 + 0.750468s^3 + 0.935494s^2 + 9.463025 \times 10^{-3}s + 4.195875 \times 10^{-3}} \tag{8.3,3}$$

As to $J(s)$, a reasonable general form for this application is

$$J(s) = \frac{k_1}{s} + k_2 + k_3 s \tag{8.3,4}$$

For obvious reasons, the three terms on the right hand side are called, respectively, *integral control, proportional control* and *rate control,* because of the way they operate on the error e. The particular form of the controlled system, here $G_{\theta\delta_e}(s)$, determines which of k_1, k_2, k_3 need to be nonzero, and what their magnitudes should be for good performance. Integral control has the characteristic of a *memory,* and steady-state errors cannot persist when it is present. Rate control has the characteristic of *anticipating* the future values of the error and thus generates *lead* in the control actuation. In using (8.3,4), we have neglected the dynamics of the elevator servo actuator and control surface, which would typically be approximated by the first-order transfer function $1/(1 + \tau s)$. Since the characteristic time of the servo actuator system, τ, is usually a small fraction of a second, and we are interested here in much longer times, this is a reasonable approximation.

For the example airplane at the chosen flight condition it turns out that we need all three terms of (8.3,4) to get a good control design. This might not always be the case. Let us first look at the use of proportional control only, in which case J is a constant gain, k_2. To select its magnitude, we use a root locus plot[2] of the system, Fig. 8.6, in which the locus of the roots of the characteristic equation of the closed loop system are plotted for variable gain k_2. We see that at a gain of about -0.5 the phugoid mode is nearly critically damped, that is, it is about to split into two real roots. At this gain, the phugoid oscillation is effectively eliminated. We note that at

[2]The term *root locus* is used throughout this chapter with the meaning ordinarily ascribed to it in the control theory literature.

the same gain the short period roots have moved in the direction of lower damping. The response of the aircraft to a unit step command in pitch angle with only proportional control is shown in Fig. 8.7*a*. It is clear that this is not an acceptable response. There is a large steady-state error (steady-state error is a feature of proportional control) and the short-period oscillation leads to excessive hunting. The steady-state error could be reduced by increasing k_2 (see Exercise 8.2), but this would further decrease the short period damping.

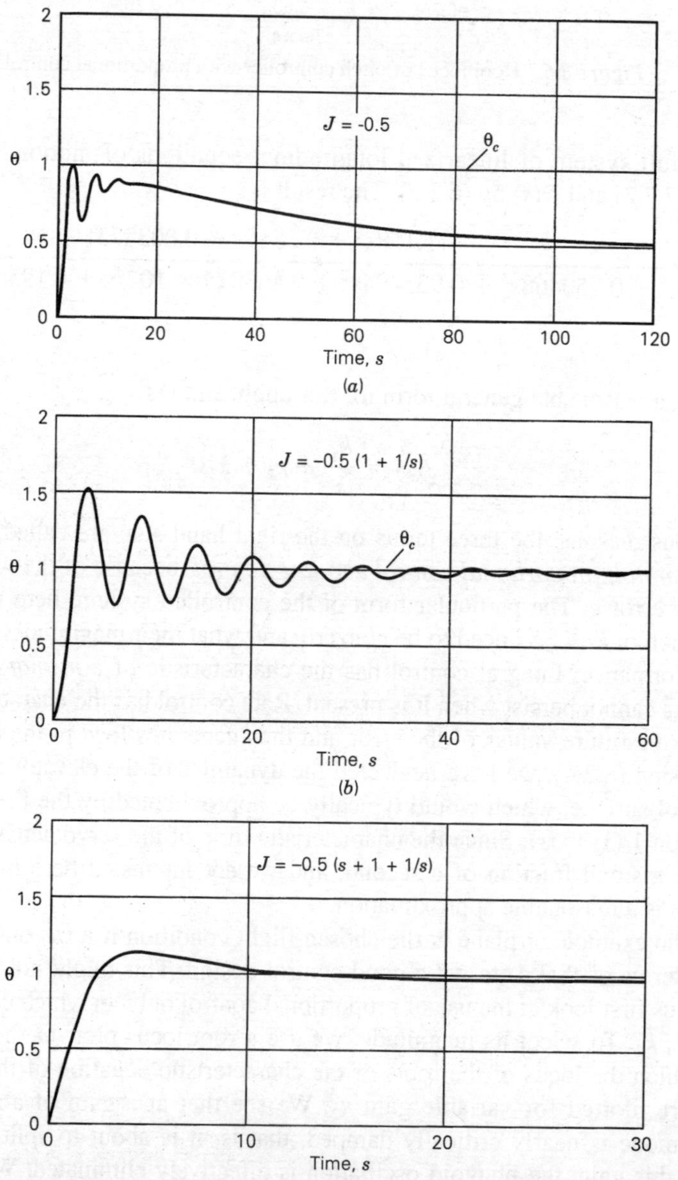

Figure 8.7 Response of pitch angle to unit step command. (*a*) With proportional control. (*b*) With proportional plus integral control. (*c*) With proportional, integral, and rate control.

We digress briefly to explore the reason for the damping behavior. It was noted previously (Sec. 6.8 and Exercise 6.4) that the term of next-to-highest degree in the characteristic equation gives "the sum of the dampings." That is, the coefficient of s^3 in (8.3,3) is the sum of the real parts of the short-period and phugoid roots. Now when $J = k_2$ the closed loop characteristic equation (8.3,2) becomes $D(s) + k_2N(s)$. That is we add a second degree numerator to a fourth degree denominator, *leaving the coefficient of s^3 unchanged.* Thus any increase in the phugoid damping can only come at the expense of that of the short-period mode. This is exactly what is seen in Fig. 8.6. The shifts of the two roots in the real direction are equal and opposite.

To eliminate the steady-state error, we use integral control and choose

$$J = -0.5\left(1 + \frac{1}{s}\right) \tag{8.3,5}$$

The result is shown in Fig. 8.7*b*. The steady-state error has been eliminated, but the short-period oscillation is now even less damped. Now the damping of the short-period mode is governed principally by M_q [see (6.3,14)], so in order to improve it we should provide a synthetic increase to M_q. A signal proportional to q is readily obtained from a pitch-rate gyro. Since $q = \dot{\theta}$ in the system model we are using, we accomplish this by adding a third term to J:

$$J = -0.5\left(s + 1 + \frac{1}{s}\right) \tag{8.3,6}$$

The result, shown in Fig. 8.7*c* is an acceptable controller, with little overshoot and no steady-state error. A commanded pitch attitude change is accomplished in about 10 sec. Note that in this illustration, all the constants in (8.3,4) are the same, that is, -0.5. Fine-tuning of these could be used to modify the behavior to reduce the overshoot or speed up the response. Throughout this maneuver, the elevator angle remains less than its steady-state value (which it approaches asymptotically), so that the gains used are indeed much smaller than the elevator control is physically capable of providing (see Exercise 8.2).

The preceding analysis does not reveal the underlying physics of why k_2 damps the phugoid. This can be understood as follows. An angle θ in the low frequency phugoid implies vertical velocity (i.e., $\dot{h} = V\theta$). Now a positive elevator angle proportional to a slowly changing θ implies a negative increment in angle of attack and hence in the lift as well. Thus k_2 leads to a vertical force (downward) 180° out of phase with the vertical velocity (upward), exactly what is required for damping.

Although the phugoid oscillation has been suppressed quite successfully by the strategy employed above, it should be remarked that in the example case neither the speed nor the altitude has been controlled. As a consequence, the speed drifts rather slowly back to its original value, and the altitude to a new steady state.

Finally it should be noted that a controller design that is correct for one flight condition, in this case high speed at high altitude, may not be acceptable at all speeds and altitudes, for example, landing approach. In the real world of AFCS design, this problem leads, as in most engineering design, to compromises between conflicting requirements. If the economics of the airplane justifies it, *gain scheduling* can be adopted; that is, the control gains are made to be functions of speed, altitude, and configuration.

8.4 Speed Controller

The phugoid makes its presence known not only in the form of transient perturbations from a steady state, but also in maneuvers, as illustrated in Sec. 7.7. We saw there for example that in changing from level to climbing flight by opening the throttle (Fig. 7.21) there results a protracted, weakly damped approach to the new state that would take more than 10 min to complete. Transitions from one value of γ to another are obviously not made in this manner, and the pilot suppresses the oscillation in this case as well. Provided that the correct θ is known for the climb condition, the same technique as discussed above would work, that is, control operating on pitch-attitude error. We illustrate an alternative concept that does not require any knowledge of the final correct pitch attitude, but that uses speed error alone. It is not self-evident how speed should be controlled, in the light of the discussion in Sec. 7.1. We saw there that both elevator and throttle influence the speed, but that the short- and long-term effects of each of these controls are quite different—the throttle principally affects the speed only in the short term. For a change of steady-state speed, the elevator must be used. Clearly, a sophisticated speed control might use both. We shall see in this example, however, that when the primary aim is to suppress the phugoid, which is a very long period oscillation, the goal can be achieved with the elevator alone. Figure 8.8 shows the system.

The command is the speed u_c and the feedback signal is the actual speed u. For output we choose speed and flight-path angle, that is, $\mathbf{y} = [u \;\; \gamma]^T$. The control vector is $\mathbf{c} = [\delta_e \;\; \delta_p]^T$ of which only the elevator is in the feedback loop. Since the controlled variable is u, which does not change appreciably in the short-period mode, we can use the phugoid approximation for the aircraft transfer function matrix $\mathbf{G}(s)$, which is the (2×2) matrix of transfer functions from $\mathbf{c}$ to $\mathbf{y}$:

$$\mathbf{G}(s) = \begin{bmatrix} G_{u\delta_e} & G_{u\delta_p} \\ G_{\gamma\delta_e} & G_{\gamma\delta_p} \end{bmatrix} \tag{8.4,1}$$

Two of the elements of $\mathbf{G}$ are implicit in (7.7,7), since $G_{\gamma\delta} = G_{\theta\delta} - G_{\alpha\delta}$ where δ stands for either δ_e or δ_p. The remaining two are (see Exercise 8.4)

$$\begin{aligned} G_{u\delta_p} &= -\frac{X_{\delta_p}}{m}\frac{u_0 M_w s}{f(s)} \\ G_{\gamma\delta_p} &= \frac{X_{\delta_p}}{m}\frac{M_w\dfrac{Z_u}{m} - M_u\dfrac{Z_w}{m}}{f(s)} \end{aligned} \tag{8.4,2}$$

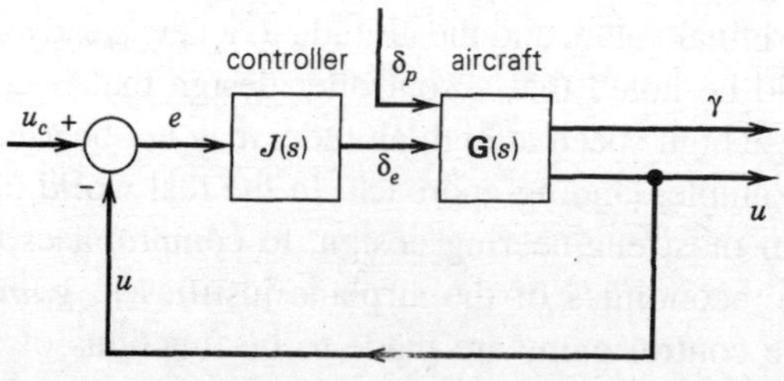

Figure 8.8 Speed controller.

Ultimately we shall want to calculate the time responses of u and γ to a throttle input δ_p. So the transfer functions we need are the two corresponding closed loop transfer functions. If we denote these by $\mathring{G}_{u\delta_p}(s)$ and $\mathring{G}_{\gamma\delta_p}(s)$, respectively, we find that they are given in terms of the aircraft transfer functions by (see Exercise 8.4)

$$\begin{aligned} \mathring{G}_{u\delta_p} &= \frac{G_{u\delta_p}}{1 + G_{u\delta_e}J} && (\delta_p \rightarrow u) \\ \mathring{G}_{\gamma\delta_p} &= G_{\gamma\delta_p} - G_{\gamma\delta_e}J\mathring{G}_{u\delta_p} && (\delta_p \rightarrow \gamma) \end{aligned} \tag{8.4,3}$$

Each of the aircraft transfer functions in (8.4,3) can, as usual, be expressed as a ratio of two polynomials, for example:

$$G_{u\delta_e} = \frac{N_{u\delta_e}}{f(s)}, \text{ etc.}$$

and

$$J = \frac{N_j}{D_j} \tag{8.4,4}$$

When this is done (8.4,3) becomes

$$\mathring{G}_{u\delta_p} = \frac{D_jN_{u\delta_p}}{fD_j + N_jN_{u\delta_e}} \tag{8.4,5}$$

$$\mathring{G}_{\gamma\delta_p} = \frac{fD_jN_{\gamma\delta_p} + N_j(N_{\gamma\delta_p}N_{u\delta_e} - N_{u\delta_p}N_{\gamma\delta_e})}{f(fD_j + N_jN_{u\delta_e})} \tag{8.4,6}$$

We know that the denominator of a transfer function is the characteristic polynomial. We also know that a linear invariant system of the kind under discussion can have only one independent characteristic equation. Thus we have an apparent paradox, since the denominator of (8.4,6) is not the same as that of (8.4,5), having the extra factor $f(s)$. Now it can be shown (see Exercise 8.6) that f is a factor of the bracketed term in the numerator of (8.4,6), and hence that it divides out of the right side and leaves the same characteristic polynomial as in (8.4,5).

As indicated above, the second-order phugoid approximation should be expected to be reasonable for this case. We shall therefore use it to choose the gains in $J(s)$, but at the end will check the solution for suitability with the exact fourth-order equations. To this end we examine the effect of $J(s)$ on the characteristic equation, that is, on

$$f(s)D_j + N_jN_{u\delta_e} = 0 \tag{8.4,7}$$

$f(s)$ is given by (6.3,9):

$$f(s) = As^2 + Bs + C$$

$N_{u\delta_e}$ is given by (7.7,7):

$$N_{u\delta_e} = a_1s + a_0$$

and for $J(s)$ we use

$$J(s) = k_1 + k_2s \tag{8.4,8}$$

so that $D_j = 1$ and $N_j = k_1 + k_2s$. Note that the k_2s term implies a signal proportional to acceleration. Such a signal could be obtained from an x-axis accelerometer or by

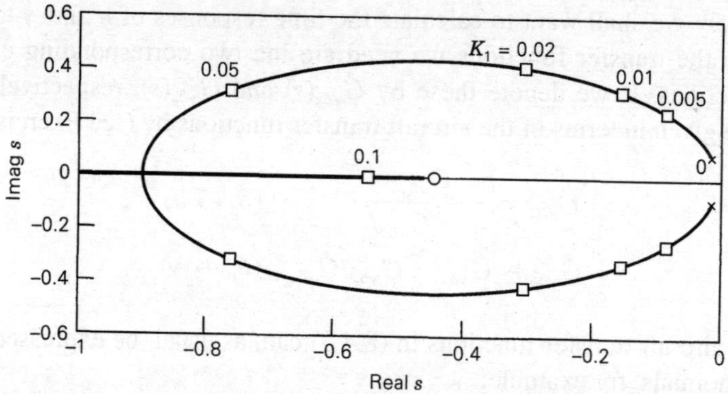

Figure 8.9 Speed controller. Root locus plot of $G_{u\delta_e}$. Phugoid approximation.

differentiating the signal from the speed sensor. The closed loop characteristic equation then becomes:

$$A's^2 + B's + C' = 0 \quad (a)$$

where

$$A' = A + a_1k_2 \quad (b) \quad (8.4,9)$$

$$B' = B + a_1k_1 + a_0k_2 \quad (c)$$

$$C' = C + a_0k_1 \quad (d)$$

The numerical values of the constants for the example jet transport are

$$\begin{aligned} A &= 2.721 \times 10^7 \\ B &= 2.633 \times 10^5 \\ C &= 1.376 \times 10^5 \\ a_1 &= 8.218 \times 10^8 \\ a_0 &= 3.653 \times 10^8 \end{aligned} \quad (8.4,10)$$

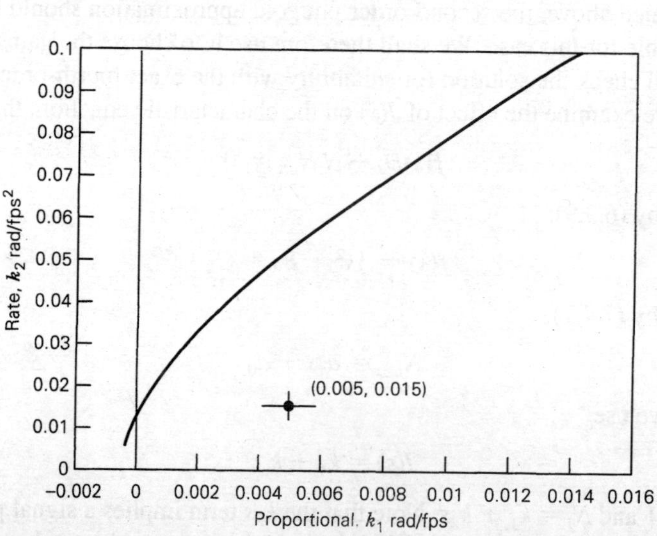

Figure 8.10 Speed controller gain relation for $\zeta = 1$. Phugoid approximation.

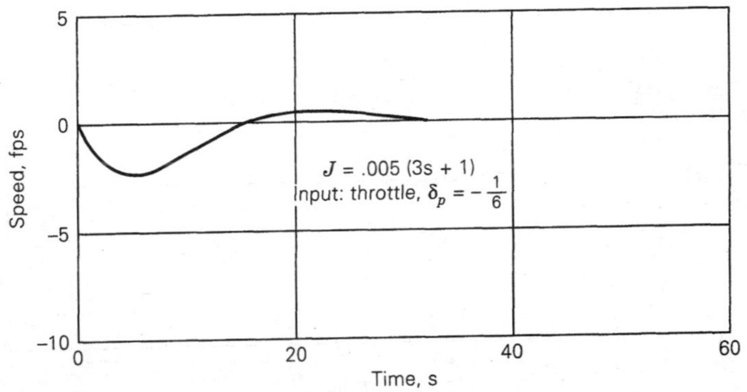

Figure 8.11 Speed controller—phugoid approximation. Speed response to throttle input.

To assess what range of values of k_1 and k_2 would be appropriate, we use three guides:

1. A reasonable elevator angle for, say, a 10 fps (3.048 m/s) speed error
2. The root locus plot for $G_{u\delta_e}$
3. The graph of k_2 vs. k_1 for critical damping

(1) The first of these is arrived at by noting that 1° of elevator for 10 fps speed error gives a k_1 of 0.0017 rad/fps. (2) The root locus plot is shown on Fig. 8.9 and indicates that the open loop roots can be moved very appreciably with a proportional gain as low as 0.005. (3) For the third guide, we note that critical damping corresponds to $B'^2 - 4A'C' = 0$. With the aid of (8.4,9) and (8.4,10) this leads to an algebraic relation between k_1 and k_2 that is solved for the graph shown on Fig. 8.10. The useful range of gains is the space below the curve, which corresponds to damped oscillations. The farther from the curve, the more overshoot would be expected in the response. We have for illustration arbitrarily chosen the gains indicated by the point marked on the graph, without regard for whether it is optimum. When used to calculate the response of airplane speed to application of a negative step in thrust, with the phugoid approximation, the result is as shown in Fig. 8.11. The throttle input corre-

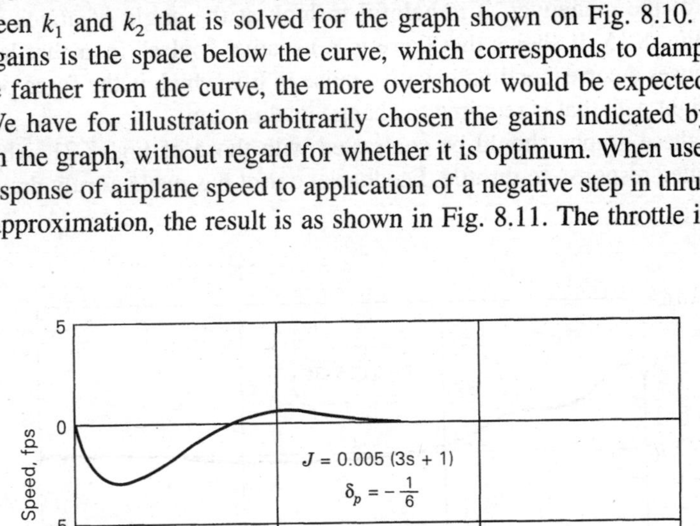

Figure 8.12 Speed controller—exact equations. Speed response.

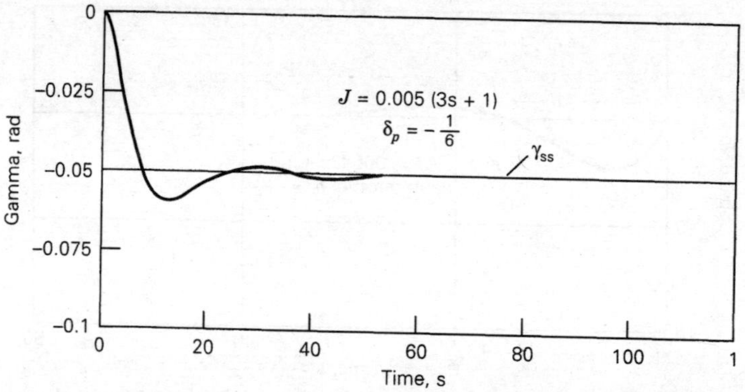

Figure 8.13 Speed controller—exact equations. Gamma response.

sponds to a steady-state descent angle of a little less than 3°. The maximum speed error, which is seen to be less than 3 fps at an initial speed of 774 fps, would probably not be perceptible to the pilot. This suggests that the chosen gains are probably not too small. The maximum elevator angle during this maneuver is less than 2° (see Fig. 8.14) so the gains are not excessive either.

To assess the performance of the controller with certainty, it is necessary to use the exact equations. The full matrix **A** for this example is (6.2,1), and **B** is (7.6,4). The most important elements of the solution are displayed in Figs. 8.12 to 8.14. The result for the speed in Fig. 8.12 confirms that the phugoid approximation is indeed good enough for preliminary design. Figure 8.13 demonstrates that the steady-state flight path angle is reached, with a small overshoot, in about 20 s. Figure 8.14 demonstrates that the elevator angle required to achieve this is small. To understand the physics of the maneuver, it is helpful to look at the angle of attack variation, graphed in Fig. 8.15. It shows that there is a negative "pulse" in α that lasts about 10 s. This causes a corresponding negative pulse in lift, which is the force perpendicular to the flight path that is required to change its direction.

Finally, these graphs should be contrasted with those of Fig. 7.21, which show the uncontrolled response to throttle. Feedback control has made a truly dramatic difference!

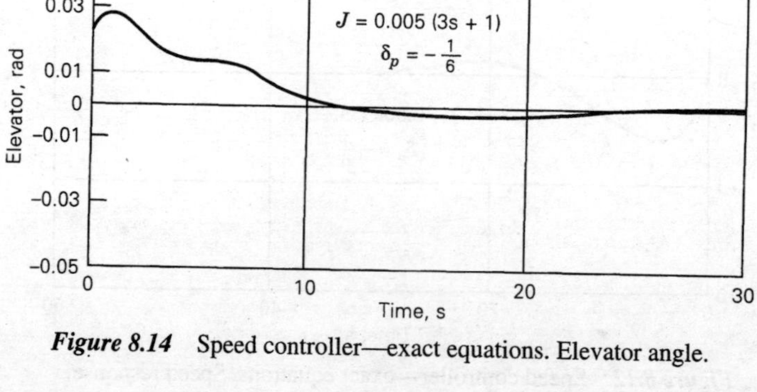

Figure 8.14 Speed controller—exact equations. Elevator angle.

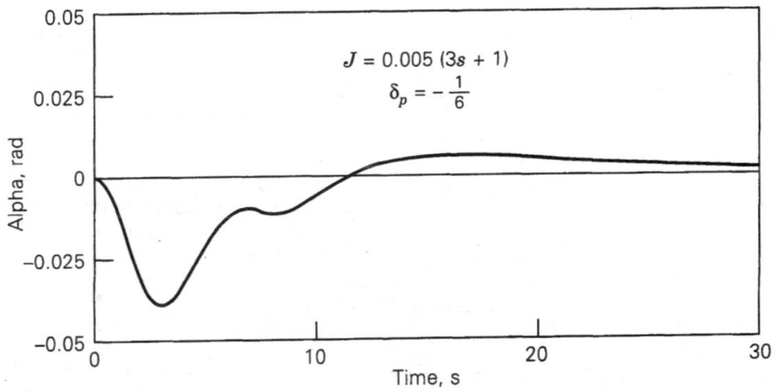

Figure 8.15 Speed controller—exact equations. Angle of attack response.

8.5 *Altitude and Glide Path Control*

One of the most important problems in the control of flight path is that of following a prescribed line in space, as defined for example by a radio beacon, or when the airplane flies down the ILS glide slope. We discuss this case by considering first a simple approximate model that reveals the main features, and then examining a more realistic, and hence more complicated case.

FLIGHT AT EXACTLY CONSTANT HEIGHT—SPEED STABILITY

The first mathematical model we consider can be regarded as that corresponding to horizontal flight when a "perfect" autopilot controls the angle of attack in such a way as to keep the height error exactly zero. The result will show that the speed variation is stable at high speeds, but unstable at speeds below a critical value near the minimum drag speed. Neumark (1950) recounts that this criterion was first discovered *in 1910* by Painlevé, and that it was at first accepted by aeronautical engineers and scientists, but later, on the basis of the theory of the phugoid, which showed no such effect, was rejected as false. In fact, to the extent that pilots can control height error by elevator control alone, that is, to the extent that they approximate the ideal autopilot we have postulated, the instability at low speed will be experienced in manual flight. Since speed variation is the most noticeable feature of this phenomenon, it is commonly referred to as *speed stability.*

We could analyze this case by applying (4.9,18) to the stated flight condition. However, it is both simpler and more illuminating to proceed directly from first principles. The airplane is flying on a horizontal straight line at variable speed V. It is implied that α is made to vary, by controlling δ_e, in such a way that the lift is kept exactly equal to the weight at all times. The equation of motion is clearly

$$m\dot{V} = T - D \tag{8.5,1}$$

where T is the horizontal component of the thrust, and D is the drag. Since the speed cannot change very rapidly, then neither does α, and we can safely ignore any effects of q and $\dot{\alpha}$ on lift and drag. Consequently, T and D are simply the thrust and drag or-

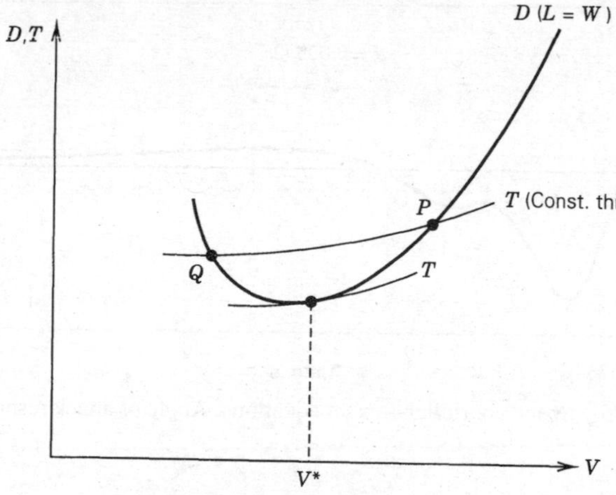

Figure 8.16 Performance graph.

dinarily used in performance analysis, as displayed in Fig. 8.16. We denote the reference thrust and drag by T_0 and D_0 and define the stability derivatives

$$T_V = \partial T/\partial V \quad \text{and} \quad D_V = \partial D/\partial V$$

so that

$$T - D = (T_0 + T_V\Delta V) - (D_0 + D_V\Delta V)$$

Since $V = V_0 + \Delta V$, and $T_0 = D_0$, (8.5,1) becomes

$$m\Delta\dot{V} = (T_V - D_V)\Delta V \tag{8.5,2}$$

This first-order differential equation has the solution

$$\Delta V = ae^{\lambda t}$$

with

$$\lambda = (T_V - D_V)/m \tag{8.5,3}$$

T_V and D_V are the slopes of the tangents to the thrust and drag curves at their intersection. If they intersect at a point such as P in Fig. 8.16, then $T_V < D_V$, λ is negative, and the motion is stable. If, on the other hand, the flight condition is at a point such as Q, the reverse is the case. λ is then >0, and the motion is unstable. If when flying at point Q there is an initial error in the speed, then it will either increase until it reaches the stable point P or it will decrease until the airplane stalls. The stable and unstable regimes are bounded by the speed V^*, which is where the thrust curve is tangent to the drag curve. V^* will be the same as V_{md} of Fig. 7.1 if $T_V = 0$. Hence the appellation "back side of the polar" is used to describe the range $V < V^*$, with reference to the portion of the aircraft polar (the graph of C_L vs. C_D) for which C_L is greater than that for maximum L/D.

Although we have analyzed only the case of horizontal flight, the result is similar for other straight-line flight paths, climbing or descending (see Exercise 8.8). Flight in the unstable regime can indeed occur when an airplane is in a low speed climb or landing approach. This speed instability is therefore not entirely academic, but can present a real operational problem, depending on by what means and how tightly the aircraft is constrained to follow the prescribed flight path. An important point insofar as AFCS design is concerned is that for speeds less than V^* *it is not possible to lock* *exactly onto a straight-line flight path, and at the same time provide stability, using the elevator control alone,* no matter how sophisticated the controller! To achieve stability it is mandatory to use a second control. This would most commonly be the throttle, but in principle spoilers that control the drag could also be used.

EXAMPLE—AN ALTITUDE CONTROLLER

In view of the above, we illustrate an altitude controller that also incorporates control of speed, using once again our example jet airplane. This time we make the system model more realistic by including first-order lag elements for the two controls: that for the elevator is mainly associated with its servo actuator (time constant 0.1 s); and that for the throttle with the relatively long time lag inherent in the build up of thrust of a jet engine following a sudden movement of the throttle (time constant 3.5 s). Another feature that is incorporated to add realism to the example is a thrust limiter. Because transport aircraft inherently respond slowly to changes in thrust, the gains chosen to give satisfactory response for very small perturbations in speed will lead to a demand for thrust outside the engine envelope for larger speed errors. We have therefore included a nonlinear feature that limits the thrust to the range $0 \leq T \leq 1.1T_0$. This contains the implicit assumptions (quite arbitrary) that the airplane, flying near its ceiling, has 10% additional thrust available, and that idling engines correspond to zero thrust.

At the same time this example illustrates an alternative approach to generating the analytical model of the system, in terms of its differential equations. In the previous illustrations we have, by contrast, used what may be termed "transfer function algebra" to arrive at transfer functions of interest, and then used these to obtain whatever results were desired. The end result of the modeling to follow is a system of differential equations that is then integrated to get time solutions. Since the limiter is inherently a nonlinear element, it is in any case not possible to include it in a transfer function based analysis.

The system block diagram is shown in Fig. 8.17. The commanded speed and altitude are the reference values u_0 and h_0, so that the two corresponding error signals are $-\Delta u$ and $-\Delta h$. Note that h is the negative of z_E used in Chap. 4. The inner loop for θ is that previously studied in Sec. 8.3, with the $J_e(s)$ modified to account for the

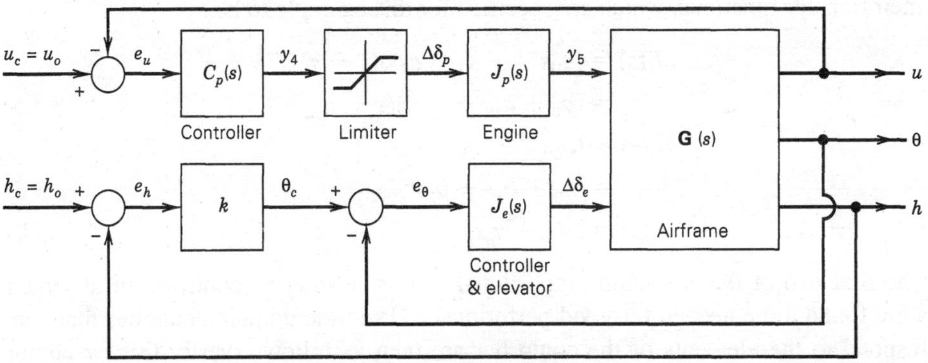

Figure 8.17 Altitude-hold controller.

elevator servo actuator. The logic of the outer loop that controls h warrants explanation. If there is an initial error in h, say the altitude is too low, then in order to correct it, the airplane's flight path must be deflected upward. This requires an increase in angle of attack to produce an increase in lift. The angle of attack and the resulting lift could of course be produced by using an angle of attack vane as sensor, and no doubt an angle of attack commanded to be a function of height error would be very effective. It might be preferred, however, to use the vertical gyro as the source of the signal, and since short-term changes in θ are effectively changes in α, then much the same result is obtained by using θ as the commanded variable. We have chosen to use stability axes, so that in the steady state, when Δh is zero, the correct value of θ is also zero. Thus, in summary, the system commands a pitch angle that is proportional to height error and the inner loop uses the elevator to make the pitch angle follow the command. While all this is going on the speed will be changing because of both gravity and drag changes. The quickest and most straightforward way of controlling the speed is with the throttle, and the third loop accomplishes that. (The symbols y_4 and y_5 denote the inputs to the limiter and the airframe, and are elements of the state vector derived below.)

The Differential Equations

The basic matrix differential equation of the airframe, with $\theta_0 = 0$, and $\Delta z_E = -\Delta h$ is obtained from (4.9,18) and (7.6,4). Up to this point, we have neglected engine dynamics and in effect regarded thrust as proportional to δ_p. The matrix $\mathbf{B}$ of (7.6,4) is structured in that way. To accommodate the facts that δ_p actually represents the throttle setting, not the thrust, and that the two are dynamically connected, we need to introduce two new symbols, y_5 and $\mathbf{c}^*$. The quantity y_5, when multiplied by X_{δ_p}, etc. yields the aerodynamic force and moment increments ΔX_c, etc., and $\mathbf{c}^*$ is defined below. The differential equation of the airframe is then

$$\dot{\mathbf{x}} = \mathbf{A}\mathbf{x} + \mathbf{B}\mathbf{c}^* \tag{8.5,4}$$

where

$$\mathbf{x} = [\Delta u \quad w \quad q \quad \theta \quad \Delta h]^T$$
$$\mathbf{c}^* = [\delta_e \quad y_5]^T$$
$$\mathbf{A} = [a_{ij}]$$
$$\mathbf{B} = [b_{ij}]$$

To obtain the differential equations of the three control elements, we begin with their transfer functions, which are specified for this example to be

$$\begin{aligned} J_e(s) &= (a_0 s^{-1} + a_1 + a_2 s)(1 + \tau_e s)^{-1} \\ &= (a_0 + a_1 s + a_2 s^2)(s + \tau_e s^2)^{-1} \end{aligned} \tag{8.5,5}$$

$$\begin{aligned} C_p(s) &= (b_0 s^{-1} + b_1 + b_2 s) \\ &= (b_0 + b_1 s + b_2 s^2)/s \end{aligned} \tag{8.5,6}$$

$$J_p(s) = 1/(1 + \tau_p s) \tag{8.5,7}$$

The first two of these contain proportional + rate + integral controls, all of which were found to be needed for good performance. The time domain equations that correspond to the elements of the controller are then as follows (verify this by taking their Laplace transforms):

$$
\begin{aligned}
\theta_c &= -k\Delta h && (a)\\
\tau_e\Delta\ddot{\delta}_e + \Delta\dot{\delta}_e &= a_0 e_\theta + a_1\dot{e}_\theta + a_2\ddot{e}_\theta && (b)\\
\dot{y}_4 &= b_0 e_u + b_1\dot{e}_u + b_2\ddot{e}_u && (c)\\
\tau_p\dot{y}_5 &= \Delta\delta_p - y_5 && (d)
\end{aligned}
\tag{8.5,8}
$$

After substituting the expressions for the two error signals, (8.5,8) yield three equations for the controls

$$
\begin{aligned}
\tau_e\Delta\ddot{\delta}_e + \Delta\dot{\delta}_e &= -ka_2\Delta\ddot{h} - a_2\ddot{\theta} - ka_1\Delta\dot{h} - a_1\dot{\theta} - ka_0\Delta h - a_0\theta && (a)\\
\dot{y}_4 &= -b_2\Delta\ddot{u} - b_1\Delta\dot{u} - b_0\Delta u && (b)\\
\tau_p\dot{y}_5 &= \Delta\delta_p - y_5 && (c)
\end{aligned}
\tag{8.5,9}
$$

For convenient integration we want a system of first-order equations and therefore have to do something about the second derivatives in (8.5,9). Since $\dot{\theta} = q$ we can replace $\ddot{\theta}$ with $\dot{q}$. For the other second derivatives, we introduce three new variables, as follows:

$$
\begin{aligned}
y_1 &= \Delta\dot{u} && (a)\\
y_2 &= \Delta\dot{h} && (b)\\
y_3 &= \Delta\dot{\delta}_e && (c)
\end{aligned}
\tag{8.5,10}
$$

With these definitions, (8.5,9*a* and *b*) can now be rewritten in terms of first derivatives as

$$
\begin{aligned}
\tau_e\dot{y}_3 + y_3 &= -(a_2\dot{q} + ka_2\dot{y}_2 + a_1 q + a_0\theta + ka_0\Delta h + ka_1 y_2)\\
\dot{y}_4 &= -(b_2\dot{y}_1 + b_0\Delta u + b_1 y_1)
\end{aligned}
\tag{8.5,11}
$$

The state vector now consists of the original five variables from (8.5,4) plus the two control variables $\Delta\delta_e$ and $\Delta\delta_p$, plus the five y_i defined above, making 12 in all. We therefore require 12 independent equations. From the foregoing equations (8.5,4) (8.5,10), (8.5,11), and (8.5,9*c*) we can get 11 of the required differential equations. That for y_1 is obtained from (8.5,10*a*) by differentiating the first component of (8.5,4) and that for y_2 by differentiating the fifth. The result of that operation is

$$
\begin{aligned}
\dot{y}_1 &= a_{12}\dot{w} + a_{11}y_1 + a_{14}q + b_{11}y_3 + b_{12}\dot{y}_5\\
\dot{y}_2 &= -\dot{w} + u_0 q
\end{aligned}
\tag{8.5,12}
$$

Finally, the 11 independent differential equations are assembled as follows:

$$
\begin{aligned}
\Delta\dot{u} &= y_1\\
\dot{w} &= a_{21}\Delta u + a_{22}w + a_{23}q + b_{21}\Delta\delta_e\\
\dot{q} &= a_{31}\Delta u + a_{32}w + a_{33}q + b_{31}\Delta\delta_e\\
\dot{\theta} &= q\\
\Delta\dot{h} &= y_2\\
\Delta\dot{\delta}_e &= y_3\\
\dot{y}_5 &= (-y_5 + \Delta\delta_p)/\tau_p\\
\dot{y}_1 &= a_{11}y_1 + a_{12}\dot{w} + a_{14}q + b_{11}y_3 + b_{12}\dot{y}_5\\
\dot{y}_2 &= u_0 q - \dot{w}\\
\dot{y}_3 &= -(a_2\dot{q} + ka_2\dot{y}_2 + a_1 q + a_0\theta + ka_0\Delta h + ka_1 y_2 + y_3)/\tau_e\\
\dot{y}_4 &= -(b_2\dot{y}_1 + b_0\Delta u + b_1 y_1)
\end{aligned}
\tag{8.5,13}
$$

Solutions

The above equations contain first derivatives on the right side as well as on the left side and hence are not in the canonical form. This is no impediment to numerical integration, however, since if the derivatives are calculated in the sequence given, each one that appears on the right has already been calculated in one of the preceding equations by the time it is needed. The twelfth and final relation needed is that which describes the limiter, in the form

$$\Delta\delta_p = f(y_4)$$

From the values of C_{D_0} and C_{L_0} given in Sec. 6.2, we find that $D_0 = T_0 = 0.0657W$. In Sec. 7.6 it was given that $\delta_p = 1$ corresponds to a thrust of $0.3W$. It follows that zero thrust corresponds to $y_4 = -0.0657/.3 = -0.219$. The nonlinear relationship for y_4 is therefore implemented in the computing program by a program fragment equivalent to

$$\begin{aligned} &\Delta\delta_p = y_4 \\ &\text{IF } y_4 < -0.219 \text{ THEN } \quad \Delta\delta_p = -0.219 \\ &\text{IF } y_4 > 0.10 \text{ THEN } \quad \Delta\delta_p = 0.10 \end{aligned} \tag{8.5,14}$$

where the maximum engine thrust has been assumed to be 10% greater than cruise thrust. Equations (8.5,13 and 8.5,14) are convenient for numerical integration. We have calculated a solution using simple Euler integration of the equations for the example jet transport with the matrices **A** and **B** given in Secs. 6.2 and 7.6, and with the following control parameters: $\tau_e = 0.1$; $\tau_p = 3.5$; $k = 0.0002$; $a_0 = a_1 = a_2 = -0.5$; $b_0 = 0.005$; $b_1 = 0.08$; $b_2 = 0.16$. Figure 8.18 shows the performance obtained in response to an initial height error of 500 ft.

It is seen that the height error is reduced to negligible proportions quickly, in about 20 s, accompanied by a theta pulse of similar duration and peak magnitude about 7°. Even with extreme throttle action, the speed takes more than 2 min to recover its reference value. This length of time is inherent in the physics of the situation and cannot be shortened significantly by changes in the controller design. On the other hand, there is no operational requirement for more rapid speed adjustment when cruising at 40,000 ft.

The peak elevator angle needed is less than 3°, but the thrust drops quickly to zero, stays there for about 30 s, then increases rapidly to its maximum. Toward the end of the maneuver the throttle behaves linearly and reduces the speed error smoothly to zero.

8.6 Lateral Control

There are five lateral state variables that can be used readily as a source of feedback signals—$\{v, p, r, \phi, \psi\}$; v from a sideslip vane or other form of aerodynamic sensor, p and r from rate gyros, and ϕ, ψ from vertical and directional gyros. Lateral acceleration is also available from an accelerometer. These signals can be used to drive the two lateral controls, aileron and rudder. Thus there is a possibility of many feedback loops. The implementation of some of these can be viewed simply as synthetic modification of the inherent stability derivatives. For example, p fed back to aileron modi-

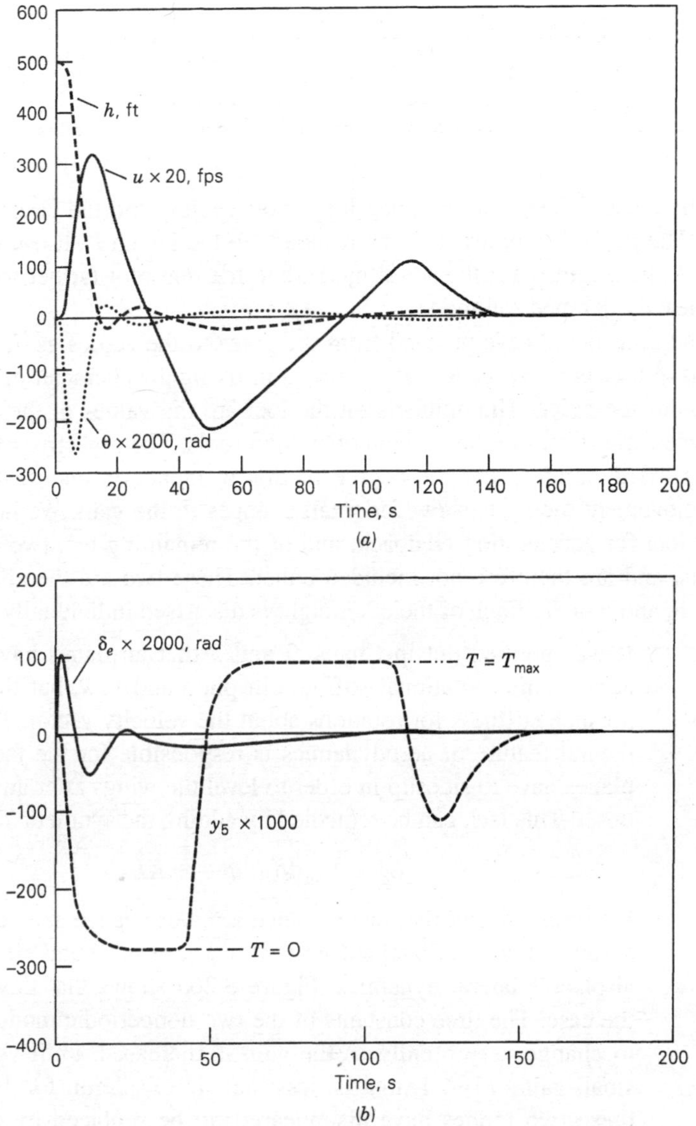

Figure 8.18 Altitude-hold controller. (*a*) Height, speed, and pitch angle. (*b*) Elevator and throttle controls.

fies L_p (roll damper), r to the rudder modifies N_r (yaw damper), and v to rudder modifies the yaw stiffness N_v, and so on. It is a helpful and instructive preliminary to a detailed study of particular lateral control objectives to survey some of these possible control loops. We could do this analytically by examining the approximate transfer functions given in Chap. 7. However, we prefer here to do this by way of example, using the now familiar jet transport, and using the full system model. We treat each loop as in Fig. 8.19, as a negative feedback with a perfect sensor and a perfect actuator, so that the loop is characterized by the simple constant gain K. For each case we present a root locus plot with the gain as parameter (Fig. 8.20) (All the root loci are symmetrical about the real axis; for some, only the upper half is shown). As is con-

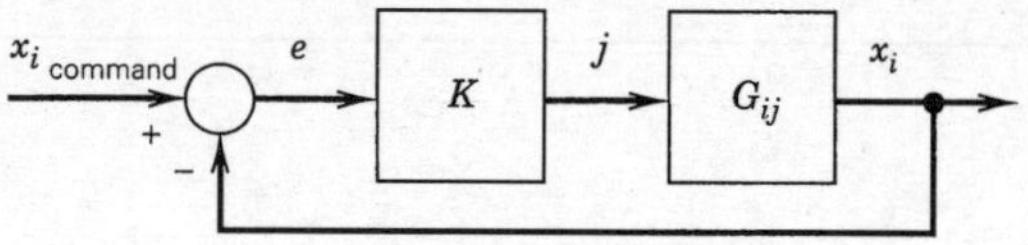

Figure 8.19 Representative loop.

ventional, the crosses designate the open loop roots (poles) and the circles the open loop zeros. The pair of complex roots corresponds to the Dutch Roll oscillation; the real root near the origin is for the spiral mode; and the real root farther to the left is that of the heavily damped roll mode.

Since the root loci always proceed from the poles to the zeroes as $|K|$ increases, the locations of the zeros can be just as important in fixing the character of the loci as the locations of the poles. The numbers on the loci are the values of the gain. Zero gain of course corresponds to the original open loop roots. The objective of control is to influence the dynamics, and the degree of this influence is manifested by the amount of movement the roots show for small changes in the gain. We have not included root loci for acceleration feedback, and of the remaining ten, two show very small effects, and are therefore not included either. These two are the aileron feedbacks: $v \rightarrow \delta_a$ and $r \rightarrow \delta_a$. Each of the other eight is discussed individually below.

$\phi \rightarrow \delta_a$ It was pointed out in Chaps. 2 and 3 that airplanes have inherent aerodynamic rotational stiffness in pitch and yaw, but that there is no such stiffness for rotations about the velocity vector. This fundamental feature of aerodynamics is responsible for the fact that airplanes have to sideslip in order to level the wings after an initial roll upset. This lack can be remedied by adding the synthetic derivative

$$L_\phi = L_{\delta_a}\, \mathrm{d}\delta_a/\mathrm{d}\phi = -KL_{\delta_a}$$

We might expect that making such a major change as adding a new aerodynamic rotational stiffness would have profound effects on the airplane's lateral dynamics. Figure 8.20*a* shows that this is indeed the case. The time constants of the two nonperiodic modes are seen to change very rapidly as the gain is increased, until with even a small gain, $|K| < 1$, that is, less than 1° of aileron for 1° of bank, these two modes have disappeared, to be replaced by a low frequency, heavily damped oscillation. The Dutch Roll remains virtually unaffected by the aileron feedback for any modest gain.

$p \rightarrow \delta_a$ This root locus is shown in Fig. 8.20*b*. The largest effect is on the roll mode, as might be expected, where a positive gain of unity (corresponding to a *decrease* in $|L_p|$) results in a substantial reduction in the magnitude of the large real root. This is accompanied by an increase in the spiral stability and a slight reduction in the Dutch Roll damping. A negative gain, (an *increase* in $|L_p|$) increases the Dutch Roll damping, shortens the roll mode time constant and causes a slight reduction in the magnitude of the spiral root (the latter not visible in the figure).

$\psi \rightarrow \delta_a$ Because ψ is the integral of r, the transfer functions for ψ have an s factor in the denominator, and hence a pole at the origin. This is

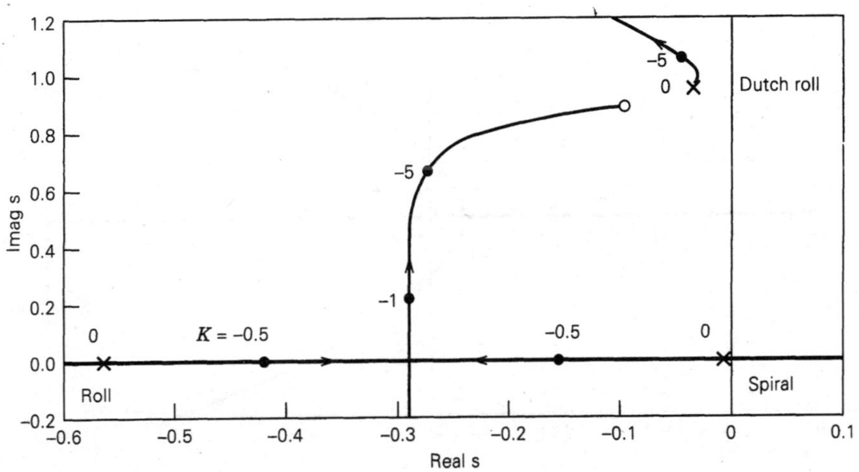

(a) $\phi \to \delta_a$

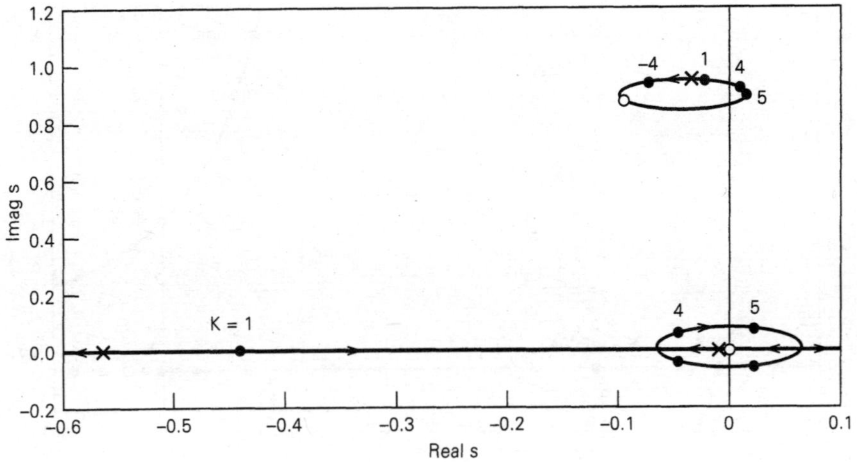

(b) $p \to \delta_a$

Figure 8.20 Root loci. (*a*) $\phi \to \delta_a$. (*b*) $p \to \delta_a$. (*c*) $\psi \to \delta_a$. (*d*) $v \to \delta_r$. (*e*) $p \to \delta_r$. (*f*) $r \to \delta_r$. (*g*) $\phi \to \delta_r$. (*h*) $\psi \to \delta_r$. (*i*) STOL airplane; $r \to \delta_r$.

seen in Fig. 8.20*c*. The expansion theorem (A.2,10) shows that the zero root of the characteristic equation leads to a constant in the solution for ψ. This is consistent with the fact that the reference direction for ψ is arbitrary.

The feedback of ψ to aileron has little relative effect on the Dutch Roll and rolling modes. Its main influence is seen on the spiral and zero roots, which are quite sensitive to this feedback. For negative gain (stick left for yaw to the right) these two modes rapidly combine into an oscillatory mode that goes unstable by the time the gain is -0.5 (0.5° aileron for 1° yaw). For all positive gains, there is an unstable divergence.

$v \to \delta_r$ This feedback (Fig. 8.20*d*) represents rudder angle proportional to sideslip, with positive gain corresponding to an increase in N_v. Note

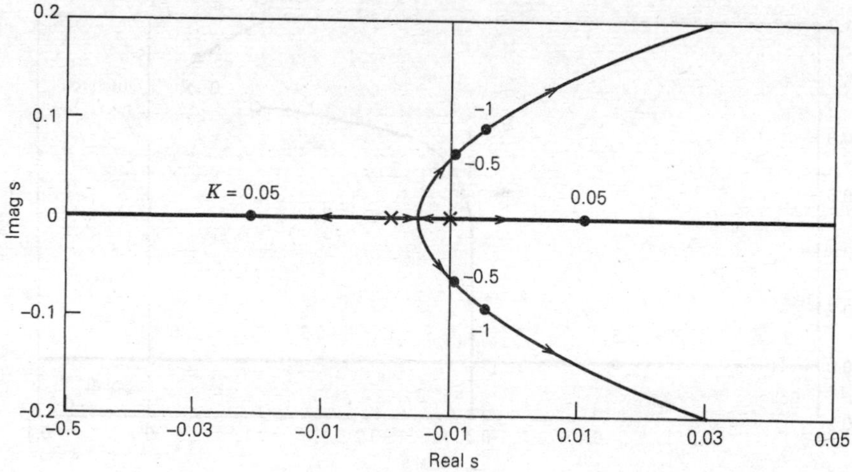

(c) $\psi \rightarrow \delta_a$

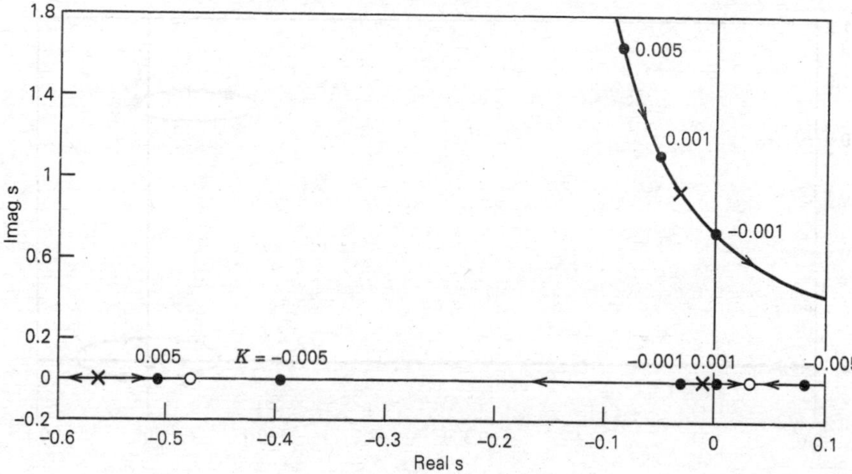

(d) $v \rightarrow \delta_r$

Figure 8.20 *(Continued)*

that a gain of 0.001 for v corresponds to $\delta_r/\beta = -0.774$. The principal effect is to increase the frequency of the Dutch Roll while simultaneously decreasing the spiral stability, which rapidly goes unstable as the gain is increased. The reverse is true for negative gain. The roll mode remains essentially unaffected.

$p \rightarrow \delta_r$ Roll rate fed back to the rudder has a large effect on all three modes. For positive gain (right rudder for roll to the right) the damping of the Dutch Roll is increased quite dramatically—it is quadrupled for a gain of about 0.2° rudder/deg/s of roll rate. This is counterintuitive (see Exercise 8.10). At the same time, the damping of the roll mode is very much diminished, and that of the spiral mode is increased. With further increase in gain the two nonperiodic modes combine to form an oscillation, which can go unstable at a gain of about 0.4.

$r \rightarrow \delta_r$ The large effects shown in Fig. 8.20*f* for the yaw damper case are what would be expected. As an aid in assessing the damping perfor-

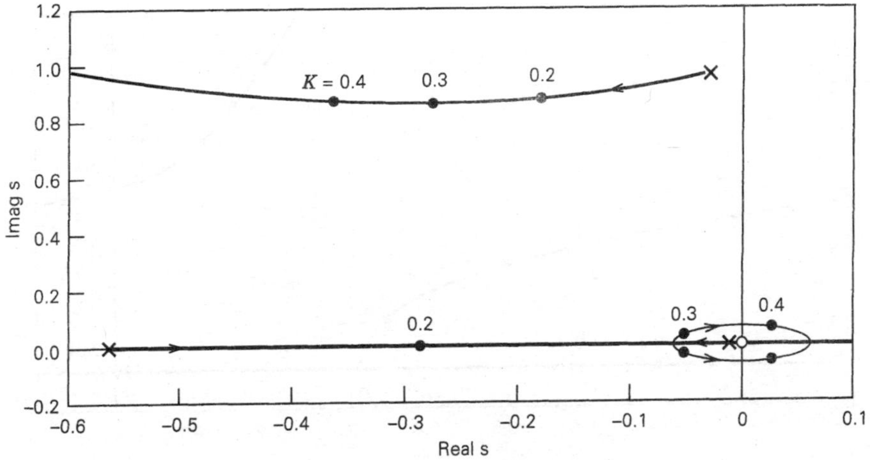

(e) $p \rightarrow \delta_r$

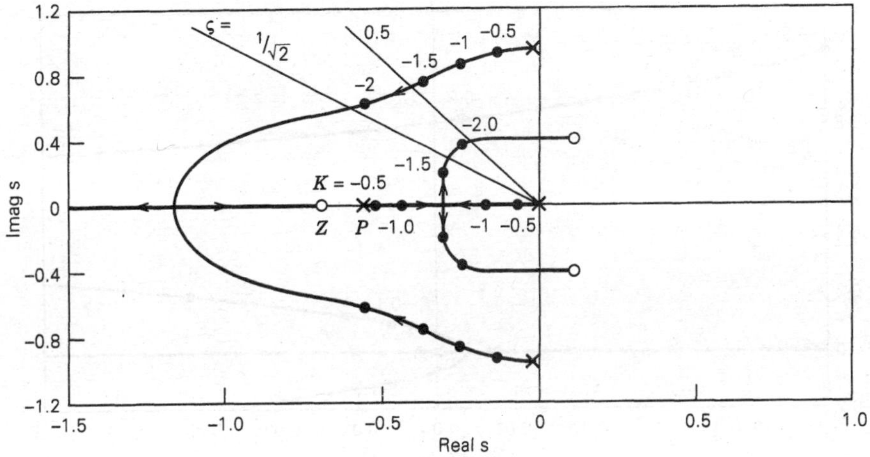

(f) $r \rightarrow \delta_r$

Figure 8.20 *(Continued)*

mance, two lines of constant relative damping ζ are shown on this figure. Negative gain corresponds to left rudder when yawing nose right. A very large increase in Dutch Roll damping is attained with a gain of -1, at which point there is a commensurate gain in the spiral damping. There is some loss in damping of the roll mode. The beneficial effects of yaw-rate feedback are clearly evident from this figure. The behavior for larger negative gains, beyond about -1.4, is especially interesting. For this airplane at this flight condition, the two real roots combine to form a new oscillation, the damping of which rapidly deteriorates with further increase in negative gain. This feature complicates the choice of gain for the yaw damper. The pattern shown is not the only one possible. Figure 8.20i is the corresponding root locus for the STOL airplane of Sec. 8.9, flying at 10,000 ft and 200 k. It illustrates the importance of the location of the zeroes of the closed loop transfer function. For the jet transport the real zero, z is to the left of the real roll mode root p. For the

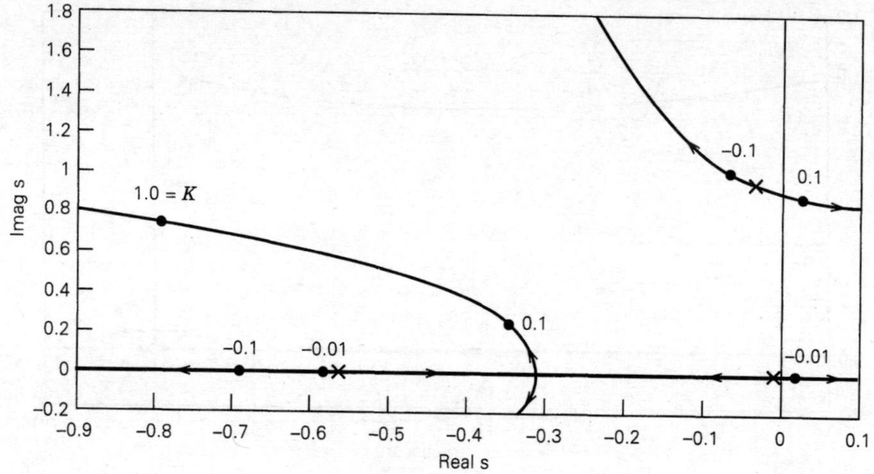

(g) $\phi \to \delta_r$

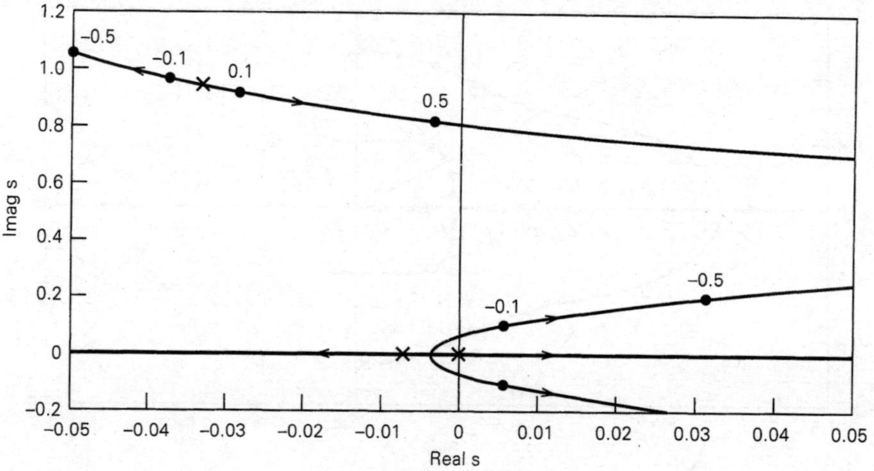

(h) $\psi \to \delta_r$

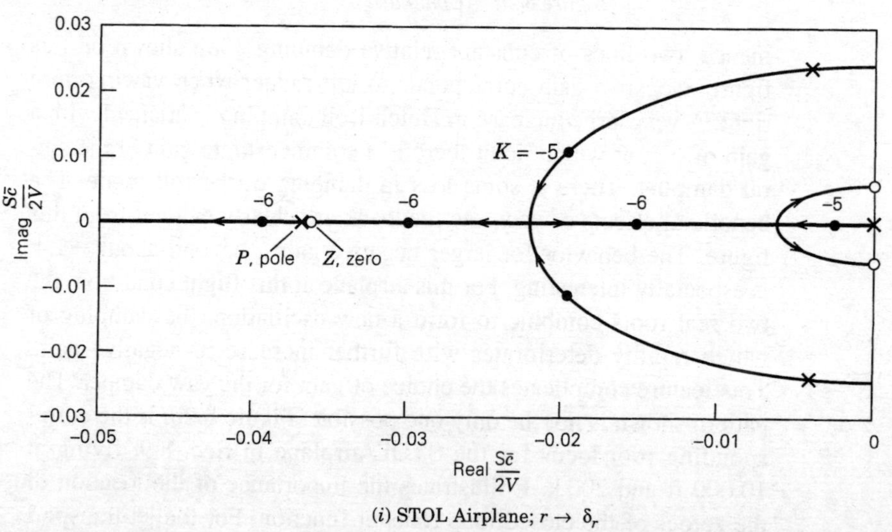

(i) STOL Airplane; $r \to \delta_r$

Figure 8.20 *(Continued)*

STOL airplane the reverse is the case. The direction of the locus emanating from this root is therefore opposite in the two cases, with a consequent basic difference in the qualitative nature of the dynamics. For the STOL airplane the Dutch Roll root splits into a real pair, one of which then combines with the spiral root to form a new low-frequency oscillation. In viewing Fig. 8.20*i* it should be noted that it was drawn for the nondimensional system model, and hence the numerical values for the roots and the gains are not directly comparable with those of Fig. 8.20*f*.

$\phi \rightarrow \delta_r$ Feeding back bank angle to the rudder produces mixed results (Fig. 8.20*g*). When the gain is negative, the spiral mode is rapidly driven unstable. On the other hand if it is positive, to improve the spiral, the Dutch Roll is adversely affected.

$\psi \rightarrow \delta_r$ The consequence of using heading to control the rudder is also equivocal. If the gain is positive (heading right induces right rudder) the null mode becomes divergent. If the gain is negative the two nonperiodic modes form a new oscillation at quite small gain that quickly becomes unstable.

8.7 Yaw Damper

Yaw dampers are widely used as components of stability augmentation systems (SAS); we saw the potential beneficial effects in Fig. 8.20*f*. At first glance the yaw damper would appear to be a very simple application of feedback control principles—just use Fig. 8.20*f* as a guide, select a reasonable gain, and add a model for the servo actuator/control dynamics. However, it is not really that simple. There is another important factor that has to be taken into account—namely that during a steady turn, the value of r is not zero. If, in that situation, the yaw damper commands a rudder angle because it senses an r, the angle would no doubt not be the right one needed for a coordinated turn. In fact during a right turn the yaw damper would always produce left rudder, whereas right rudder would usually be required (see Fig. 7.24). This characteristic of the yaw damper is therefore undesirable. To eliminate it, the usual method is to introduce a high-pass or "washout" filter, which has zero gain in the steady state and unity gain at high frequency. The zero steady state gain eliminates the feedback altogether in a steady turn. The system that results is pictured in Fig. 8.21, where the meaning of the filter time constant is illustrated. For the servo actuator/rudder combination of this large airplane we assume a first order system of time constant 0.3 sec.

The closed loop transfer function for Fig. 8.21 is readily found to be [see (8.2,1)]

$$\mathring{G}_{rr_c} = \frac{JG_{r\delta_r}}{1 + WJG_{r\delta_r}} \tag{8.7,1}$$

This transfer function was used to calculate a number of transient responses to illustrate the effects of $J(s)$ and $W(s)$.

As a reference starting point, Fig. 8.22 shows the open loop response ($W = 0$) to

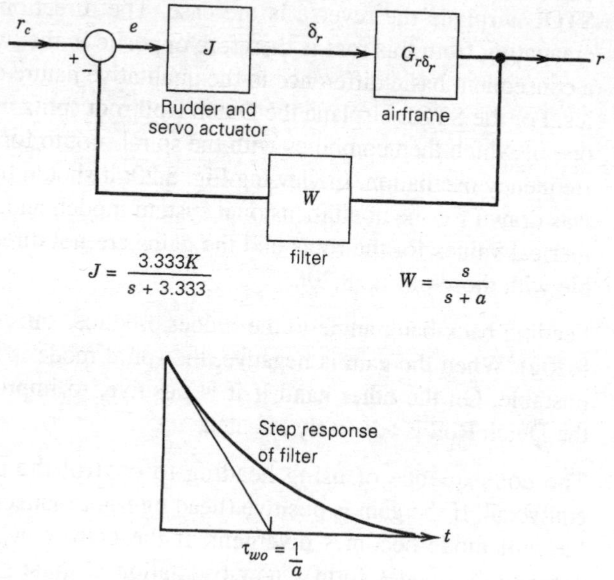

Figure 8.21 Yaw damper.

a unit impulse of yaw rate command r_c. It is evident that there is a poorly damped oscillatory response (the Dutch Roll) that continues for about 2 min and is followed by a slow drift back to zero (the spiral mode). Fig. 8.22*a* shows that the control dynamics (i.e., $J(s)$) has not had much effect on the response.

Figure 8.23 shows what happens when the yaw damper is turned on with the same input as in Fig. 8.22. It is seen that the response is very well damped with either of the two gains shown, which span the useful range suggested by Fig. 8.20*f*, and that the spiral mode effect has also been suppressed.

It remains to choose a time constant for the washout filter. If it is too long, the washout effect will be insufficient; if too short, it may impair the damping performance. To assist in making the choice, it is helpful to see how the parameter $a = 1/\tau_{wo}$ affects the lateral roots. Figure 8.24 shows the result for a gain of $K = -1.6$. The roots in this case consist of those shown in Fig. 8.20*f* plus an additional small real root associated with the filter. It is seen that good damping can be realized for values of a up to about 0.3, that is, for time constant τ down to about 3 s. This result is very dependent on the gain that is chosen. While the oscillatory modes are behaving as displayed, the real roots are also changing—the roll root decreasing in magnitude from -2.31 at $a = 0$ to -1.95 at $a = 0.32$. The new small real root starts at the origin when $a = 0$ and moves slowly to the left, growing to -0.00464 at $a = 0.32$. When the filter time constant is 5 s the small root is -0.0038, corresponding to an aperiodic mode with $t_{\text{half}} = 182$ s. It is instructive to compare the performance of the yaw damper with and without the filter for an otherwise identical case. This is done in Fig. 8.25. It is seen that the main difference between them comes from the small real root, which after 5 min has reduced the yaw rate to about 5% of its peak value. This slow decay is unlikely to present a problem since the airplane heading is inevitably controlled, either by a human or automatic pilot. In either case, the residual r would rapidly be eliminated (see Sec. 8.8).

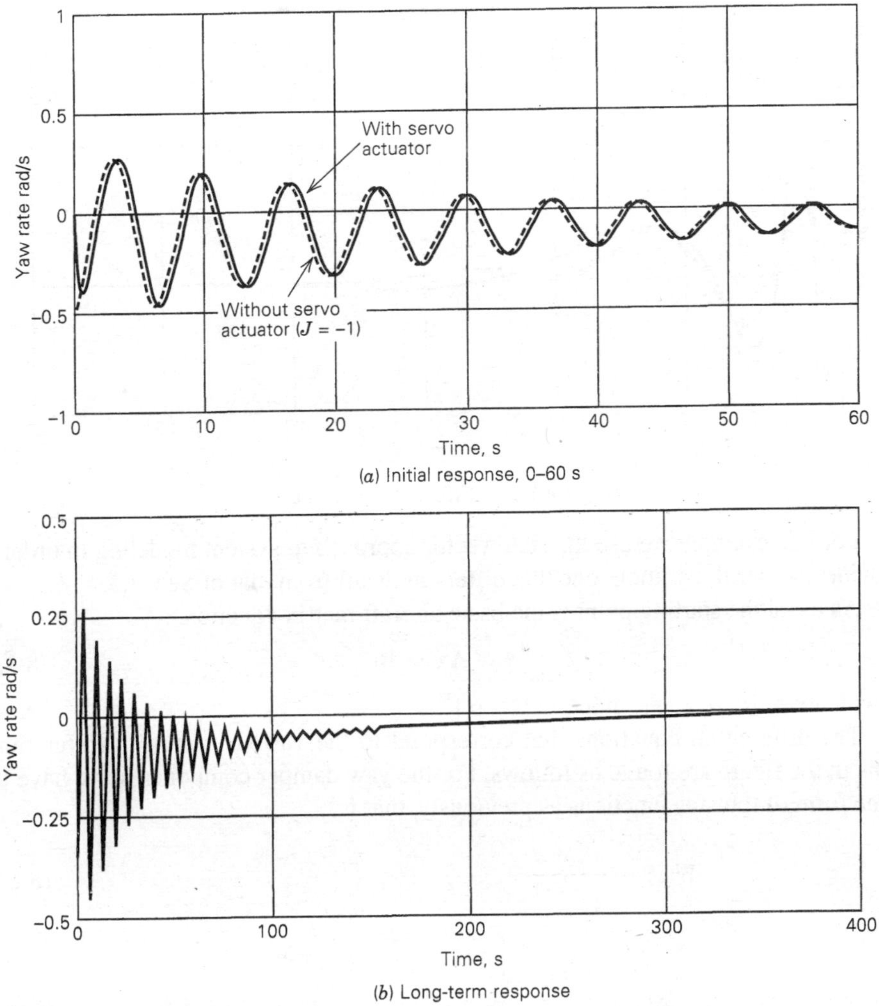

Figure 8.22 Yaw rate impulse response—open loop, $W = 0$. (*a*) Initial response, 0–60 sec. (*b*) Long-term response.

8.8 Roll Controller

This example is of another common component of an AFCS, a control loop that maintains the wings level when flying on autopilot, or that can be commanded to roll the airplane into a turn and hold it there. We shall see in this particular case that the resulting turn is virtually truly banked, even though no special provision has been made to control sideslip.

The block diagram of the system is shown in Fig. 8.26. It incorporates the yaw damper described in the previous section and adds two additional loops. The outer loop commands ϕ. The ϕ error is converted to a roll rate command by J_p, and it is the roll rate error that is then used to drive the aileron servo actuator. If the roll rate followed the command instantaneously, without lag, the bank angle response would be exponential (i.e., $\dot{\phi} \propto \phi$). In reality of course this ideal behavior is not achieved because of the airframe and servo dynamics.

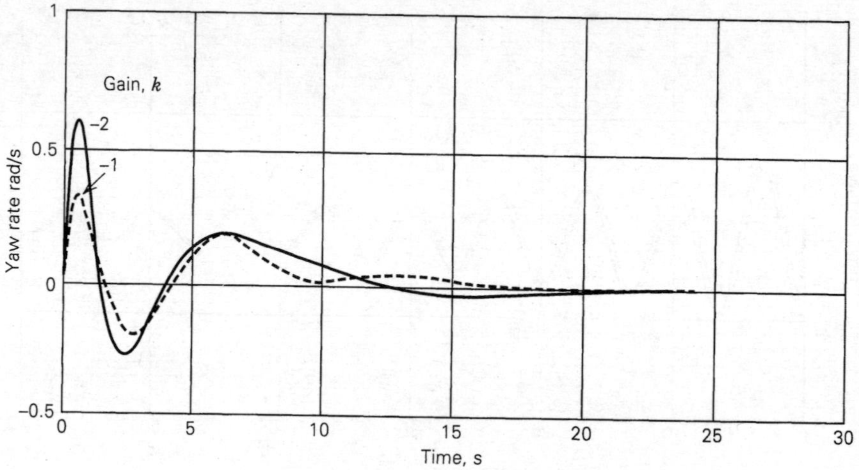

Figure 8.23 Effect of yaw damper.

For this example we use the state vector approach to system modeling in order to provide another illustration, one that differs in detail from that of Sec. 8.5.

As usual the starting point is the basic aircraft matrix equation,

$$\dot{\mathbf{x}} = \mathbf{Ax} + \mathbf{Bc} \tag{8.8,1}$$

in which $\mathbf{x} = [v \ p \ r \ \phi]^T$ and $\mathbf{c} = [\delta_a \ \delta_r]^T$.

The differential equations that correspond to the various control transfer functions in the figure are found as follows. For the yaw damper components, we have the same form of transfer functions as previously, that is,

$$W(s) = \frac{s}{s + \dfrac{1}{\tau_{w0}}} \tag{8.8,2}$$

and

$$J_r(s) = \frac{K_r/\tau_r}{s + \dfrac{1}{\tau_r}} \tag{8.8,3}$$

$$\therefore \quad \bar{\delta}_r = -J_r W \bar{r} = -\frac{K_r s/\tau_r}{\left(s + \dfrac{1}{\tau_{w0}}\right)\left(s + \dfrac{1}{\tau_r}\right)} \bar{r} \tag{8.8,4}$$

From (8.8,4) we get the differential equation

$$\ddot{\delta}_r = \frac{K_r}{\tau_r}\dot{r} - \left(\frac{1}{\tau_r} + \frac{1}{\tau_{w0}}\right)\dot{\delta}_r - \frac{1}{\tau_r \tau_{w0}} \tag{8.8,5}$$

For J_p we use the constant K_p, and for J_a we use a first order servoactuator

$$J_a = \frac{K_a/\tau_a}{s + \dfrac{1}{\tau_a}} \tag{8.8,6}$$

The relation between $\bar{\delta}_a$, $\bar{p}$, and $\bar{\phi}_c$ is seen from the diagram to be

$$\bar{\delta}_a = J_a(\bar{p}_c - \bar{p}) = J_a J_p(\bar{\phi}_c - \bar{\phi}) - J_a \bar{p} \tag{8.8,7}$$

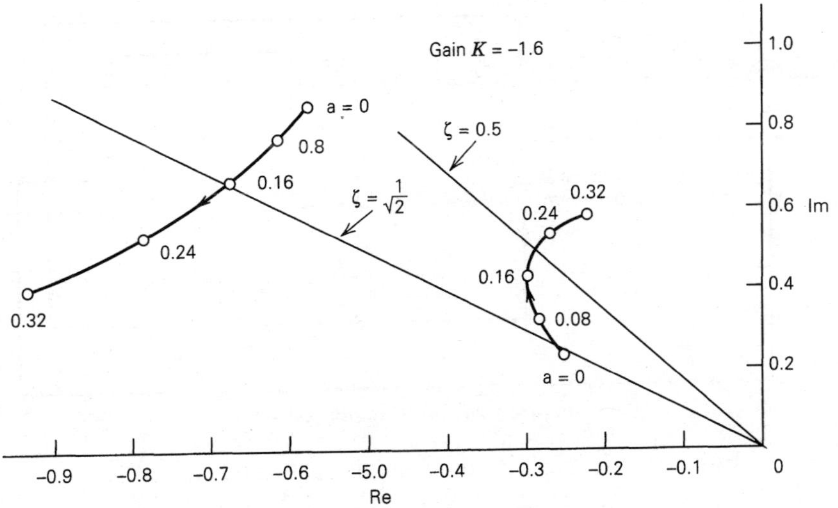

Figure 8.24 System poles for varying washout time constant.

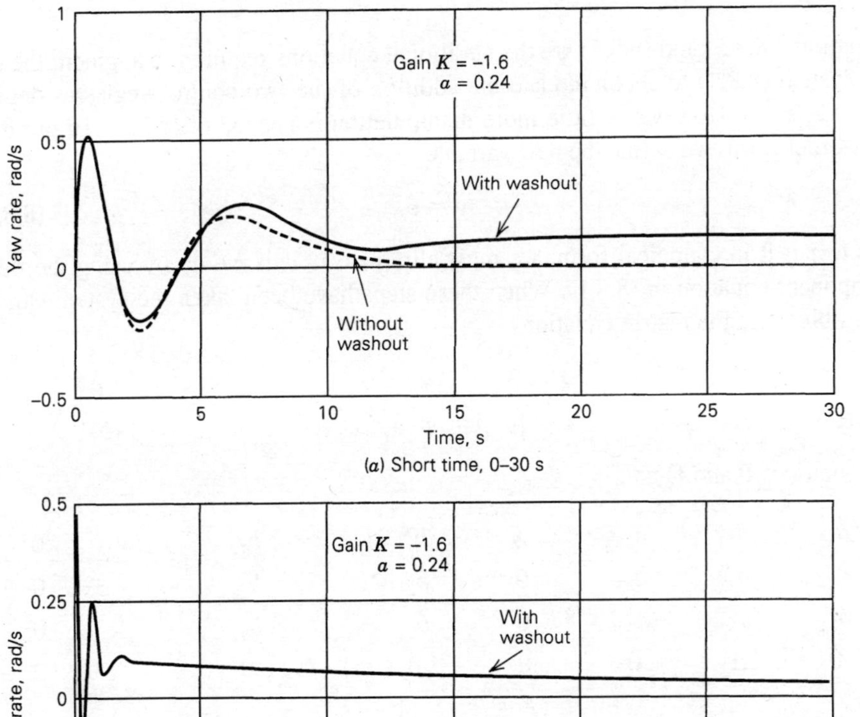

Figure 8.25 Effect of washout filter on yaw damper performance. (*a*) Short time, 0–30 sec. (*b*) Long time.

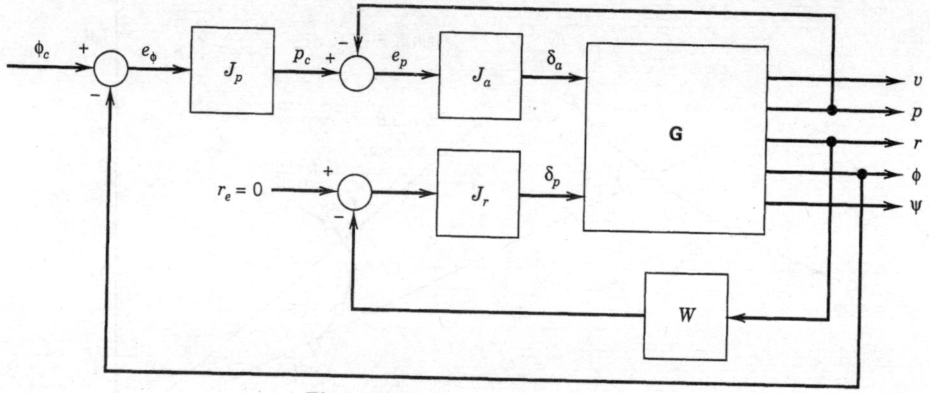

Figure 8.26 Roll control system.

When we substitute for J_a and J_p the differential equation that results is

$$\dot{\delta}_a = -\frac{K_aK_p}{\tau_a}\phi - \frac{K_a}{\tau_a}p - \frac{1}{\tau_a}\delta_a + \frac{K_aK_p}{\tau_a}\phi_c \tag{8.8,8}$$

Equations (8.8,5) and (8.8,8) are the additional equations required to augment the basic system (8.8,1) to accommodate the addition of the two control angles as dependent variables. However, a little more manipulation is needed of (8.8,5). To put it in first-order form, we define the new variable

$$y = \dot{\delta}_r \tag{8.8,9}$$

and to put it in canonical form, we must eliminate $\dot{r}$. This we do by using the third component equation in (8.8,1). When these steps have been taken the system can be assembled into the matrix equation

$$\dot{\mathbf{z}} = \mathbf{Pz} + \mathbf{Q}\phi_c \qquad (a)$$

where

$$\mathbf{z} = [v \; p \; r \; \phi \; \delta_a \; \delta_r \; y]^{\mathrm{T}} \qquad (b) \qquad (8.8,10)$$

The matrices **P** and **Q** are:

$$\mathbf{P} = \begin{bmatrix} a_{11} & a_{12} & a_{13} & g & b_{11} & b_{12} & 0 \\ a_{21} & a_{22} & a_{23} & 0 & b_{21} & b_{22} & 0 \\ a_{31} & a_{32} & a_{33} & 0 & b_{31} & b_{32} & 0 \\ 0 & 1 & 0 & 0 & 0 & 0 & 0 \\ 0 & -\dfrac{K_a}{\tau_a} & 0 & -\dfrac{K_aK_p}{\tau_a} & -\dfrac{1}{\tau_a} & 0 & 0 \\ 0 & 0 & 0 & 0 & 0 & 0 & 1 \\ \dfrac{K_r}{\tau_r}a_{31} & \dfrac{K_r}{\tau_r}a_{32} & \dfrac{K_r}{\tau_r}a_{33} & 0 & \dfrac{K_r}{\tau_r}b_{31} & \left(\dfrac{K_r}{\tau_r}b_{32} - \dfrac{1}{\tau_r\tau_{w0}}\right) & -\left(\dfrac{1}{\tau_r} + \dfrac{1}{\tau_{w0}}\right) \end{bmatrix}$$

$$\mathbf{Q} = [0 \; 0 \; 0 \; 0 \; (K_aK_p/\tau_a) \; 0 \; 0]^{\mathrm{T}} \tag{8.8,12}$$

Equation (8.8,10) was solved by numerical integration for two cases, with the results shown on Figs. 8.27 and 8.28. The various gains and time constants used were se-

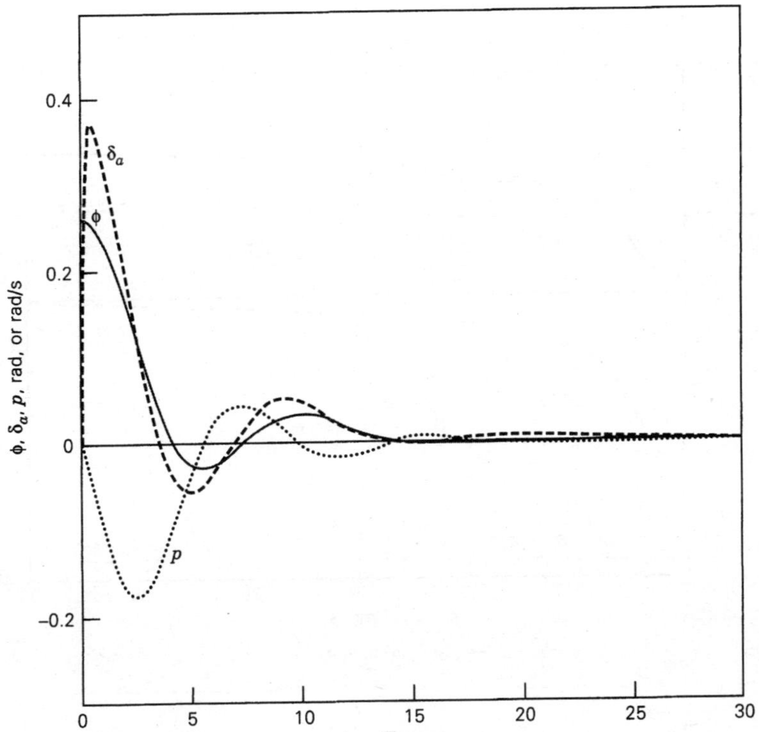

(*a*) ϕ, p, and δ_a

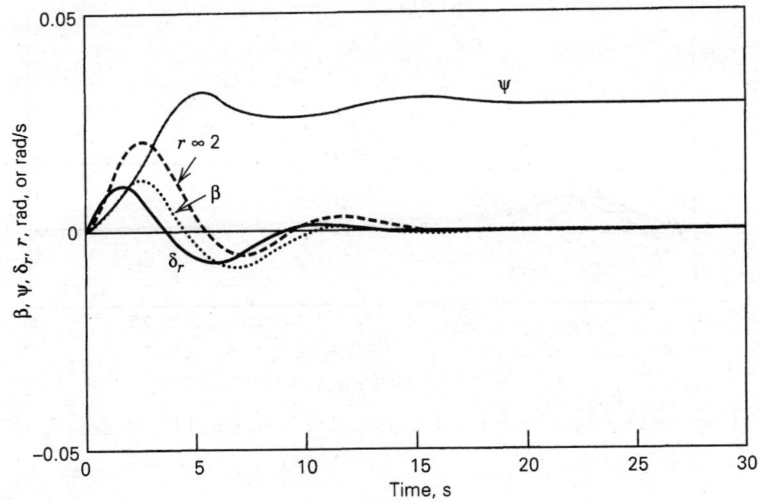

(*b*) β, r, ψ, δ_r

Figure 8.27 Response of roll controller to initial ϕ of 0.262 rad (15°). (*a*) ϕ, p, and δ_a. (*b*) β, r, ψ, δ_r.

lected somewhat arbitrarily, as follows:

$$K_p = 1.5 \;\; K_a = -1.0 \;\; K_r = -1.6; \; \tau_a = .15 \;\; \tau_r = .30 \;\; \tau_{w0} = 4.0$$

On the first of these figures, response to an initial bank error, we see that all the state variables experience a reasonably well damped oscillatory decay, and that the maxi-

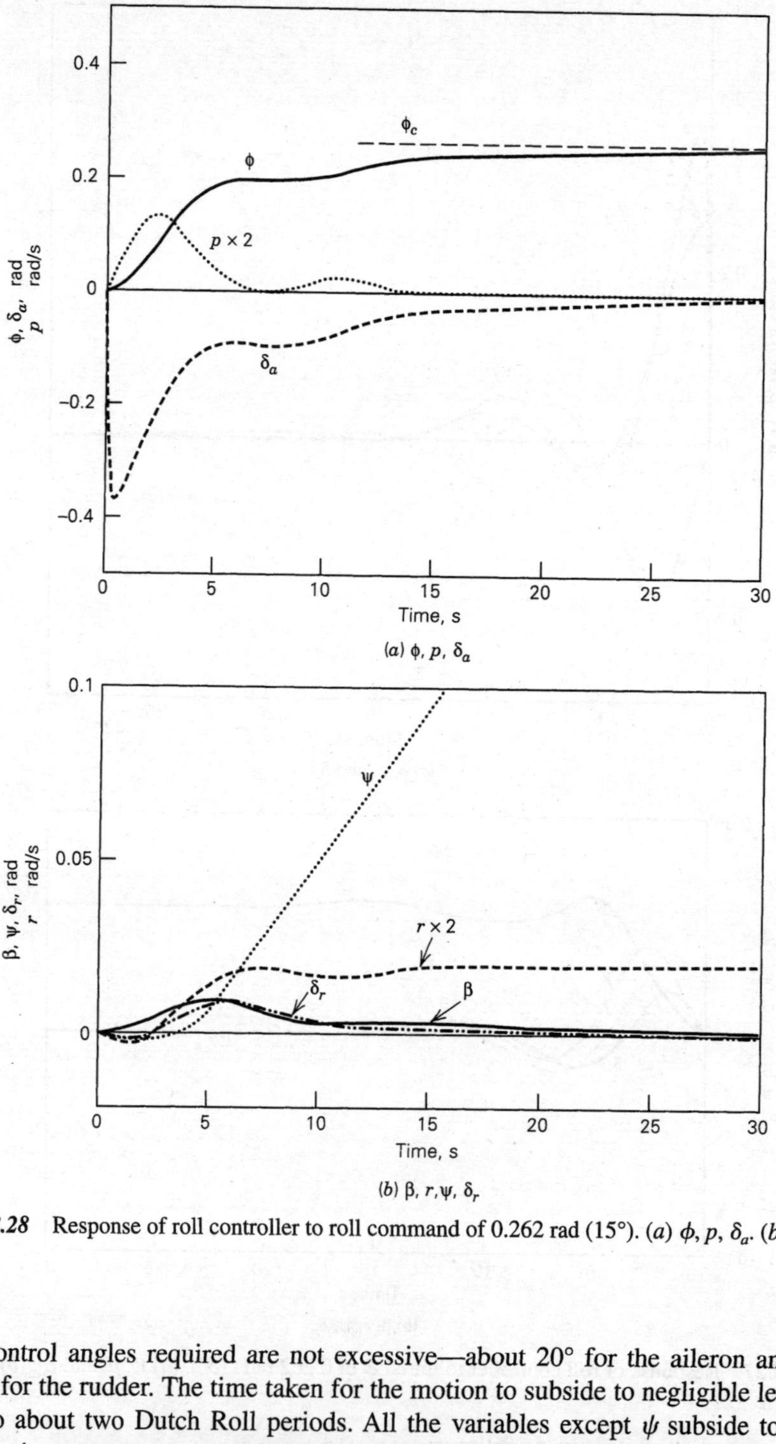

Figure 8.28 Response of roll controller to roll command of 0.262 rad (15°). (*a*) ϕ, p, δ_a. (*b*) β, r, ψ, δ_r.

mum control angles required are not excessive—about 20° for the aileron and less than 1° for the rudder. The time taken for the motion to subside to negligible levels is equal to about two Dutch Roll periods. All the variables except ψ subside to zero, whereas ψ asymptotes to a new steady state. When level flight is reestablished, the airplane has changed its heading by about 1.8°.

The second figure shows the response to a 15° bank command. The new steady state is approached with a damped oscillation that takes about 15 s to decay. The steady state is clearly a turn to the right, in which r has a constant value and ψ is increasingly linearly. All the other variables, including the two control angles, are very small. It is especially interesting that the sideslip angle is almost zero. Clearly this controller has the capability to provide the bank angle needed for a coordinated turn. (The angle of attack and lift would of course have to be increased.)

8.9 Gust Alleviation

For the final example in this chapter, we turn to a study of the application of automatic controls to reduce the response of an airplane to atmospheric turbulence (Byrne, 1983). This is obviously a useful goal for many flight situations, the benefits including increased passenger comfort, reduction of pilot workload, and possibly reductions in structural loading and fatigue, and in fuel consumption. The case reported here is for a STOL airplane, which is especially vulnerable to turbulence, since the relatively low operating speed makes it more responsive to turbulence, and because its duty cycle requires it to spend relatively more time at low altitudes where turbulence is more intense.

The numerical data used in the study were supplied by the de Havilland Aircraft Co., and although it does not apply to any particular airplane, it is representative of the class.

In this situation, where random turbulence produces random forces and moments on the airplane, which in turn result in random motion, the methods of analysis we have used in the foregoing examples, being essentially deterministic, are not applicable. Random processes have to be described by statistical functions. Let $f(t)$ represent such a random function. Two of the key statistical properties that characterize it are the *spectrum function,* derived from a Fourier analysis of $f(t)$ and the closely related *correlation function* (Etkin, 1972). The spectrum function or *spectral density,* as it is frequently called, is denoted $\Phi_{ff}(\omega)$. The area under the $\Phi_{ff}(\omega)$ curve that is contained between the two frequencies ω_1 and ω_2 is equal to the contribution to $\frac{1}{2}\overline{f^2}$ (where $\overline{f^2}$ is the mean-squared-value of f)[3] that comes from all the frequencies in the band $\omega_1 \Rightarrow \omega_2$ that are contained in the Fourier representation of f.

A BASIC THEOREM

When a system with transfer function $G(s)$ is subjected to an input with spectrum function $\Phi_{ii}(\omega)$ the spectral density $\Phi_{rr}(\omega)$ of the response $r(t)$ is given by

$$\Phi_{rr}(\omega) = \Phi_{ii}(\omega)|G(i\omega)|^2 \tag{8.9,1}$$

where $G(i\omega)$ is the frequency response function defined in Sec. 7.5. There is a generalization of (8.9,1) available for multiple inputs (Etkin, 1972, p. 94). Figure 8.29 shows the relationships expressed in (8.9,1) for a second-order system of moderate damping.

In the case at hand the motion studied is the lateral motion, and the forces needed are Y, L, and N. The gust vector $\mathbf{g}$ (Fig. 8.1) that is the source of these forces has elements that represent aspects of the motion of the atmosphere. It has four components

$$\mathbf{g} = [v_g \ p_g \ r_{1g} \ r_{2g}]^T$$

[3]The factor $\frac{1}{2}$ comes from the use of *two-sided* spectra.

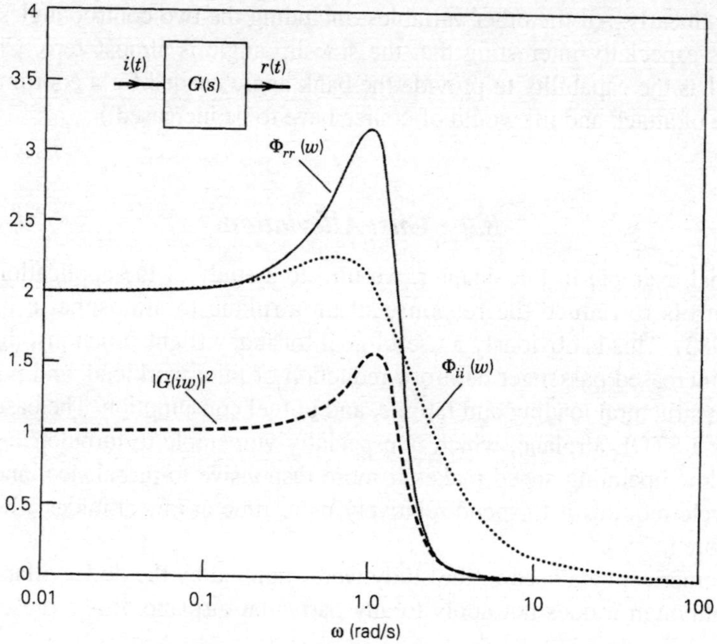

Figure 8.29 Response to random input.

v_g is the y-component of the turbulent velocity, p_g is the lateral gradient of the z-component, $p_g = \partial w_g/\partial y$, and $r_{1g} = -\partial u_g/\partial y$, $r_{2g} = \partial v_g/\partial x$. Each of these inputs is capable of producing aerodynamic actions on the airframe. v_g acts just like v in producing forces and moments like $Y_v v$ and $L_v v$; p_g acts like p and produces moments like $L_p p$ and $N_p p$; and r_{1g} and r_{2g} also produce forces and moments as a result of their effects on the relative wind at the wing and tail. The details of this theory can be found in Etkin (1981). The incremental force and moments can be expressed in terms of the gust components by the equation

$$\begin{bmatrix} \Delta Y \\ \Delta L \\ \Delta N \end{bmatrix}_g = \mathbf{Fg} \tag{8.9,1}$$

Where $\mathbf{F}$ is a (3×4) matrix of "gust derivatives." The basic differential equation of the system is then (8.8,1) with an added term to account for the turbulence, that is,

$$\dot{\mathbf{x}} = \mathbf{Ax} + \mathbf{Bc} + \mathbf{Tg} \tag{8.9,2}$$

$$\mathbf{Tg} = \begin{bmatrix} \dfrac{\Delta Y_g}{m} \\ \dfrac{\Delta L_g}{I'_x} + I'_{zx}\Delta N_g \\ I'_{zx}\Delta L_g + \dfrac{\Delta N_g}{I'_z} \\ 0 \end{bmatrix}$$

In order to alleviate the response of the airplane, it is necessary to be specific and choose an output that is to be minimized, and then to find what control actions will

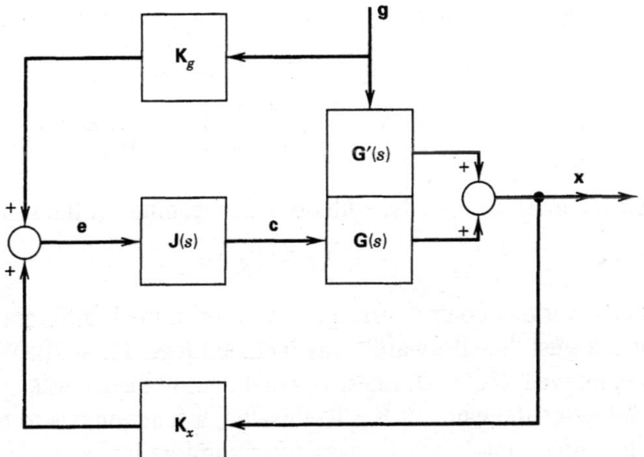

Figure 8.30 Gust alleviation system.

be successful in doing so. In the cited study, the output chosen to be minimized was a passenger comfort index (see below).

A block diagram of the system considered is given in Fig. 8.30. The state vector is the set $\mathbf{x} = [v \;\; p \;\; r \;\; \phi]^T$ and the control vector is $\mathbf{c} = [\delta_a \;\; \delta_r]^T$. The model includes full state feedback via the (2×4) gain matrix $\mathbf{K}_x$, control servo actuators described by the (2×2) matrix $\mathbf{J}$, and also includes the possibility of using measurements of the turbulent motion to influence the controls via the (2×4) gain matrix $\mathbf{K}_g$.

We now proceed to complete the differential equation of the system. We start with the servo actuator transfer function $\mathbf{J}(s)$, which is given, as in our previous examples, by first-order elements:

$$\mathbf{J} = \begin{bmatrix} \dfrac{1/\tau_a}{s + 1/\tau_a} & 0 \\ 0 & \dfrac{1/\tau_r}{s + 1/\tau_r} \end{bmatrix} \tag{8.9,3}$$

Equation (8.9,3) corresponds to the pair of differential equations

$$\begin{aligned} \dot{\delta}_a &= (e_a - \delta_a)/\tau_a \\ \dot{\delta}_r &= (e_r - \delta_r)/\tau_r \end{aligned} \tag{8.9,4}$$

or

$$\dot{\mathbf{c}} = -\mathbf{P}\mathbf{c} + \mathbf{P}\mathbf{e} \tag{8.9,5}$$

where

$$\mathbf{P} = \begin{bmatrix} \dfrac{1}{\tau_a} & 0 \\ 0 & \dfrac{1}{\tau_r} \end{bmatrix} \tag{8.9,6}$$

We see from Fig. 8.30 that $\mathbf{e} = \mathbf{K}_g\mathbf{g} + \mathbf{K}_x\mathbf{x}$, so that (8.9,5) becomes

$$\dot{\mathbf{c}} = \mathbf{P}\mathbf{K}_x\mathbf{x} - \mathbf{P}\mathbf{c} + \mathbf{P}\mathbf{K}_g\mathbf{g} \tag{8.9,7}$$

We can now combine (8.9,2) and (8.9,7) into the augmented differential equation of the system:

$$\begin{bmatrix} \dot{\mathbf{x}} \\ \dot{\mathbf{c}} \end{bmatrix} = \begin{bmatrix} \mathbf{A} & \mathbf{B} \\ \mathbf{PK}_x & -\mathbf{P} \end{bmatrix} \begin{bmatrix} \mathbf{x} \\ \mathbf{c} \end{bmatrix} + \begin{bmatrix} \mathbf{T} \\ \mathbf{PK}_g \end{bmatrix} \mathbf{g} \tag{8.9,8}$$

This can be written more compactly, with obvious meanings of the symbols, as

$$\dot{\mathbf{z}} = \hat{\mathbf{A}}\mathbf{z} + \hat{\mathbf{T}}\mathbf{g} \tag{8.9,9}$$

In the cited study, various control strategies were examined, differentiated primarily by whether or not gust "feedforward" was included (i.e., $\mathbf{K}_g \neq 0$). When only state feedback was employed, ($\mathbf{K}_g = 0$) linear optimal control theory was used to ascertain the optimum values of the gains in $\mathbf{K}_x$. To this end, a function has to be chosen to be minimized. The choice made was a passenger comfort index made up of a linear combination of sideways seat acceleration along with angular accelerations $\dot{p}$ and $\dot{r}$. The seat acceleration depends on how far the seat is from the CG, so an average was used for this quantity. The optimum that resulted entailed the feedback of each of the four state variables to each of the two controls, a very complicated control system! However, it was found that there was very little difference in performance between

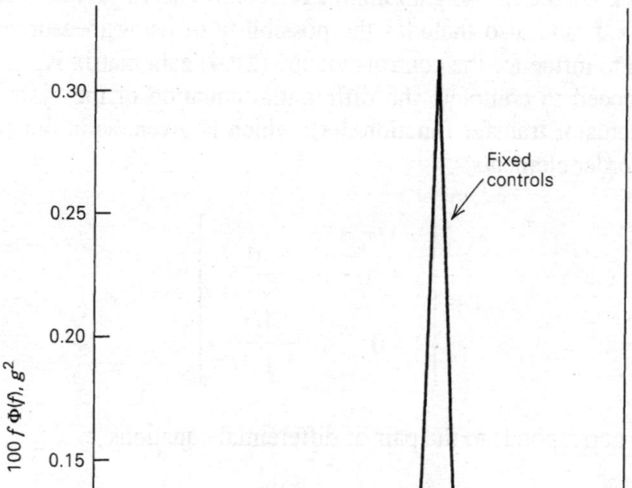

Figure 8.31 Lateral acceleration spectra. Rearmost seat, with yaw damper.

this optimum and a simple yaw damper. The result is shown on Fig. 8.31, in the form of the spectral density of sideways acceleration of the rearmost seat. This form of plot, $f\Phi(f)$ vs. $\log f$, is commonly used. The area under any portion of this curve is also equal to the mean-square contribution of that frequency band, just as with $\Phi(f)$ vs. f. Results are shown for three cases—the basic airframe with fixed controls, a conventional autopilot, and the selected yaw damper. Very substantial reduction of response to turbulence has clearly been achieved with a relatively simple control strategy.

An alternative to conventional linear optimal control theory was found to be better for the case when gust measurement is assumed to be possible. It stems from a theorem of Rynaski et al. (1979). It is seen from (8.9,2) that if one could make $\mathbf{Bc} + \mathbf{Tg} = \mathbf{0}$ then one would have completely canceled the gust input with control action, and the airplane would fly as if it were in still air! This equality presupposes that the control is given by

$$\mathbf{c} = -\mathbf{B}^{-1}\mathbf{T}\mathbf{g}$$

that is, that $\mathbf{B}$ has an inverse. This would require $\mathbf{B}$ to be a square matrix [i.e., to be (4×4)], which in turn would require that the airplane have two more independent

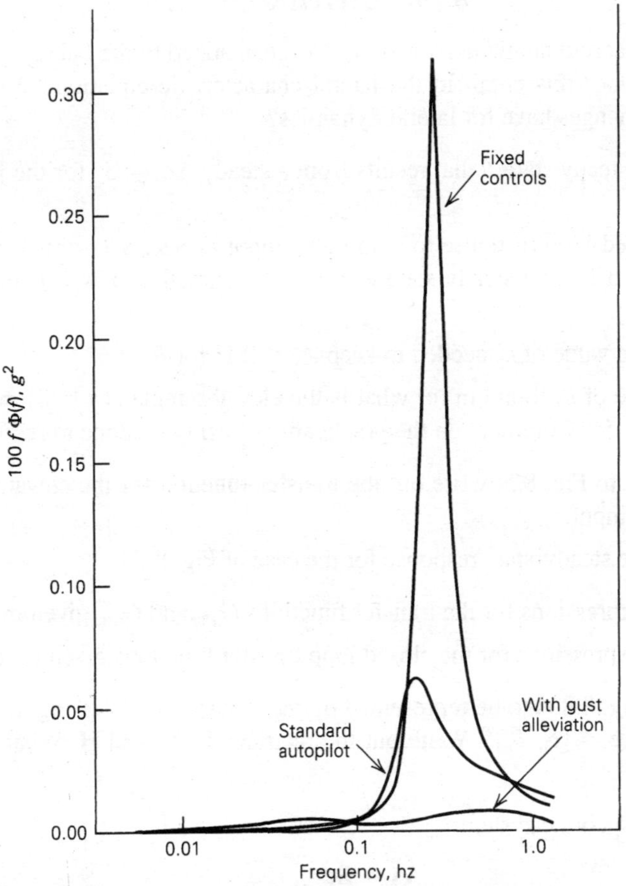

Figure 8.32 Lateral acceleration spectra. Rearmost seat, with yaw damper and gust feedforward.

(and sufficiently powerful) controls than it actually has. Although this is not beyond the realm of imagination, it was not a feasible option in the present study. However, there is available the "generalized inverse," which provides in a certain sense the best approximation to the desired control law. The generalized inverse of **B** is the inverse of the (2×2) matrix $\mathbf{B}^T\mathbf{B}$. This leads to the control law

$$\mathbf{c} = -(\mathbf{B}^T\mathbf{B})^{-1}\mathbf{B}^T\mathbf{T}\mathbf{g}$$

(The second $\mathbf{B}^T$ is needed to yield a (2×1) matrix on the right-hand side). This law still requires, however, that all four components of **g** be sensed in order to compute **c**. Sensing all components of **g** is not impossible, indeed it may not even be impractical. However, a good result can be obtained with a subset of **g** consisting only of v_g and r_{2g}, both of which can be measured with an aerodynamic yawmeter, a sideslip vane or other form of sensor. The end result of combining gust sensing in this way with the yaw damper, with the gust sensor placed an optimum distance forward of the CG, is shown in Fig. 8.32. It is clear that this control strategy has been successful in achieving a very large reduction in seat acceleration.

8.10 Exercises

8.1 Assume that an aerodynamic derivative L_ϕ has been added to the lateral force system. What changes does this entail in the lateral characteristic equation? What implications do these changes have for lateral dynamics?

8.2 (a) What is the steady state θ that results from a steady $\Delta\delta_e = 5°$ for the jet transport of Sec. 8.3?

(b) For the closed-loop response to a unit step input in Sec. 8.3, with $J = k_2$, derive an expression for the steady-state error e_{ss} as a function of k_2. (Hint: start with (8.3,1)).

(c) Calculate the value of k_2 needed to keep $e_{ss} < 0.1°$ for $\theta_c = 5°$.

(d) For the value of k_2 found in (c) what is the elevator angle at $t = 0^+$ when θ_c is a step input of 5°? Comment on the practicality of using k_2 alone to reduce e_{ss}.

8.3 (a) With respect to Fig. 8.5, write out the transfer function for the elevator angle response to θ_c input.

(b) Calculate the steady-state response for the case of Fig. 8.7*c*.

8.4 (a) Derive the expressions for the transfer functions $G_{u\delta_p}$ and $G_{\gamma\delta_p}$ given in (8.4,2).

(b) Derive the expressions for the closed loop transfer functions given in (8.4,3).

8.5 The system of Fig. 8.8 is to be represented by the block diagram of Fig. 8.1 with $\mathbf{x} = [u \;\; w \;\; q \;\; \theta]^T$ and $\mathbf{c} = [\delta_e \;\; \delta_p]^T$. Write out the matrices **D**, **E**, and **H**. What are the dimensions of **J**?

8.6 Given the 2×2 algebraic system

$$\mathbf{A}\bar{\mathbf{x}} = \mathbf{B}\bar{\mathbf{c}}$$

for which

$$\frac{\bar{x}_i}{\bar{c}_j} = \frac{N_{ij}(s)}{D(s)}$$

where $D(s)$ is det **A**. Prove that

$$\det \mathbf{N} = \det \mathbf{A} \det \mathbf{B}$$

that is, that $$(N_{11}N_{22} - N_{12}N_{21}) = D(B_{11}B_{22} - B_{12}B_{21})$$

Relate this result to the elimination of f in (8.4,6).

8.7 Modify the system model of the altitude-hold autopilot (Fig. 8.17) to include a rate term in the block indicated by the constant k. Show explicitly what changes result in (8.5,13).

8.8 Modify the analysis of Sec. 8.5 that leads to (8.5,13) to include climbing or gliding flight, that is, $\theta_0 \neq 0$.

8.9 (a) Prove that in the yaw damper with washout, the steady-state yaw rate for a step in r_c is independent of the washout time constant and is given by

$$r_{ss} = r_c J(0) G_{r\delta_r}(0)$$

(b) Prove that if the washout filter is in the forward path, instead of the feedback path, then

$$r_{ss} = 0$$

regardless of the washout time constant.

8.10 A positive gain K in the $p \rightarrow \delta_r$ loop (Figs. 8.19 and 8.20*e*) implies right rudder response when the right wing dips down. On the face of it this would seem to be destabilizing. Explain, using the modal diagram Fig. 6.15 and Table 6.9, why the damping of the Dutch Roll mode is actually increased by this feedback. (Hint: r produced by a yaw damper is normally used to increase Dutch Roll damping.)

8.11 Add an outer loop to the system of Fig. 8.26 to control the heading angle ψ. Draw a new block diagram and write out the augmented system differential equation. (Hint: design the loop to command a bank angle proportional to heading error.)

8.12 Write out the full system of equations corresponding to (8.9,8).

8.13 (This is a miniresearch project). The relative locations of the pole P and the zero Z on Fig. 8.20*f* are relevant to the design of a yaw damper in that they determine the characteristic structure of the root locus. Discover, by any means, what changes to the jet transport's aerodynamics would move its zero to the right of the pole. Two suggestions: (1) Systematically perturb each of the stability derivatives and note the sensitivity of the locations of P and Z to each, and (2) use approximate transfer functions to get analytical approximations for P and Z. When you have found some changes to the stability derivatives that would produce the desired result, discuss the design changes that would be needed to achieve the altered derivatives.

8.11 Additional Symbols Introduced in Chapter 8

$\mathbf{c}$	control vector
$D(s)$	denominator of transfer function
D_V	$\partial D/\partial V$
$\mathbf{e}$	error vector
$\mathbf{F}$	matrix of gust derivatives
$\mathbf{g}$	gust vector
$\mathbf{G}'$	transfer function $\mathbf{g} \Rightarrow \mathbf{x}$
$\mathring{G}$	closed-loop transfer function
$\mathbf{H}$	feedback matrix
$\mathbf{J}$	transfer function $\mathbf{e} \Rightarrow \mathbf{c}$
K	gain constant
$N(s)$	numerator of transfer function
p_g	gust gradient $\partial w_g/\partial y$
r_{1g}	gust gradient $-\partial u_g/\partial y$
r_{2g}	gust gradient $\partial v_g/\partial x$
$\mathbf{r}$	reference vector
$\mathbf{T}$	gust input matrix
T_V	$\partial T/\partial V$
V^*	speed at which $T_V = D_V$
u_g, v_g, w_g	components of gust velocity
$\mathbf{y}$	output vector
$\mathbf{z}$	feedback vector
$\Phi(\omega)$	spectrum function
τ	time constant

APPENDIX A

Analytical Tools

A.1 Linear Algebra

In this book no formal distinction is made between vectors and matrices, the former being simply column matrices, as is common in treatments of linear algebra. In particular the familiar vectors of mechanics, such as force and velocity, are simply three-component column matrices. We use boldface letters for both matrices and vectors, for example, $\mathbf{A} = [a_{ij}]$ and $\mathbf{v} = [v_i]$. The corresponding lowercase letter defines the magnitude (or *norm*) of the vector. The transpose and inverse are denoted as usual by superscripts, for example, $\mathbf{A}^T$ and $\mathbf{A}^{-1}$. When appropriate to the context, a subscript is used to denote the frame of reference for a physical vector, for example, $\mathbf{V}_E = [u\ v\ w]^T$ denotes a vector whose components in frame F_E are (u, v, w). The three-component vectors of physics have the following properties:

Scalar product

$$c = \mathbf{a} \cdot \mathbf{b} = \mathbf{a}^T\mathbf{b} = a_1b_1 + a_2b_2 + a_3b_3 \tag{A.1,1}$$

c is a scalar, with magnitude $ab \cos \theta$, where θ is the angle between **a** and **b**

Vector product

$$\mathbf{c} = \mathbf{a} \times \mathbf{b} = \tilde{\mathbf{a}}\mathbf{b} = \begin{bmatrix} a_2b_3 - a_3b_2 \\ a_3b_1 - a_1b_3 \\ a_1b_2 - a_2b_1 \end{bmatrix} \tag{A.1,2}$$

where

$$\tilde{\mathbf{a}} = \begin{bmatrix} 0 & -a_3 & a_2 \\ a_3 & 0 & -a_1 \\ -a_2 & a_1 & 0 \end{bmatrix} \tag{A.1,3}$$

c is a vector perpendicular to the plane of **a** and **b**, with direction following the right-hand rule for the sequence **a, b, c** and has the magnitude $ab \sin \theta$, where θ is the angle ($<180°$) between **a** and **b**

Unit vectors

The basis unit vectors are **i,j,k** such that

$$[\mathbf{i}\ \ \mathbf{j}\ \ \mathbf{k}] = \begin{bmatrix} 1 & 0 & 0 \\ 0 & 1 & 0 \\ 0 & 0 & 1 \end{bmatrix} = [\delta_{ij}] = \mathbf{I} \tag{A.1,4}$$

Where δ_{ij} is the Kronecker delta.

Square matrices have the following properties:

Minor Determinant, Cofactor

The minor determinant m_{ij} of a matrix $[a_{ij}]$ is the determinant of the reduced matrix that remains after the ith row and jth column of $[a_{ij}]$ have been deleted.

The cofactor is $c_{ij} = m_{ij}(-1)^{i+j}$

Adjoint

The adjoint of a matrix is the transpose of the matrix of cofactors,

$$\text{adj } [a_{ij}] = [c_{ij}]^T \tag{A.1,5}$$

Inverse

Provided that $\det \mathbf{A} \neq 0$ the inverse is given by

$$\mathbf{A}^{-1} = \frac{\text{adj } \mathbf{A}}{\det \mathbf{A}} \tag{A.1,6}$$

A.2 The Laplace Transform

Let $x(t)$ be a known function of t for values of $t > 0$. Then the Laplace transform of $x(t)$ is defined by the integral relation

$$\bar{x}(s) = \mathcal{L}[x(t)] = \int_0^\infty e^{-st} x(t)\, dt \tag{A.2,1}$$

The integral is convergent only for certain functions $x(t)$ and for certain values of s. The Laplace transform is defined only when the integral converges. This restriction is weak and excludes few cases of interest to engineers. It should be noted that the original function $x(t)$ is converted into a new function of the transform variable s by the transformation. The two notations for the transform shown on the left-hand side of (A.2,1) will be used interchangeably. The transforms of some functions that commonly occur in problems of linear systems are listed in Table A.1.

TRANSFORMS OF DERIVATIVES

$$\mathcal{L}\left[\frac{dx}{dt}\right] = \int_0^\infty e^{-st} \frac{dx}{dt}\, dt$$

$$= \int_{t=0}^\infty e^{-st}\, dx = xe^{-st}\Big]_{t=0}^\infty + s\int_0^\infty xe^{-st}\, dt$$

When $xe^{-st} \to 0$ as $t \to \infty$ (only this case is considered), then

$$\mathcal{L}\left[\frac{dx}{dt}\right] = -x(0) + s\bar{x}(s) \tag{A.2,2}$$

Table A.1
Laplace Transforms

	$x(t)$	$\bar{x}(s)$
1	$\delta(t)$	1
2	$\delta(t-T)$	e^{-sT}
3	1 or $\mathbf{1}(t)$	$\dfrac{1}{s}$
4	$\mathbf{1}(t-T)$	$\dfrac{e^{-sT}}{s}$
5	$f(t-T)\mathbf{1}(t-T)$	$e^{-sT}\,\mathcal{L}[f(t)]$
6	t	$\dfrac{1}{s^2}$
7	$\dfrac{t^{n-1}}{(n-1)!}$	$\dfrac{1}{s^n}$
8	e^{at}	$\dfrac{1}{s-a}$
9	$\sin at$	$\dfrac{a}{s^2+a^2}$
10	$\cos at$	$\dfrac{s}{s^2+a^2}$
11	te^{at}	$\dfrac{1}{(s-a)^2}$
12	$\dfrac{t^{n-1}}{(n-1)!}\,e^{at}$	$\dfrac{1}{(s-a)^n}$
13	$e^{at}\sin bt$	$\dfrac{b}{(s-a)^2+b^2}$
14	$e^{at}\cos bt$	$\dfrac{s-a}{(s-a)^2+b^2}$
15	$\sinh at$	$\dfrac{a}{s^2-a^2}$
16	$\cosh at$	$\dfrac{s}{s^2-a^2}$
17	$e^{at}\sinh bt$	$\dfrac{b}{(s-a)^2-b^2}$
18	$e^{at}\cosh bt$	$\dfrac{s-a}{(s-a)^2-b^2}$
19	$\dot{x}(t)$	$s\bar{x}(s)-x(0)$
20	$e^{at}x(t)$	$\bar{x}(s-a)$

where $x(0)$ is the value of $x(t)$ when $t = 0$.[1] The process may be repeated to find the higher derivatives by replacing $x(t)$ in (A.2,2) by $\dot{x}(t)$, and so on. The result is

$$\mathcal{L}\left[\frac{d^n x}{dt^n}\right] = -\frac{d^{n-1}x}{dt^{n-1}}(0) - s\frac{d^{n-2}x}{dt^{n-2}}(0) - \cdots - s^{n-1}x(0) + s^n\bar{x}(s) \quad \text{(A.2,3)}$$

TRANSFORM OF AN INTEGRAL

Let the integral be

$$y = \int x(t)\, dt$$

and let it be required to find $\bar{y}(s)$. By differentiating with respect to t, we get

$$\frac{dy}{dt} = x(t)$$

thus

$$\bar{x}(s) = \mathcal{L}\left[\frac{dy}{dt}\right] = s\bar{y}(s) - y(0)$$

and

$$\bar{y}(s) = \frac{1}{s}\bar{x}(s) + \frac{1}{s}y(0) \quad \text{(A.2,4)}$$

METHODS FOR THE INVERSE TRANSFORMATION

The Use of Tables of Transforms

Extensive tables of transforms (like Table A.1) have been published that are useful in carrying out the inverse process. When the transform involved can be found in the tables, the function $x(t)$ is obtained directly.

The Method of Partial Fractions

In some cases it is convenient to expand the transform $\bar{x}(s)$ in partial fractions, so that the elements are all simple ones like those in Table A.1. The function $x(t)$ can then be obtained simply from the table. This procedure is illustrated with an example. Let the second-order system of Sec. 7.3 be initially quiescent, that is, $x(0) = 0$, and $\dot{x}(0) = 0$, and let it be acted upon by a constant unit force applied at time $t = 0$. Then $f(t) = 1$, and $\bar{f}(s) = 1/s$ (see Table A.1). Then (see (7.4,1))

$$\bar{x}(s) = \frac{1}{s(s^2 + 2\zeta\omega_n s + \omega_n^2)} \quad \text{(A.2,5)}$$

[1]To avoid ambiguity when dealing with step functions, $t = 0$ should always be interpreted as $t = 0^+$.

Let us assume that the system is aperiodic; that is, that $\zeta > 1$. Then the roots of the characteristic equation are real and equal to

$$\lambda_{1,2} = n \pm \omega' \tag{A.2,6}$$

where

$$n = -\zeta\omega_n$$
$$\omega' = \omega_n(\zeta^2 - 1)^{1/2}$$

The denominator of (A.2,5) can be written in factored form so that

$$\bar{x}(s) = \frac{1}{s(s-\lambda_1)(s-\lambda_2)} \tag{A.2,7}$$

Now let (A.2,7) be expanded in partial fractions,

$$\bar{x}(s) = \frac{A}{s} + \frac{B}{(s-\lambda_1)} + \frac{C}{(s-\lambda_2)} \tag{A.2,8}$$

By the usual method of equating (A.2,7) and (A.2,8), we find

$$A = \frac{1}{\lambda_1\lambda_2}$$
$$B = \frac{1}{\lambda_1(\lambda_1 - \lambda_2)}$$
$$C = \frac{1}{\lambda_2(\lambda_2 - \lambda_1)}$$

Therefore

$$\bar{x}(s) = \frac{1/\lambda_1\lambda_2}{s} + \frac{1/\lambda_1(\lambda_1 - \lambda_2)}{s - \lambda_1} + \frac{1/\lambda_2(\lambda_2 - \lambda_1)}{s - \lambda_2}$$

By comparing these three terms with items 3 and 8 in Table A.1, we may write down the solution immediately as

$$\begin{aligned} x(t) &= \frac{1}{\lambda_1\lambda_2} + \frac{1}{\lambda_1(\lambda_1 - \lambda_2)} e^{\lambda_1 t} - \frac{1}{\lambda_2(\lambda_1 - \lambda_2)} e^{\lambda_2 t} \\ &= \frac{1}{\omega_n^2}\left[1 + \frac{n - \omega'}{2\omega'} e^{(n+\omega')t} - \frac{n + \omega'}{2\omega'} e^{(n-\omega')t}\right] \end{aligned} \tag{A.2,9}$$

Heaviside Expansion Theorem

When the transform is a ratio of two polynomials in s, the method of partial fractions can be generalized. Let

$$\bar{x}(s) = \frac{N(s)}{D(s)}$$

where $N(s)$ and $D(s)$ are polynomials and the degree of $D(s)$ is higher than that of $N(s)$. Let the roots of $D(s) = 0$ be a_r, so that

$$D(s) = (s - a_1)(s - a_2) \cdots (s - a_n)$$

Then the inverse of the transform is

$$x(t) = \sum_{r=1}^{n} \left\{ \frac{(s - a_r)N(s)}{D(s)} \right\}_{s=a_r} e^{a_r t} \tag{A.2,10}$$

The effect of the factor $(s - a_r)$ in the numerator is to cancel out the same factor of the denominator. The substitution $s = a_r$ is then made in the reduced expression.[2]

In applying this theorem to (A.2,7), we have the three roots $a_1 = 0$, $a_2 = \lambda_1$, $a_3 = \lambda_2$, and $N(s) = 1$. With these roots, (A.2,9) follows immediately from (A.2,10).

The Inversion Theorem

The function $x(t)$ can be found formally from its transform $\bar{x}(s)$ by the application of the inversion theorem Jaeger (1949) and Carslaw and Jaeger (1947). It is given by the line integral

$$x(t) = \frac{1}{2\pi i} \lim_{\omega \to \infty} \int_{\gamma - i\omega}^{\gamma + i\omega} e^{st}\bar{x}(s)\, ds \tag{A.2,11}$$

where γ is a real number greater than the real part of all values of s for which $\bar{x}(s)$ diverges. That is, $s = \gamma$ is a straight line on the s plane lying parallel to the imaginary axis, and to the right of all the poles of $\bar{x}(s)$. This theorem can be used, employing the methods of contour integrals in the complex plane, to evaluate the inverse of the transform.

Extreme Value Theorems

Equation (A.2,2) may be rewritten as

$$-x(0) + s\bar{x}(s) = \int_0^{\infty} e^{-st}\dot{x}(t)\, dt$$

$$= \lim_{T \to \infty} \int_0^{T} e^{-st}\dot{x}(t)\, dt$$

We now take the limit $s \to 0$ while T is held constant, that is,

$$-x(0) + \lim_{s \to 0} s\bar{x}(s) = \lim_{T \to \infty} \int_0^{T} \lim_{s \to 0} e^{-st}\dot{x}(t)\, dt$$

$$= \lim_{T \to \infty} \int_0^{T} \dot{x}(t)\, dt = \lim_{T \to \infty} [x(T) - x(0)]$$

Hence

$$\lim_{s \to 0} s\bar{x}(s) = \lim_{T \to \infty} x(T) \tag{A.2,12}$$

This result, known as the *final value theorem,* provides a ready means for determining the asymptotic value of $x(t)$ for large times from the value of its Laplace transform.

[2]For the case of repeated roots, see Jaeger (1949).

In a similar way, by taking the limit $s \to \infty$ at constant T, the integral vanishes for all finite $\dot{x}(t)$ and we get the *initial value theorem.*

$$\lim_{s \to \infty} s\bar{x}(s) = x(0) \tag{A.2,13}$$

A.3 The Convolution Integral

The response of any linear system to any arbitrary input $f(t)$ can be obtained from integrals of the two basic response functions $h(t)$ and $A(t)$. $h(t)$ is the response to the unit impulse $\delta(t)$, and $A(t)$ is the response to the unit step $\mathbf{1}(t)$. The system is assumed to be initially quiescent. If not, the transient associated with nonzero initial conditions must be added to the following integrals. The response to $f(t)$ is then given by Duhamel's integral, or the convolution integral:

$$x(t) = \int_{\tau=0}^{t} h(t-\tau) f(\tau)\, d\tau \qquad (a)$$

$$x(t) = \int_{\tau=0}^{t} A(t-\tau) \dot{f}(\tau)\, d\tau \qquad (f(0)=0) \qquad (b)$$

(A.3,1)

When $f(0)$ is not zero, then there must be added to (A.3,1*b*) a term to allow for the initial step in $f(t)$; i.e.,

$$x(t) = f(0)A(t) + \int_{\tau=0}^{t} A(t-\tau)\dot{f}(\tau)\, d\tau \tag{A.3,2}$$

The physical significance of these integrals is brought out by considering them as the limits of the following sums

$$x(t) = \Sigma h(t-\tau) f(\tau)\, \Delta\tau \qquad (a)$$

$$x(t) = A(t)f(0) + \Sigma A(t-\tau)\dot{f}(\tau)\, \Delta\tau \qquad (b)$$

(A.3,3)

Typical terms of the summations are illustrated in Figs. A.1 and A.2. The summation forms are quite convenient for computation, especially when the interval $\Delta\tau$ is kept constant.

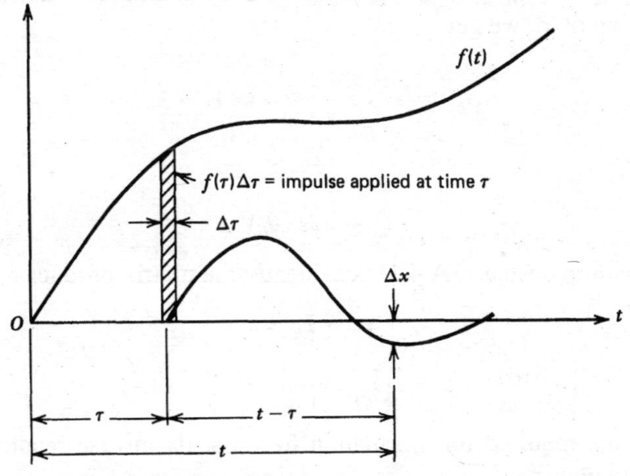

Figure A.1 Duhamel's integral, impulsive form. $\Delta x = h(t-\tau) f(\tau)\, \Delta\tau$ = response at time t to impulse at time τ.

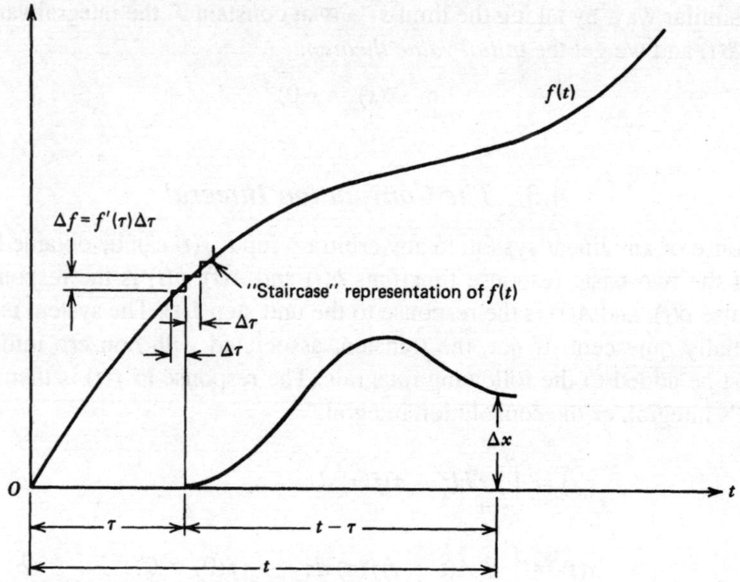

Figure A.2 Duhamel's integral, indicial form. Δf = step input applied at time τ, $\Delta x = A(t-\tau)\Delta f$ = response at time t to step input Δf.

A.4 Coordinate Transformations

TRANSFORMATION OF A VECTOR

Let **v** be a vector with the components

$$\mathbf{v}_a = \begin{bmatrix} v_{a_1} \\ v_{a_2} \\ v_{a_3} \end{bmatrix} \text{ in } F_a \quad \text{and} \quad \mathbf{v}_b = \begin{bmatrix} v_{b_1} \\ v_{b_2} \\ v_{b_3} \end{bmatrix} \text{ in } F_b$$

The component of v_{a_1} in the direction of x_{b_i} is $v_{a_1} \cos(\theta_{i1})$ where θ_{i1} denotes the angle between $O_b x_{b_i}$ and $O_a x_{a_1}$ (see Fig. A.3). Thus by adding the three components of v_{a_j} in the direction of x_{b_i} we get

$$v_{b_i} = \sum_{j=1}^{3} l_{ij} v_{a_j} \qquad i = 1 \cdots 3 \tag{A.4,1}$$

where

$$l_{ij} = \cos(\theta_{ij}) \tag{A.4,2}$$

are the nine direction cosines. (A.4,1) is evidently the matrix product

$$\mathbf{v}_b = \mathbf{L}_{ba}\mathbf{v}_a \qquad (a)$$

where (A.4,3)

$$\mathbf{L}_{ba} = [l_{ij}] \qquad (b)$$

and constitutes the required transformation formula. Its inverse readily reverses the transformation to give

$$\mathbf{v}_a = \mathbf{L}_{ba}^{-1}\mathbf{v}_b = \mathbf{L}_{ab}\mathbf{v}_b$$

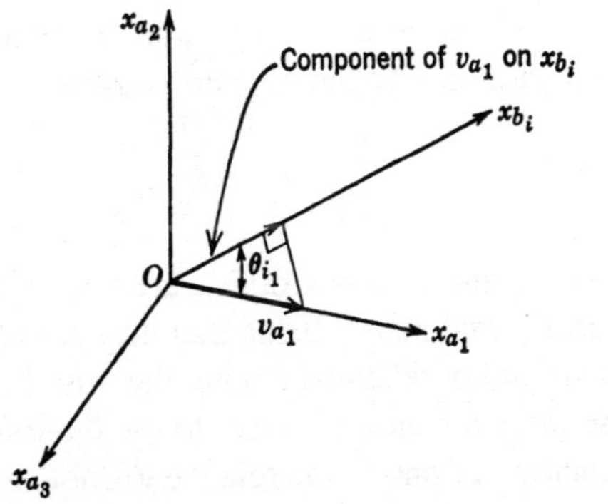

Figure A.3 Component of vector.

where

$$\mathbf{L}_{ab} \equiv \mathbf{L}_{ba}^{-1} \tag{A.4,4}$$

When a vector is successively transformed through several frames of reference, for example, F_a, F_b, F_c . . . then

$$\mathbf{v}_b = \mathbf{L}_{ba}\mathbf{v}_a$$

and

$$\mathbf{v}_c = \mathbf{L}_{cb}\mathbf{v}_b = \mathbf{L}_{cb}(\mathbf{L}_{ba}\mathbf{v}_a)$$

Since also $\mathbf{v}_c = \mathbf{L}_{ca}\mathbf{v}_a$, then it follows that

$$\mathbf{L}_{ca} = \mathbf{L}_{cb}\mathbf{L}_{ba}$$

and similarly for additional transformations.

The sequence of subscripts in the preceding expression should be noted, as it provides a convenient mnemonic for remembering these relations.

PROPERTIES OF THE L MATRIX

Since $\mathbf{v}_a$ and $\mathbf{v}_b$ are physically the same vector $\mathbf{v}$, the magnitude of $\mathbf{v}_a$ must be the same as that of $\mathbf{v}_b$, that is, v^2 is an invariant of the transformation. From (A.4,3) this requires

$$v^2 = \mathbf{v}_b^T\mathbf{v}_b = \mathbf{v}_a^T\mathbf{L}_{ba}^T\mathbf{L}_{ba}\mathbf{v}_a = \mathbf{v}_a^T\mathbf{v}_a \tag{A.4,5}$$

It follows from the last equality of (A.4,5) that

$$\mathbf{L}_{ba}^T\mathbf{L}_{ba} = \mathbf{I} \tag{A.4,6}$$

Equation (A.4,6) is known as the *orthogonality condition* on $\mathbf{L}_{ba}$. From (A.4,6) it follows that

$$|\mathbf{L}_{ba}|^2 = 1$$

and hence that $|\mathbf{L}_{ba}|$ is never zero and the inverse of $\mathbf{L}_{ba}$ always exists. In view of (A.4,6) we have, of course, that

$$\mathbf{L}_{ba}^T = \mathbf{L}_{ba}^{-1} = \mathbf{L}_{ab} \tag{A.4,7}$$

that is the inverse and the transpose are the same. Equation (A.4,6) together with (A.4,3*b*) yields a set of conditions on the direction cosines,

$$\sum_{k=1}^{3} l_{ki} l_{kj} = \delta_{ij} \tag{A.4,8}$$

It follows from (A.4,8) that the columns of $\mathbf{L}_{ba}$ are vectors that form an orthogonal set (hence the name "orthogonal matrix") and that they are of unit length.

Since (A.4,8) is a set of six relations among the nine l_{ij}, then only three of them are independent. These three are an alternative to the three independent Euler angles for specifying the orientation of one frame relative to another.

THE L MATRIX IN TERMS OF ROTATION ANGLES

The transformations associated with single rotations about the three coordinate axes are now given. In each case F_a represents the initial frame, F_b the frame after rotation, and the notation for $\mathbf{L}$ identifies the axis and the angle of the rotation (see Fig. A.4). Thus in each case

$$\mathbf{v}_b = \mathbf{L}_i(X_i)\mathbf{v}_a \tag{A.4,9}$$

By inspection of the angles in Fig. A.4, the following matrices are readily verified.

$$\mathbf{L}_1(X_1) = \begin{bmatrix} 1 & 0 & 0 \\ 0 & \cos X_1 & \sin X_1 \\ 0 & -\sin X_1 & \cos X_1 \end{bmatrix}$$

$$\mathbf{L}_2(X_2) = \begin{bmatrix} \cos X_2 & 0 & -\sin X_2 \\ 0 & 1 & 0 \\ \sin X_2 & 0 & \cos X_2 \end{bmatrix} \tag{A.4,10}$$

$$\mathbf{L}_3(X_3) = \begin{bmatrix} \cos X_3 & \sin X_3 & 0 \\ -\sin X_3 & \cos X_3 & 0 \\ 0 & 0 & 1 \end{bmatrix}$$

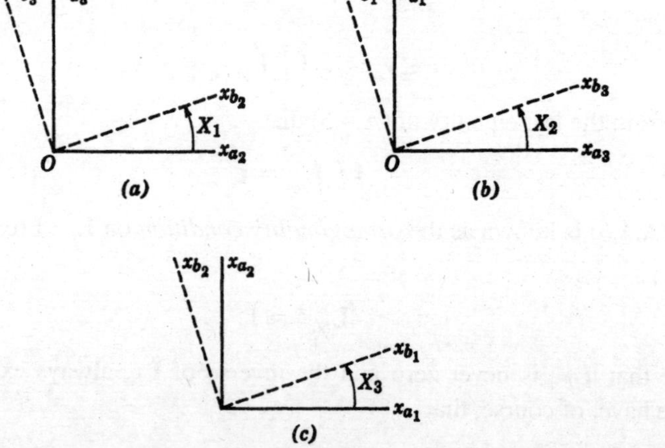

Figure A.4 The three basic rotations. (*a*) About x_{a_1}. (*b*) About x_{a_2}. (*c*) About x_{a_3}.

The transformation matrix for any sequence of rotations can be constructed readily from the above basic formulas. For the case of Euler angles, which rotate frame F_E into F_B as defined in Sec. 4.4, the matrix corresponds to the sequence $(X_3, X_2, X_1) = (\psi, \theta, \phi)$, giving

$$\mathbf{L}_{BE} = \mathbf{L}_1(\phi) \cdot \mathbf{L}_2(\theta) \cdot \mathbf{L}_3(\psi) \tag{A.4,11}$$

[The sequence of angles in (A.4,11) is opposite that of the rotations, since each transformation matrix *pre*multiplies the vector arrived at in the previous step.] The result of multiplying the three matrices is

$$\mathbf{L}_{BE} = \begin{bmatrix} \cos\theta\cos\psi & \cos\theta\sin\psi & -\sin\theta \\ \sin\phi\sin\theta\cos\psi - \cos\phi\sin\psi & \sin\phi\sin\theta\sin\psi + \cos\phi\cos\psi & \sin\phi\cos\theta \\ \cos\phi\sin\theta\cos\psi + \sin\phi\sin\psi & \cos\phi\sin\theta\sin\psi - \sin\phi\cos\psi & \cos\phi\cos\theta \end{bmatrix} \tag{A.4,12}$$

TRANSFORMATION OF THE DERIVATIVE OF A VECTOR

Consider a vector $\mathbf{v}$ that is being observed simultaneously from two frames F_a and F_b that have relative rotation—say F_b rotates with angular velocity $\boldsymbol{\omega}$ relative to F_a, which we may regard as fixed. From (A.4,3)

$$\mathbf{v}_b = \mathbf{L}_{ba}\mathbf{v}_a$$

The derivatives of $\mathbf{v}_a$ and $\mathbf{v}_b$ are of course

$$\dot{\mathbf{v}}_a = \begin{bmatrix} \dot{v}_{a_1} \\ \dot{v}_{a_2} \\ \dot{v}_{a_3} \end{bmatrix} \quad \text{and} \quad \dot{\mathbf{v}}_b = \begin{bmatrix} \dot{v}_{b_1} \\ \dot{v}_{b_2} \\ \dot{v}_{b_3} \end{bmatrix} \tag{A.4,13}$$

where $\dot{\mathbf{v}}_{a_1} = (d/dt)(v_{a_1})$, and so forth. It is important to note that $\dot{\mathbf{v}}_a$ and $\dot{\mathbf{v}}_b$ are not simply two sets of components of the *same* vector, but are actually two *different* vectors.

Now because F_b rotates relative to F_a, the direction cosines l_{ij} are changing with time, and the derivative of (A.4,3) is

$$\dot{\mathbf{v}}_b = \mathbf{L}_{ba}\dot{\mathbf{v}}_a + \dot{\mathbf{L}}_{ba}\mathbf{v}_a \tag{A.4,14}$$

or alternatively

$$\dot{\mathbf{v}}_a = \mathbf{L}_{ab}\dot{\mathbf{v}}_b + \dot{\mathbf{L}}_{ab}\mathbf{v}_b$$

the second terms representing the effect of the rotation.

Since $\dot{\mathbf{L}}$ must be independent of $\mathbf{v}$, the matrix $\dot{\mathbf{L}}_{ab}$ can readily be identified by considering the case when $\mathbf{v}_b$ is constant (see Fig. A.5.). For then, from the fundamental definitions of derivative and cross product, the derivative of $\mathbf{v}$ as seen from F_a is readily shown to be

$$\frac{d\mathbf{v}}{dt} = \boldsymbol{\omega} \times \mathbf{v} \tag{A.4,15}$$

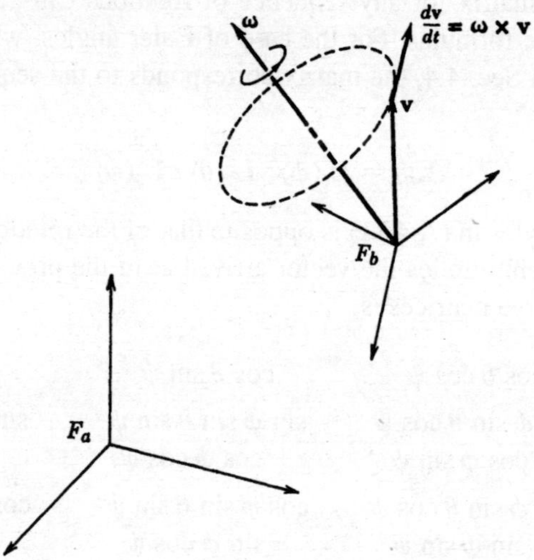

Figure A.5 Rotating vector of constant magnitude.

The matrix equivalent to (A.4,15) is

$$\dot{\mathbf{v}}_a = \tilde{\boldsymbol{\omega}}_a \mathbf{v}_a \tag{A.4,16}$$

where

$$\tilde{\boldsymbol{\omega}}_a = \begin{bmatrix} 0 & -\omega_{a_3} & \omega_{a_2} \\ \omega_{a_3} & 0 & -\omega_{a_1} \\ -\omega_{a_2} & \omega_{a_1} & 0 \end{bmatrix}$$

The corresponding result from (A.4,14) is

$$\dot{\mathbf{v}}_a = \dot{\mathbf{L}}_{ab}\mathbf{v}_b \tag{A.4,17}$$

It follows from equating (A.4,16) and (A.4,17) that

$$\dot{\mathbf{L}}_{ab}\mathbf{v}_b = \tilde{\boldsymbol{\omega}}_a \mathbf{v}_a$$

or

$$\dot{\mathbf{L}}_{ab}\mathbf{v}_b = \tilde{\boldsymbol{\omega}}_a \mathbf{L}_{ab}\mathbf{v}_b \tag{A.4,18}$$

for all $\mathbf{v}_b$. Whence

$$\dot{\mathbf{L}}_{ab} = \tilde{\boldsymbol{\omega}}_a \mathbf{L}_{ab}$$

and

$$\tilde{\boldsymbol{\omega}}_a = \dot{\mathbf{L}}_{ab}\mathbf{L}_{ba}$$

Finally if the above argument is repeated with F_b considered fixed, and F_a having angular velocity $-\boldsymbol{\omega}$, we clearly arrive at the reciprocal result

$$\dot{\mathbf{L}}_{ba} = -\tilde{\boldsymbol{\omega}}_b \mathbf{L}_{ba} \tag{A.4,19}$$

From (A.4,18) and (A.4,19), recalling that $\tilde{\boldsymbol{\omega}}$ is skew-symmetric so that $\tilde{\boldsymbol{\omega}}^T = -\tilde{\boldsymbol{\omega}}$, the reader can readily derive the result

$$\tilde{\boldsymbol{\omega}}_a = \mathbf{L}_{ab}\tilde{\boldsymbol{\omega}}_b \mathbf{L}_{ba} \tag{A.4,20}$$

From (A.4,14), (A.4,18), and (A.4,19) we have the alternative relations

$$\begin{aligned} \dot{\mathbf{v}}_b &= \mathbf{L}_{ba}\dot{\mathbf{v}}_a - \tilde{\boldsymbol{\omega}}_b\mathbf{v}_b \\ \dot{\mathbf{v}}_a &= \mathbf{L}_{ab}\dot{\mathbf{v}}_b + \tilde{\boldsymbol{\omega}}_a\mathbf{v}_a \end{aligned} \tag{A.4,21}$$

with two additional permutations made possible by (A.4,20). A particular form we shall finally want for application is that which uses the components of $\dot{\mathbf{v}}_a$ transformed into F_b, viz.

$$\mathbf{L}_{ba}\dot{\mathbf{v}}_a = \dot{\mathbf{v}}_b + \tilde{\boldsymbol{\omega}}_b\mathbf{v}_b \tag{A.4,22}$$

TRANSFORMATION OF A MATRIX

Equation (A.4,20) is an example of the transformation of a matrix, the elements of which are dependent on the frame of reference. Generally the matrix of interest $\mathbf{A}$ occurs in an equation of the form

$$\mathbf{v} = \mathbf{A}\mathbf{u} \tag{A.4,23}$$

where the elements of the (physical) vectors $\mathbf{u}$ and $\mathbf{v}$ and of the matrix $\mathbf{A}$ are all dependent on the reference frame. We write (A.4,23) for each of the two frames F_a and F_b, that is,

$$\begin{aligned} \mathbf{v}_a &= \mathbf{A}_a\mathbf{u}_a \quad (a) \\ \mathbf{v}_b &= \mathbf{A}_b\mathbf{u}_b \quad (b) \end{aligned} \tag{A.4,24}$$

and transform the second to

$$\mathbf{L}_{ba}\mathbf{v}_a = \mathbf{A}_b\mathbf{L}_{ba}\mathbf{u}_a$$

Premultiplying by $\mathbf{L}_{ab}$ we get

$$\mathbf{v}_a = \mathbf{L}_{ab}\mathbf{A}_b\mathbf{L}_{ba}\mathbf{u}_a \tag{A.4,25}$$

By comparison with (A.4,24*a*) we get the general result

$$\mathbf{A}_a = \mathbf{L}_{ab}\mathbf{A}_b\mathbf{L}_{ba} \tag{A.4,26}$$

A.5 Computation of Eigenvalues and Eigenvectors

Some software packages provide for the calculation of eigenvalues and eigenvectors of matrices directly. The software used for many of the computations in this book is the *Student Version of Program CC*,[3] which does not do this. However, these important system properties can readily be obtained from it, as shown in the following.

Program CC is oriented to the calculation of transfer functions and presents them in various forms; one is the *pole-zero* form. Any transfer function of the system, for example, that from elevator angle to pitch rate, when displayed in this form, will show the eigenvalues in the denominator. That is how we obtained the eigenvalues presented in Chap. 6.

[3]Available from Systems Technology, Inc., 13766 South Hawthorne Blvd., Hawthorne, CA, 90250-7083 U.S.

For the eigenvectors, we turn to the expansion theorem (A.2,10). Consider a case where the input to the system is $\delta_c = \delta(t)$, Dirac's delta function. The response of the ith component of the state vector to this input in the mode corresponding to eigenvalue λ is

$$x_i(t) = \left[\frac{(s-\lambda)N_i(s)}{D(s)}\right]_{s=\lambda} e^{\lambda t} \tag{A.5,1}$$

The ratio of this component to x_1 for the same input $\delta(t)$ is

$$\frac{x_i(t)}{x_1(t)} = \frac{N_i(\lambda)}{N_1(\lambda)} \tag{A.5,2}$$

This ratio gives the ith component of the eigenvector for the mode associated with λ. Any component can be chosen for reference instead of x_1, as illustrated in Figs. 6.3 and 6.15.

A.6 Velocity and Acceleration in an Arbitrarily Moving Frame

Since in many applications, we want to express the position, inertial velocity, and inertial acceleration of a particle in components parallel to the axes of moving frames, we need general theorems that allow for arbitrary motion of the origin, and arbitrary angular velocity of the frame. These theorems are presented below.

Let $F_M(Oxyz)$ be any moving frame with origin at O and with angular velocity $\boldsymbol{\omega}$ relative to F_I. Let $\mathbf{r} = \mathbf{r}_0 + \mathbf{r}'$ be the position vector of a point P of F_M (see Fig. A.6). Let the velocity and acceleration of P relative to F_I be $\mathbf{v}$ and $\mathbf{a}$. Then in F_I

$$\begin{aligned} \mathbf{v}_I &= \dot{\mathbf{r}}_I \\ \mathbf{a}_I &= \ddot{\mathbf{r}}_I \end{aligned} \tag{A.6,1}$$

We want expressions for the velocity and acceleration of P in terms of the components of $\mathbf{r}'$ in F_M. Expanding the first of (A.6,1)

$$\begin{aligned} \mathbf{v}_I &= \dot{\mathbf{r}}_{0I} + \dot{\mathbf{r}}'_I \\ &= \mathbf{v}_{0I} + \dot{\mathbf{r}}'_I \end{aligned} \tag{A.6,2}$$

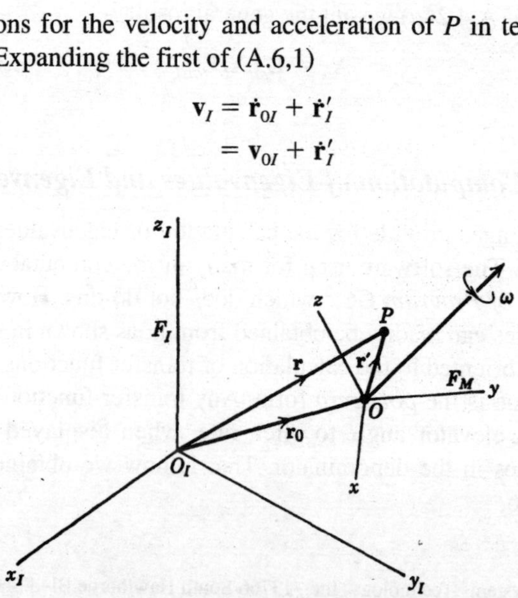

Figure A.6 Moving coordinate system.

where $\mathbf{v}_0$ is the velocity of O relative to F_I. The velocity components in F_M are given by

$$\mathbf{v}_M = \mathbf{L}_{MI}\mathbf{v}_I = \mathbf{L}_{MI}(\mathbf{v}_{0I} + \dot{\mathbf{r}}'_I) = \mathbf{v}_{0_M} + \mathbf{L}_{MI}\dot{\mathbf{r}}'_I$$

From the rule for transforming derivatives (A.4,22)

$$\mathbf{L}_{MI}\dot{\mathbf{r}}'_I = \dot{\mathbf{r}}'_M + \tilde{\boldsymbol{\omega}}_M\mathbf{r}'_M \tag{A.6,3}$$

whence

$$\mathbf{v}_M = \mathbf{v}_{0_M} + \dot{\mathbf{r}}'_M + \tilde{\boldsymbol{\omega}}_M\mathbf{r}'_M \tag{A.6,4}$$

The first term of (A.6,4) is the velocity of O relative to F_I, the second is the velocity of P as measured by an observer fixed in F_M, and the last is the "transport velocity," that is the velocity relative to F_I of the point of F_M that is momentarily coincident with P. The total velocity of P relative to F_I is the sum of these three components. Following traditional practice in flight dynamics, we denote

$$\boldsymbol{\omega}_M = \begin{bmatrix} p \\ q \\ r \end{bmatrix}, \qquad \mathbf{r}'_M = \begin{bmatrix} x \\ y \\ z \end{bmatrix} \tag{A.6,5}$$

(When necessary, subscripts are added to the components to identify particular moving frames.)

The scalar expansion of (A.6,4) is then

$$\begin{aligned} v_x &= v_{0_x} + \dot{x} + qz - ry \\ v_y &= v_{0_y} + \dot{y} + rx - pz \\ v_z &= v_{0_z} + \dot{z} + py - qx \end{aligned} \tag{A.6,6}$$

These expressions then give the components, parallel to the moving coordinate axes, of the velocity of P relative to the *inertial* frame.

On differentiating $\mathbf{v}_I$ and using (A.6,4) we find the components of inertial acceleration parallel to the F_M axes to be

$$\begin{aligned} \mathbf{a}_M &= \mathbf{L}_{MI}\dot{\mathbf{v}}_I = \dot{\mathbf{v}}_M + \tilde{\boldsymbol{\omega}}_M\mathbf{v}_M \\ &= \dot{\mathbf{v}}_{0_M} + \ddot{\mathbf{r}}'_M + \dot{\tilde{\boldsymbol{\omega}}}_M\mathbf{r}'_M + \tilde{\boldsymbol{\omega}}_M\dot{\mathbf{r}}'_M + \tilde{\boldsymbol{\omega}}_M\mathbf{v}_{0_M} + \tilde{\boldsymbol{\omega}}_M\dot{\mathbf{r}}'_M + \tilde{\boldsymbol{\omega}}_M\tilde{\boldsymbol{\omega}}_M\mathbf{r}'_M \\ &= \mathbf{a}_{0_M} + \ddot{\mathbf{r}}'_M + \dot{\tilde{\boldsymbol{\omega}}}_M\mathbf{r}'_M + 2\tilde{\boldsymbol{\omega}}_M\dot{\mathbf{r}}'_M + \tilde{\boldsymbol{\omega}}_M\tilde{\boldsymbol{\omega}}_M\mathbf{r}'_M \end{aligned} \tag{A.6,7}$$

where $\mathbf{a}_{0_M} = \dot{\mathbf{v}}_{0_M} + \tilde{\boldsymbol{\omega}}_M\mathbf{v}_{0_M} = \mathbf{L}_{MI}\dot{\mathbf{v}}_{0_I}$ is the acceleration of O relative to F_I.

The total inertial acceleration of P is seen to be composed of the following parts:

- $\mathbf{a}_{0_M}$: the acceleration of the origin of the moving frame
- $\ddot{\mathbf{r}}'_M$: the acceleration of P as measured by an observer fixed in the moving frame
- $\dot{\tilde{\boldsymbol{\omega}}}_M\mathbf{r}'_M$: the "tangential" acceleration owing to rotational acceleration of the frame F_M
- $2\tilde{\boldsymbol{\omega}}_M\dot{\mathbf{r}}'_M$: the Coriolis acceleration
- $\tilde{\boldsymbol{\omega}}_M\tilde{\boldsymbol{\omega}}_M\mathbf{r}'_M$: the centripetal acceleration

Three of the five terms vanish when the frame F_M has no rotation, and only $\ddot{\mathbf{r}}'_M$ remains if it is inertial. Note that the Coriolis acceleration is perpendicular to $\boldsymbol{\omega}_M$ and

$\dot{\mathbf{r}}'_M$, and the centripetal acceleration is directed along the perpendicular from P to $\boldsymbol{\omega}$. The scalar expansion of (A.6,7) gives the required inertial acceleration components of P as

$$\begin{aligned} a_x &= a_{0_x} + \ddot{x} + 2q\dot{z} - 2r\dot{y} - x(q^2 + r^2) + y(pq - \dot{r}) + z(pr + \dot{q}) \\ a_y &= a_{0_y} + \ddot{y} + 2r\dot{x} - 2p\dot{z} + x(pq + \dot{r}) - y(p^2 + r^2) + z(qr - \dot{p}) \\ a_z &= a_{0_z} + \ddot{z} + 2p\dot{y} - 2q\dot{x} + x(pr - \dot{q}) + y(qr + \dot{p}) - z(p^2 + q^2) \end{aligned} \tag{A.6,8}$$

APPENDIX B

Data for Estimating Aerodynamic Derivatives

This appendix contains a limited amount of data on stability and control derivatives. It is not intended to be used for design. That requires much more detail than could possibly be provided here. It is intended to display some representative orders of magnitude and trends, and to provide numerical data that teachers and students can use for exercises. All the data pertain to subsonic flight of rigid airplanes. Much of the information comes from either the USAF Datcom (USAF, 1978) or from the data sheets of the Royal Aeronautical Society of Great Britain (now out of print), which is also the source for some of the Datcom data. We have taken some liberties in extracting and presenting this information, but have not changed any essential content. For information about derivatives at transonic and supersonic speeds and for geometries different from those covered in the following, the reader is referred to the USAF Datcom. When estimating derivatives, reference should also be made to Tables 5.1 and 5.2.

B.1 Lift-Curve Slope, $C_{L\alpha}$

NOTATION

A	aspect ratio
c	chord
$C_{L\alpha}$	lift-curve slope of wing alone
$C_{l\alpha}$	two-dimensional (airfoil) lift-curve slope
$(C_{l\alpha})_{\text{theory}}$	theoretical value of $C_{l\alpha}$
K	an empirical factor
M	Mach number
R	Reynolds number, $Vc\rho/\mu$
β	Prandtl-Glauert compressibility factor, $\sqrt{1-\mathrm{M}^2}$
κ	$\beta C_{l\alpha}/2\pi$
$\Lambda_{c/2}$	sweepback angle of midchord line

NOTES

- The source of the data for airfoils and wings is USAF datcom. It applies to rigid straight-tapered wings at subsonic speeds and small angle of attack.
- The section lift-curve slope is given by

$$C_{l\alpha} = \frac{1.05}{\beta} K(C_{l\alpha})_{\text{theory}} \tag{B.1,1}$$

where K is given in Fig. B.1,1*a* and $(C_{l\alpha})_{\text{theory}}$ in Fig. B.1,1*b*. Y_{90} and Y_{99} are the airfoil thicknesses, in percent of chord, at 90% and 99% of the chord back from the leading edge, as illustrated, and the trailing edge angle is defined in terms of these thicknesses by

$$\tan \tfrac{1}{2}\phi'_{TE} = \frac{\tfrac{1}{2}(Y_{90} - Y_{99})}{9} \tag{B.1,2}$$

- The lift-curve slope $C_{l\alpha}$ of the wing alone is given in Fig. B.1,2. The inset equation is seen to approach the theoretically correct limits of $\pi A/2$ as $A \to 0$ and 2π as $\{A \to \infty, \kappa \to 1, \Lambda \to 0, \beta \to 1\}$.
- Figure B.1,3 gives some theoretical values of the body effect on $C_{L\alpha}$ for unswept wings in mid-wing combination with an infinite circular cylinder body. For values of $A < 1$, the theory also applies to delta wings with pointed tips.

In Fig. B.1,3*a* the wing angle of attack is the same as that of the fuselage; that is, $\epsilon = 0$. In this case the lift of the wing-body combination increases to a maximum value, then decreases with increasing body diameter. Where there is a wing setting, i.e., $\epsilon \neq 0$, and $\alpha_B = 0$ (Fig. B.1,3*b*), the lift of the combination decreases with increasing σ.

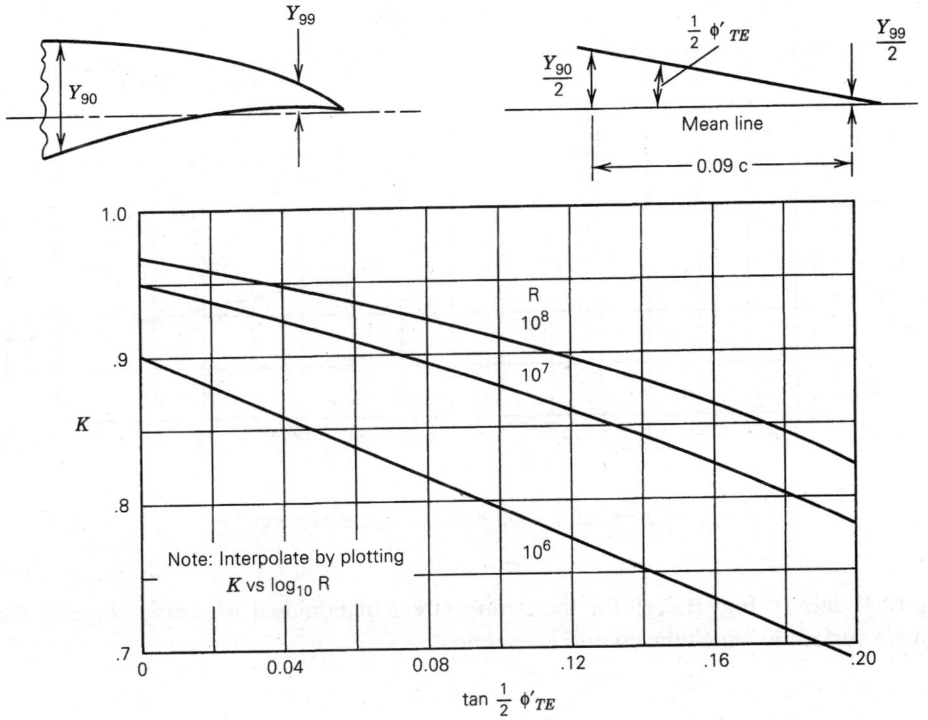

(a)

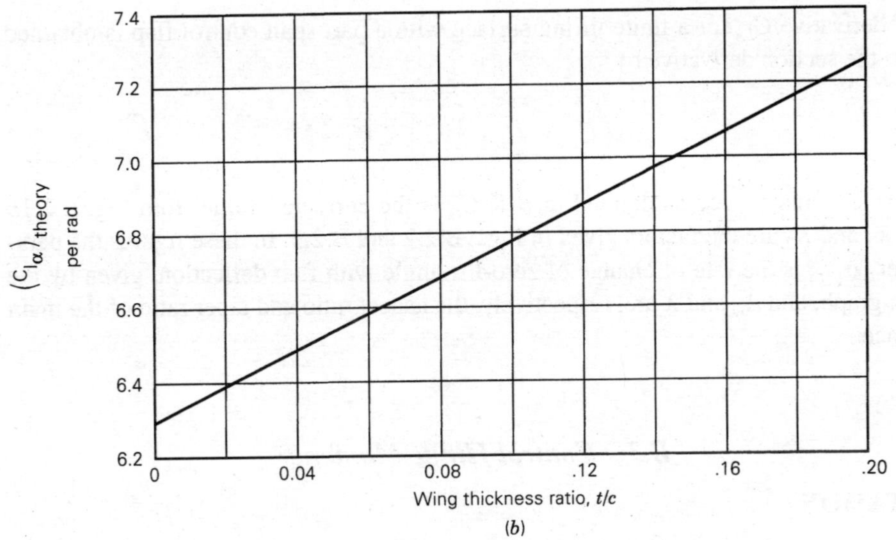

(b)

Figure B.1,1 Two-dimensional lift-curve slope.

B.2 Control Effectiveness, C_{L_δ}

SECTION DATA

Figure B.2,1*a* presents theoretical values of the two-dimensional control derivative C_{l_δ} for simple flaps in incompressible flow. These values can be corrected by the em-

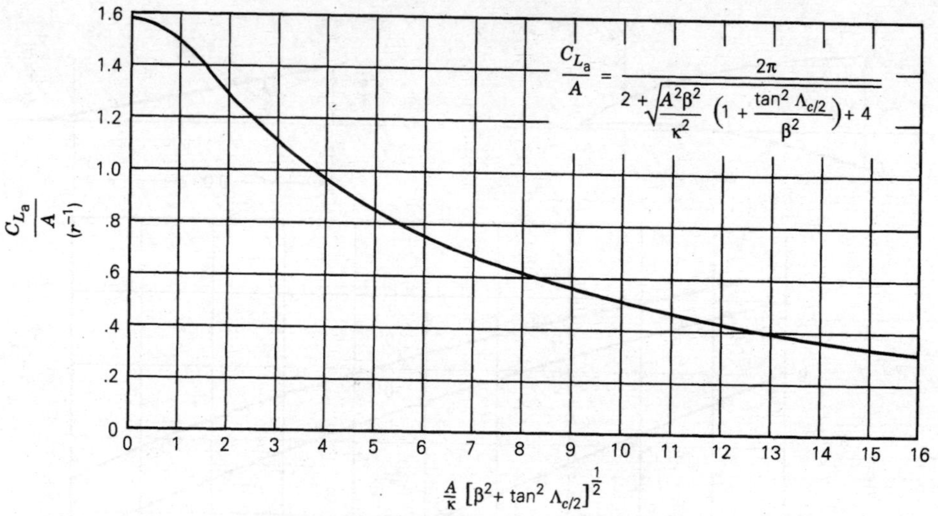

Figure B.1,2 Subsonic wing lift-curve slope.

pirical data of Fig. B.2,1*b* for the strong effect of nonideal lift-curve slope of the main surface to which the control is attached.

SURFACE DATA

The derivative C_{L_δ} for a finite lifting surface with a part span control flap is obtained from the section derivative by

$$C_{L_\delta} = C_{l_\delta}\left(\frac{C_{L_\alpha}}{C_{l_\alpha}}\right)K_1K_2$$

where C_{L_α} and C_{l_α} are as defined in B.1, C_{l_δ} is the corrected value from Fig. B.2,1*b* and K_1 and K_2 are the factors given in Figs. B.2,2 and B.2,3. In these figures the parameter $(\alpha_\delta)_{C_l}$ is the rate of change of zero-lift angle with flap deflection, given by the inset graph, and A_w and λ are, respectively, the aspect ratio and taper ratio of the main surface.

B.3 *Control Hinge Moments*

NOTATION

τ	trailing-edge angle defined by the tangents to the upper and lower surfaces at the trailing edge
$(b_1)_{0T}$	theoretical rate of change of hinge-moment coefficient with angle of attack for incompressible inviscid two-dimensional flow
$(b_1)_0$	actual rate of change of hinge-moment coefficient with angle of attack for incompressible two-dimensional flow

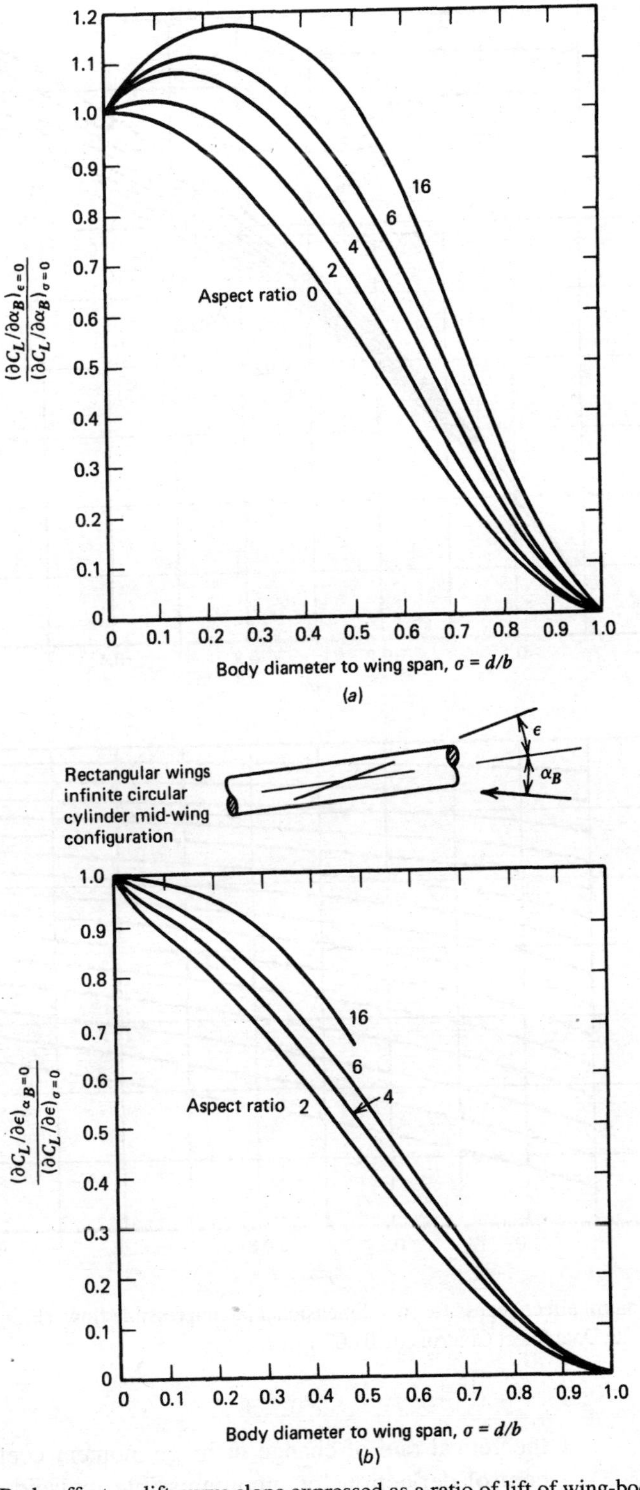

Figure B.1,3 Body effect on lift-curve slope expressed as a ratio of lift of wing-body combination to lift of wing alone. (From "Lift and Lift Distribution of Wings in Combination with Slender Bodies of Revolution," by H. J. Luckert, *Can. Aero. J.*, December 1955.)

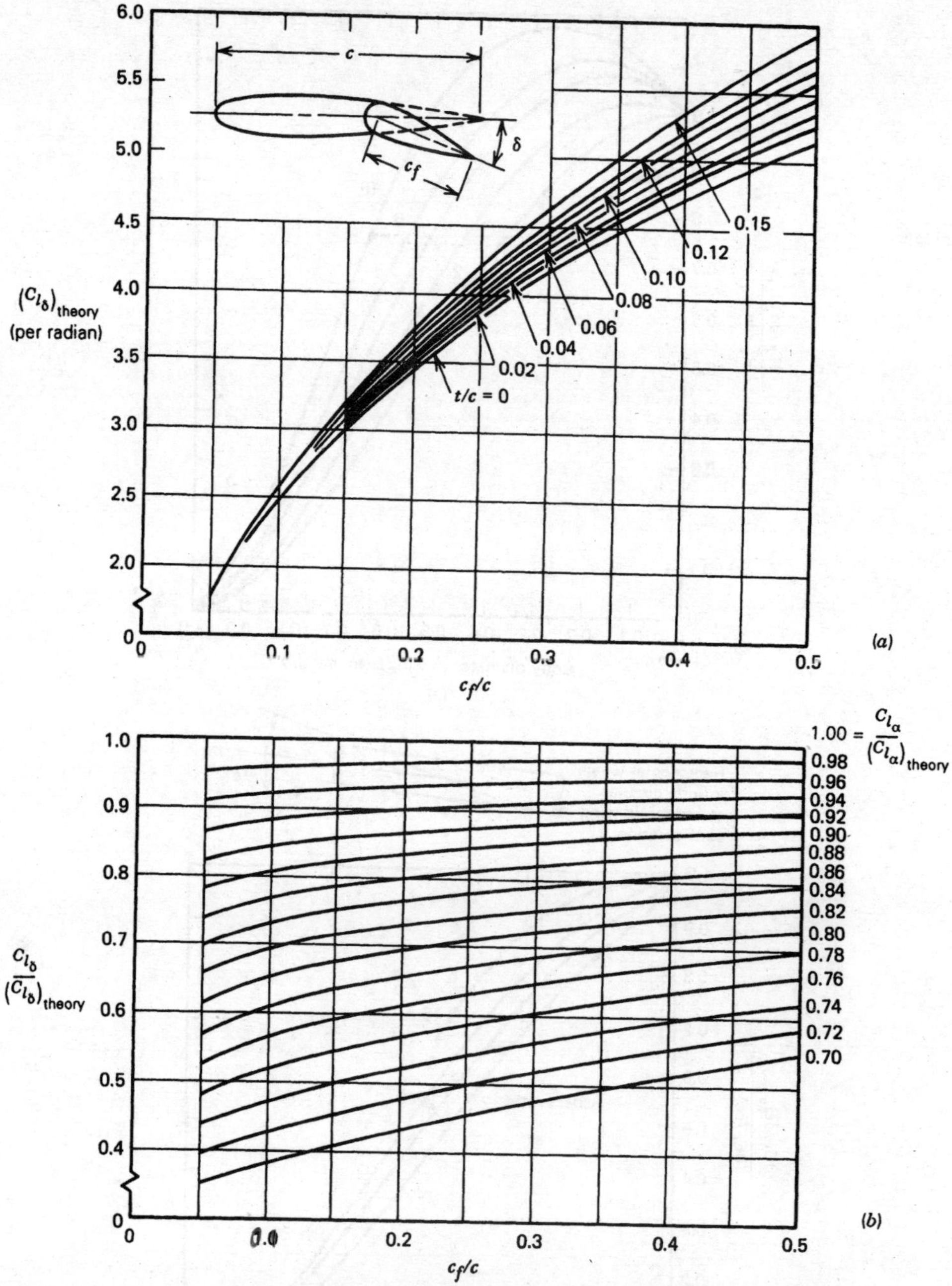

Figure B.2,1 Control effectiveness for two-dimensional incompressible flow. (From Royal Aeronautical Society Data Sheet Controls 01.01.03.)

$(b_2)_{0T}$	theoretical rate of change of hinge-moment coefficient with control deflection for incompressible inviscid two-dimensional flow
$(b_2)_0$	actual rate of change of hinge-moment coefficient with control deflection for incompressible two-dimensional flow

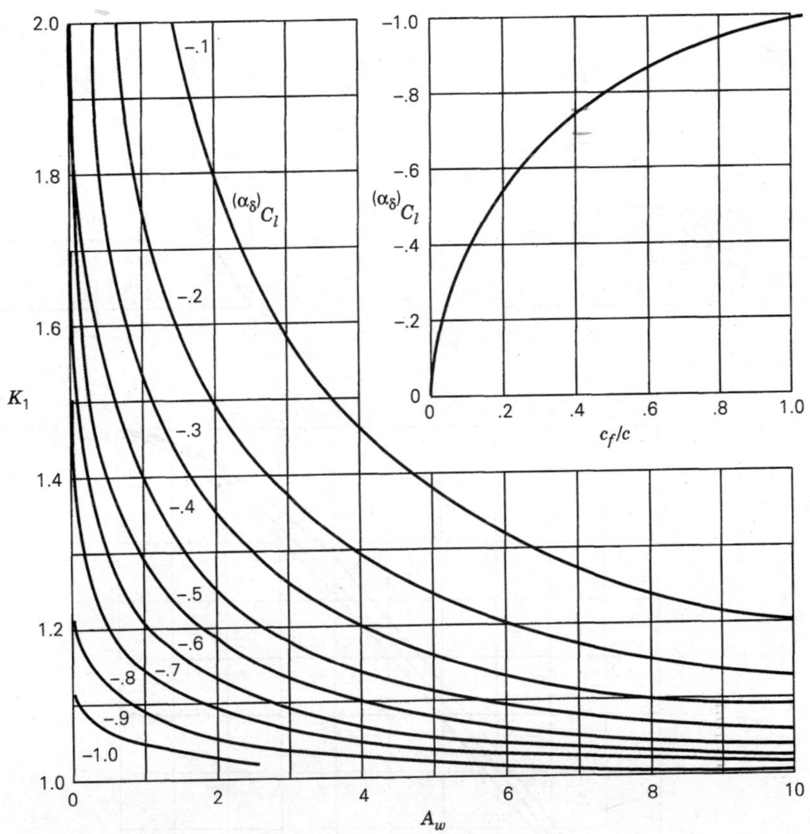

Figure B.2,2 Flap-chord factor.

$(b_1)_{0_{bal}}$, $(b_2)_{0_{bal}}$	rates of change of control hinge-moment coefficients with incidence and control-surface deflection, respectively, in two-dimensional flow for control surfaces with sealed gap and nose balance
F_1, α_i/δ	induced angle of attack correction to $(b_1)_0$ and $(b_2)_0$, respectively, where F_1 is the value of $(\alpha_i/\delta)\ [C_{l_\alpha}/C_{l_\delta}]$ when $c_f = c$
F_2, $\Delta(b_2)$	stream-line curvature correction to $(b_1)_0$ and $(b_2)_0$, respectively, where F_2 is the value of $\Delta(b_2)$ when $c_f = c$
F_3	factor to F_2 and $\Delta(b_2)$ allowing for nose balance
Balance	ratio of control-surface area forward of hinge line to control-surface area behind hinge line

NOTES

Figures B.3,1 and B.3,2

The curves of Fig. B.3,1 were derived for a standard series of airfoils with plain controls for which $\tan(\frac{1}{2})\tau = t/c$ (referred to by an asterisk). To correct for airfoils

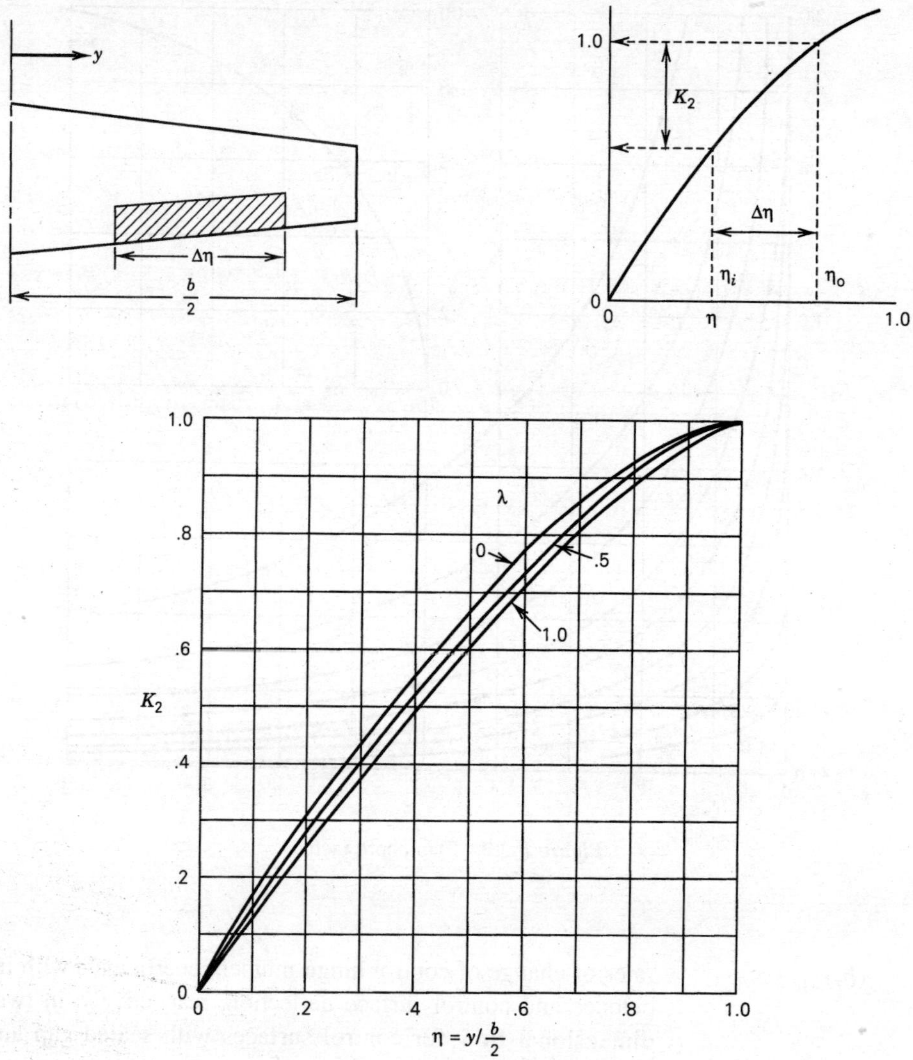

Figure B.2,3 Span factor for inboard flaps.

with tan $(\frac{1}{2})\tau$ different from *t/c*, values of $(b_1)^*_{0T}$, $(C_{l_\alpha})^*_{\text{theory}}$ and $C^*_{l_\alpha}$ are calculated for the given *t/c* ratio; then $(b_1)_0$ is calculated from

$$(b_1)_0 = (b_1)^*_0 + 2[(C_{l_\alpha})^*_{\text{theory}} - C^*_{l_\alpha}](\tan(\tfrac{1}{2}\tau) - t/c). \qquad \text{(B.3,1)}$$

Values of $(C_{l_\alpha})^*_{\text{theory}}$ and $C^*_{l_\alpha}$ may be obtained as in Appendix B.1.

The curves apply for values of angle of attack and control deflection for which there is no flow separation over the airfoil; for these conditions $(b_1)_0$ can be estimated to within ±0.05. The data refer to sealed gaps but may be used if the gap is not greater than 0.002*c*.

The above discussion also applies to the data given in Fig. B.3,2 for $(b_2)_0$. The subscript 1 in Eq. B.3,1 becomes a subscript 2, α becomes δ, and values of $(C_{l_\delta})^*_{\text{theory}}$ and $C^*_{l_\delta}$ may be obtained from Appendix B.2.

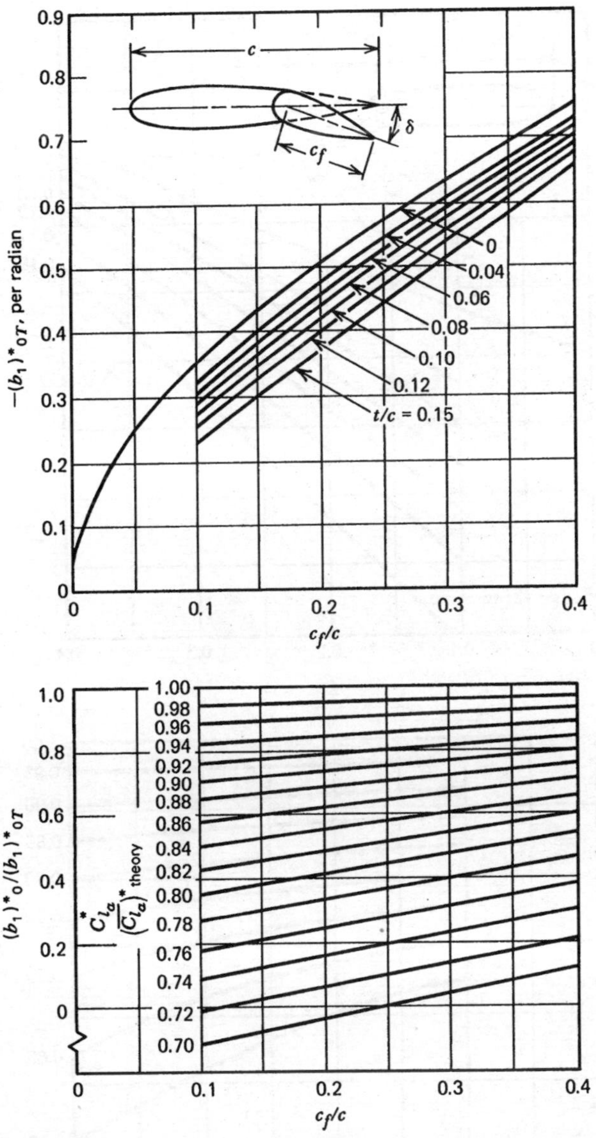

Figure B.3,1 Rate of change of hinge-moment coefficient with angle of attack for a plain control in incompressible two-dimensional flow. (From Royal Aeronautical Society Data Sheet Controls, 04.01.01.)

Figure B.3,3

The effect of nose balance on $(b_1)_0$ and $(b_2)_0$ can be estimated from the curves given on this figure. The data were obtained from wind-tunnel tests on airfoils with control-chord/airfoil-chord ratio of 0.3. Relatively small changes in nose and trailing-edge shape, and airflow over the control surface, may have a large effect on hinge moments for balanced control surfaces, so that estimates of nose-balance effect will be fairly inaccurate. If the control-surface gap is unsealed, the hinge-moment coefficients of plain and nosebalanced controls will generally become more positive.

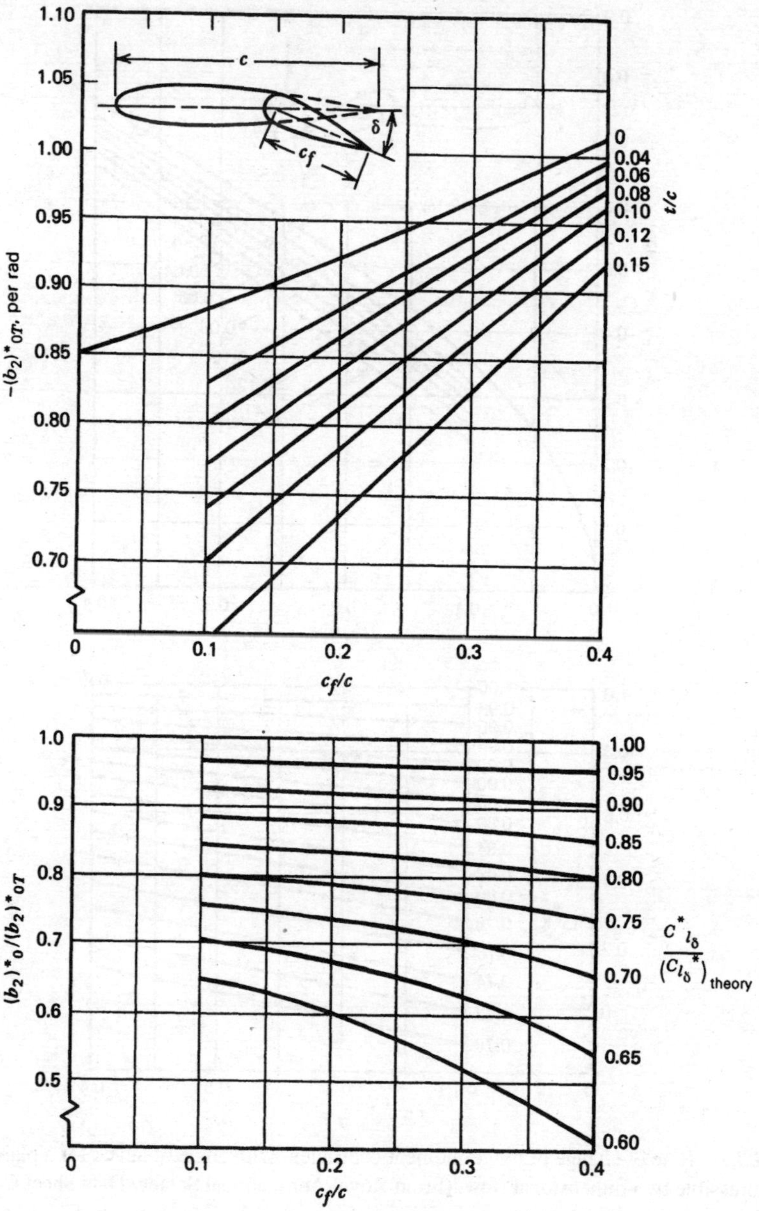

Figure B.3,2 Rate of change of hinge-moment coefficient with control deflection for a plain flap in two-dimensional flow. (From Royal Aeronautical Society Data Sheet Controls 04.01.02.)

Figure B.3,4

Two-dimensional hinge-moment coefficients for control surfaces with nose balance can be corrected for finite aspect ratio of the main surface using the factors given in the curves and the following equations:

$$b_1 = (b_1)_0(1 - F_1) + F_2F_3C_{l_\alpha} \tag{B.3,2}$$

$$b_2 = (b_2)_0 - (\alpha_i/\delta)(b_1)_0 + \Delta(b_2)F_3C_{l_\delta} \tag{B.3,3}$$

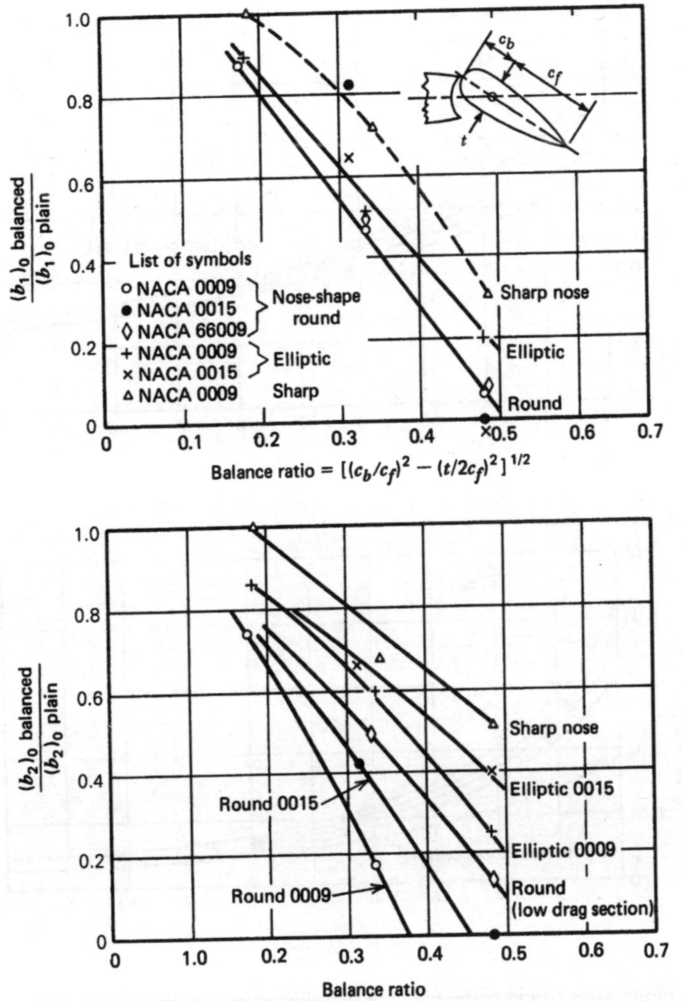

Figure B.3,3 Effect of nose balance on two-dimensional plain-control hinge-moment coefficients. (From Royal Aeronautical Society Data Sheet 04.01.03.)

For plain control surfaces the above equations are used with $F_3 = 1$. $(b_1)_0$ and $(b_2)_0$ can be obtained from Fig. B.3,1 and B.3,2, respectively, for plain controls. For nose-balanced controls, the two-dimensional coefficients $(b_1)_0$ and $(b_2)_0$ must include the effect of nose-balance. Values of C_{l_α} can be obtained from Sec. B.1, and those for C_{l_δ} from Fig. B.2,1.

Lifting-surface theory was applied to unswept wings with elliptic spanwise lift distribution to derive the factors. Full-span control surfaces were assumed together with constant ratios of c_f/c and constant values of $(b_1)_0$ and $(b_2)_0$ across the span. The factors apply to wings with taper ratios of 2 to 3 if c_f/c, $(b_1)_0$ and $(b_2)_0$ do not vary by more than $\pm 10\%$ from their average values.

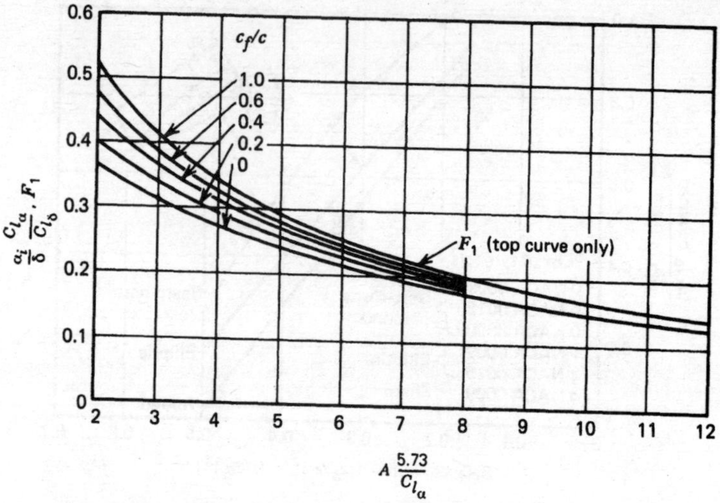

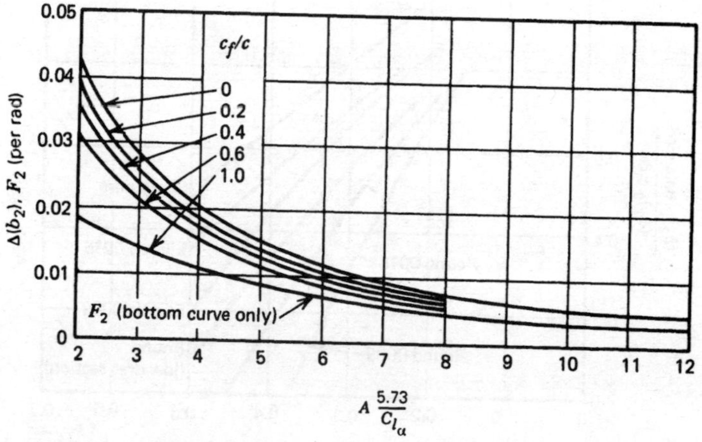

Figure B.3,4 Finite-aspect-ratio corrections for two-dimensional plain and nose-balanced control hinge-moment coefficients (C_{l_α} per radian). (From Royal Aeronautical Society Data Sheet Controls 04.01.05.)

B.4 Tab Effectiveness, b_3

The data and method that follows is taken from the USAF Datcom. It provides estimates of b_3 [see (2.5,1)] for two-dimensional subsonic attached flow over airfoils with a control surface and tab. Corrections for part-span tabs can be made by multiplying the result for two dimensions by the ratio of the control surface area spanned by the tab to the total control surface area (both areas being measured aft of the hinge line).

$$b_3 = \left(\frac{\partial C_{h_f}}{\partial \delta_t}\right)_{\alpha,\delta_f} = \left(\frac{\partial C_{h_f}}{\partial \delta_t}\right)_{C_l,\delta_f} - \left(\frac{\partial C_{h_f}}{\partial C_l}\right)_{\delta_t,\delta_f}\left(\frac{\partial C_l}{\partial \alpha}\right)_{\delta_t,\delta_f}\left(\frac{\partial \alpha}{\partial \delta_t}\right)_{C_l,\delta_f} \tag{B.4,1}$$

where

$\left(\dfrac{\partial C_{h_f}}{\partial \delta_t}\right)_{c_l,\delta_f}$ is the change in control section hinge-moment coefficient due to tab deflection, measured at constant values of lift and flap deflection. This value is obtained from Fig. B.4,1.

$\left(\dfrac{\partial C_{h_f}}{\partial C_l}\right)_{\delta_t,\delta_f}$ is the change in control section hinge-moment coefficient due to lift variation, measured at constant values of tab and flap deflection. This value is obtained from Fig. B.4,2.

$\left(\dfrac{\partial C_l}{\partial \alpha}\right)_{\delta_t,\delta_f}$ is the section lift-curve slope of the primary panel (wing, horizontal tail, etc.) at constant values of tab and flap deflection. This value can be obtained from (B.1,1).

$\left(\dfrac{\partial \alpha}{\partial \delta_t}\right)_{c_l,\delta_f}$ is the rate of change of angle of attack due to a change in tab deflection in the linear range at constant values of lift and flap deflection. This value can be obtained from Fig. B.4,3.

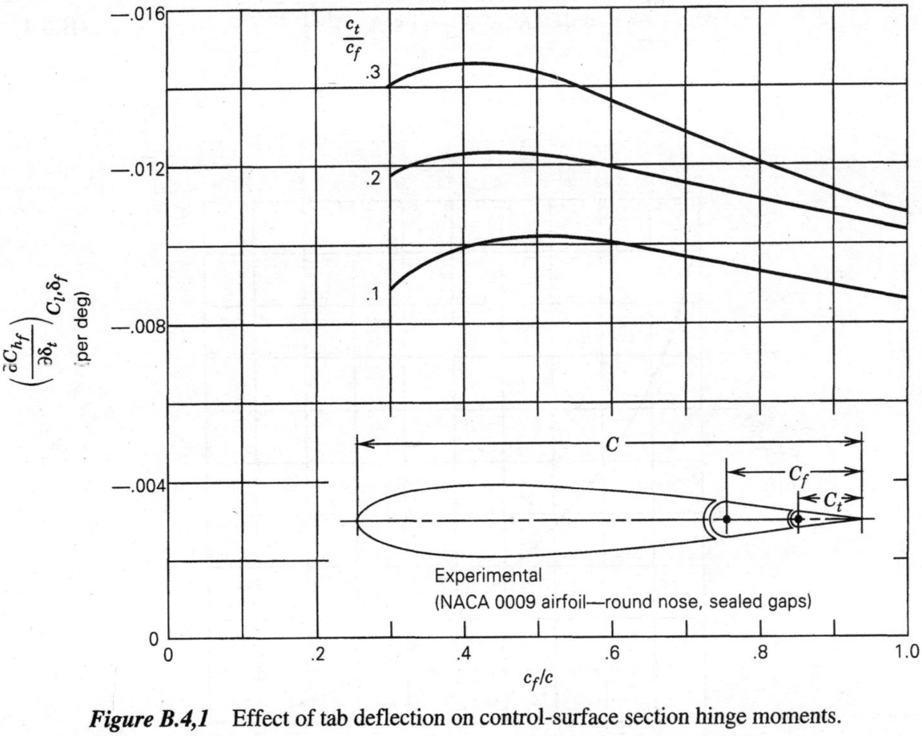

Figure B.4,1 Effect of tab deflection on control-surface section hinge moments.

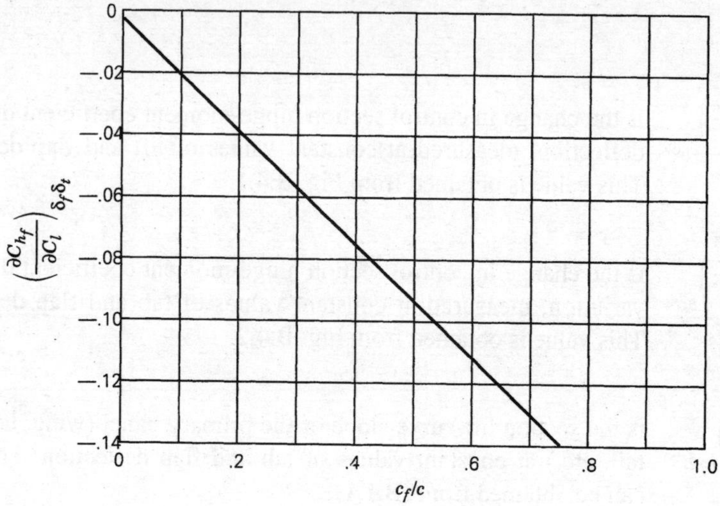

Figure B.4,2 Effect of section lift coefficient on flap section hinge moments.

B.5 Downwash, $\frac{\partial \epsilon}{\partial \alpha}$

The method and data that follow are taken from the USAF Datcom. The average low-speed downwash gradient at the horizontal tail is given by

$$\frac{\partial \epsilon}{\partial \alpha} = 4.44\ [K_A K_\lambda K_H (\cos \Lambda_{c/4})^{1/2}]^{1.19} \tag{B.5,1}$$

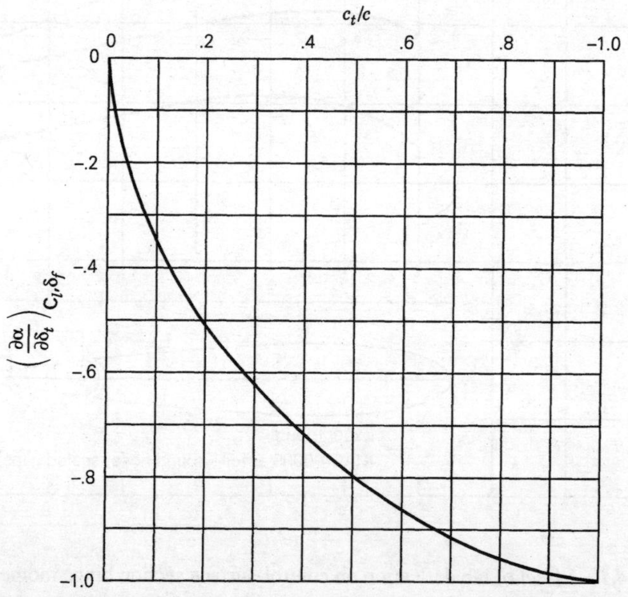

Figure B.4,3 Rate of change of angle of attack due to a change in tab deflection.

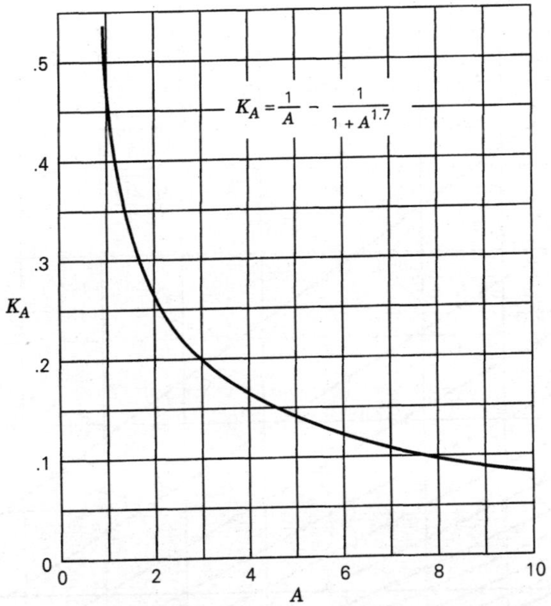

Figure B.5,1 Wing aspect-ratio factor.

where K_A, K_λ, and K_H are wing-aspect-ratio, wing-taper-ratio, and horizontal-tail-location factors obtained from Figs. B.5,1, B.5,2, and B.5,3, respectively. $\Lambda_{c/4}$ is the sweepback angle of the wing $\frac{1}{4}$ chord line.

At higher subsonic speeds the effect of compressibility is approximated by

$$\left(\frac{\partial \epsilon}{\partial \alpha}\right)_M = \left(\frac{\partial \epsilon}{\partial \alpha}\right)_{\text{low speed}} \frac{(C_{L_\alpha})_M}{(C_{L_\alpha})_{\text{low speed}}} \tag{B.5,2}$$

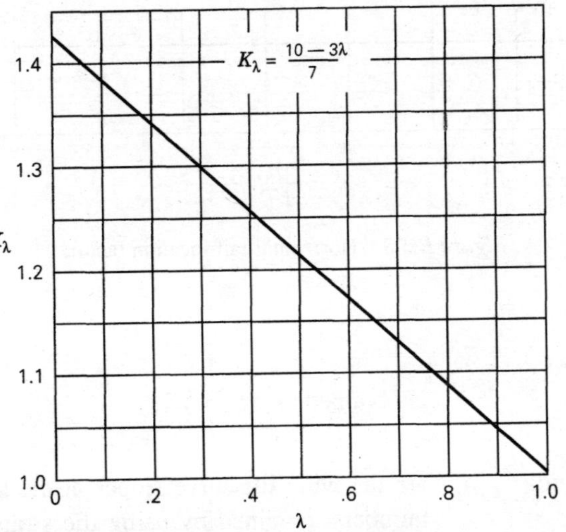

Figure B.5,2 Wing taper-ratio factor.

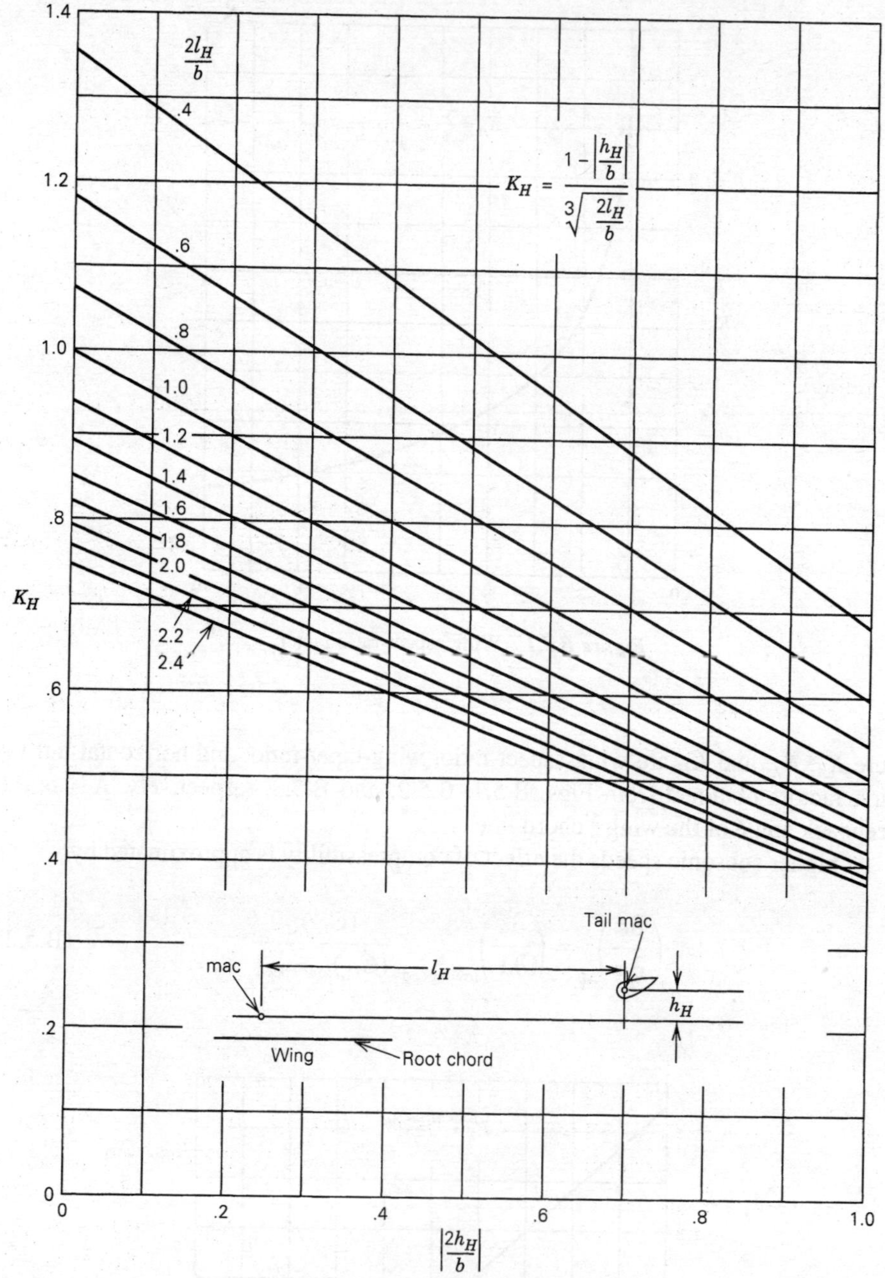

Figure B.5,3 Horizontal-tail-location factor.

where

$\left(\dfrac{\partial \epsilon}{\partial \alpha}\right)_{\text{low speed}}$ is obtained using (B.5,1)

$(C_{L_\alpha})_{\text{low speed}}$ and $(C_{L_\alpha})_M$ are the wing lift-curve slopes at the appropriate Mach numbers, obtained by using the straight-tapered-wing method of Sec. B.1

B.6 Effect of Bodies on Neutral Point and C_{m0}

NOTATION

c	local wing chord at center line of fuselage or nacelle
$\bar{c}$	mean aerodynamic chord
w	maximum width of fuselage or nacelle
S	gross wing area
Δh_n	shift of neutral point due to fuselage or nacelle as a fraction of $\bar{c}$, positive aft
S_B	area of planform of body
S_{BF}	area of planform of body, forward of $0.25\bar{c}$
c'	root chord of wing without fillets
$(C_{m0})_B$	increment to C_{m_0} due to a body at zero lift
θ	reflex angle of fillet, i.e., angle between wing root chord and lower surface of fillet for upswept fillets, or the upper surface for downswept fillets, positive as indicated in Fig. B.8,2
λ	fillet lift-increment ratio, i.e., $C_{l_\delta}/C_{l_\alpha}$, considering the fillet to be a flap of chord 1_f

NOTES

Figure B.6,1: Δh_n

The data for estimating Δh_n presented in this graph were derived from wind-tunnel tests. The forward shift in neutral point is mainly dependent on the length and width of the body forward of the wing. The values of Δh_n given by the curves are accurate to within $\pm 0.01\bar{c}$, and are about 5% higher for low-wing, and the same amount lower for high-wing configurations. The data are inapplicable if the wing is clear of the body. Separate values should be computed for fuselages and nacelles, and the results added to obtain the total neutral-point shift.

Figure B.6,2: $(C_{m0})_B$

The curves given in this figure apply to stream-line bodies of circular or near circular cross section with midwing configurations. For high- or low-wing configurations a positive or negative $\Delta(C_{m0})_B = 0.004$ is added, respectively, to the value derived from the curves. The curves apply only for angles of attack up to about 15° for stream-line bodies where the pitching moment of the body varies linearly with angle of attack.

In the wind-tunnel tests from which the data were derived, the wings had straight trailing edges at the wing-body junction. Fillets have a large effect on C_{m0}, however, especially if θ is large. The following equation may be used to estimate the fillet effect if $0.12 < l_f/c < 0.5$ and $0.03 < S_f/b < 0.075$:

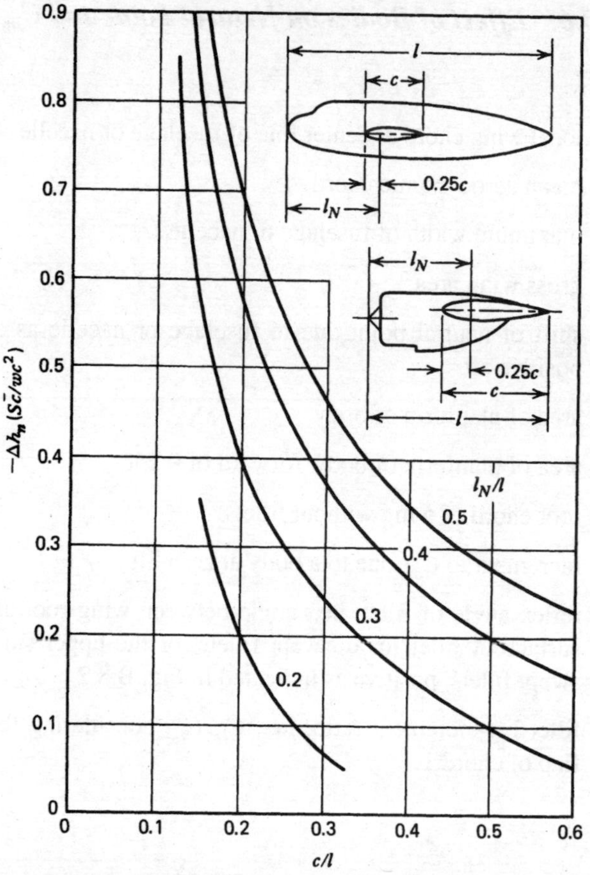

Figure B.6,1 Effect of a fuselage or nacelle on neutral-point position. (From Royal Aeronautical Society Data Sheet Aircraft 08.01.01.)

$$C_{m0} \text{ due to fillets} = [0.046 + 0.08(dC_m/dC_L)_w\lambda\theta - 0.2(c + l_f)/c](w + S_f)/b$$

The value obtained from the curves, the fillet effect, and the effect due to wing position are added to determine $(C_{m0})_B$.

B.7 Propeller and Slipstream Effects

PROPELLER NORMAL FORCE

The following method of estimating the propeller normal force is due to Ribner (1944). The normal force is expressed in terms of the derivative $\partial C_{N_p}/\partial \alpha_p$ (see Sec. 3.4), which is given by

$$\partial C_{N_p}/\partial \alpha_p = f C_{Y'_{\psi 0}}$$

The factor f is the same for all propellers, and is given in Fig. B.7,1 as a function of $T_c = T/\rho V^2 d^2$. The value of $C_{Y'_{\psi 0}}$ varies with the propeller and its operating condition. The values for a particular propeller family are given in Fig. B.7,2. Extrapolation to

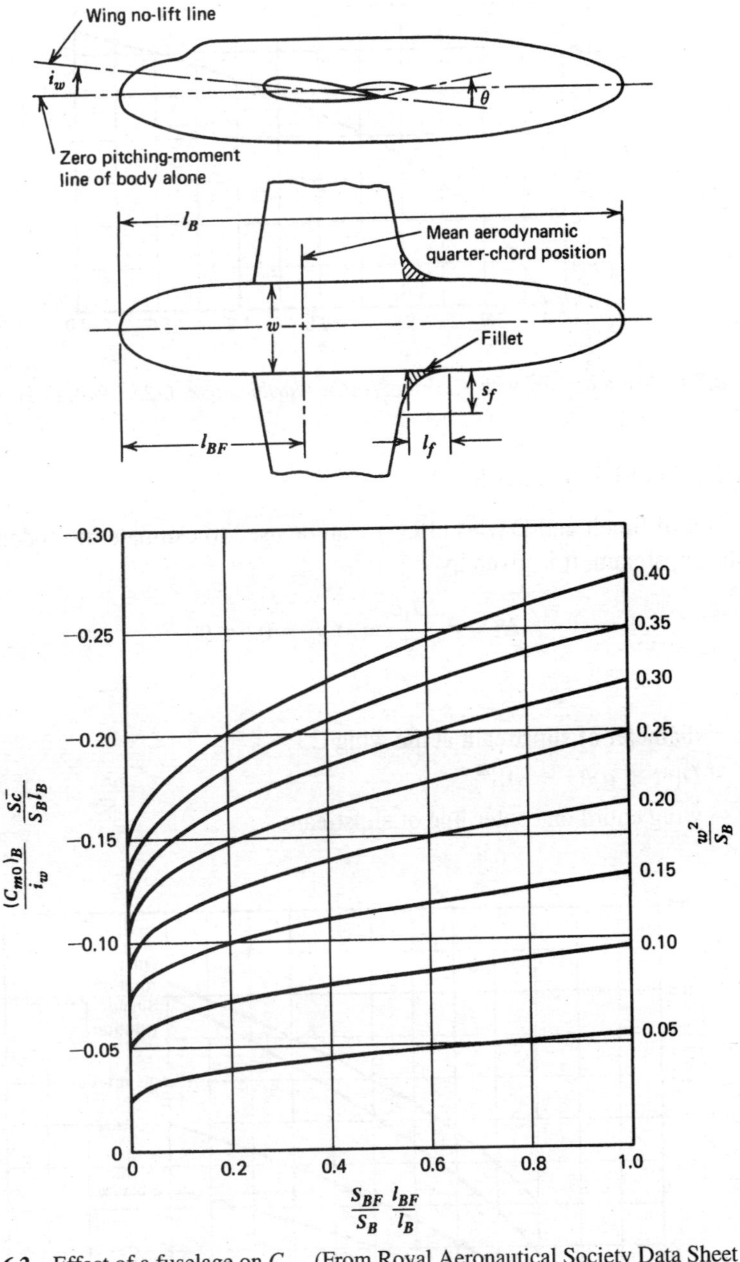

Figure B.6,2 Effect of a fuselage on C_{m_0}. (From Royal Aeronautical Society Data Sheet Aircraft 08.01.07.)

other propellers can be made by means of Fig. B.7,3, on the basis of the "side-force-factor," SFF. This is a geometrical propeller parameter, given approximately by

$$SFF = 525[(b/D)_{0.3} + (b/D)_{0.6}] + 270(b/D)_{0.9}$$

where (b/D) is the ratio of blade width to propeller diameter, and the subscript is the relative radius at which this ratio is measured.

Also given in Ribner (1944) are some curves which are useful for estimating the upwash or downwash at the propeller plane. These are reproduced in Fig. B.7,4.

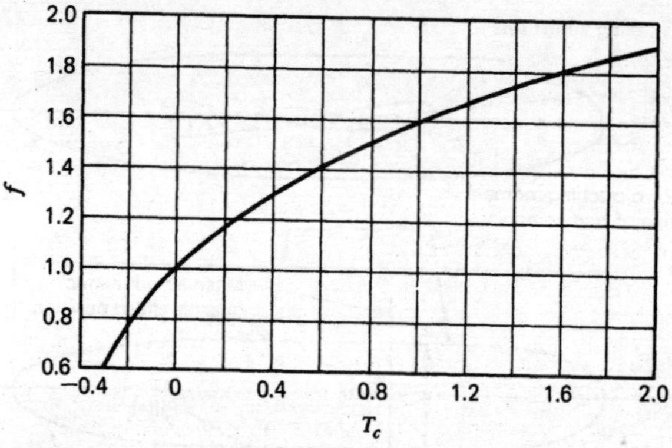

Figure B.7,1 Variation of f with T_c. (From *NACA Wartime Rept. L-25,* 1944, by H. S. Ribner.)

LIFT DUE TO SLIPSTREAM

The method of Smelt and Davies (1937)[‡] can be used to estimate the added wing lift due to the slipstream. It is given by

$$\Delta C_L = \frac{D_1 c}{S} s(\lambda C_{L_0} - 0.6 a_0 \theta)$$

where

D_1 = diameter of slipstream at the wing C.P.
$= D[(1 + a)/(1 + s)]^{1/2}$
c = wing chord on center line of slipstream

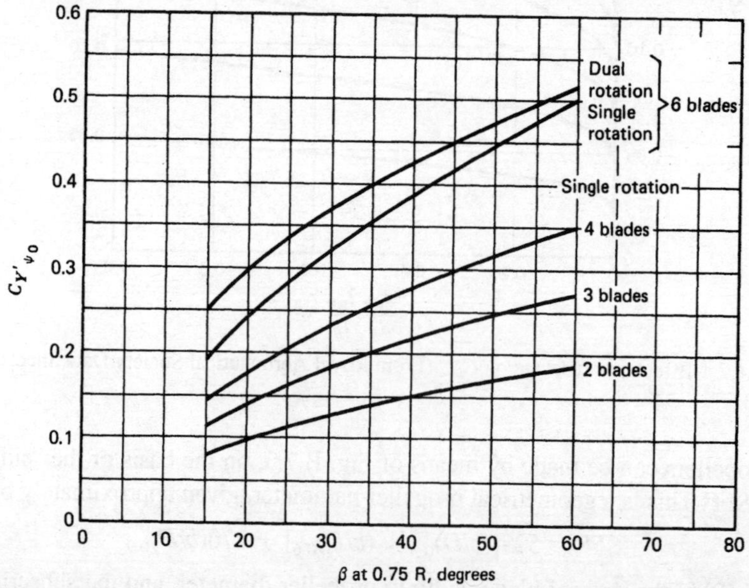

Figure B.7,2 Variation of $C_{Y'_{\psi 0}}$ with blade angle. (From *NACA Wartime Rept. L-25,* 1944, by H. S. Ribner.)

[‡]See also Ribner and Ellis (1972).

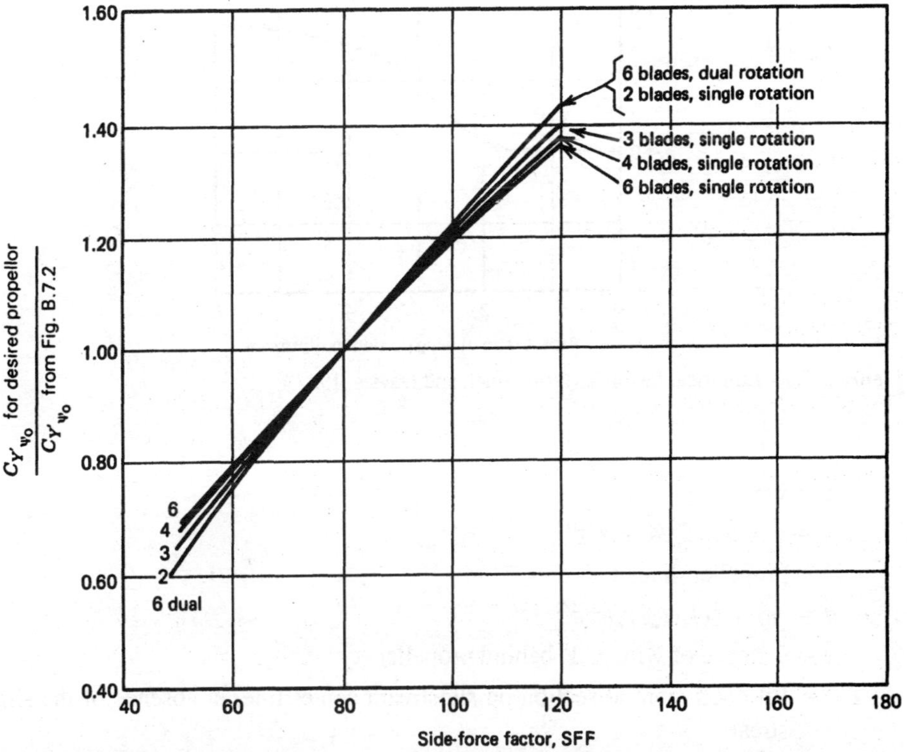

Figure B.7,3 Ratio of normal force derivatives. (From *NACA Wartime Rept. L-25,* 1944, by H. S. Ribner.)

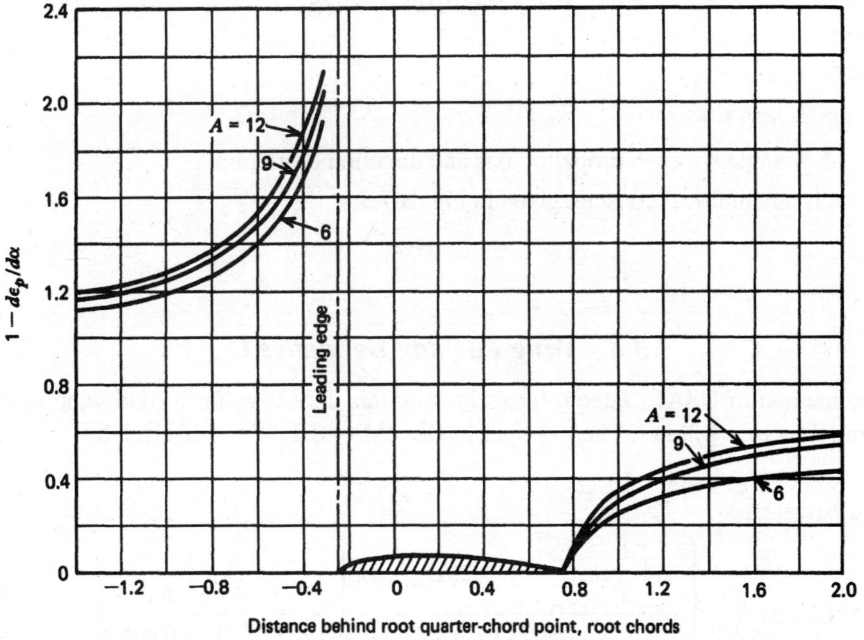

Figure B.7,4 Value of $1 - d\epsilon/d\alpha$ on longitudinal axis of elliptic wing for aspect ratios 6, 9, and 12. (From *NACA Wartime Rept. L-25,* 1944, by H. S. Ribner.)

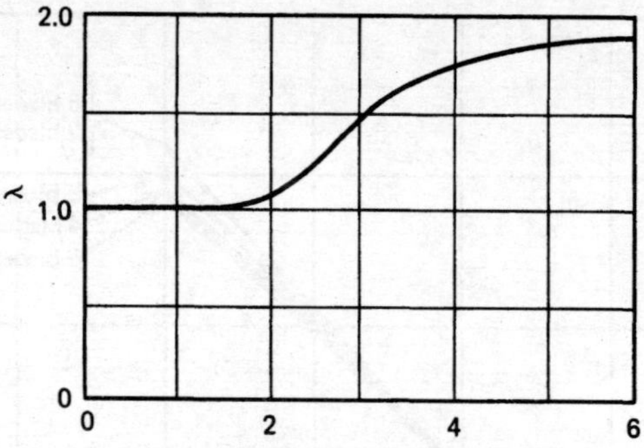

Figure B.7,5 Empirical factor λ. (From Smelt and Davies, 1937.)

S = wing area

$s = a + ax/(D^2/4 + x^2)^{1/2}$

D = propeller diameter

$a = -\frac{1}{2} + \frac{1}{2}(1 + 8T_c/\pi)^{1/2}$

x = distance of wing C.P. behind propeller

C_{L_0} = lift coefficient at section on slipstream center line, in absence of the slipstream

a_0 = two-dimensional lift-curve slope of wing section

θ = angle of downwash of slipstream at wing C.P. calculated from the equation

$$1/\theta^{0.8} = 0.016x/D + 1/\theta_0{}^{0.8}$$

where

$\theta_0 = a\phi/(1 + a)$

ϕ = angle between propellor axis and direction of motion.

λ is an empirical constant given in Fig. B.7,5.

B.8 Wing Pitching Derivative C_{m_q}

The method of USAF Datcom for estimating this derivative for a rigid wing in subsonic flow is as follows. The low-speed value ($\mathbf{M} \approx 0.2$) of C_{m_q} is given by

$$(C_{m_q})_{\mathbf{M}\approx 0.2} =$$

$$-0.7C_{l_\alpha} \cos \Lambda_{c/4} \left\{ \frac{A\left[\frac{1}{2}(h_{n_w} - h) + 2(h_{n_w} - h)^2\right]}{A + 2\cos \Lambda_{c/4}} + \frac{1}{24}\left(\frac{A^3 \tan^2 \Lambda_{c/4}}{A + 6\cos \Lambda_{c/4}}\right) + \frac{1}{8} \right\} \tag{B.8,1}$$

where

C_{l_α} is the wing section lift curve slope from Sec. B.1 (per rad).

$\Lambda_{c/4}$ is the sweepback angle of the wing $\frac{1}{4}$ chord line.

For higher subsonic speeds the derivative is obtained by applying an approximate compressibility correction.

$$(C_{m_q})_{\mathbf{M}>0.2} = \left[\frac{\dfrac{A^3 \tan^2 \Lambda_{c/4}}{AB + 6 \cos \Lambda_{c/4}} + \dfrac{3}{B}}{\dfrac{A^3 \tan^2 \Lambda_{c/4}}{A + 6 \cos \Lambda_{c/4}} + 3} \right] (C_{m_q})_{\mathbf{M}\approx 0.2} \tag{B.8,2}$$

where A is aspect ratio, and

$$B = \sqrt{1 - \mathbf{M}^2 \cos^2 \Lambda_{c/4}} \tag{B.8,3}$$

B.9 Wing Sideslip Derivatives C_{l_β}, C_{n_β}

The methods that follow are simplified versions of those given in USAF Datcom. They apply to rigid straight-tapered wings in subsonic flow.

The derivative C_{l_β}:

For $A \geq 1.0$:

$$C_{l_\beta} = C_L\left[\left(\frac{C_{l_\beta}}{C_L}\right)_{\Lambda_{c/2}} K_{M_\Lambda} + \left(\frac{C_{l_\beta}}{C_L}\right)_A\right] + \Gamma\left(\frac{C_{l_\beta}}{\Gamma} K_{M_\Gamma}\right) + \theta \tan \Lambda_{c/4} \frac{\Delta C_{l_\beta}}{\theta \tan \Lambda_{c/4}} \text{ (per deg)} \tag{B.9,1}$$

For $A < 1.0$:

$$C_{l_\beta} = C_L\left[-\frac{1}{57.3}\frac{2}{3}\frac{1}{A}\right] - \Gamma\left(\frac{A}{6}\right) \text{ (per deg)} \tag{B.9,2}$$

where

$\left(\dfrac{C_{l_\beta}}{C_L}\right)_{\Lambda_{c/2}}$ is the wing-sweep contribution obtained from Fig. B.9,1.

K_{M_Λ} is the compressibility correction to the sweep contribution, obtained from Fig. B.9,2.

$\left(\dfrac{C_{l_\beta}}{C_L}\right)_A$ is the aspect-ratio contribution, including taper-ratio effects, obtained from Fig. B.9,3.

$\dfrac{C_{l_\beta}}{\Gamma}$ is the dihedral effect for uniform geometric dihedral, obtained from Fig. B.9,4.

Γ is the dihedral angle in degrees.

K_{M_Γ} is the compressibility correction factor to the uniform-geometric-dihedral effect, obtained from Fig. B.9,5.

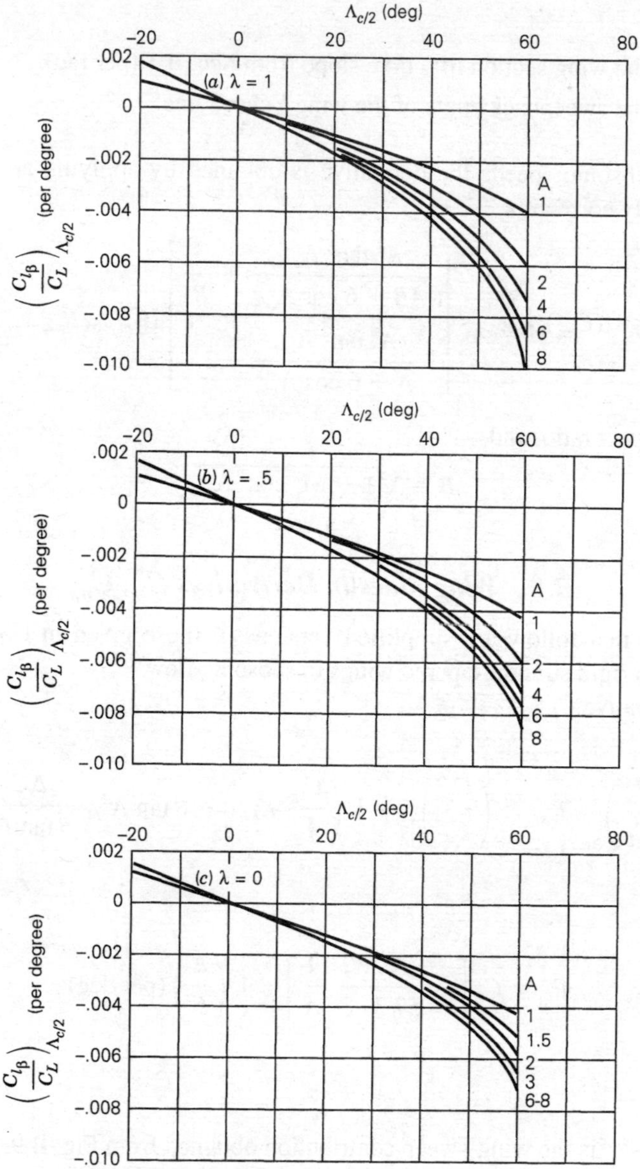

Figure B.9,1 Wing sweep contribution to C_{l_β}.

$\dfrac{\Delta C_{l_\beta}}{\theta \tan \Lambda_{c/4}}$ is the wing-twist correction factor, obtained from Fig. B.9,6.

θ is the wing-twist between the root and tip stations, negative for washout (see Fig. B.9,6).

$\Lambda_{c/2}$ is the sweepback angle of the midchord line.

$\Lambda_{c/4}$ is the sweepback angle of the $\frac{1}{4}$ chord line.

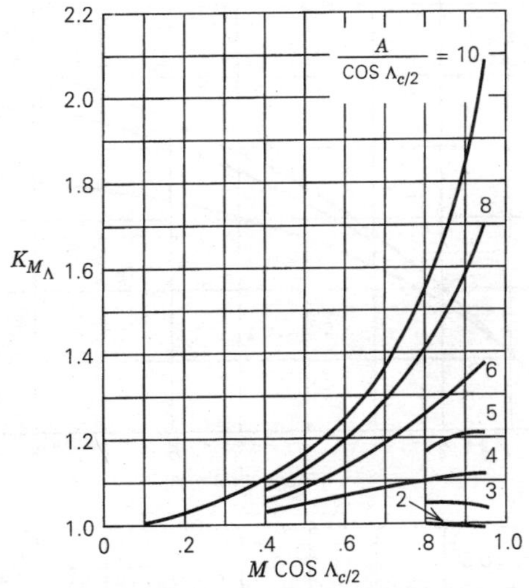

Figure B.9,2 Compressibility correction factor to sweep contribution to wing C_{l_β}.

The derivative C_{n_β}:

$$\left.\frac{C_{n_\beta}}{C_L^2}\right)_{\text{low speed}} = \frac{1}{57.3}\left[\frac{1}{4\pi A} - \frac{\tan \Lambda_{c/4}}{\pi A(A + 4\cos\Lambda_{c/4})}\left(\cos\Lambda_{c/4} - \frac{A}{2} - \frac{A^2}{8\cos\Lambda_{c/4}} + 6(h_{n_w} - h)\frac{\sin\Lambda_{c/4}}{A}\right)\right] \quad \text{(per deg)} \qquad \text{(B.9,3)}$$

For subcritical speeds, the low-speed derivative can be modified by the Prandtl-Glauert rule to yield approximate corrections for the first-order three-dimensional effects of compressibility:

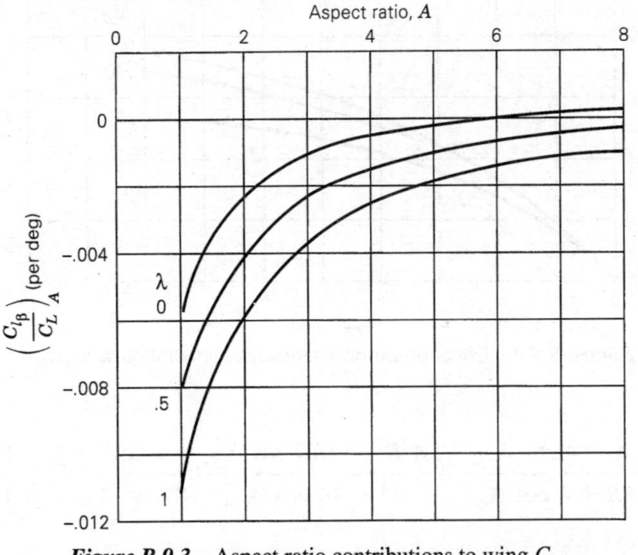

Figure B.9,3 Aspect ratio contributions to wing C_{l_β}.

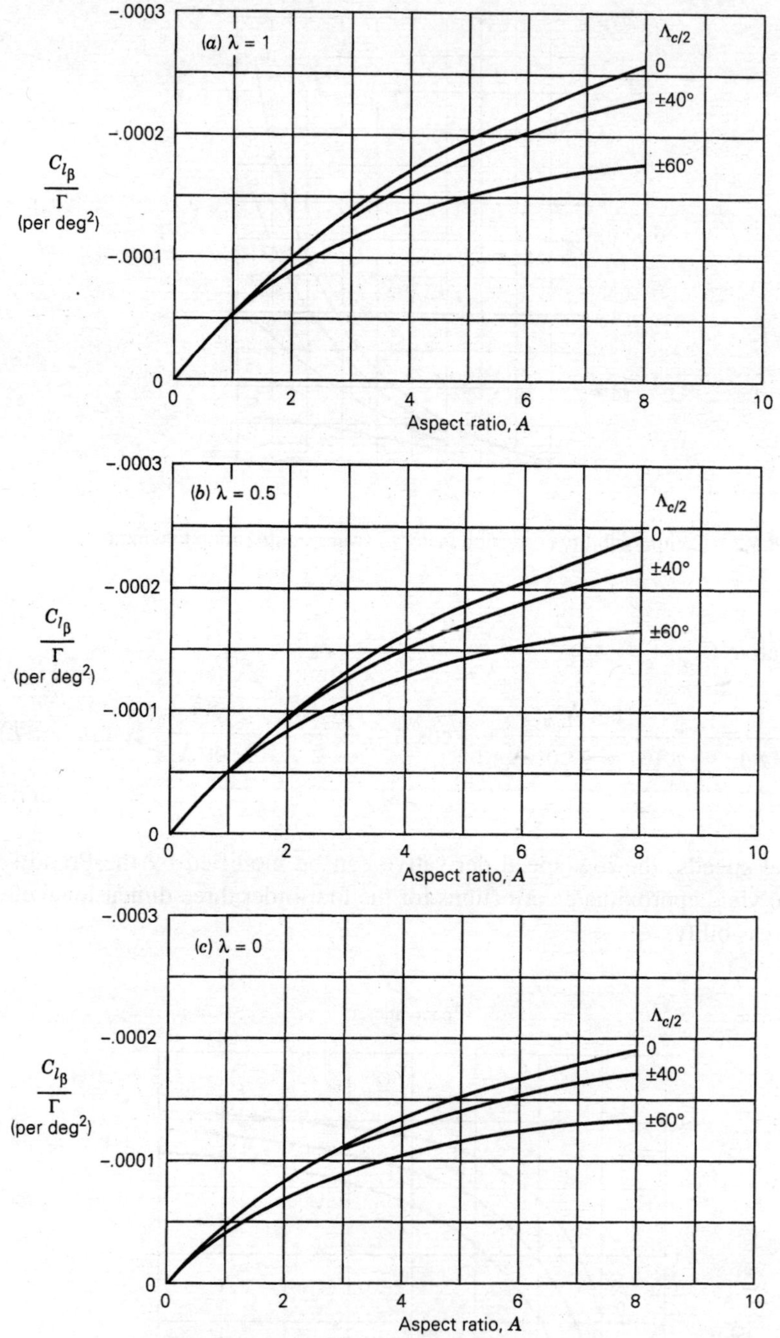

Figure B.9,4 Effect of uniform geometric dihedral on wing C_{l_β}.

$$\left(\frac{C_{n_\beta}}{C_L^2}\right)_{\mathrm{M}} = \left(\frac{A + 4\cos\Lambda_{c/4}}{AB + 4\cos\Lambda_{c/4}}\right)\left(\frac{A^2B^2 + 4AB\cos\Lambda_{c/4} - 8\cos^2\Lambda_{c/4}}{A^2 + 4A\cos\Lambda_{c/4} - 8\cos^2\Lambda_{c/4}}\right)\left(\frac{C_{n_\beta}}{C_L^2}\right)_{\text{low speed}} \tag{B.9,4}$$

where B is given by (B.8,3).

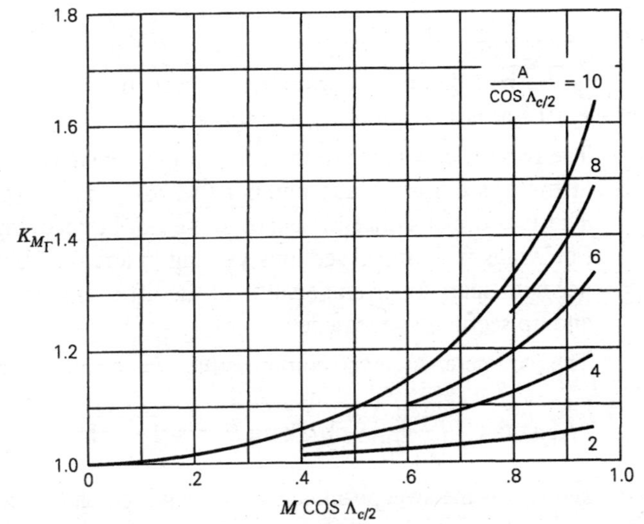

Figure B.9,5 Compressibility correction to dihedral effect on wing C_{l_β}.

B.10 Wing Rolling Derivatives C_{l_p}, C_{n_p}

The following methods are simplified versions of those given in USAF Datcom. They apply to rigid straight-tapered wings in subsonic flow, in the linear range of C_L vs. α.

The derivative C_{l_p}:

$$C_{l_p} = \left(\frac{\beta C_{l_p}}{\kappa}\right)_{C_L=0} \left(\frac{\kappa}{\beta}\right) \frac{(C_{l_p})_\Gamma}{(C_{l_p})_{\Gamma=0}} + (\Delta C_{l_p})_{\text{drag}} \tag{B.10,1}$$

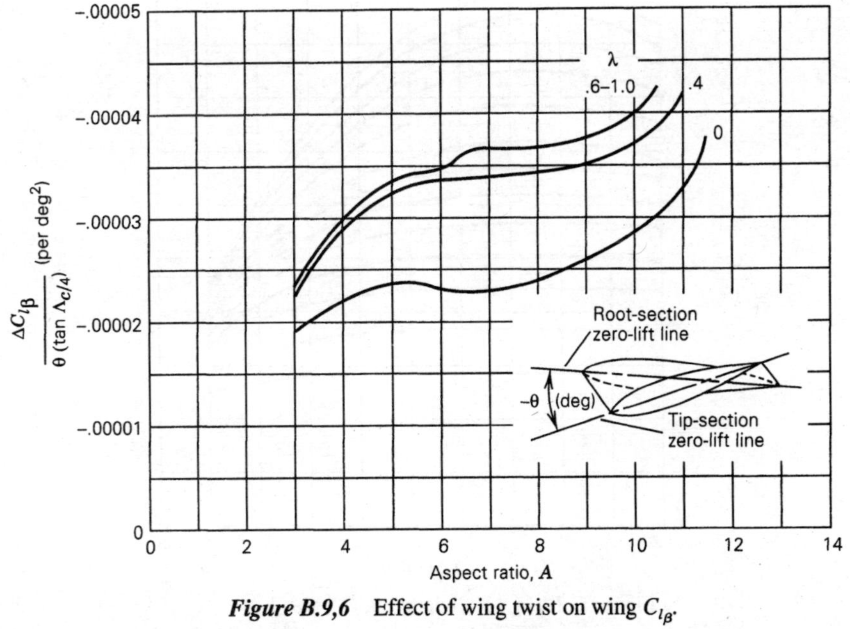

Figure B.9,6 Effect of wing twist on wing C_{l_β}.

where

$\left(\dfrac{\beta C_{l_p}}{\kappa}\right)_{C_L=0}$ is the roll-damping parameter at zero lift, obtained from Fig. B.10,1 as a function of Λ_β and $\beta A/\kappa$.

The parameter κ is the ratio of the two-dimensional lift-curve slope at the appropriate Mach number to $2\pi/\beta$; that is, $(C_{l_\alpha})_M/(2\pi/\beta)$. The two-dimensional lift-curve slope is obtained from Sec. B.1. For wings with airfoil sections varying in a reasonably linear manner with span, the average value of the lift-curve slopes of the root and tip sections is adequate.

The parameter Λ_β is the compressible sweep parameter given as

$$\Lambda_\beta = \tan^{-1}\left(\frac{\tan \Lambda_{c/4}}{\beta}\right), \text{ where } \beta = \sqrt{1 - \mathbf{M}^2}.$$

and $\Lambda_{c/4}$ is the sweepback angle of the wing $\frac{1}{4}$ chord line.

$\dfrac{(C_{l_p})_\Gamma}{(C_{l_p})_{\Gamma=0}}$ is the dihedral-effect parameter given by

$$\frac{(C_{l_p})_\Gamma}{(C_{l_p})_{\Gamma=0}} = \left[1 - 2\frac{z}{b/2}\sin\Gamma + 3\left(\frac{z}{b/2}\right)^2 \sin^2\Gamma\right] \tag{B.10,2}$$

where

Γ is the geometric dihedral angle, positive for the wing tip above the plane of the root chord.

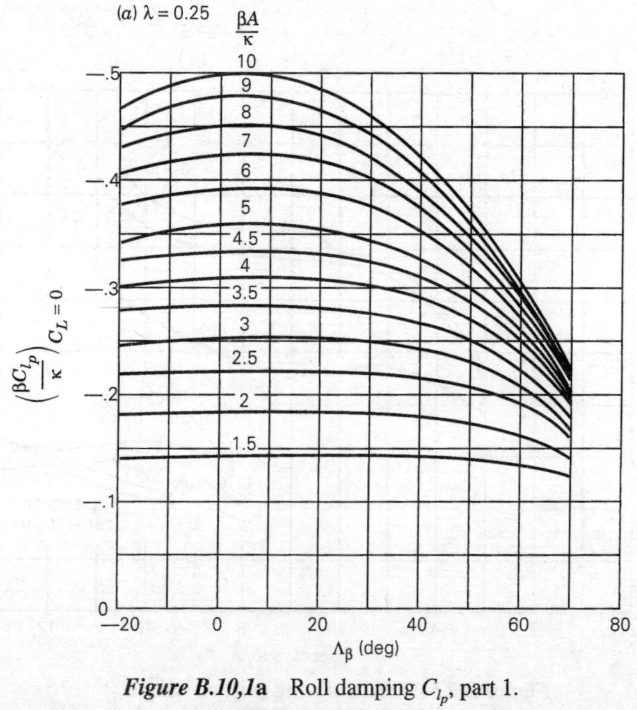

Figure B.10,1a Roll damping C_{l_p}, part 1.

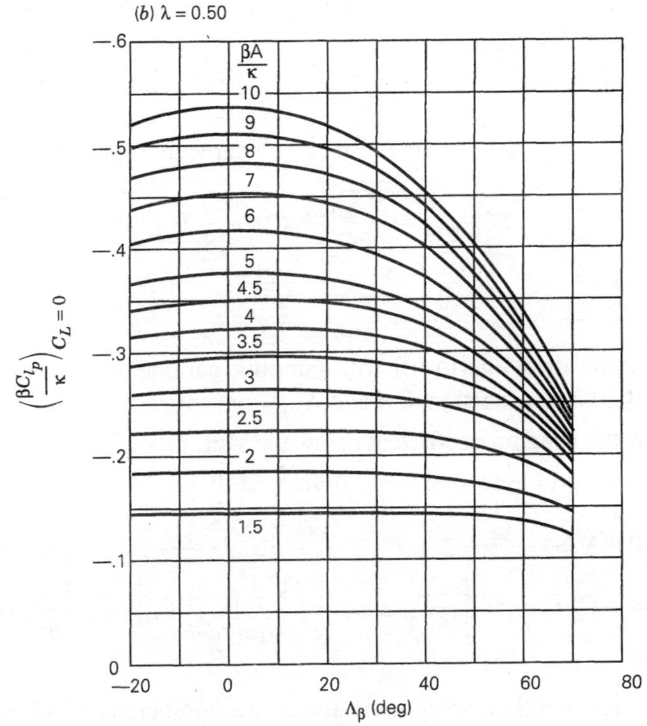

Figure B.10,1b Roll damping C_{l_p}, part 2.

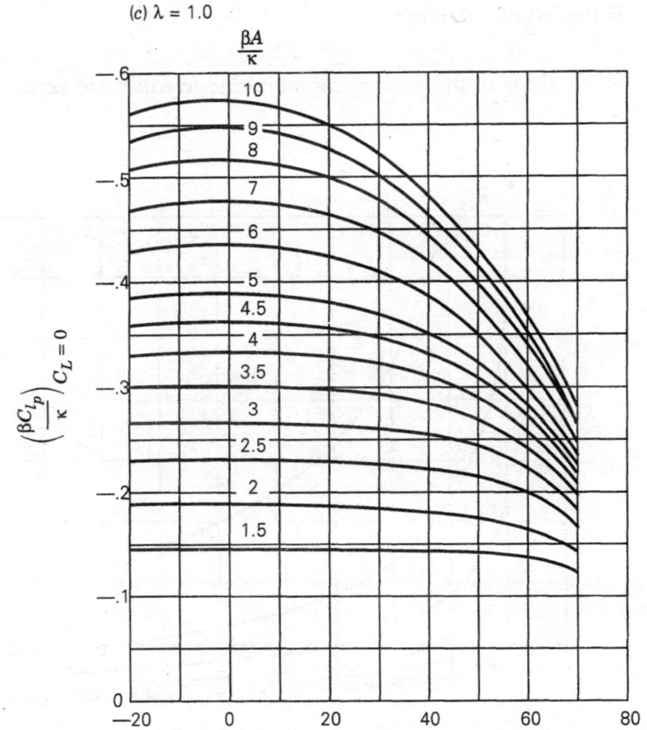

Figure B.10,1c Roll damping C_{l_p}, part 3.

z is the vertical distance between the CG and the wing root chord, positive for the CG above the root chord.

b is the wing span.

$(\Delta C_{l_p})_{\text{drag}}$ is the increment in the roll-damping derivative due to drag, given by

$$(\Delta C_{l_p})_{\text{drag}} = \frac{(C_{l_p})_{C_{DL}}}{C_L^2} C_L^2 - \frac{1}{8} C_{D0} \qquad \text{(B.10,3)}$$

where

$\frac{(C_{l_p})_{C_{DL}}}{C_L^2}$ is the drag-due-to-lift roll-damping parameter obtained from Fig. B.10,2 as a function of A and $\Lambda_{c/4}$.

C_L is the wing lift coefficient below the stall.

C_{D0} is the profile or total zero-lift drag coefficient.

The derivative C_{n_p}:

$$C_{n_p} = -C_{l_p} \tan\alpha - \left[-C_{l_p} \tan\alpha - \left(\frac{C_{n_p}}{C_L}\right)_{\substack{C_L=0\\M}} C_L \right] + \left(\frac{\Delta C_{n_p}}{\theta}\right)\theta \qquad \text{(B.10,4)}$$

where

C_{l_p} is the roll-damping derivative at the appropriate Mach number estimated above

α is the angle of attack.

C_L is the lift coefficient.

$\left(\frac{C_{n_p}}{C_L}\right)_{\substack{C_L=0\\M}}$ is the slope of the yawing moment due to rolling at zero lift given by

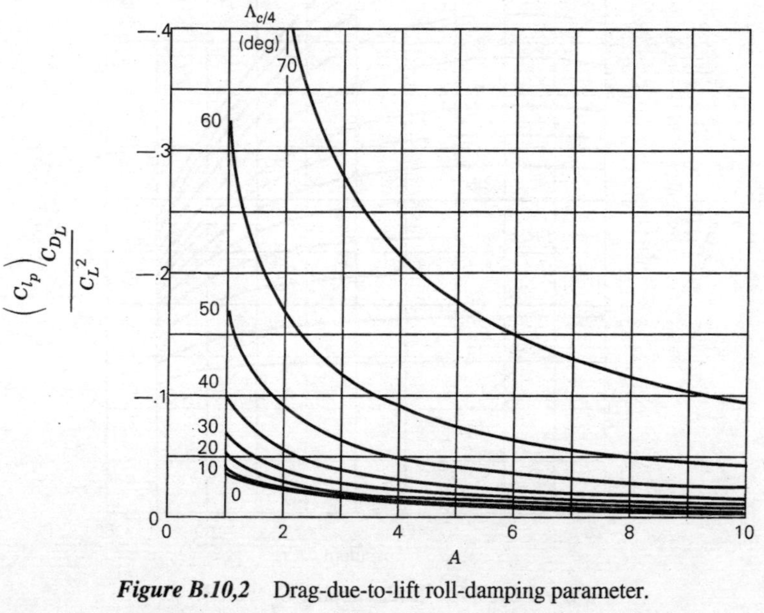

Figure B.10,2 Drag-due-to-lift roll-damping parameter.

$$\left.\frac{C_{n_p}}{C_L}\right)_{\substack{C_L=0\\M}} = \left(\frac{A + 4\cos\Lambda_{c/4}}{AB + 4\cos\Lambda_{c/4}}\right)\left[\frac{AB + \frac{1}{2}(AB + \cos\Lambda_{c/4})\tan^2\Lambda_{c/4}}{A + \frac{1}{2}(A + \cos\Lambda_{c/4})\tan^2\Lambda_{c/4}}\right]\left(\frac{C_{n_p}}{C_L}\right)_{\substack{C_L=0\\M=0}} \tag{B.10,5}$$

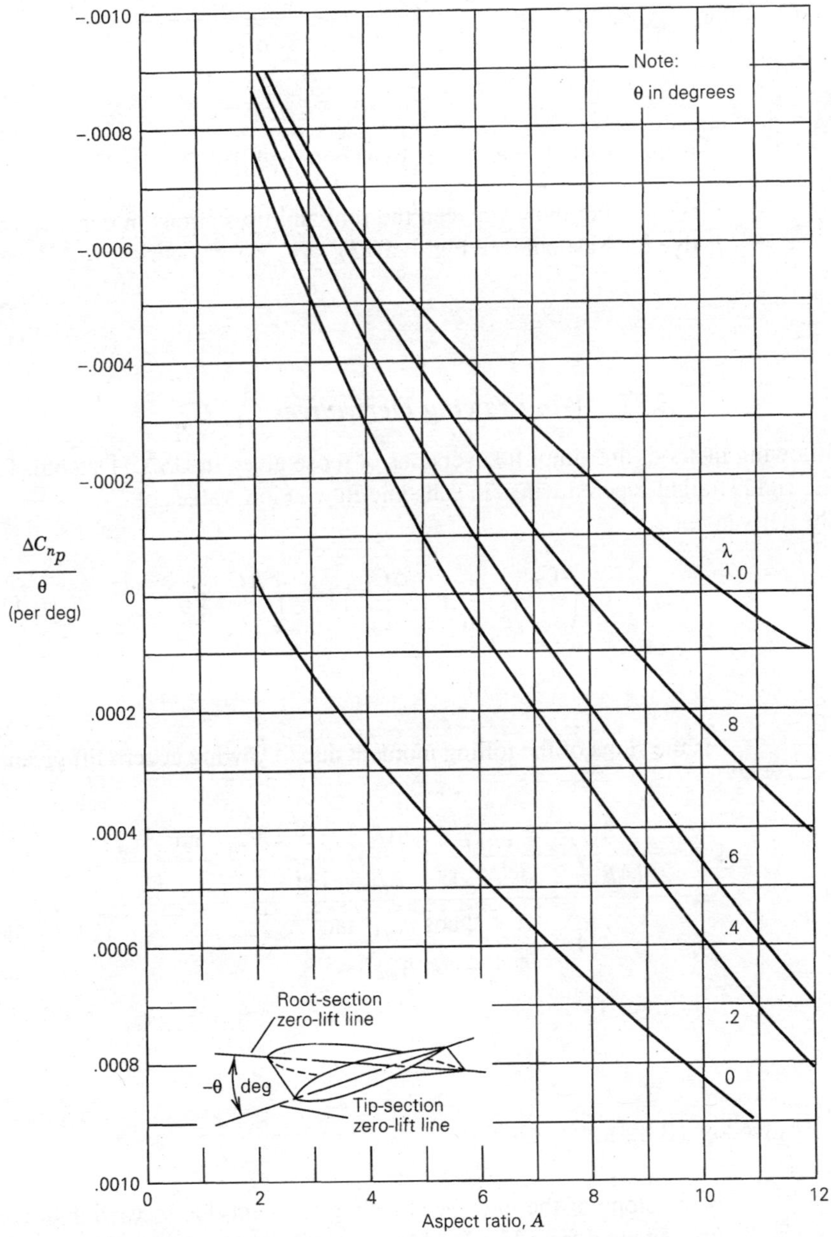

Figure B.10,3 Effect of wing twist on wing rolling derivative C_{n_p}.

where

B is given by (B.8,3).

$\Lambda_{c/4}$ is the sweepback angle of the $\frac{1}{4}$ chord line.

$\left(\dfrac{C_{n_p}}{C_L}\right)_{\substack{C_L=0\\M=0}}$ is the slope of the low-speed yawing moment due to rolling at zero lift given by

$$\left(\frac{C_{n_p}}{C_L}\right)_{\substack{C_L=0\\M=0}} = -\frac{1}{6}\,\frac{A + 6(A + \cos\Lambda_{c/4})\left((h_{n_w} - h)\dfrac{\tan\Lambda_{c/4}}{A} + \dfrac{\tan^2\Lambda_{c/4}}{12}\right)}{A + 4\cos\Lambda_{c/4}} \tag{B.10,6}$$

$\dfrac{\Delta C_{n_p}}{\theta}$ is the effect of linear wing twist obtained from Fig. B.10,3.

θ is the wing twist between the root and tip stations in degrees, negative for washout (see Fig. B.10,3).

B.11 Wing Yawing Derivatives C_{l_r}, C_{n_r}

The following methods are simplified versions of those given in USAF Datcom. They apply to rigid straight-tapered wings in subsonic flow at low values of C_L.

The derivative C_{l_r}:

$$C_{l_r} = C_L\left(\frac{C_{l_r}}{C_L}\right)_{\substack{C_L=0\\M}} + \left(\frac{\Delta C_{l_r}}{\Gamma}\right)\Gamma + \left(\frac{\Delta C_{l_r}}{\theta}\right)\theta \tag{B.11,1}$$

where

$\left(\dfrac{C_{l_r}}{C_L}\right)_{\substack{C_L=0\\M}}$ is the slope of the rolling moment due to yawing at zero lift given by

$$\left(\frac{C_{l_r}}{C_L}\right)_{\substack{C_L=0\\M}} = \frac{1 + \dfrac{A(1-B^2)}{2B(AB + 2\cos\Lambda_{c/4})} + \dfrac{AB + 2\cos\Lambda_{c/4}}{AB + 4\cos\Lambda_{c/4}}\,\dfrac{\tan^2\Lambda_{c/4}}{8}}{1 + \dfrac{A + 2\cos\Lambda_{c/4}}{A + 4\cos\Lambda_{c/4}}\,\dfrac{\tan^2\Lambda_{c/4}}{8}}\left(\frac{C_{l_r}}{C_L}\right)_{\substack{C_L=0\\M=0}} \tag{B.11,2}$$

where

B is given by (B.8,3).

$\left(\dfrac{C_{l_r}}{C_L}\right)_{\substack{C_L=0\\M=0}}$ is the slope of the low-speed rolling moment due to yawing at zero lift, obtained from Fig. B.11,1 as a function of aspect ratio, sweep of

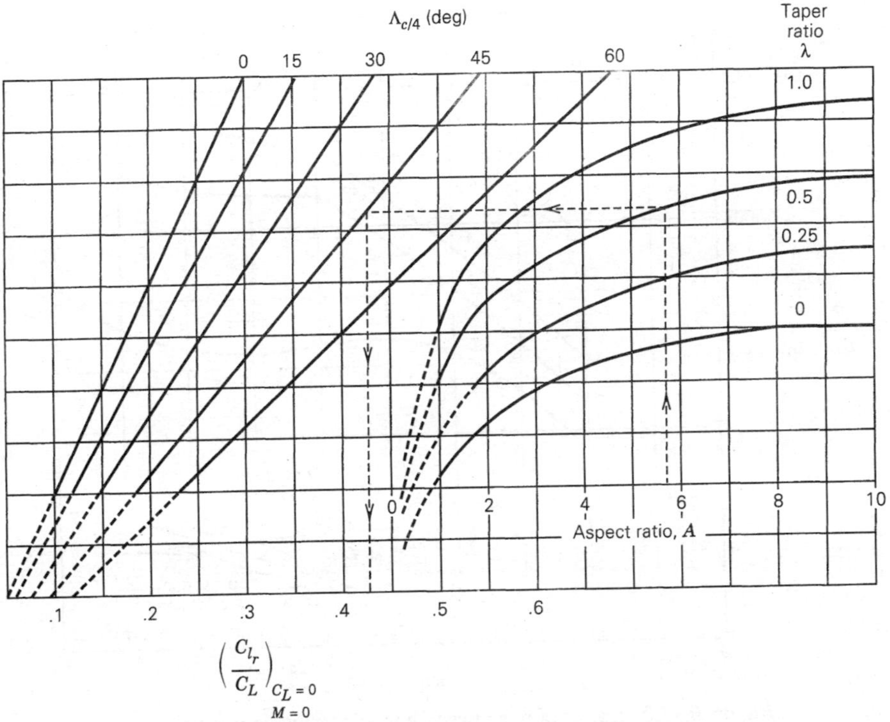

Figure B.11,1 Wing yawing derivative C_{l_r}.

the quarter-chord, and taper ratio. (B.11,2) modifies the low-speed value by means of the Prandtl-Glauert rule to yield approximate corrections for the first-order three-dimensional effects of compressible flow up to the critical Mach number.

$\frac{\Delta C_{l_r}}{\Gamma}$ is the increment in C_{l_r} due to dihedral, given by

$$\frac{\Delta C_{l_r}}{\Gamma} = \frac{1}{12} \frac{\pi A \sin \Lambda_{c/4}}{A + 4 \cos \Lambda_{c/4}} \tag{B.11,3}$$

Γ is the geometric dihedral angle in radians, positive for the wing tip above the plane of the root chord.

$\frac{\Delta C_{l_r}}{\theta}$ is the increment in C_{l_r} due to wing twist obtained from Fig. B.11,2.

θ is the wing twist between the root and tip sections in degrees, negative for washout (see Fig. B.11,2).

The derivative C_{n_r}:

$$C_{n_r} = \left(\frac{C_{n_r}}{C_L^2}\right) C_L^2 + \left(\frac{C_{n_r}}{C_{D_0}}\right) C_{D_0} \tag{B.11,4}$$

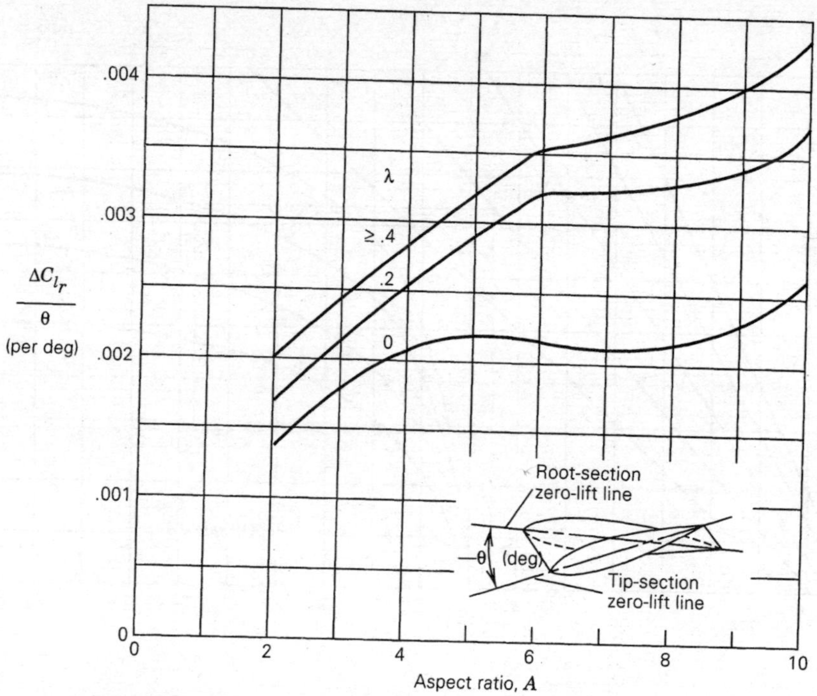

Figure B.11,2 Effect of wing twist on wing yawing derivative C_{l_r}.

where

C_L is the wing lift coefficient.

$\frac{C_{n_r}}{C_L^2}$ is the low-speed drag-due-to-lift yaw-damping parameter obtained from Fig. B.11,3 as a function of wing aspect ratio, taper ratio, sweepback, and CG position.

$\frac{C_{n_r}}{C_{D0}}$ is the low-speed profile-drag yaw-damping parameter obtained from Fig. B.11,4 as a function of the wing aspect ratio, sweep-back, and CG position.

C_{D0} is the wing profile drag coefficient evaluated at the appropriate Mach number. For this application C_{D_0} is assumed to be the profile drag associated with the theoretical ideal drag due to lift and is given by

$$C_{D_0} = C_D - \frac{C^2_L}{\pi A}$$

where C_D is the total drag coefficient at a given lift coefficient.

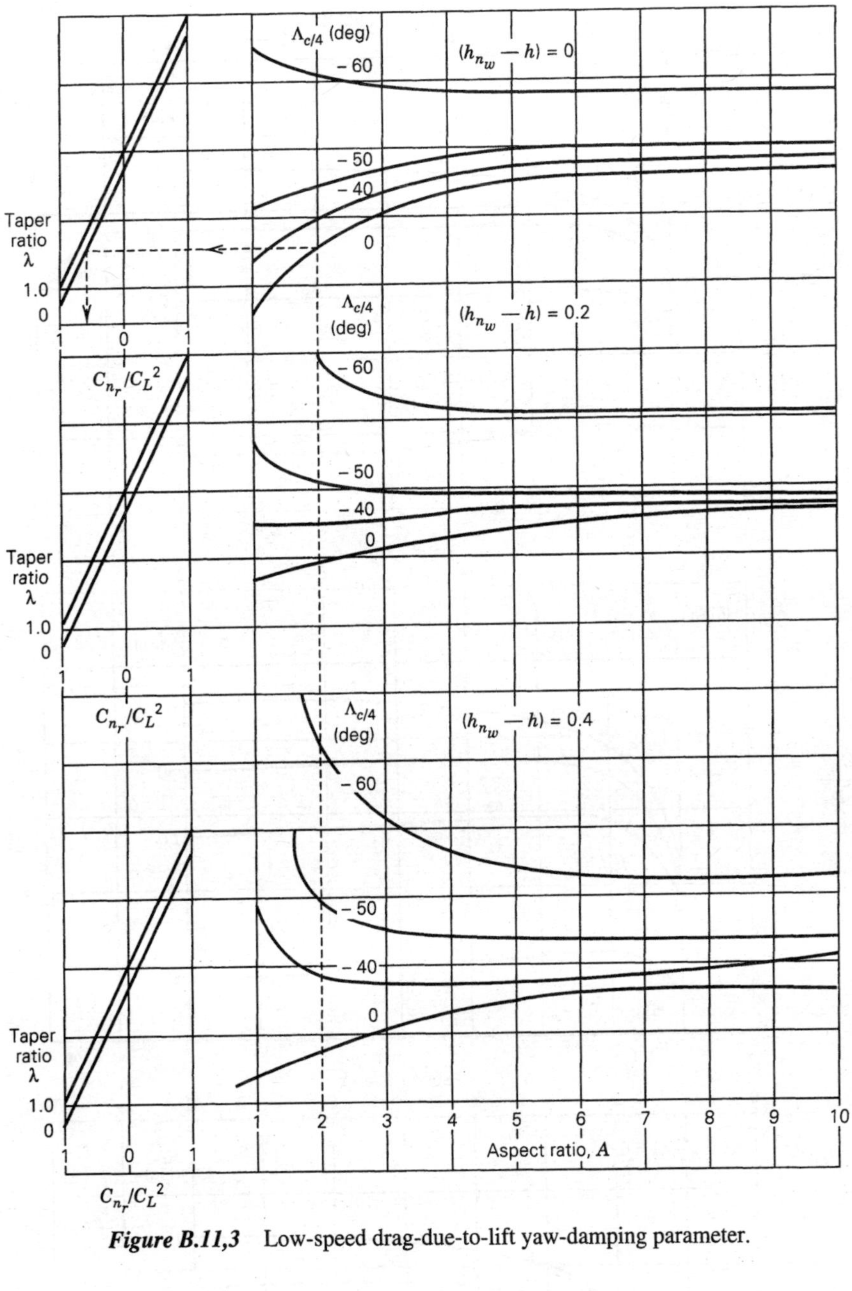

Figure B.11,3 Low-speed drag-due-to-lift yaw-damping parameter.

B.12 Changes in Inertias and Stability Derivatives with Change of Body Axes

A matrix **A** that connects two vectors **u** and **v** as in (A.4,23) transforms between two reference frames as in (A.4,26). We now apply this rule to the inertia matrix and to matrices of stability derivatives.

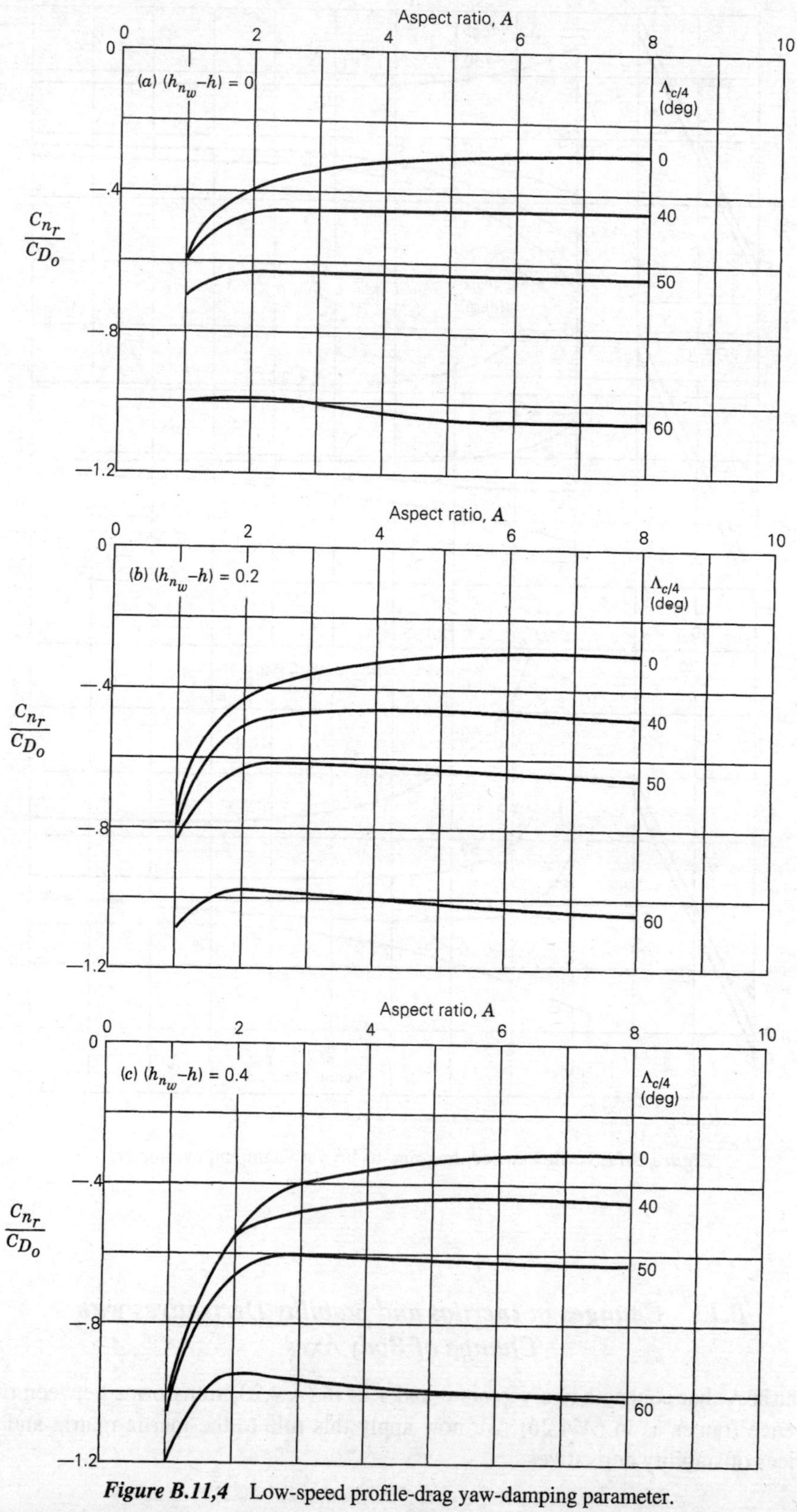

Figure B.11,4 Low-speed profile-drag yaw-damping parameter.

TRANSFORMATION OF INERTIAS

The inertia matrix **I** connects angular momentum with angular velocity [see (4.3,4) and (4.3,5)] via

$$\mathbf{h} = \mathbf{I}\boldsymbol{\omega}$$

and hence belongs to the class of matrices covered by (A.4,26). It follows that for two sets of body axes, denoted F_{B_1} and F_{B_2} connected by the transformation $\mathbf{L}_{12}$, the inertias in frame F_{B_2} can be obtained from those in F_{B_1} by

$$\mathbf{I}_2 = \mathbf{L}_{21}\mathbf{I}_1\mathbf{L}_{12} \tag{B.12,1}$$

If the two frames are two sets of body axes such that x_{B_1} is rotated about y_{B_1} through angle ξ to bring it to x_{B_2}, then (see Appendix A.4)

$$\mathbf{L}_{21} = \begin{bmatrix} \cos\xi & 0 & -\sin\xi \\ 0 & 1 & 0 \\ \sin\xi & 0 & \cos\xi \end{bmatrix} \tag{B.12,2}$$

The inertias in frame F_{B_2}, denoted by an asterisk, are then obtained from those in F_{B_1}, with the usual assumption of symmetry about the xz plane, by the relations

$$\begin{aligned} I_x^* &= I_x\cos^2\xi + I_z\sin^2\xi + I_{zx}\sin 2\xi \\ I_z^* &= I_x\sin^2\xi + I_z\cos^2\xi - I_{zx}\sin 2\xi \\ -I_{zx}^* &= \tfrac{1}{2}(I_x - I_z)\sin 2\xi + I_{zx}(\sin^2\xi - \cos^2\xi) \end{aligned} \tag{B.12,3}$$

TRANSFORMATION OF STABILITY DERIVATIVES

All of the stability derivatives with respect to linear and angular velocities and velocity derivatives can be expressed as sums of expressions of the form of (A.4,23). That is, with the usual assumptions about separation of longitudinal and lateral motion, we can write

$$\begin{bmatrix} \Delta X \\ \Delta Y \\ \Delta Z \end{bmatrix} = \begin{bmatrix} X_u & 0 & X_w \\ 0 & Y_v & 0 \\ Z_u & 0 & Z_w \end{bmatrix}\begin{bmatrix} \Delta u \\ v \\ \Delta w \end{bmatrix} + \begin{bmatrix} 0 & X_q & 0 \\ Y_p & 0 & Y_r \\ 0 & Z_q & 0 \end{bmatrix}\begin{bmatrix} p \\ q \\ r \end{bmatrix} + \begin{bmatrix} 0 & 0 & 0 \\ 0 & 0 & 0 \\ 0 & 0 & Z_{\dot{w}} \end{bmatrix}\begin{bmatrix} \Delta\dot{u} \\ \dot{v} \\ \Delta\dot{w} \end{bmatrix} \tag{B.12,4}$$

$$\begin{bmatrix} \Delta L \\ \Delta M \\ \Delta N \end{bmatrix} = \begin{bmatrix} 0 & L_v & 0 \\ M_u & 0 & M_w \\ 0 & N_v & 0 \end{bmatrix}\begin{bmatrix} \Delta u \\ v \\ \Delta w \end{bmatrix} + \begin{bmatrix} L_p & 0 & L_r \\ 0 & M_q & 0 \\ N_p & 0 & N_r \end{bmatrix}\begin{bmatrix} p \\ q \\ r \end{bmatrix} + \begin{bmatrix} 0 & 0 & 0 \\ 0 & 0 & 0 \\ 0 & 0 & M_{\dot{w}} \end{bmatrix}\begin{bmatrix} \Delta\dot{u} \\ \dot{v} \\ \Delta\dot{w} \end{bmatrix} \tag{B.12,5}$$

Each of the six matrices of derivatives above transforms according to the rule (A.4,26). When **L** is given by (B.12,2) we have the transformation from an initial set of body axes (unprimed) to a second set (primed) as follows:

Longitudinal

$$
\begin{aligned}
(X_u)' &= X_u \cos^2 \xi - (X_w + Z_u) \sin \xi \cos \xi + Z_w \sin^2 \xi \\
(X_w)' &= X_w \cos^2 \xi + (X_u - Z_w) \sin \xi \cos \xi - Z_u \sin^2 \xi \\
(X_q)' &= X_q \cos \xi - Z_q \sin \xi \\
(X_{\dot{u}})' &= Z_{\dot{w}} \sin^2 \xi \quad (1) \\
(X_{\dot{w}})' &= -Z_{\dot{w}} \sin \xi \cos \xi \quad (1) \\
(Z_u)' &= Z_u \cos^2 \xi - (Z_w - X_u) \sin \xi \cos \xi - X_w \sin^2 \xi \\
(Z_w)' &= Z_w \cos^2 \xi + (Z_u + X_w) \sin \xi \cos \xi + X_u \sin^2 \xi \\
(Z_q)' &= Z_q \cos \xi + X_q \sin \xi \\
(Z_{\dot{u}})' &= -Z_{\dot{w}} \sin \xi \cos \xi \quad (1) \\
(Z_{\dot{w}})' &= Z_{\dot{w}} \cos^2 \xi \\
(M_u)' &= M_u \cos \xi - M_w \sin \xi \\
(M_w)' &= M_w \cos \xi + M_u \sin \xi \\
(M_q)' &= M_q \\
(M_{\dot{u}})' &= -M_{\dot{w}} \sin \xi \quad (1) \\
(M_{\dot{w}})' &= M_{\dot{w}} \cos \xi \quad (1)
\end{aligned}
\qquad \text{(B.12,6)}
$$

Lateral

$$
\begin{aligned}
(Y_v)' &= Y_v \\
(Y_p)' &= Y_p \cos \xi - Y_r \sin \xi \\
(Y_r)' &= Y_r \cos \xi + Y_p \sin \xi \\
(L_v)' &= L_v \cos \xi - N_v \sin \xi \\
(L_p)' &= L_p \cos^2 \xi - (L_r + N_p) \sin \xi \cos \xi + N_r \sin^2 \xi \\
(L_r)' &= L_r \cos^2 \xi - (N_r - L_p) \sin \xi \cos \xi - N_p \sin^2 \xi \\
(N_v)' &= N_v \cos \xi + L_v \sin \xi \\
(N_p)' &= N_p \cos^2 \xi - (N_r - L_p) \sin \xi \cos \xi - L_r \sin^2 \xi \\
(N_r)' &= N_r \cos^2 \xi + (L_r + N_p) \sin \xi \cos \xi + L_p \sin^2 \xi
\end{aligned}
\qquad \text{(B.12,7)}
$$

(1) For consistency of assumptions, the derivatives with respect to $\dot{u}$ and $(X_{\dot{w}})'$ are usually ignored.

APPENDIX C

Mean Aerodynamic Chord, Mean Aerodynamic Center, and $C_{m_{ac_w}}$

C.1 Basic Definitions

In the normal flight range, the resultant aerodynamic forces acting on any lifting surface can be represented as a lift and drag acting at the mean aerodynamic center ($\bar{x}$, $\bar{y}$, $\bar{z}$), together with a pitching couple $C_{m_{ac_w}}$ which is independent of angle of attack (see Fig. 2.8).

The pitching moment of a wing is nondimensionalized by the use of the mean aerodynamic chord $\bar{c}$.

Both the m.a. center and the m.a. chord lie in the plane of symmetry of the wing. However, in determining them it is convenient to work with the half-wing.

These quantities are defined by (see Fig. C.1)[1]

$$\bar{c} = \frac{2}{S}\int_0^{b/2} c^2\,dy \tag{C.1,1}$$

$$\bar{x} = \frac{2}{C_L S}\int_0^{b/2} C_{l_a} cx\,dy \tag{C.1,2}$$

$$\bar{y} = \frac{2}{C_L S}\int_0^{b/2} C_{l_a} cy\,dy = \eta_{cp}\frac{b}{2} \tag{C.1,3}$$

$$\bar{z} = \frac{2}{C_L S}\int_0^{b/2} C_{l_a} cz\,dy \tag{C.1,4}$$

where b = wing span

c = local chord

C_L = total lift coefficient

C_{l_a} = local additional lift coefficient, proportional to C_L

C_{l_b} = local basic lift coefficient, independent of C_L

$C_l = C_{l_b} + C_{l_a}$ = total local lift coefficient

[1]The coordinate system used applies only to this appendix.

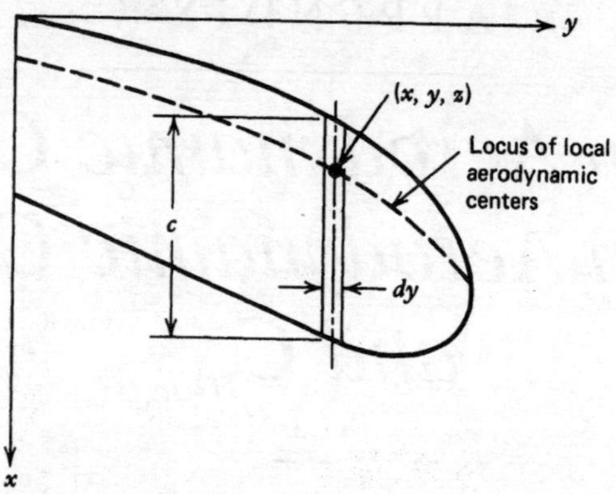

Figure C.1 Local aerodynamic center coordinates.

m_{ac} = pitching moment, per unit span, about aerodynamic center (Fig. C.4)

S = wing area

y = spanwise coordinate of local aerodynamic center measured from axis of symmetry

x = chordwise coordinate of local aerodynamic center measured aft of wing apex

z = vertical coordinate of local aerodynamic center measured from xy plane

η_{cp} = lateral position of the center of pressure of the additional load on the half-wing as a fraction of the semispan

The coordinates of the m.a. center depend on the additional load distribution; hence the position of the true m.a. center will vary with wing angle of attack if the form of the additional loading varies with angle of attack. For a wing that has no aerodynamic twist, the m.a. center of the half-wing is also the center of pressure of the half-wing. If there is a basic loading (i.e., at zero overall lift, due to wing twist), then $(\bar{x}, \bar{y}, \bar{z})$ is the center of pressure of the additional loading.

The height and spanwise position of the local aerodynamic centers may be assumed known, and hence $\bar{y}$ and $\bar{z}$ for the half-wing can be calculated once the additional spanwise loading distribution is known. However, in order to calculate $\bar{x}$, the fore-and-aft position of each local aerodynamic center must be known first. If all the local aerodynamic centers are assumed to lie on the nth-chord line (assumed to be straight), then

$$\bar{x} = nc_r + \bar{y} \tan \Lambda_n \quad \text{(C.1,5)}$$

where c_r = wing root chord

Λ_n = sweepback of nth-chord line, degrees

Ideal two-dimensional flow theory gives $n = \frac{1}{4}$ for subsonic speeds and $n = \frac{1}{2}$ for supersonic speeds.

The m.a. chord is located relative to the wing by the following procedure:

1. In (C.1,2) replace C_{l_a} by C_L, and for x use the coordinates of the $\frac{1}{4}$-chord line.
2. The value of $\bar{x}$ so obtained (the mean quarter-chord point) is the $\frac{1}{4}$-point of the m.a. chord.

The above procedure and the definition of $\bar{c}$ (see C.1,1) are used for *all* wings.

C.2 Comparison of m.a. Chord and m.a. Center for Basic Planforms and Loading Distributions

In Table C.1 taken from (Yates, 1952), values of m.a. chord and $\bar{y}$ are given for some basic planforms and loading distributions.

In the general case the additional loading distribution and the spanwise center-of-pressure position can be obtained by methods such as those of De Young and Harper (1948), Weissinger (1947), and Stanton-Jones (1950). For a trapezoidal wing with the local aerodynamic centers on the nth-chord line, the chordwise location of the mean aerodynamic center from the leading edge of the m.a. chord expressed as a fraction of the m.a. chord h_{n_w} is given by

$$h_{n_w} = n + \frac{3(1+\lambda)^2}{8(1+\lambda+\lambda^2)}\left[\eta_{cp} - \frac{1+2\lambda}{3(1+\lambda)}\right] A \tan \Lambda_n \qquad \text{(C.2,1)}$$

Table C.1

Planform	*Additional Loading Distribution*	*M.A.C.* $\bar{c}$	$\bar{y}$
Constant taper and sweep (trapezoidal)	Any	$\frac{2c_r}{3}\frac{1+\lambda+\lambda^2}{1+\lambda}$	$\eta_{cp}\cdot\frac{b}{2}$
Constant taper and sweep (trapezoidal)	Proportional to wing chord (uniform C_{l_a})	$\frac{2c_r}{3}\frac{1+\lambda+\lambda^2}{1+\lambda}$	$\frac{b}{2}\cdot\frac{1+2\lambda}{3(1+\lambda)}$
Constant taper and sweep (trapezoidal)	Elliptic	$\frac{2c_r}{3}\frac{1+\lambda+\lambda^2}{1+\lambda}$	$\frac{b}{2}\cdot\frac{4}{3\pi}$
Elliptic (with straight sweep of line of local a.c.)	Any	$\frac{c_r}{3}\cdot\frac{8}{\pi}$	$\eta_{cp}\cdot\frac{b}{2}$
Elliptic (with straight sweep of line of local a.c.)	Elliptic (uniform C_{l_a})	$\frac{c_r}{3}\cdot\frac{8}{\pi}$	$\frac{b}{2}\cdot\frac{4}{3\pi}$
Any (with straight sweep of line of local a.c.)	Elliptic	$\frac{2}{S}\int_0^{b/2} c^2\,dy$	$\frac{b}{2}\cdot\frac{4}{3\pi}$

where A = aspect ratio, b^2/S

λ = taper ratio, c_t/c_r

c_t = wing-tip chord

The length of the chord through the centroid of area of a trapezoidal half-wing is equal to $\bar{c}$. For the same wing with uniform spanwise lift distribution (i.e., C_{l_a} = const) and local aerodynamic centers on the nth-chord line, the m.a. center also lies on the chord through the centroid of area. The chord through the centroid of area of a wing having an elliptic planform is not the same as $\bar{c}$, but the m.a. center for elliptic loading and the centroid of area both lie on the same chord (see Yates, 1952).

C.3 m.a. Chord and m.a. Center for Swept and Tapered Wings (Subsonic)

The ratio $\bar{c}/c_r$ is plotted against λ in Fig. C.2 for straight tapered wings with streamwise tips. The spanwise position of the m.a. center of the half-wing (or the center of pressure of the additional load) for uniform spanwise loading is also given in Fig. C.2. These functions are given in Table C.1.

The m.a. chord is located by means of the distance x of the leading edge of the m.a. chord aft of the wing apex:

$$x = \frac{b}{2} \cdot \frac{1}{3} \frac{1 + 2\lambda}{1 + \lambda} \tan \Lambda_0$$

$$= \frac{1 + 2\lambda}{12} c_r A \tan \Lambda_0 \tag{C.3,1}$$

where Λ_0 = sweepback of wing leading edge, degrees.

The sweepback of the leading edge is related to the sweep of the nth-chord line Λ_n by the relation

$$A \tan \Lambda_0 = A \tan \Lambda_n + 4n \frac{1 - \lambda}{1 + \lambda} \tag{C.3,2}$$

Using (C.3,2) and the expression for $\bar{c}/c_r$, x can be obtained in terms of $\bar{c}$ and Λ_n from

$$\frac{x}{\bar{c}} = \frac{(1 + 2\lambda)(1 + \lambda)}{8(1 + \lambda + \lambda^2)} \left[A \tan \Lambda_n + 4n \frac{1 - \lambda}{1 + \lambda} \right] \tag{C.3,3}$$

The fractional distance of the m.a. center aft of the leading edge of the m.a. chord, h_{n_w}, is given for swept and tapered wings at low speeds and small incidences in Fig. C.3. The dotted lines show the aerodynamic-center position for wings with unswept trailing edges. The curves have been obtained from theoretical and experimental data. The curves apply only within the linear range of the curve of wing lift against pitching moment, provided that the flow is subsonic over the entire wing. The probable error of h_{n_w} given by the curves is within 3%.

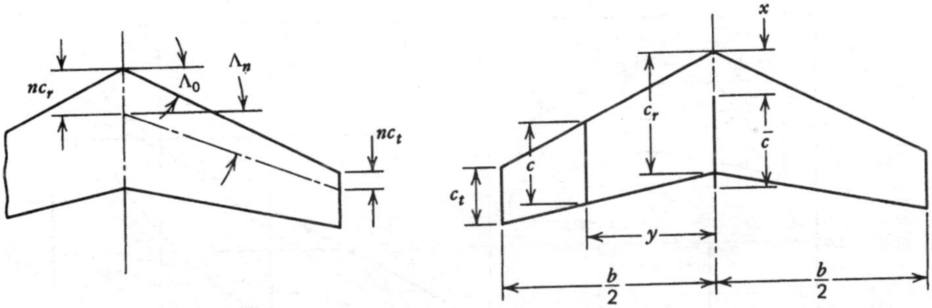

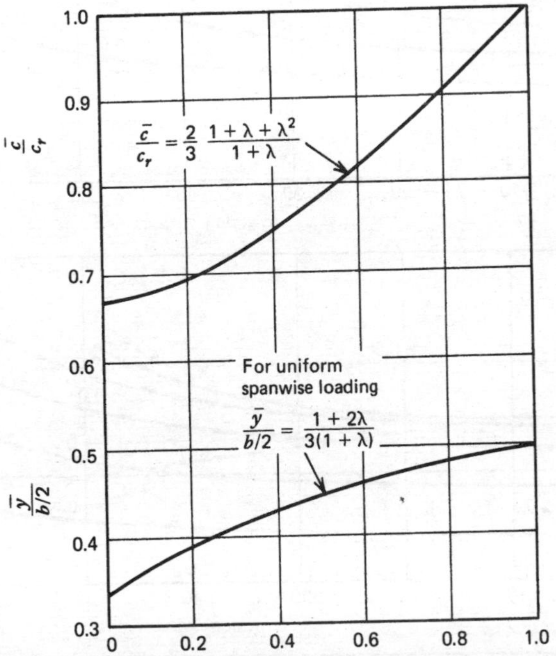

Figure C.2 Mean aerodynamic chord for straight tapered wings; and spanwise position of mean aerodynamic center for uniform spanwise loading (i.e., constant C_{L_α}). (From "Notes on the Mean Aerodynamic Chord and the Mean Aerodynamic Center of a Wing" by A. H. Yates, *J. Roy. Aero. Soc.*, June 1952.)

C.4 C_{mac_w}

The total load on each section of a wing has three parts as illustrated by Fig. C.4*a*. The resultant of the local additional lift l_a, is the lift L_a acting through the m.a. center (Fig. C.4*b*).

The resultant of the distribution of the local basic lift l_b is a pitching couple whenever the line of aerodynamic centers is not straight and perpendicular to x. This couple is given by

$$M_1 = 2\int_0^{b/2} (x - \bar{x}) l_b \, dy = 2\int_0^{b/2} x l_b \, dy$$

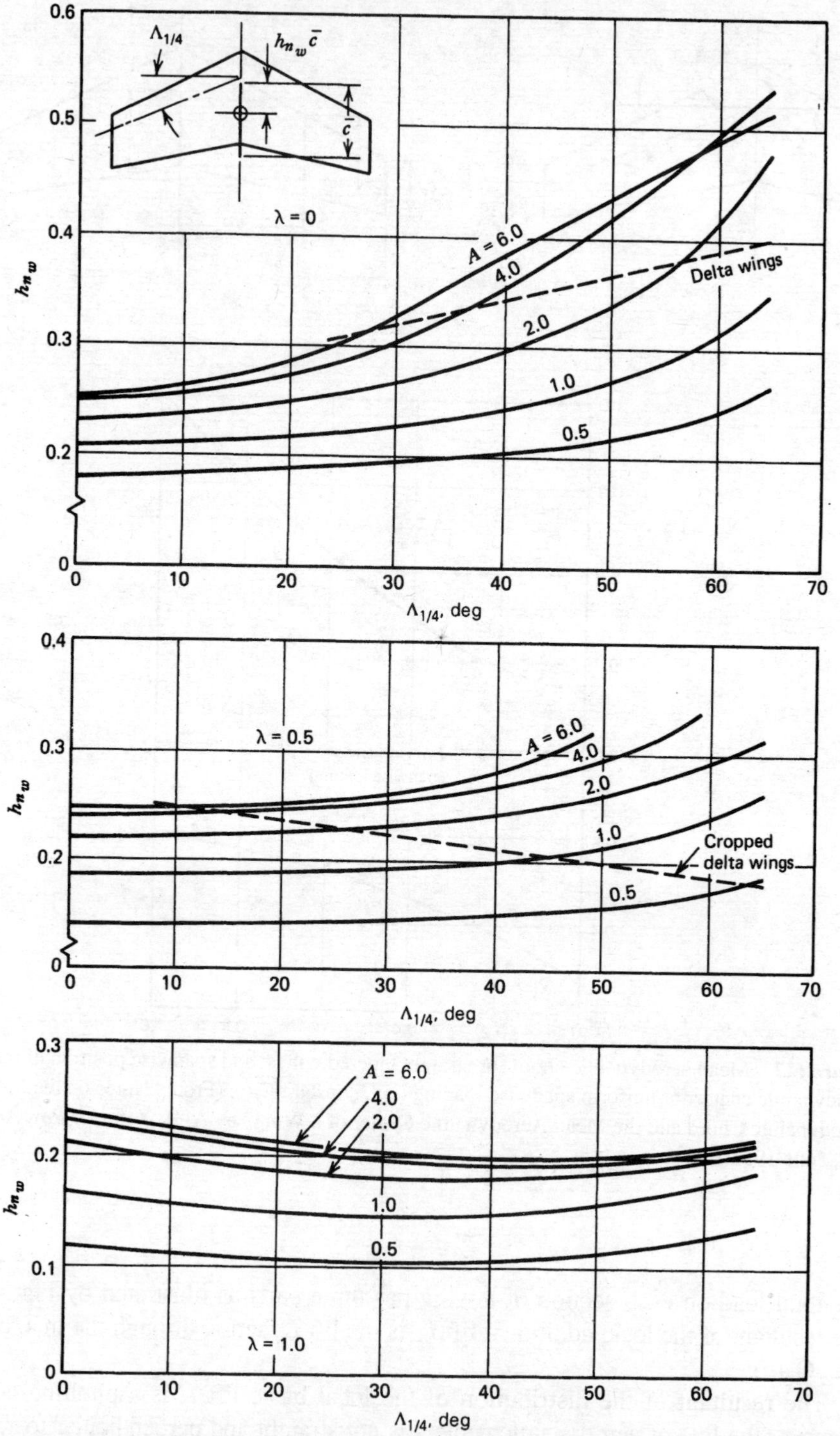

Figure C.3 Chordwise position of the mean aerodynamic center of swept and tapered wings at low speeds expressed as a fraction of the mean aerodynamic chord. (From Royal Aeronautical Data Sheet Wings 08.01.01.)

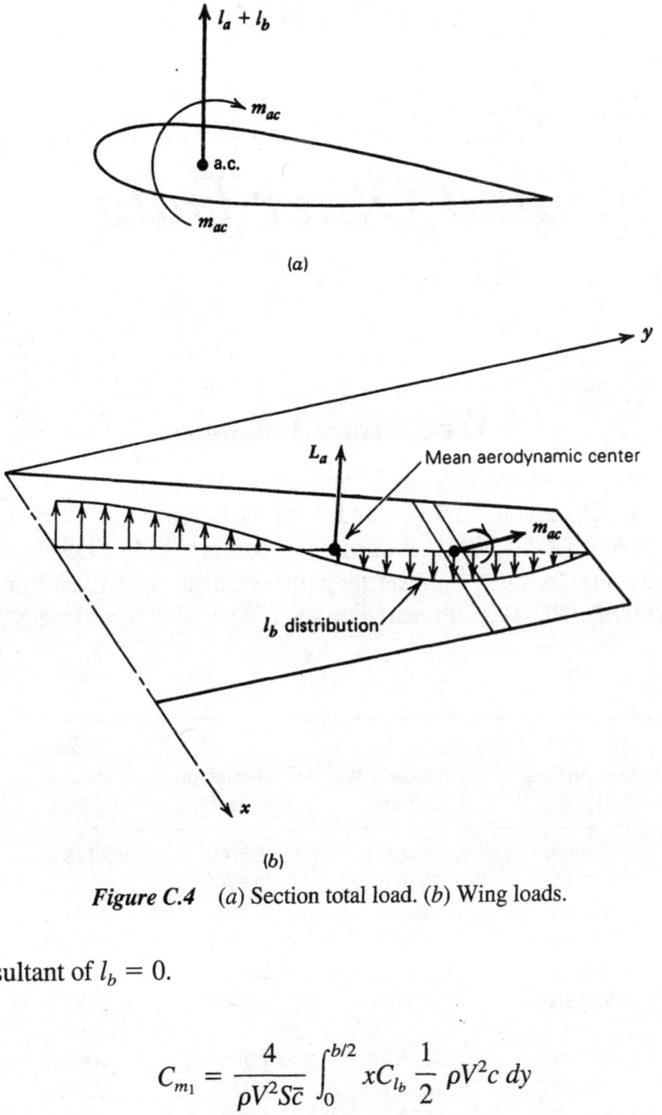

Figure C.4 (*a*) Section total load. (*b*) Wing loads.

since the resultant of $l_b = 0$.

Then

$$C_{m_1} = \frac{4}{\rho V^2 S\bar{c}} \int_0^{b/2} xC_{l_b} \frac{1}{2} \rho V^2 c \, dy$$

$$= \frac{2}{S\bar{c}} \int_0^{b/2} C_{l_b} xc \, dy \tag{C.4,1}$$

The resultant of the m_{ac} distribution is given by

$$C_{m_2} = \frac{2}{S\bar{c}} \int_0^{b/2} C_{m_{ac}} c^2 \, dy \tag{C.4,2}$$

The total pitching-moment coefficient about the m.a. center is then

$$C_{m_{ac_w}} = C_{m_1} + C_{m_2} = \text{const} \tag{C.4,3}$$

If $C_{m_{ac}}$ is constant across the span, and equals C_{m_2}, then (C.4,2) also becomes the defining equation for $\bar{c}$.

APPENDIX D

The Standard Atmosphere and Other Data

The Standard Atmosphere

The tables that follow are derived from *The ARDC Model Atmosphere, 1959,* by Minzner, R. A., Champion, K. S. W., and Pond, H. L. Air Force Cambridge Research Center Report No. TR-59-267, U.S. Air Force, Bedford, MA, 1959. The values in the tables are the same, for most engineering purposes, as those derived from *U.S. STANDARD ATMOSPHERE, 1976.* Prepared by the USAF, NASA, and the NOAA.

English Units[a]

Altitude h, *ft*	*Temperature* T, °*R*	*Pressure* P, *lb/ft*^{2}	*Density* ρ, *lb sec*^{2}*/ft*^{4}	*Speed of sound, ft/sec*	*Kinematic viscosity, ft*^{2}*/sec*
0	518.69	2116.2	2.3769^{-3}	1116.4	1.5723^{-4}
1,000	515.12	2040.9	2.3081	1112.6	1.6105
2,000	511.56	1967.7	2.2409	1108.7	1.6499
3,000	507.99	1896.7	2.1752	1104.9	1.6905
4,000	504.43	1827.7	2.1110	1101.0	1.7324
5,000	500.86	1760.9	2.0482	1097.1	1.7755
6,000	497.30	1696.0	1.9869^{-3}	1093.2	1.8201^{-4}
7,000	493.73	1633.1	1.9270	1089.3	1.8661
8,000	490.17	1572.1	1.8685	1085.3	1.9136
9,000	486.61	1512.9	1.8113	1081.4	1.9626
10,000	483.04	1455.6	1.7556	1077.4	2.0132
11,000	479.48	1400.0	1.7011^{-3}	1073.4	2.0655^{-4}
12,000	475.92	1346.2	1.6480	1069.4	2.1196
13,000	472.36	1294.1	1.5961	1065.4	2.1754
14,000	468.80	1243.6	1.5455	1061.4	2.2331
15,000	465.23	1194.8	1.4962	1057.4	2.2927
16,000	461.67	1147.5	1.4480^{-3}	1053.3	2.3544^{-4}
17,000	458.11	1101.7	1.4011	1049.2	2.4183
18,000	454.55	1057.5	1.3553	1045.1	2.4843
19,000	450.99	1014.7	1.3107	1041.0	2.5526
20,000	447.43	973.27	1.2673	1036.9	2.6234

English Units[a] *(Continued)*

Altitude h, *ft*	*Temperature* T, °*R*	*Pressure* P, *lb/ft*^{2}	*Density* ρ, *lb sec*^{2}*/ft*^{4}	*Speed of sound, ft/sec*	*Kinematic viscosity, ft*^{2}*/sec*
21,000	443.87	933.26	1.2249^{-3}	1032.8	2.6966^{-4}
22,000	440.32	894.59	1.1836	1028.6	2.7724
23,000	436.76	857.24	1.1435	1024.5	2.8510
24,000	433.20	821.16	1.1043	1020.3	2.9324
25,000	429.64	786.33	1.0663	1016.1	3.0168
26,000	426.08	752.71	1.0292^{-3}	1011.9	3.1044^{-4}
27,000	422.53	720.26	9.9311^{-4}	1007.7	3.1951
28,000	418.97	688.96	9.5801	1003.4	3.2893
29,000	415.41	658.77	9.2387	000.13	3.3870
30,000	411.86	629.66	8.9068	994.85	3.4884
31,000	408.30	601.61	8.5841^{-4}	990.54	3.5937^{-4}
32,000	404.75	574.58	8.2704	986.22	3.7030
33,000	401.19	548.54	7.9656	981.88	3.8167
34,000	397.64	523.47	7.6696	977.52	3.9348
35,000	394.08	499.34	7.3820	973.14	4.0575
36,000	390.53	476.12	7.1028^{-4}	968.75	4.1852^{-4}
37,000	389.99	453.86	6.7800	968.08	4.3794
38,000	389.99	432.63	6.4629	968.08	4.5942
39,000	389.99	412.41	6.1608	968.08	4.8196
40,000	389.99	393.12	5.8727	968.08	5.0560
41,000	389.99	374.75	5.5982^{-4}	968.08	5.3039^{-4}
42,000	389.99	357.23	5.3365	968.08	5.5640
43,000	389.99	340.53	5.0871	968.08	5.8368
44,000	389.99	324.62	4.8493	968.08	6.1230
45,000	389.99	309.45	4.6227	968.08	6.4231
46,000	389.99	294.99	4.4067^{-4}	968.08	6.7380^{-4}
47,000	389.99	281.20	4.2008	968.08	7.0682
48,000	389.99	268.07	4.0045	968.08	7.4146
49,000	389.99	255.54	3.8175	968.08	7.7780
50,000	389.99	243.61	3.6391	968.08	8.1591
51,000	389.99	232.23	3.4692^{-4}	968.08	8.5588^{-4}
52,000	389.99	221.38	3.3072	968.08	8.9781
53,000	389.99	211.05	3.1527	968.08	9.4179
54,000	389.99	201.19	3.0055	968.08	9.8792
55,000	389.99	191.80	2.8652	968.08	1.0363^{-3}
56,000	389.99	182.84	2.7314^{-4}	968.08	1.0871^{-3}
57,000	389.99	174.31	2.6039	968.08	1.1403
58,000	389.99	166.17	2.4824	968.08	1.1961
59,000	389.99	158.42	2.3665	968.08	1.2547
60,000	389.99	151.03	2.2561	968.08	1.3161

English Units[a] *(Continued)*

Altitude h, *ft*	*Temperature* T, °*R*	*Pressure* P, *lb/ft*^{2}	*Density* ρ, *lb sec*^{2}*/ft*^{4}	*Speed of sound, ft/sec*	*Kinematic viscosity, ft*^{2}*/sec*
61,000	389.99	143.98	2.1508^{-4}	968.08	1.3805^{-3}
62,000	389.99	137.26	2.0505	968.08	1.4481
63,000	389.99	130.86	1.9548	968.08	1.5189
64,000	389.99	124.75	1.8636	968.08	1.5932
65,000	389.99	118.93	1.7767	968.08	1.6712
66,000	389.99	113.39	1.6938^{-4}	968.08	1.7530^{-3}
67,000	389.99	108.10	1.6148	968.08	1.8387
68,000	389.99	102.06	1.5395	968.08	1.9286
69,000	389.99	98.253	1.4678	968.08	2.0230
70,000	389.99	93.672	1.3993	968.08	2.1219
71,000	389.99	89.305	1.3341^{-4}	968.08	2.2257^{-3}
72,000	389.99	85.142	1.2719	968.08	2.3345
73,000	389.99	81.174	1.2126	968.08	2.4486
74,000	389.99	77.390	1.1561	968.08	2.5683
75,000	389.99	73.784	1.1022	968.08	2.6938

[a]Note: the notation xxx^{-n} means $xxx \times 10^{-n}$.

SI Units

Altitude h, *m*	*Temperature* T, *K*	*Pressure* P *N/m*^{2}	*Density* ρ *kg/m*^{3}	*Speed of Sound m/s*	*Kinematic Viscosity, m*^{2}*/s*
0	288.16	1.01325^{+5}	1.2250	340.29	1.4607^{-5}
300	286.21	9.7773^{+4}	1.1901	339.14	1.4956
600	284.26	9.4322	1.1560	337.98	1.5316
900	282.31	9.0971	1.1226	336.82	1.5687
1,200	280.36	8.7718	1.0900	335.66	1.6069
1,500	278.41	8.4560	1.0581	334.49	1.6463
1,800	276.46	8.1494	1.0269	333.32	1.6869
2,100	274.51	7.8520	9.9649^{-1}	332.14	1.7289
2,400	272.57	7.5634	9.6673	330.96	1.7721
2,700	270.62	7.2835	9.3765	329.77	1.8167
3,000	268.67	7.0121	9.0926	328.58	1.8628
3,300	266.72	6.7489	8.8153	327.39	1.9104
3,600	264.77	6.4939	8.5445	326.19	1.9595
3,900	262.83	6.2467	8.2802	324.99	2.0102
4,200	260.88	6.0072	8.0222	323.78	2.0626
4,500	258.93	5.7752	7.7704	322.57	2.1167
4,800	256.98	5.5506	7.5247	321.36	2.1727
5,100	255.04	5.3331	7.2851	320.14	2.2305
5,400	253.09	5.1226	7.0513	318.91	2.2903
5,700	251.14	4.9188^{+4}	6.8234^{-1}	317.69	2.3522^{-5}

SI Units (*Continued*)

Altitude h, *m*	*Tempera-ture* T, *K*	*Pressure* P *N/m²*	*Density* ρ *kg/m³*	*Speed of Sound* *m/s*	*Kinematic Viscosity,* *m²/s*
6,000	249.20	4.7217	6.6011	316.45	2.4161
6,300	247.25	4.5311	6.3845	315.21	2.4824
6,600	245.30	4.3468	6.1733	313.97	2.5509
6,900	243.36	4.1686	5.9676	312.72	2.6218
7,200	241.41	3.9963	5.7671	311.47	2.6953
7,500	239.47	3.8299	5.5719	310.21	2.7714
7,800	237.52	3.6692	5.3818	308.95	2.8503
8,100	235.58	3.5140	5.1967	307.68	2.9320
8,400	233.63	3.3642	5.0165	306.41	3.0167
8,700	231.69	3.2196	4.8412	305.13	3.1046
9,000	229.74	3.0800	4.6706	303.85	3.1957
9,300	227.80	2.9455	4.5047	302.56	3.2903
9,600	225.85	2.8157	4.3433	301.27	3.3884
9,900	223.91	2.6906	4.1864	299.97	3.4903
10,200	221.97	2.5701	4.0339	298.66	3.5961
10,500	220.02	2.4540	3.8857	297.35	3.7060
10,800	218.08	2.3422	3.7417	296.03	3.8202
11,100	216.66	2.2346	3.5932	295.07	3.9564
11,400	216.66	2.1317	3.4277	295.07	4.1474
11,700	216.66	2.0335^{+4}	3.2699^{-1}	295.07	4.3475^{-5}
12,000	216.66	1.9399	3.1194	295.07	4.5574
12,300	216.66	1.8506	2.9758	295.07	4.7773
12,600	216.66	1.7654	2.8388	295.07	5.0078
12,900	216.66	1.6842	2.7081	295.07	5.2494
13,200	216.66	1.6067	2.5835	295.07	5.5026
13,500	216.66	1.5327	2.4646	295.07	5.7680
13,800	216.66	1.4622	2.3512	295.07	6.0462
14,100	216.66	1.3950	2.2430	295.07	6.3378
14,400	216.66	1.3308	2.1399	295.07	6.6434
14,700	216.66	1.2696	2.0414	295.07	6.9637
15,000	216.66	1.2112	1.9475	295.07	7.2995
15,300	216.66	1.1555	1.8580	295.07	7.6514
15,600	216.66	1.1023	1.7725	295.07	8.0202
15,900	216.66	1.0516	1.6910	295.07	8.4068
16,200	216.66	1.0033	1.6133	295.07	8.8119
16,500	216.66	9.5717^{+3}	1.5391	295.07	9.2366
16,800	216.66	9.1317	1.4683	295.07	9.6816
17,100	216.66	8.7119	1.4009	295.07	1.0148^{-4}
17,400	216.66	8.3115	1.3365	295.07	1.0637
17,700	216.66	7.9295^{+3}	1.2751^{-1}	295.07	1.1149^{-4}
18,000	216.66	7.5652	1.2165	295.07	1.1686
18,300	216.66	7.2175	1.1606	295.07	1.2249
18,600	216.66	6.8859	1.1072	295.07	1.2839
18,900	216.66	6.5696	1.0564	295.07	1.3457

Other Data

Conversion Factors

Multiply	*By*	*To Get*
Pounds (lb)	4.448	Newtons (N)
Feet (ft)	0.3048	Meters (m)
Slugs	14.59	Kilograms (kg)
Slugs per cubic foot (slugs/ft^3)	515.4	Kilograms per cubic meter (kg/m^3)
Miles per hour (mph)	0.4471	Meters per second (m/s)
Knots (kt)	0.5151	Meters per second (m/s)
Knots (kt)	1.152	Miles per hour (mph)

Gravity

$g = 32.2\ \text{ft/s}^2 = 9.81\ \text{m/s}^2$ at sea level.

APPENDIX E

Data for the Boeing 747-100

The Boeing 747 is a highly successful, large, four-engined turbofan transport aircraft. The model 100 first entered service in January 1970, and since then it has continued to be developed through a series of models and special versions. As of May 1990, only versions of the model 400 were being marketed. By the year 1994, close to 800 Boeing 747s were in operation around the world, and the aircraft was still in production.

The data for the Boeing 747-100 contained in this appendix are based on Heffley and Jewell (1972). A three-view drawing of the aircraft is given in Fig. E.1. A body axis system F_B is located with origin at the CG and its x-axis along the fuselage reference line (FRL). The CG is located at 0.25 $\bar{c}$ (i.e., $h = 0.25$), and this is the location that applies for the tabulated data. The thrust line (TL) makes an angle of 2.5° with respect to the FRL as shown.

Three flight cases are documented in the data tables. They all represent straight and level steady-state flight at a fixed altitude. Case I has the aircraft in its landing configuration with 30° flaps, landing gear down, and an airspeed 20% above the stalling speed. Cases II and III represent two cruising states with the flaps retracted and the gear up.

The data in Table E.1 define the flight conditions that apply to the three cases. It should be noted that the moments and product of inertia are given relative to the body frame F_B shown in Fig. E.1. Here the weight and inertias for Case I are smaller than those for the other two cases because the amount of fuel on board during landing is less than that during the cruise. If the data are to be applied to a reference frame different from F_B (e.g., to stability axes F_S) then the given inertias will have to be transformed according to (B.12,3). Note that F_B can be rotated into F_S by a single rotation of ξ about the y-axis. Values for ξ are contained in Table E.1.

The dimensional derivatives corresponding to F_B of Fig. E.1 are contained in Tables E.2 to E.4. Since F_B can be rotated into the stability axes F_S by a single rotation of ξ about the y-axis, it follows that the transformations of (B.12,6 and B.12,7) can be used to obtain the derivatives corresponding to F_S. Values for ξ are contained in Table E.1.

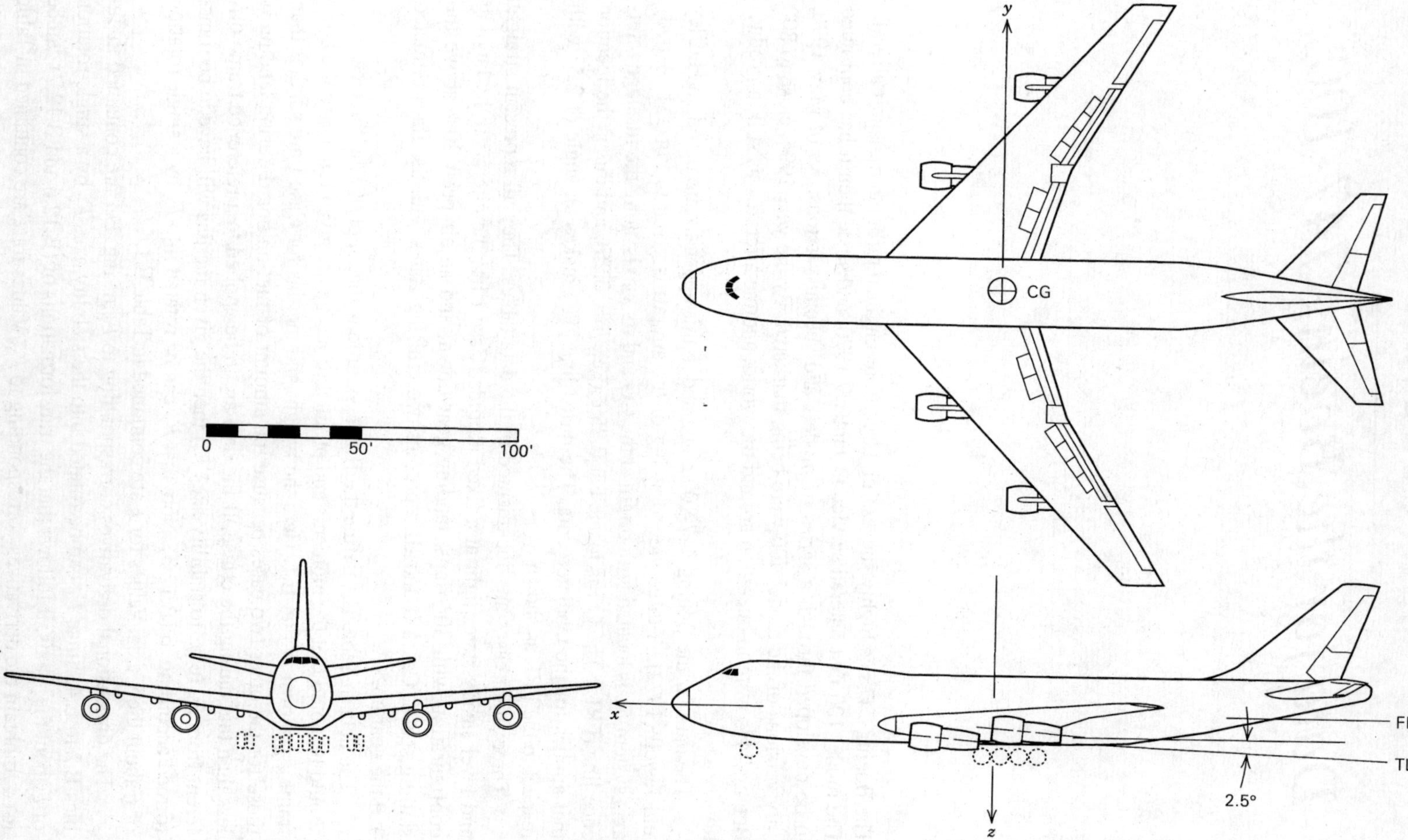

Figure E.1 Boeing 747-100.

Table E.1
Boeing 747-100 Data

(S = 5,500 ft^2, b = 195.68 ft, $\bar{c}$ = 27.31 ft, h = 0.25)

	Case I	*Case II*	*Case III*
Altitude (ft)	0	20,000	40,000
M	0.2	0.5	0.9
V (ft/s)	221	518	871
W (lb)	5.640×10^5	6.366×10^5	6.366×10^5
I_x (slug·ft^2)	1.42×10^7	1.82×10^7	1.82×10^7
I_y (slug·ft^2)	3.23×10^7	3.31×10^7	3.31×10^7
I_z (slug·ft^2)	4.54×10^7	4.97×10^7	4.97×10^7
I_{zx} (slug·ft^2)	8.70×10^5	9.70×10^5	9.70×10^5
ξ (degrees)	−8.5	−6.8	−2.4
C_D	0.263	0.040	0.043

Table E.2
Boeing 747-100 Dimensional Derivatives

Case I (**M** = 0.2)
Longitudinal

	X (lb)	*Z (lb)*	*M (ft·lb)*
u (ft/s)	-3.661×10^2	-3.538×10^3	3.779×10^3
w (ft/s)	2.137×10^3	-8.969×10^3	-5.717×10^4
q (rad/s)	0	-1.090×10^5	-1.153×10^7
$\dot{w}$ (ft/s^2)	0	5.851×10^2	-7.946×10^3
δ_e (rad)	1.680×10^4	-1.125×10^5	-1.221×10^7

Lateral

	Y (lb)	*L (ft·lb)*	*N (ft·lb)*
v (ft/s)	-1.559×10^3	-8.612×10^4	3.975×10^4
p (rad/s)	0	-1.370×10^7	-6.688×10^6
r (rad/s)	0	4.832×10^6	-1.014×10^7
δ_a (rad)	0	-3.200×10^6	-1.001×10^6
δ_r (rad)	5.729×10^4	1.034×10^6	-6.911×10^6

Table E.3
Boeing 747-100 Dimensional Derivatives

Case II (**M** = 0.5)

Longitudinal

	X (lb)	*Z (lb)*	*M (ft·lb)*
u (ft/s)	-4.883×10^1	-1.342×10^3	8.176×10^3
w (ft/s)	1.546×10^3	-8.561×10^3	-5.627×10^4
q (rad/s)	0	-1.263×10^5	-1.394×10^7
$\dot{w}$ (ft/s²)	0	3.104×10^2	-4.138×10^3
δ_e (rad)	3.994×10^4	-3.341×10^5	-3.608×10^7

Lateral

	Y (lb)	*L (ft·lb)*	*N (ft·lb)*
v (ft/s)	-1.625×10^3	-7.281×10^4	4.404×10^4
p (rad/s)	0	-1.180×10^7	-2.852×10^6
r (rad/s)	0	6.979×10^6	-7.323×10^6
δ_a (rad)	0	-2.312×10^6	-7.555×10^5
δ_r (rad)	1.342×10^5	3.073×10^6	-1.958×10^7

Table E.4
Boeing 747-100 Dimensional Derivatives

Case III (**M** = 0.9)

Longitudinal

	X (lb)	*Z (lb)*	*M (ft·lb)*
u (ft/s)	-3.954×10^2	-8.383×10^2	-2.062×10^3
w (ft/s)	3.144×10^2	-7.928×10^3	-6.289×10^4
q (rad/s)	0	-1.327×10^5	-1.327×10^7
$\dot{w}$ (ft/s²)	0	1.214×10^2	-5.296×10^3
δ_e (rad)	1.544×10^4	-3.677×10^5	-4.038×10^7

Lateral

	Y (lb)	*L (ft·lb)*	*N (ft·lb)*
v (ft/s)	-1.198×10^3	-2.866×10^4	5.688×10^4
p (rad/s)	0	-8.357×10^6	-5.864×10^5
r (rad/s)	0	5.233×10^6	-7.279×10^6
δ_a (rad)	0	-3.391×10^6	4.841×10^5
δ_r (rad)	7.990×10^4	2.249×10^6	-2.206×10^7

References

References Cited

AGARD. Aeroelastic Effects from a Flight Mechanics Standpoint. *Conference Proceedings,* no. 46, March 1970.

AGARD. *Flight Test Manual,* Vol. II, Stability and Control. AGARD, 1959.

AGARD. Rotorcraft System Identification. *Advisory Report No. 280, NATO,* 1991.

ANSI/AIAA. Recommended Practise—Atmospheric and Space Flight Vehicle Coordinate Systems. *AIAA,* Washington, DC, 1992.

Bairstow, L. *Applied Aerodynamics,* 2nd ed. Longmans, Green & Co., 1939.

Bisplinghoff, Raymond L. *Principles of Aeroelasticity.* John Wiley & Sons, Inc., New York, 1962.

Bollay, W. Aerodynamic Stability and Automatic Control. *Jour. Aero. Sci.* vol. 18, No. 9, 1951.

Bryan, G. H. *Stability in Aviation.* Macmillan Co., London, 1911.

Bur. Aero. The Human Pilot. Vol. III of *Bur. Aero. Rept. AF-6-4,* 1954.

Byrne, J. E. Design of a Lateral Ride Comfort Control System for STOL Aircraft. *University of Toronto UTIAS Tech. Note No. 247,* 1983.

Campbell, J. P., and McKinney, M. O. Summary of Methods for Calculating Dynamic Lateral Stability and Response and for Estimating Lateral Stability Derivatives. *NACA Rept. 1098,* 1952.

Carslaw, H. S., and Jaeger, J. C. *Operational Methods in Applied Mathematics,* 2nd ed. Oxford University Press, London, 1947.

Cooper, G. E., and Harper, R. P., Jr. The Use of Pilot Rating in the Evaluation of Aircraft Handling Qualities. *NASA TN D-5153,* 1969.

Davenport, A. G. Rationale for Determining Design Wind Velocities. *J. Struct. Div. ASCE,* vol. 86, 1960.

De Young, J., and Harper, C. W. Theoretical Symmetric Span Loadings at Subsonic Speeds for Wings Having Arbitrary Planforms. NACA Rept. 921, 1948.

Dowell, Earl H., et al., ed. A Modern Course in Aeroelasticity. *Solid Mechanics and Its Applications,* vol. 32. Dordrecht, Boston: Kluwer Academic, 1994.

Duncan, W. J. *Control and Stability of Aircraft.* Cambridge University Press, Cambridge, 1952.

Ellis, N. D. Aerodynamics of Wing-Slipstream Interaction. *Univ. of Toronto UTIAS Rept. 169,* 1971.

Etkin, B. Aerodynamic Transfer Functions: An Improvement on Stability Derivatives for Unsteady Flight. *Univ. of Toronto UTIA Rept. 42,* 1956.

Etkin, B. *Dynamics of Flight—Stability and Control.* 1st ed., John Wiley & Sons, Inc., New York, 1959.

Etkin, B., and Hughes, P. C. Explanation of the Anomalous Spin Decay of Satellites with Long Flexible Antennae. *J. Spacecraft,* vol. 4, no. 9, Sept. 1967.

Etkin, B. *Dynamics of Atmospheric Flight.* John Wiley & Sons, Inc., New York, 1972.

Etkin, B. Turbulent Wind and Its Effect on Flight. *AIAA Jour. of Aircraft,* vol. 18, no. 5, 1981.

Etkin, B. *Dynamics of Flight, Stability and Control.* 2nd ed. John Wiley & Sons, Inc., New York, 1982.

Etkin, B., and Etkin, D. A. Critical Aspects of Trajectory Prediction: Flight in Non-uniform Wind, in AGARDograph No. 301, v. 1 (1990).

Filotas, L. T. Approximate Transfer Functions for Large Aspect Ratio Wings in Turbulent Flow. *J. Aircraft,* vol. 8, no. 6, June 1971.

Flax, A. H., and Lawrence, H. R. The Aerodynamics of Low-Aspect Ratio Wing-Body Combinations. *Cornell Aero Lab. Rept. CAL-37,* 1951.

Fung, Y. C. *The Theory of Aeroelasticity.* John Wiley & Sons, Inc., New York, 1955.

Gates, S. B. Proposal for an Elevator Manoeuvreability Criterion. *ARC R & M 2677,* 1942.

Heffley, R. K., and Jewell, N. F. Aircraft Handling Qualities Data. *NASA CR 2144,* Dec. 1972.

Heppe, R. R., and Celinker, L. Airplane Design Implications of the Inertia Coupling Problem. *IAS Preprint 723,* 1957.

Hughes, P. C. *Spacecraft Attitude Dynamics.* John Wiley & Sons, Inc., New York, 1986.

ICAO. Airworthiness of Aircraft. Anex 8, Chap. 2; para 2.3; *Flying Qualities,* 1991.

Jaeger, J. C. *An Introduction to the Laplace Transformation.* Methuen & Co., London, 1949.
Kleinman, D. L., Baron, S., and Levison, W. H. An Optimal Control Model of Human Response. Part I: Theory and Validation. *Automatica,* vol. 6, pp. 357–369, 1970.
Kuethe, A. M., and Chow, C. Y. *Foundations of Aerodynamics.* 3rd ed., John Wiley & Sons, Inc., New York, 1976.
Kuhn, R. E. Semiempirical Procedure for Estimating Lift and Drag Characteristics of Propeller-Wing-Flap Configurations for V/STOL Airplanes. *NASA Memo 1-16-59L,* Feb. 1959.
Lanchester, F. W. *Aerodonetics.* A. Constable & Co. Ltd., London, 1908.
Langhaar, H. L. Dimensional Analysis and Theory of Models. John Wiley & Sons, Inc., New York, 1951.
Maine, R. E., and Iliff, K. W. Application of Parameter Estimation to Aircraft Stability and Control. *NASA Reference Publication no. 1168,* 1986.
Margason, R. J., Hammond, A. D., and Gentry, G. L. Longitudinal Stability and Control Characteristics of a Powered Model of a Twin-Propeller Deflected-Slipstream STOL Airplane Configuration. *NASA TN D-3438,* 1966.
McLean, D. *Automatic Flight Control Systems.* Prentice Hall International (U.K.) Ltd., 1990.
McCormick, B. W. *Aerodynamics, Aeronautics, and Flight Mechanics.* John Wiley & Sons, Inc., New York, 1994.
McLaughlin, M. D. A Theoretical Investigation of the Short-Period Dynamic Longitudinal Stability of Airplanes Having Elastic Wings of 0° to 60° Sweepback. *NACA TN 3251,* 1954.
McRuer, D. T., and Krendel, E. S. Mathematical Models of Human Pilot Behavior. *AGARDograph M-188,* Nov. 1973.
McRuer, D., Ashkenas, I., and Graham, D. *Aircraft Dynamics and Automatic Control.* Princeton Univ. Press, Princeton, NJ, 1973.
Michael, W. H., Jr. Analysis of the Effects of Wing Interference on the Tail Contributions to the Rolling Derivatives. *NACA Rept. 1086,* 1952.
Miele, A. *Flight Mechanics—I. Theory of Flight Paths.* Addison-Wesley Pub. Co., Reading, MA, 1962.
Miles, J. W. On the Compressibility Correction for Subsonic Unsteady Flow. *J. Aero. Sci.,* vol. 17, no. 3, 1950.
Miles, J. W. Unsteady Flow Theory in Dynamic Stability. *J. Aero. Sci.,* vol. 17, no. 1, 1950.
Millikan, C. B. The Influence of Running Propellors on Airplane Characteristics. *J. Aero Sci.,* vol. 7, no. 3, 1940.
Milne, R. D. Dynamics of the Deformable Airplane. *ARC R & M 3345,* 1964.
Neumark, S. Longitudinal Stability, Speed and Height. *RAE Rept. Aero 2265,* 1948; also *Aircraft Engineering,* Nov. 1950.
Nyquist, H. Regeneration Theory. *Bell System Tech. Jour.* vol. 11, Jan. 1932.
O'Hara, F. Handling Criteria. *J. Roy. Aero. Soc.,* vol. 71, no. 676, pp. 271–291, 1967.
Perkins, C. D. Development of Airplane Stability and Control Technology. *J. Aircraft,* vol. 7, no. 4, 1970.
Phillips, W. H. Effect of Steady Rolling on Longitudinal and Directional Stability. *NACA TN. 1627,* 1948.
Pinsker, W. J. G. Charts of Peak Amplitudes in Incidence and Sideslip in Rolling Manoeuvres Due to Inertia Cross Coupling. *RAE Rept. No. Aero 2604,* April 1958.
Priestley, E. A General Treatment of Static Longitudinal Stability with Propellors, with Application to Single Engine Aircraft. *ARC R & M 2732,* 1953.
Reid, L. D. The Measurement of Human Pilot Dynamics in a Pursuit-Plus-Disturbance Task. *Univ. of Toronto UTIAS Rept. 138,* 1969.
Ribner, H. S. Notes on the Propellor and Slip-Stream in Relation to Stability. *NACA Wartime Rept. L25,* 1944.
Ribner, H. S. Formulas for Propellors in Yaw and Charts of the Side-Force Derivative. *NACA Rept. 819,* 1945.
Ribner, H. S. Field of Flow About a Jet and Effect of Jets on Stability of Jet-Propelled Airplanes. *NACA Wartime Rept. L213,* 1946.

Ribner, H. S., and Ellis, N. D. Aerodynamics of Wing-slipstream Interaction. *CASI Transactions,* vol. 5, no. 2, 1972.

Rodden, W. P. An Aeroelastic Parameter for Estimation of the Effect of Flexibility on the Lateral Stability and Control of Aircraft. *J. Aero. Sci.,* vol. 23, no. 7, 1956.

Rodden, W. P., and Giesing, J. P. Application of Oscillatory Aerodynamic Theory to Estimation of Dynamic Stability Derivatives. *J. Aircraft,* vol. 7, no. 3, 1970.

Routh, E. J. *Dynamics of a System of Rigid Bodies,* 6th ed. Macmillan & Co., London, 1905.

Rynaski, E. G., Andrisani, D. III, Weingarten, N. Active Control for the Total-in-Flight Simulator. *NASA CR 3118,* 1979.

Schlichting, H., and Truckenbrodt, E. *Aerodynamics of the Airplane.* McGraw Hill International Book Co., New York, 1979.

Smelt, R., and Davies, H. Estimation of Increase in Lift Due to Slipstream. *ARC R & M 1788,* 1937.

Stanton-Jones, R. An Empirical Method for Rapidly Determining the Loading Distributions on Swept Back Wings. Coll. Aero. (Cranfield) Rept. 32, 1950.

Stevens, B. L., and Lewis, F. L. *Aircraft Control and Simulation.* John Wiley and Sons, Inc., New York, 1992.

Synge, J. L., and Griffith, B. A. *Principles of Mechanics.* McGraw-Hill Pub. Co., Inc., New York, 1942.

Theodorsen, Th. General Theory of Aerodynamic Instability and the Mechanism of Flutter. *NACA Rept. 496,* 1934.

Tobak, M. On the Use of the Indicial Function Concept in the Analysis of Unsteady Motions of Wings and Wing-Tail Combinations. *NACA Rept. 1188,* 1954.

Truitt, R. W. *Hypersonic Aerodynamics.* Ronald Press, New York, 1959.

USAF. Stability and Control Datcom. *Air Force Flight Dynamics Lab., U.S. Air Force,* 1960 (revised 1978).

USAF. Flying Qualities of Piloted Airplanes. *USAF Spec.* MIL-F-8785C, 1980.

USAF. Flying Qualities of Piloted Vehicles. USAF MIL-STD-1797A, 1990.

Weil, J., and Sleeman, W. C., Jr. Prediction of the Effects of Propellor Operation on the Static Longitudinal Stability of Single-Engine Tractor Monoplanes with Flaps Retracted. *NACA Rept. 941,* 1949.

Weissinger, J. The Lift Distribution of Swept Back Wings. NACA TM 1120, 1947.

Yates, A. H. Notes on the Mean Aerodynamic Chord and the Mean Aerodynamic Centre of a Wing. *J. Roy. Aero. Soc.,* vol. 56, p. 461, June 1952.

Additional Reading

Anderson, J. D. *Introduction to Flight.* 2nd ed. McGraw Hill, Inc., New York, 1985.

Bertin, J. J., and Smith, M. J. *Aerodynamics for Engineers.* 2nd ed. Prentice Hall, NJ, 1989.

Blakelock, J. H. *Automatic Control of Aircraft and Missiles.* 2nd ed. John Wiley & Sons, Inc., New York, 1991.

McCormick, B. W. *Aerodynamics, Aeronautics, and Flight Mechanics.* John Wiley & Sons, Inc., New York, 1994.

Nelson, R. C. *Flight Stability and Automatic Control.* McGraw Hill Inc., New York, 1989.

Shevell, R. S. *Fundamentals of Flight.* 2nd ed. Prentice Hall, NJ, 1989.

INDEX

$ Job 25 A 11/21/2010 14:39:15